Collaborative Design

Springer
London
Berlin
Heidelberg
New York
Barcelona
Hong Kong
Milan
Paris
Singapore
Tokyo

Stephen A.R. Scrivener, Linden J. Ball and Andrée Woodcock (Eds.)

Collaborative Design

Proceedings of CoDesigning 2000

With 124 Figures

Springer

Stephen A.R. Scrivener, DipAD, HDFA, PhD
VIDe Research Centre, Coventry School of Art & Design, Coventry University, Priory Street, Coventry, CV1 5FB

Linden J. Ball, BSc, PhD
Institute of Behavioural Sciences, University of Derby, Mickleover, Derby, DE3 5GX

Andrée Woodcock, BSc, MSc
VIDe Research Centre, Coventry School of Art & Design, Coventry University, Priory Street, Coventry, CV1 5FB

ISBN 1-85233-341-3 Springer-Verlag London Berlin Heidelberg

British Library Cataloguing in Publication Data
Collaborative design : proceedings of CoDesigning 2000
Woodcock
1.Design - Computer network resources - Congresses
2.Communication in design - Congresses 3.Communication in design - Computer network resources - Congresses 4.Design - Environmental aspects - Congresses
I.Scrivener, Stephen A. R. II.Ball, Linden J. III. Woodcock, Andree
745.2'0285'4678
ISBN 1852333413

Library of Congress Cataloging-in-Publication Data
Collaborative design / Stephen A.R. Scrivener, Linden J. Ball and Andrée Woodcock (eds.)
p. cm.
Includes bibliographical references and index.
ISBN 1-85233-341-3 (alk. paper)
1. Engineering design. 2. Artistic collaboration. I. Scrivener, Stephen A. R. II. Ball, Linden J., 1963- III. Woodcock, Andrée, 1959-
TA174.C5725 2000
620'.0042--dc21 00-034504

Printed in Great Britain

Typesetting: Camera ready by editors
Printed and bound at the Athenæum Press Ltd., Gateshead, Tyne & Wear
34/3830-543210 Printed on acid-free paper SPIN 10771205

Preface

This book comprises the proceedings of CoDesign 2000, held at Coventry between September 11th and 13th, 2000. The conference 'Call for Proposals' argued the following:

> *'In recent years, the view of design as an essentially individual creative activity has come under increasing question. Instead, for a variety of reasons, design is being viewed, studied and developed as a collective, collaborative, even community process.'*

In truth, the collective nature of design has been recognised for many years and is well developed in some design domains, such as socio-technical and human-computer system design, architecture and engineering design. This developing interest in the collaborative, social and organisational aspects of design is reflected in the fact that most design-related conferences now have one or more sessions devoted to it. Nevertheless, it seemed both timely and appropriate to initiate a conference focused specifically on the 'co' in CoDesign.

In framing the conference, the organisers intended that the 'co' in CoDesign should convey multiple meanings, to include, for example, collaborative, cooperative, concurrent, user-centred, participatory, socio-technical and community design. In other words, any development where design as a group process is explored. With this in mind, the conference aimed to be inclusive rather than prescriptive, and no specific sub-themes were identified. Instead the 'Call for Proposals' acknowledged that any research concerned with exploring design as a collaborative process was of interest to CoDesigning 2000.

In the event the response to the 'Call for Proposals' was excellent with some two hundred abstracts being submitted, the great majority being relevant to CoDesigning. In due course, around one hundred papers were submitted for peer review, of which the forty-nine included in this book were accepted for publication (many others were accepted as poster papers and appear in the adjunct proceedings published by Coventry University).

Although themes were not prescribed, they were easily identifiable in the papers accepted for publication. Consequently, we have organised the proceedings into Parts around these themes. For us, the salience of these themes is one indicator of coherence in the material presented. Other sub-themes and issues are evident within each Part and we have drawn out and elaborated these in the introduction and summary to each Part. In short, we believe that these proceedings provide a timely and useful contribution to the development of an aspect of design that is becoming of increasing importance to an ever widening range of design domains.

S.A.R. Scrivener, L.J. Ball and A. Woodcock
Coventry, UK,
September 2000

Contents

Part II Issues in Developing Integrated Collaborative Design Approaches

Part III Studies of Collaborative Design

Part IV Mediated Communication

Part V Computer-mediated Collaborative Design

List of Contributors

Ali-MacLachlan I., Product Design Engineering, Glasgow School of Art, Glasgow, Scotland, UK, i.ali-maclachlan@gsa.ac.uk

Anderl R., Department of Computer Integrated Design, University of Technology Darmstadt, Darmstadt, Germany, anderl@dik.tu-darmstadt.de

Ashton P., Historical & Professional Studies, School of Art & Design, Staffordshire University, Stoke-on-Trent, England, UK, p.a.ashton@staffs.ac.uk

Athavankar U.A., Industrial Design Center, Indian Institute of Technology, Powai, Mumbai, India, uaa@idc.iitb.ernet.in

Austin S.A., Civil & Building Engineering, Loughborough University, Loughborough, Leicestershire. LE11 3TU, England, UK, s.a.austin@lboro.ac.uk

Ball L.J., Cognitive and Behavioural Sciences Research Group, Institute of Behavioural Sciences, University of Derby, Mickleover, Derby, DE3 5GX, England, UK, l.j.ball@derby.ac.uk

Bhamra T., Enterprise Integration, Cranfield University, Bedford, England, UK, t.bhamra@cranfield.ac.uk

Brown R., School of Architecture, University of East London, London, England, UK, lbal@compuserve,com

Bruce M., Management & Design Centre, Department of Textiles, UMIST, Manchester, England, UK, margaret.bruce@umist.ac.uk

Burningham C., Human Centred Systems Design Research Group, Department of Psychology, University of Plymouth, Plymouth, England, UK

Burns A.D., Enterprise Integration, Cranfield University, Cranfield, Bedford, MK43 2AL, England, UK, a.d.burns@cranfield.ac.uk

Burr J., Mads Clausen Institute for Product Innovation, University of Southern Denmark, DK-6400, Sønderborg, Denmark, buur@mci.sdu.dk

Caldwell N.H.M., Engineering Design Centre, Engineering Department, Cambridge University, Cambridge, England, UK, nhmc1@eng.cam.ac.uk

Chen C.D., VIDe Research Centre, School of Art & Design, Coventry University, Coventry, England,UK, cdchen@cdchen.freeserve.co.uk

Cooper R., Design & Innovation Research Group, Institute for Social Research, University of Salford, Salford, M3 6EQ, England, UK, r.cooper@salford.ac.uk

Cross N., Department of Design & Innovation, Open University, Milton Keynes, England, UK, n.g.cross@open.ac.uk
Culverhouse P.F., Human Centred Systems Design Research Group, School of Electronic & Electrical Engineering, University of Plymouth, Plymouth, UK, pculverhouse@plym.ac.uk
Daly L., Management & Design Centre, Department of Textiles, UMIST, Manchester, England, UK, lucy.daly@umist.ac.uk
Dandavate U., Sonic Rim, Columbus, Ohio, USA, uday@sonicrim.com
Demirbilek O., University of New South Wales, Faculty of the Built Environment, Industrial Design, Sydney, Australia, oyad@fbe.unsw.edu.au
Demirkan H., Bilkent University, Interior Architecture & Environmental Design, Ankara, Turkey, demirkan@bilkent.edu.tr
Deshmukh H.S., Industrial Design Center, Indian Institute of Technology, Powai, Mumbai, India, hsdes8pi@ccs.iitb.ernet.in
Erol R., Design & Innovation Research Unit, Sheffield Hallam University, Sheffield, S11 8UZ, England, UK, r.a.erol@shu.ac.uk
Evans S., Enterprise Integration, Cranfield University, Cranfield, Bedford, MK43 2AL, England, UK, s.evans@cranfield,ac.uk
Fadel G., Department of Mechanical Engineering, Clemson University, Clemson, South Carolina, USA, gfadel@ces.clemson.edu
Felton A.J., School of Engineering & the Built Environment, University of Wolverhampton, Priorslee Campus, Telford, Shropshire, TF2 9NT, England, UK, a.j.felton@wlv.ac.uk
Filson A., The National Centre for Product Design & Development Research, University of Wales Institute, Cardiff (UWIC), Cardiff, Wales, UK, afilson-pdr@uwic.ac.uk
Fowles R.A., The Welsh School of Architecture, Cardiff University, Bute Building, Cardiff, Wales, UK, fowlesra@cf.ac.uk
Gabriel G., Architectural & Design Science, University of Sydney, Australia, gerard@arch.usyd.edu.au
Gierhardt H., Lehrstuhl für Próduktentwicklung, Technische Universitaet München, Boltzmannstraße 15, 85748 Garching, Munich, Germany, gierhardt@km.mw.tum.de
Gill N., Industrial Design Center, Indian Institute of Technology, Powai, Mumbai, India, neel8pi@css.iitb.ernet.in
Goeller H., Design, The Ohio State University, Columbus, Ohio, USA, goeller.1@osu.edu
Hakkarainen K., Department of Psychology, University of Helsinki, Helsinki, Finland, kai.hakkarainen@helsinki.fi
Hargreaves G., Industrial Design, Massey University College of Design, Fine Arts & Music, Wellington, New Zealand, g.w.hargreaves@massey.ac.nz
Harun R., Management & Design Centre, Department of Textiles, UMIST, Manchester, England, UK, richelle.harun@stud.umist.ac.uk
Harwood R., Head of Department of Textile Design & Production, De Montfort University, Leicester, England, UK, rjh@dmu.ac.uk
Heeley J., Design & Innovation Group, Salford University, Salford, England, UK, jheeley@salford.ac.uk
Huxor A., Centre for Electronic Arts, Middlesex University, Barnet, Herts, England, UK, a.huxor@mdx.ac.uk
Illman M.E., Centre for Design Research, School of Design, University of Northumbria, Newcastle upon Tyne, NE1 8ST, England, UK
Jagodzinski A.P., Human Centred Systems Design Research Group, School of Computing, University of Plymouth, Plymouth, England, UK, peterj@soc.plym.ac.uk

Kälviäinen M., Kuopio Academy of Crafts & Design, Keopio, Finland, mkalviai@koto.pspt.fi
Kirby P., The Martin Centre, Department of Architecture, Cambridge University, Cambridge, Cambridgeshire, CB2 2GB, England, UK
Kolarevic B., Digital Design Research Lab (DDRL), Architecture Department, University of Pennsylvania, Philadelphia, USA, branko@pobox.upenn.edu
Lahti H., Craft Science, Department of Education, University of Helsinki, Helsinki, Finland, henna.lahti@helsinki.fi
Lee L.-C., VIDe Research Centre, School of Art & Design, Coventry University, Coventry, England, UK, ray@leelc.freeserve.co.uk
Lewis A., DERC, University of Wales Institute, Cardiff, Wales, UK, alewis@uwic.ac.uk
Lindemann H., Lehrstuhl für Produktentwicklung, Technische Universitaet München, Boltzmannstraße 15, 85748 Garching, Munich, Germany, lindem@pe.mw.tum.de
Lloyd P., Design Methodology, Industrial Design Engineering, Technical University Delft, Delft, The Netherlands, p.lloyd@io.tudelft.nl
Lofthouse V., School of Industrial & Manufacturing Science, Cranfield University, Cranfield, Bedfordshire, England, UK, v.a.lofthouse@cranfield.ac.uk
Logan G., Rehabilitation Engineering Centre, Royal Brisbane Hospital, Brisbane, Australia, logang@health.qld.gov.au
Luck R., Research Group for Inclusive Environments, Construction Management, University of Reading, Reading, England, UK, r.a.c.luck@reading.ac.uk
MacMillan S., The Martin Centre, Department of Architecture, Cambridge University, Cambridge, Cambridgeshire, CB2 2GB, England, UK, sgm24@cam.ac.uk
Maher M.L., Architectural & Design Science, University of Sydney, Australia, mary@arch,usyd.edu.au
Mazijoglou M., VIDe Research Centre, School of Art & Design, Coventry University, Coventry, England, UK, m.mazijoglou@coventry.ac.uk
McDonagh-Philp D., Design & Technology, Loughborough University, Loughborough, England, UK, d.c.mcdonagh-philp@lboro.ac.uk
Milne A.J., Centre for Design Research, Design Division, Stanford University, Stanford, CA, USA, amilne@stanford.edu
Molyneux, R.W., 28 Shepherd's Walk, Hassocks, West Sussex, BN6 8EB, England, UK
Moreau-Yates D., School of Architecture, University of East London, London, England, UK, studio@fbmarchitects.demon.co.uk
Muukkonen H., Department of Psychology, University of Helsinki, Helsinki, Finland, hanni.muukkonen@helsinki.fi
Ng E.Y.Y., Architecture Department, Chinese University of Hong Kong, Hong Kong, RoC, edwardng@cuhk.edu.hk
Noble A., Photographic Design, Massey University, Wellington, New Zealand, a.noble@massey.co.nz
Oostra M.A.R., Building Technology, Architecture Department, Delft University of Technology, Delft, The Netherlands, m.oostra@bk.tudelft.nl
Ormerod T.C., Department of Psychology, Lancaster University, Fylde College, Lancaster, LA1 4YF, England, UK, t.ormerod@lancs.ac.uk
Parsons R., Human Centred Systems Design Research Group, School of Computing, University of Plymouth, Plymouth, England, UK
Pederson J., User Centred Design, Innovation Department, Danfoss Drives A/5, DK-6300 Graasten, Denmark, jpedersen@danfoss.com

Pontecorvo M.S., Emergent Design, San Francisco, California, USA, ponte@emergent-design.com
Porter J.M., Department of Design & Technology, Loughborough University, Loughborough, Leicestershire, England, UK, j.m.porter@lboro.ac.uk
Porter S., School of Art & Design, Coventry University, Priory Street, Coventry, England, UK, s.porter@coventry.ac.uk
Press M., Design & Innovation Research Unit, Cultural Studies, Sheffield Hallam University, Sheffield, England, UK, m.p.press@shu.ac.uk
Radcliffe D., Mechanical Engineering, The University of Queensland, Brisbane, Australia, radcliffe@mech.uq.edu.au
Reed S.E., Human Centred Systems Design Research Group (HCSDRG), Department of Psychology, University of Plymouth, Plymouth, England, UK, sreed@plym.ac.uk
Regli W.C., GICL, Math & Computer Science, Drexel University, Philadelphia, USA, regli@drexel.edu
Reid F.J.M., Human Centred Systems Design Research Group (HCSDRG), Department of Psychology, University of Plymouth, Plymouth, England, UK, freid@plym.ac.uk
Robinson C., Visual Communication Design, Massey University, Wellington, New Zealand, claire_robinson@design.whp.ac.nz
Rodgers P.A., Design Department, Napier University, Edinburgh, EH10 5DT, Scotland, UK, p.rodgers@napier.ac.uk
Rowley F.J.B., Centre for Design Research, School of Design, University of Northumbria, Newcastle upon Tyne, NE1 8ST, England, UK
Sanders E.B.-N., Sonic Rim, 1086 N4th, Columbus, Ohio, USA, liz@sonicrim.com
Scrivener S.A.R., VIDe Research Centre, School of Art & Design, Coventry University, Coventry, England, UK, s.scrivener@coventry.ac.uk
Seitamaa-Hakkarainen P., Crafts Science, Savonlinna Department of Teacher Education, University of Joensuu, Savonlinna, Finland, pirita.seitamaa-hakkarainen@joensuu.fi
Sevy J., GICL, Math & Computer Science, Drexel University, Philadelphia, USA, jsevy@mcs.drexel.edu
Simoff S., Architectural & Design Science, University of Sydney, Australia, simeon@arch.usyd.edu.au
Spence R., The Martin Centre, Department of Architecture, Cambridge University, Cambridge, Cambridgeshire, CB2 2GB, England, UK
Steele J., Civil & Building Engineering, Loughborough University, Loughborough, Leicestershire, LE11 3TU, England, UK, j.l.steele@lboro.ac.uk
Steiner D., New Product Development, ACCO Brands Inc, Lincolnshire, Illinois, USA, denise_steiner@acco.com
Svengren L., School of Business, Department of Marketing, Stockholm University, Stockholm, Sweden, lsv@fek.su.se
Thomas M., Design & Innovation Research Group, University of Salford, Salford, M3 6EQ, England, UK, m.thomas4@salford.ac.uk
Tierney L., Design & Innovation Group, Research Centre for Art & Design, University of Salford, Salford, England, UK, l.r.tierney@salford.ac.uk
To C.K.M., Management Panel, Institute of Textiles & Clothing, The Hong Kong Polytechnic University, Hong Kong, tctokm@polyu.edu.hk
Turner S., HCI Research Group, School of Computing, Napier University, Edinburgh, Scotland, UK, s.turner@dcs.napier.ac.uk
Tyler D., Clothing Design & Technology, Manchester Metropolitan University, Manchester, England, UK, d.tyler@mmu.ac.uk

Valkenburg R.C., Industrial Design Engineering, Product Innovation & Management, Delft University of Technology, Delft, The Netherlands, r.valkenburg@io.tudelft.nl

van der Lugt R., Product Innovation Management, Sub-faculty of Industrial Design Engineering, Delft University of Technology, Delft, The Netherlands, r.vanderlugt@io.tudelft.nl

van der Veen G.J., Centre for Design Research, School of Design, University of Northumbria, Newcastle upon Tyne, NE1 8ST, England, UK

van der Veer G., FEW/IMSE, Division of Mathematics & Computer Science, Vrije Universiteit Amsterdam, Amsterdam, The Netherlands, gerrit@cs.vu.nl

van Welie M., FEW/IMSE, Division of Mathematics & Computer Science, Vrije Universiteit Amsterdam, Amsterdam, The Netherlands, martijn@cs.vu.nl

Wagner I., Institute for Technology Assessment & Design, Vienna University of Technology, Vienna, Austria, iwagner@pop.tuwien.ac.at

William C., Sonic Rim, Columbus, OH, USA, colin@sonicrim.com

Woodcock A., VIDe Research Centre, School of Art & Design, Coventry University, Coventry, England, UK, a.woodcock@coventry.ac.uk

Yair K., Design & Innovation Research Unit, Cultural Studies, Sheffield Hallam University, Sheffield, England, UK, k.e.yair@shu.ac.uk

Young R.A., Centre for Design Research, School of Design, University of Northumbria, Newcastle upon Tyne, NE1 8ST, England, UK, robert.young@unn.ac.uk

Zaychik V., GICL, Math & Computer Science, Drexel University, Philadelphia, USA, zaychik@drexel.edu

Keynote

Chapter 1

Generative Tools for Co-designing

E. B.-N. Sanders

'Dreams are the stuff the future is made of.'
(Jensen, 1999)

1.1 Preview

Whose dreams are 'the stuff the future is made of'? The dreams of CEOs? Technologists?Business people?Or the dreams of everyday people?

In this paper I will talk about a journey toward a future being made from the dreams of everyday people. I will describe how we are learning to catalyse, capture and collect their dreams and aspirations. I'll do so by showing examples of 'tools' we give people so they can express themselves visually and verbally. Then I will tell how we are beginning to work with people, using their dreams and aspirations to inform as well as inspire the design development process.

1.2 Imagine

Today's world is made of the dreams of the CEO, technologists and business strategists. It is technologically driven, materialistically delivered and fast. Those of us involved in bringing new products and services to market reduce people to the roles they play in the design development process. We see and think of them as 'consumers' when they shop, as 'customers' when they buy and as 'users' when they interact with the objects of their desire. Rarely do we integrate the roles that people play to come to see them in a holistic manner. Rarely do we take the time to get to know them as people.

We are beginning now to veer away from this reductionistic, product-focused world. We are heading into a new space where relationships between people matter more than products and where human experience is what matters most of all. The new space is people-centred, not just user-centred. At this point in time, information and communication technologies have given people the ability to say what they think and to demand what they want, when they want it. And people are beginning to speak up.

Imagine a world whose future is made of the dreams of everyday people. This new world could offer a people-centred array of experiences instead of the mind-numbing, consumer-driven array of choices we have today. It could connect us more deeply to what really is important in our lives. It could be a more humanistic world. The design challenges we face in achieving this dream are immense (Alexander and Tromble, 2000), particularly in the digital domain (Joy, 2000).

To reach this new space, we will need to bring everyday people into the centre of the design development process, respecting their ideas and desires. People-centred design is a process of discovering possibilities and opportunities with people that addresses their needs and aspirations for experience. It must be a continual process. It is a process that has already begun (Levine *et al.*, 2000).

1.3 The Journey

We are on a journey toward a future being made from the dreams of everyday people. The part of the journey that I have travelled started with pre-school children. In the mid 80's I was asked to be the 'user advocate' on a project to design a headset product (using voice-recognition technology) for pre-school children. (Couch and Sanders, 1989). It clearly was a technology in search of a need. The product would ask the child questions and the child would respond with either 'yes' or 'no' in an attempt to win the game. Our task was to design the product form since the software design had already been completed.

The industrial designers I collaborated with were quite confident in their ability to serve as user advocates for adult users of consumer products but were not at all confident in their ability to represent pre-schoolers (two to six years old). They wanted answers to many questions. For example:

- The technology required that the microphone be at most one inch from the child's mouth. Would pre-schoolers find wearing a headset acceptable?
- Would they wear it long enough to play the game?
- How heavy could the headset be? How heavy should it be?
- What was the size range of heads among pre-schoolers?
- What should it look like so pre-schoolers would be attracted to it?
- What would their parents think about the idea of a headset product?
- What color or colors should it be? Does this differ for boys and girls?
- How can we determine one color palette that is liked both by boys and girls?

Since 'the literature' did not exist to answer any of these questions, we set out to collect our own data, enlisting the help of a local pre-school (being the parent of a recent pre-school graduate helped). Not knowing exactly which methods would give us the most useful results, we tried several ways to ask each question. The approach had to be very quick because we let each child leave the session when they wanted to go back outside to play. Knowing that many pre-schoolers are not verbally proficient, we invented exercises that did not require verbal output skills. The children could respond by selecting, pointing, drawing, coloring, and/or constructing. For example, on the colour issues, we looked at:

- colour palettes created by the children;
- pictures of the headset that the children coloured themselves;
- which of nine painted headsets they liked the best;
- what they said their favourite colour was;
- what colours they were wearing that day.

We ended up with an award-winning (from the design industry) product. But the product did not do well in the marketplace. Kids and their parents' thought that it looked *'awesome'* but the children said the games it played were *'boring'*. *'It asks me the same questions every time I play this game, Mommy!'* Parents did not return to buy more software and the product was eventually taken out of the marketplace.

As it turns out, the best 'product' of this experiment was the creation of the first set of generative research tools that inspired others to follow. More importantly, it was the beginning of our understanding of the participatory process. We found that pre-school children (as well as their teachers and parents) could be useful partners in the design development process if we give them appropriate tools with which to express themselves. We learned that pre-school teachers were a particularly valuable source of information about pre-school children. Their impromptu responses to the questions about colour replicated the data we obtained from the children.

We also learned that products must be designed holistically, *i.e.*, with hardware and software together. This last lesson is one that is only now being recognised and acted upon in the design development process.

1.4 The Landscape Grows

We created new tools and methods by applying the participatory approach on subsequent design projects with children. But the application of this approach to products for 'grown-ups' was slower to emerge. Apparently, designers who felt that they can represent users saw little need to bring everyday people into the design process as participants. So the first participatory design projects with adult participants featured either ill people or technically knowledgeable people in the design development of medical products and instruments.

Over time, however, we applied these new methods and tools to products for all types of people as well as to many other design domains including:

- information/communication design (e.g., packaging and user assistance materials);
- user interface design (e.g., for computer and communications devices);
- environment design (e.g., retail environments);
- design of branding strategies.

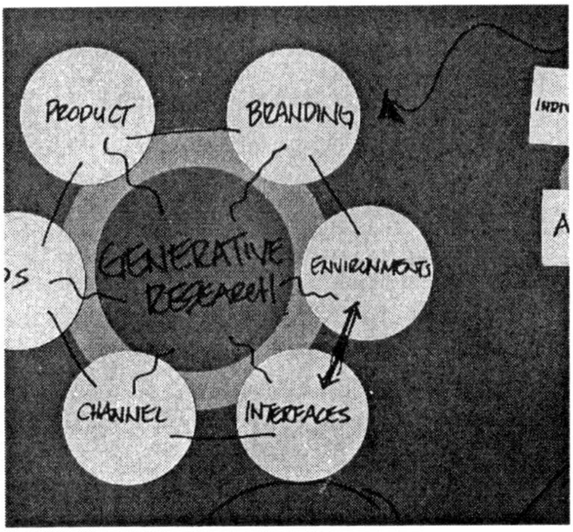

Figure 1.2. Tools for thinking: 'how do you expect your work to change in the future?'

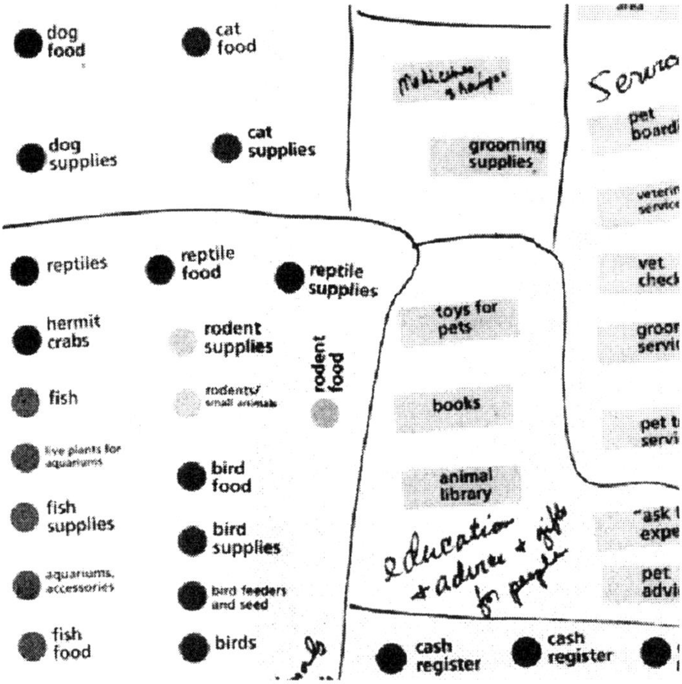

Figure 1.3. Tools for mapping: 'make a pet store that meets your needs as a dog (cat, horse, *etc.*) owner'

Figure 1.4. Tools for visioning: 'what will your work environment feel like in the future?'

Figure 1.5. Tools for feeling: 'use pictures and words to show a health-related experience in your past.'

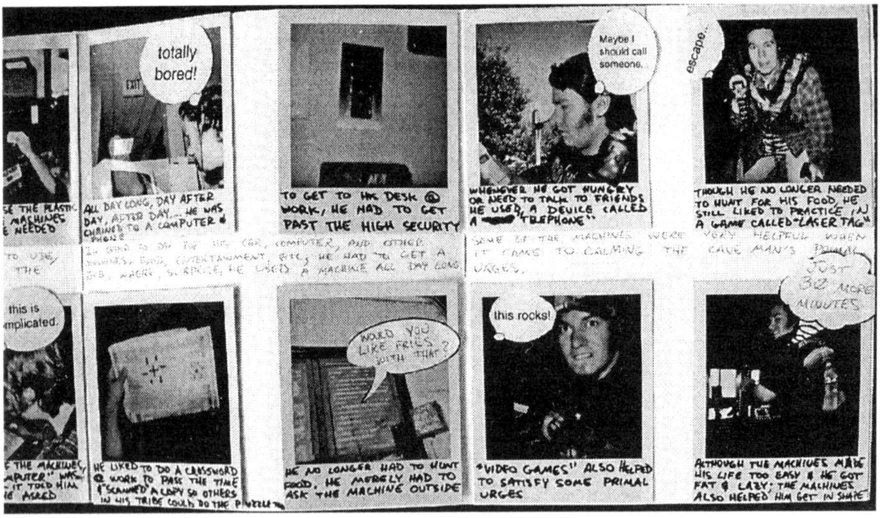

Figure 1.6. Tools for storytelling: 'tell us a story about your life with consumer products at home.'

Figure 1.7. Tools for dreaming: 'use shapes and stickers to make spaces for your ideal home experience.'

What do all these tools have in common? They take advantage of the visual ways we have of sensing, knowing, remembering and expressing. The tools give access and expression to the emotional side of experience and acknowledge the subjective perspective. They reveal the unique personal histories people have that contribute to the content and quality of their experiences. These are qualities useful to those of us involved in making people-centred decisions.

1.7 New Spaces

Other uses of the participatory methods and tools have emerged in the last few years. We are beginning to use the generative tools to seed a mindset of collective creativity in group workshops. We call them Strategic Visioning Workshops. In these workshops we use an assortment of large toolkits to enable a group of people to work together to express their ideas and dreams. Visual toolkits allow people the time and space to listen to each other's ideas and dreams, thereby facilitating more effective collaboration. The transformation that takes place when a group of people goes from a verbal exchange of ideas to a collective and visually expressive mode is remarkable. It is invariably positive and can often be quite therapeutic for the participants.

In addition to collaborative visioning, we have found that the generative tools are useful for collaborative thinking, mapping, dreaming and storytelling

1.8 Who Designs Now?

The use of generative tools for co-designing has implications for what it means to be a designer and a design researcher today. We can see that these roles are becoming mutually interdependent and are converging to the point where they are blurring. Who creates the tools for the new design language? Who interprets the results of this new design language? Designers and researchers will need to work together to explore the implications of this new language.

How does the emergence of the new tools change the nature of design education? Designers will be trained to go beyond the individualised expression of visual communication and learn how to become involved in the creation and construction of generative tools. The creative spirit of people in the design profession is essential for the exploration, development and analysis of the new tools.

Collective generativity is beginning to replace individual creativity. Mutual respect is essential. The use of generative tools for codesigning requires a new mindset on the part of designers and design researchers, *i.e.*, respect not only for each other, but also respect for the people who are served by design. This will perhaps be the most difficult challenge.

1.9 Looking Out Further

The emergence of a participatory culture and people-centred design will affect *what* companies design and produce in the future, with results more relevant and meaningful to people's lives. It will hopefully also influence the strategies that companies follow for designing and producing in the future (Lasn, 1999).

A collection of dreams is emerging. A collision is imminent. The aftermath will be new, experientially defined categories driven by a convergence of perspectives.

It will be a world made up of the dreams of everyday people. Maybe we can help it to grow (Brand, 1999).

1.10 References and Related Readings

Alexander C, Tromble M (2000) Code for creation: a steering function for genetic engineering. Limn 5

Barlow JP (1994) The economy of ideas. Wired, March 85

Brand S (1999) The clock of the long now. Basic Books

Chomsky N (1965) Aspects of the theory of syntax. Mouton

Couch J, Sanders EB-N (1989) Voyager: not just another talking teacher toy. Innovation, Winter

Jensen R (1999) The dream society: how the coming shift from information to imagination will transform your business. McGraw-Hill, New York

Joy B (2000) Design for the digital revolution. Fortune, March 6

Kelly K (1998) New rules for the new economy. Viking, New York

Krippendorff K (1996) A trajectory of artificiality and new principles of design. Unpublished manuscript

Lasn K (1999) Culture jam: the uncooling of America. Eagle Brook, New York

Leinbach CO (1999) About cake and the top ten things (more or less) Dorothy must have been thinking when she said Toto I've a feeling were not industrial anymore. Unpublished manuscript

Levine R, Locke C, Searls D, Weinberger D (2000) The Cluetrain manifesto: the end of business as usual. Perseus Books

McCloud S (1994) Understanding comics. Kitchen Sink Press

Pine BJ, Gilmore JH (1999) The experience economy: work is theatre and every business a stage. Harvard Business School Press, Boston

Polanyi M (1983) The tacit dimension. Peter Smith, Gloucester MA

Reed E S (1996) The necessity of experience. Yale University Press, New Haven

Sanders EB-N (1992) Converging perspectives: product development research for the 1990s. Design Management Journal, Fall

Sanders EB-N (1999) Postdesign and participatory culture. Paper presented at the Useful and critical: The position of research in design conference, University of Art and Design, Helsinki

Schlossberg E (1998) Interactive excellence: defining and developing new standards for the twenty-first century. The Ballantine Publishing Group, New York

Part I

Collaborative Design Methods

Part I Collaborative Design Methods

Introduction

This opening Part contains contributions relating to the promotion of team working in its broadest sense. These concern the better understanding of the process itself, the development of tools and techniques to support design team members, and the development of methods which promote user contributions at all stages in design (especially in terms of participatory design).

Goeller serves to place design in its global context, by looking at how a designer from one culture may design for people in another. In a global market a multitude of possible interpretations and misinterpretations can occur. Designers need to take into account these variations in interpretations and the diversity of human value systems by including users' physical and psychological needs early in the design process. This paper explores how these might be captured.

The need for users to be taken into account from the start of the design process is further considered by Porter and Porter. They focus on the relationship between ergonomics and design, and how the development of tools, especially when tailored to meet the requirements of designers can facilitate the consideration of human factors by designers. They outline three such tools (an ergonomics checklist, the extended use of package drawings and a questionnaire approach to identifying pleasure in product use), developed to meet the needs of designers working on projects of short duration.

Young, Van Der Veen, Illman and Rowley's contribution, whilst focusing on the role of the designer within the product creation process, considers an area of growing interest - sensorial attributes. The authors draw on case studies to look at evidence for cooperation and non-cooperation in relation to the integration of sensorial attributes within different companies. Their conclusions consider the extent to which experiences and emotional factors are integrated into the product-creation process. These indicate that co-design and communication in this field is vague, companies tend to focus on one sense, and then the physical rather than emotional attributes.

Oostra uses game theory to explain the role of the architect in traditional building and product development process, concentrating on the ways in which the building game may be manipulated by architects to favour the product development game.

The architecture theme is continued by Fowles in relation to group working in community architecture and ecological design. He considers different levels of user participation, the relationship between the architect and the user, and delineates ways in which 'partnership, balance and equity' may be achieved through symmetry, role interchangeability, transparency and ownership, and how involvement may lead to personal development.

Fowles' paper also moves the section on collaborative methods to case studies of participatory design in action. For many, the design process, although including user consultation, may still not lead to emancipation, empowerment and inclusion in design decisions. Luck presents an architectural design case study which uses semi-structured interviews and steering groups to enable the greater representation of the needs of physically impaired people.

Noble and Robinson consider participatory action research in the context of a 'live' community-based project. More especially they look at a model of participatory design which honours the knowledge and experience of the participants in the genre of documentary photography. Pederson and Buur outline new methods of extending user-centred design through the use of metaphors allowing participants to improvise social interaction and design collaboration. A case study on the design of treatment works illustrates how metaphors of game playing and film making can be used for the discussion of goals and activities through to embodiment and user interaction of the ideas during film making.

The case study presented by Dandavate, Steiner and William on the development of office products shows how user-centred design methods were used to identify a new product user, the Mobile Professional. A range of techniques was employed by the design team to understand what it means to be a Mobile Professional. For example, observational techniques used to document activities and facts that users were unconscious of or unable to articulate, were combined with participatory approaches to understand the user's goals, aspirations, and the emotional and cognitive context of their lives.

Again, looking at methods to facilitate team working, van Welie and van der Veer consider the use and function of different forms of representation at different stages of the design process. Their design method, DUTCH (Designing for Users and Tasks from Concepts to Handles), combines both formal representations (such ass NUAN) to give precision when required for product detail and specification, and informal representations (e.g., scenario play, videoclips) to allow for creative interpretation and communication of ideas to designers and clients.

The Part closes with a return to the effects of globalisation, although this time in relation to the ultilisation of worldwide resources (e.g., manufacturing capability and design competence) by multinational companies for competitive advantage. In their paper, Lindemann, Anderl, Gierhardt and Fadel present an on-going case study of a distributed, international, 24-hour design and development project involving three design teams. Distribution Methodology is used for identifying typical problems and developing solution activities.

Chapter 2

Mind that User's Mind: Incorporating Cultural Difference to User-centred Design Approaches

H. Goeller

2.1 Introduction

With today's ability to be in every part of the world at the push of a button, the ability to see, hear, learn about what is going on elsewhere is greater than ever. We are also increasingly able to fabricate pictures through the manipulation of visual information.. With this surge to a potentially worldwide 'connectedness' comes the opportunity, if not responsibility, for the design profession to consider its implications. While it might not be of any consequence if individuals are able to mentally connect with or believe in the validity of the pictures presented, when it comes to the acceptance of concepts and design ideas the judgements individuals make are rather important. Judgements are made based on references which are personal interpretations of individually selected visual, acoustical, haptic and organisation-related information. This information is said to be evaluated in respect to its adequacy and relevancy to the viewer. With today's potential for a product or space to be marketed globally the variety of possible interpretations has multiplied, as has the option and danger for mis-interpretation. Not knowing the context, not knowing the references used for interpretation makes it impossible to predict how an object will be judged or used. For many years, in educating the next generations of designers we have been advocating the importance of the careful study of the user along with the study of the existing or prospective context. In developing design objectives for a particular project we have been stressing the incorporation of data that relates to both the prospective users' physical as well as psychological needs, and asking for user participation early on in the design process, in order to detect the design implications effected by a diversity of possible interpretations. For the success of a proposed design concept the mental connections made by the audience, clients and costumers are crucial. While we accept that a designer's own innate references serve as guidance, we also need to accept that the chances they are shared are rather slim. In addition, if we acknowledge the fact that a design solution is only as good as the designer's ability to project, envision and translate identified needs, how can this ability be improved? Therefore, it seems essential to develop a consistent approach to obtaining and evaluating this kind of data, and to find ways to organise the process of translating this information to derive acceptable and meaningful design solutions. The intuitive approach would be to focus on meaning: a designer's intuitive interpretation of objectives, we learned, in most cases only addressed and successfully 'reached' a user group of the same level of visual literacy, and, maybe even more importantly, a user group with the same cultural background.

2.2 The Study of Cultural Context and Shared Values

Surya Vanka (1990) tried to address some of these issues by asking the question: how can we improve the intuitive acceptance of products and how can we overcome cultural stereotyping? This question lead to another: what does a designer educated in one culture - with the task to design a product or a space for a foreign market - know about the needs of people of another culture. With companies being increasingly interested in developing products for the international market, which criteria will a designer use to direct the design decisions? For a 'modern movement' advocate the answer might come easy: there should be one international design language - we all are of the same kind, functionally, anthropometrically, emotionally. Graffiti in the trains of New York, however, pictures on trucks crossing rural Columbia, decorations in tents of Siberian nomads, patterns in Arabic hallways, the geometric purity of Japanese buildings, speak to us of human life interpreted differently.

If they are manifestations of self-expression, they also are indications of an innate quest to give meaning to what we do, and who we are. If we could decipher these manifestations they could serve as references. Our notion that to offer one solution to fit everyone was supported by another aspect. That is, while exposure to, and appreciation for, cultural differences seems to increase, so does the need for personal identification; the faster information and ideas are exchanged, the stronger the need seems to become to cling to established personal values, as one's only true and reliable references. That would mean that in order to make a product recognisable, likeable, memorable, comfortable for the prospective user of a different culture, we need to find a way to help designers respond to the terms of that culture.

We started to search for specific factors that exist in each culture, and that could be used as references for design. The intent was to identify 'manifestations of meaning' that might allow us to establish correlations to culture-specific values. Table 2.1 shows our recommendation for consideration and steps to take for exposing culturally meaningful information.

Table 2.1. Vanka's process model

Typical starting point for project

DEFINE	Define the problem, product, project, client, designer, user, market *Definition Techniques*	Define the language and vocabulary that the client, designer are familiar with *Definition Techniques*	Articulate the philosophy, beliefs, ideology, concerns of the client and designer *Definition Techniques*
FILTER	Determine the design knowledge relevant to the problem *Filters*	Determine the domains within the user's culture that need to be investigated *Filters*	Determine the knowledge required to undersatnd the problem *Filters*
LEARN	Acquire the design knowledge needed to design the product *Principles*	Understand cultural relationships *Principles & Patterns*	Acquire higher level of knowledge pertinent to problem and method *Patterns*
SEARCH	Search for insights into design methods and innovations appropriate to problem *Principles*	Assimilate a vocabulary and an internally consistent model of the user's culture *Search Domains*	Search for insights into philosophical approaches within and outside product and culture *Patterns*
EXPLORE	Use design, cultural and philosophical knowledge to generate design concepts *Principles*	Extrapolate and add to cultural vocabulary and internally consistent model *Search Domains*	Investigate new philosophical ideas *Patterns*
CONSTRUCT	Use design knowledge to synthesize product *Principles*	Organize and structure the new vocabulary and internally consistent model *Search Domains*	Articulate the expanded philosophy; add or modify patterns *Patterns*
EVALUATE	Use design knowledge to evaluate the effectiveness of the product *Principles*	Evaluate the internally consistent model and vocabulary *Principles & Patterns*	Evaluate the exapaned philosophy ethically, ecologically, spiritually and pragmatically *Principles*

PRODUCT LANGUAGE PHILOSOPHY

It should be pointed out that behind a simple-sounding statement, such as 'understand cultural relationship' lurks a rather involved process of gathering information from very different sources. Each cultural context consists of an array of fluctuating factors that belong to different domains. These factors range from being very stable to being rapidly changing; 1. geography, climate and anthropographic data are factors that can be considered stable; 2. linguistic, social and religious expressions and patterns of behaviour are gradually changing; 3. artistic, political and commercial issues or principles are factors of rapid change. Examples collected in each domain and in different categories, including colours, textures and pattern, and also notions, artefacts and events will reveal a culture's 'deep structure' and the values shared by that particular group. This information can then be selectively applied as referential information, and can serve as the base for developing a cultural vocabulary. This approach, however, is time consuming.

It is questionable whether normal project budgets will support it. Our conviction, however, became even stronger that designing should not be an isolated artistic process, but the result of a synthesis of knowledge gained through the analysis of information obtained from and through all participants.

2.3 Do Shared Lead Values to an Appreciation of the Same Design?

To gain a better understanding of how a particular design concept is judged by different people, another study tried to identify whether shared values lead to the appreciation of the same design style. One fundamental aspect of a shared culture is shared values. Value-forming factors in a culture are affected by change; therefore, a group of about the same age from the same educational background and area would embrace similar values and preferences.

Could there be 'innate taste' categories, which might lead to a consistency in appreciating and selecting the same design style? The study by Gibson (1990) considered those rather different interpretations of what is seen as beautiful or 'tasteless' which often challenge a designer-client relationship. While a designer has to trust his own innate interpretation, designing without taking into consideration a different interpretation has often lead to conflicts arising too late in the design process to make changes. Simply expecting a user to adapt is an attitude no longer nurtured by education.

In this study we learned that even assumed shared values did not automatically result in 'shared taste' or the appreciation of the same design style. The study used examples of different tpyes of furniture, chairs, tables, lighting fixtures, and building facades, and established four categories of design styles. These categories reflected an application of decorational elements ranging from direct and obvious - termed 'literal' application (such as flowers painted on the surface of a plate) to indirect, hardly visible, 'abstract' interpretation of a form determining factor (such as a bowl vaguely resembling a flower). See Figure 2.1 for a contour sketch of the examples used in the category chairs and tables. While this study revealed that indeed quite consistent appreciation categories did exist within this group of prospective consumers, it differed greatly from individual to individual. These responses additionally enforced the notion that the assumption that appreciations are shared cannot be made, and that given no choice, a user might reluctantly and only temporarily accept an otherwise well-functioning solution.

Figure 2.1. An example of chairs and tables

2.4 Incorporating User-relevant Data

At the Department of Design the goal to prepare future designers for the challenges of the profession includes the incorporation of systematic approaches to better understand the needs and aspirations of clients, users, and consumers. Of particular concern is to improve the designer client relationship along with developing design concepts that are not the result of a designer's personal taste but which take into consideration the array of possible variations in interpretation and diversity of human value systems. My colleagues, Reinhart Butter, Professor of Industrial Design, and Klaus Krippendorff, Professor of Communications have been advocating a new approach to design for several years. What was termed by them, and became known as, the study of 'product semantics' took into consideration that products not only need to visually convey a clear message, but that its visual appearance along with the user interface and interrelationship needs to be predetermined by the designer. This well-published systematic approach to designing products enables a designer to incorporate information that is meaningful to users. It has completely changed how we teach design. What has become increasingly important in interior design is the systematic and conscious identification of individual interpretations, as we must aim to create environments that the inhabitants can identify with: spaces which evoke emotional judgement and have personal meaning. The French sociologist Pierre Bourdieu stated that the first level of understanding refers to what something looks and feels like, and the second to what it reminds us of. The latter becomes the most powerful component in rendering a judgement. Powerful because it relates to both conventions and cultural symbolism. This unconscious connection serves as a personal emotional pilot that is able to overrule logic, and may resist facts.

How can we make sure that these connections are respected, insofar as they might lead to solutions that we as designers might not anticipate or even appreciate? In respect of the fact that we all are processing and interpreting information based on past experiences, it is 'designing for experience' that attracts us most at the moment. New design solutions expect visions. To formulate the design parameters for such a vision, different approaches can be taken. One group of =professional designers, most of them graduates of The Ohio State University, under the direction of Liz Sanders, employ various methods which focus on users' experience. Prior to any conceptualisation or project specification attempt, users' expectations are studied. One of the research methodologies used - the 'life cycle scenario' - presents a collection of images and words to selected potential consumers. The pictures and words serve as stimuli and help the participants to express their thoughts, ideas and feelings. In this approach psychologists, business strategists and designers are working together to reveal unconscious expectations which allow a direct look at culture-related personal determination of needs. The information gathered is remarkable as it lends itself not only to collection of behaviour-based data but also to discovering new avenues for design.

If we share experiences, if we share the same or similar sources of information, we must have shared references. If employed well by a designer these could prompt a strong reaction of recognition and identification. Figure 2.2 shows one of the steps we take in interior design as part of the design process that is geared to determine - early on - any manifestations of meaning shared among the group. This is done to make the design student aware of his own connotations, and also to reveal variations. This approach involving clients and users utilises meditation in the form of high-concentration sessions to search for references that may exist in the subconscious of the participants. Figure 2.2 shows the results of one of these sessions searching for what is perceived as 'natural'. The search is directed to unconsciously-stored information that is linked to shapes, forms, patterns, rhythms, material, textures, sounds, taste, smell, colour and light, and organisation or interfaces. All participants are advised to document their responses using pictures rather than words, because words will force the person summarising the results to re-interpret. Once the connotations are revealed and compared to identify the most frequent and therefore strongest ones, the question of how this information can be used is still a major challenge for a designer. As in the study described above, it will be the designer's responsibility to decide if these connotations should become references and should be, for example, used in an immediately obvious way (direct application), or be integrated such that the user will not necessarily see but rather 'feel' them (abstracted application). From the educational viewpoint, both the intellectual and the less visible but still sensual interpretation of the learned data is emphasised. When the first ideas are formulated the user participates again in order to test the adequacy of the designer's translation, the level of recognition, and associations made. This process still has some weakness due to the designer's ability to manipulate the selection. Only manifestations that make sense to the designer might get selected. Another drawback, especially in designing spaces, lies in the wide range of options in applying the resulting visual elements.

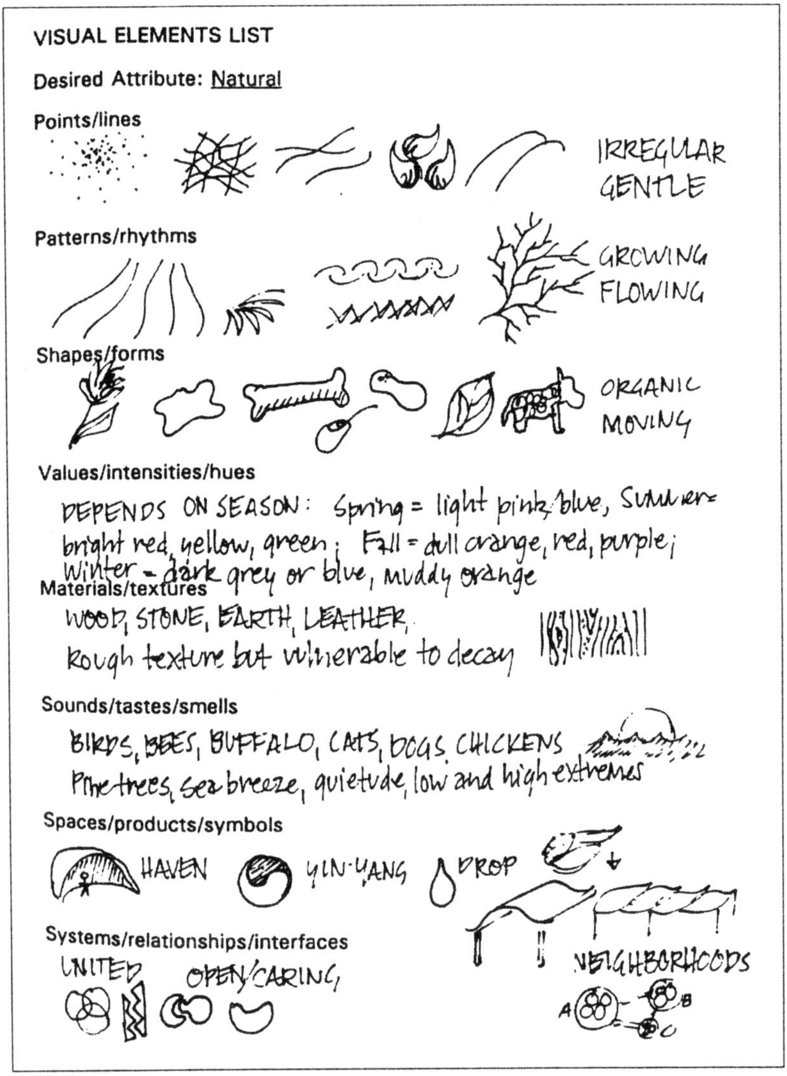

Figure 2.2. Worksheet identifying manifestations of meaning

The tendency to try to use some of the obvious references in a rather unimaginative way or, due to the complexity of the task, to let some go, is rather strong. Mike Brokamp (1993), also a graduate advisee of mine, suggested an additional step. If we can make use of the basic elements that are already unconsciously used by humans to store space-related information, a user-friendly space organisation and orientation might be greatly enhanced. For interior and environmental design, cognitive maps are sources of information of great value. If the participants' expectations linked to an existing mental pattern could be visualised, an organisational strategy tool could be developed that could support the intuitive knowledge of what is going on in a particular space. Figure 2.3 shows the visual elements used by a group of doctors and nurses discussing spatial relationships and options for the organisation of an operating room.

Group Consensus Study Results

Cognitive Element	Group 1	Group 2	Group 3
Store	□	□	□
Walls	~	□	□
Corridor	~	~	□
Gathering Place	○	○	○
Info Center	□	○	○

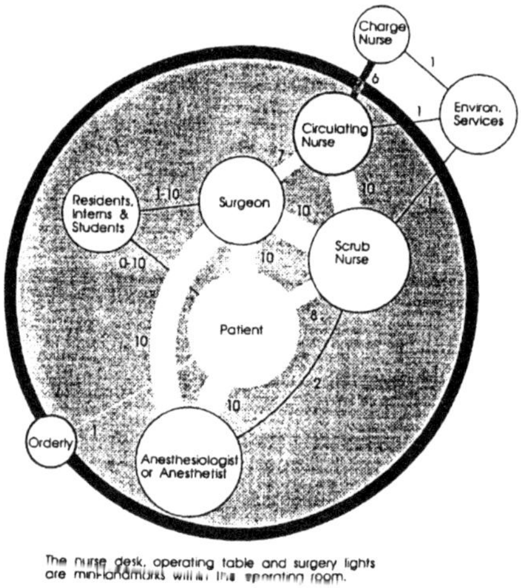

Figure 2.3. Group code concensus (above), cognitive landmarks map (below)

2.5 Will It Make a Difference?

The argument could be made that given more options to access information faster than ever before, as mentioned earlier, the need for developing culture sensitive design solutions might decrease. With increased choices consumers are no longer forced to adapt to products that do not suit or please them. Culture as a regulating and decision-influencing factor might have less of an impact, as the more accessible the market becomes globally, the more we will find variety. As a result the strength of innate cultural identity-related factors could diminish, and a designer presenting his own individual style will do no harm. It is only a question of marketing and the costs included in the product-development that might be still a concern.

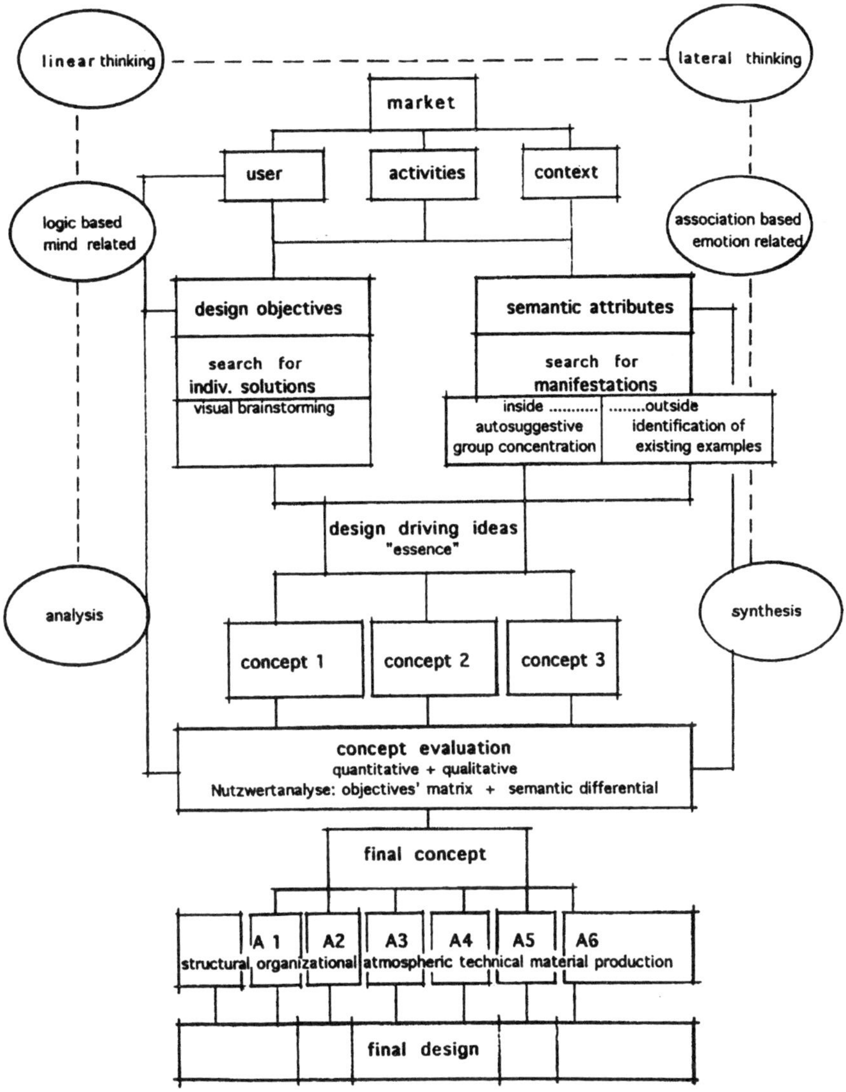

Figure 2.4. Goeller design process

Recent developments, however, show that new kinds of groups sharing similar values are forming across existing cultures. The term culture is now used more often referring to potentially shared goals rather than ethnic values supporting this shift. The success of chain restaurants and concept stores world-wide suggests that it might not matter which country an environment is designed for, as it is the projected image that consumers (are supposed to) identify with. Maybe in the future the variety of lifestyles will displace certain aspects of design which are stillof major concern. The aim of better access, better response to, and better communication of design options that respect and encourage diversity, is still important today.

Figure 2.4 shows the process and project steps introduced to my students in interior design. The sometimes rather unclear part of how, and on what basis, a particular design solution is obtained is made visible by this process, as the steps call for the systematic gathering and documentation of project-context information.

This process also demands working in close contact with the client and potential users at different points of the project's developmental stages: at the beginning for defining the context and design objectives, in developing the semantic attributes and manifestations of meaning, and at various times for evaluating interpretations and design options. The different steps within this process seek to strengthen the designer's thought processes and ideation capability by stressing linear and lateral thinking alike. Both logic and emotional factors need to be addressed to enable a designer to make a convincing statement in the attempt to project tasks and relationships and to envision environments that provide individually meaningful experiences. The projects developed in this way will, hopefully take into consideration the potential, uniqueness, and future of the prospective users. With the opportunity to create virtual environments, we can easily test to what degree the spaces created will have shared interpretations. Each designer's responsibility and goal must be to create design solutions that establish strong and meaningful connotations in the mind of the user, as only through can be as unconscious identification, which is essential for the user's emotional and functional stability to come about.

2. 7 References

Brokamp M (1993) Development of a computational model for defining meaning for interior space design attributes. MA Thesis, Department of Industrial Design, The Ohio State University

Gibson K (1990) Manifestations of meaning, the design process redefined. MA Thesis, Department of Industrial Design, The Ohio State University

Krippendorff K, Butter R (1989) Product Semantics. Design Issues 5:2

Vanka S (1990) A methodology for cross-cultural semantic design. MA Thesis, Department of Industrial Design, The Ohio State University

Chapter 3

Co-designing: Designers and Ergonomics

C.S. Porter and J.M. Porter

3.1 Introduction

It is clear from the literature, and practice, that the most successful co-designing between ergonomists and designers happens in contexts where they are able to work together in a team and ergonomics and design are able to happen concurrently (Haslegrave and Holmes, 1994; Porter and Porter, 1998). However, this 'ideal' is often not achievable in small design houses as a consequence of financial and other constraints. It is, likewise, often not achievable for many design students working on short design projects with little or no support from an ergonomics 'expert.' The way in which this problem is dealt with may be by ignoring, or by paying 'lip service' to the identification of ergonomics issues and problems and the appropriate ways of dealing with them. Hasdogen (1995) supports this view; in her investigation of human models used by product designers and ergonomists in the product design industry, she also found the product designers critical of the ergonomics models and methods available to them.

In a previous paper (Porter and Porter, 1999) we examined, through the literature and a review of good practice, the possible reasons for the lack of uptake (in some areas) of ergonomics data and methods. We identified that the communication of ergonomics information to designers, and the use of this information by designers, may fail for a number of reasons, including:

- communication at an inappropriate point in the design process;
- communication difficulties between ergonomists and designers caused mainly by education and practice differences;
- communication of ergonomics information, data and methods in an inappropriate fashion by ergonomists.

The literature supports these reasons, identifies further reasons and also demonstrates an increasing interest in this area. Burns and Vicente (1997) note that ergonomics data and methods are too onerous for designers to use. In their study of engineering designers (who are arguably better equipped to deal with science information than arts-based designers) they found there is a perceived high cost associated with the finding of appropriate ergonomics information amongst the mass of that potentially provided. In a later publication (Vicente *et al.*, 1998) they described designers as 'conservative searchers, probably as a consequence of adapting to domain constraints'. It is their firm conviction that current handbooks are not the answer to the problems of communication of ergonomics information to designers at any point in the design process.

There are many data gaps in the information provided, particularly where emerging technologies are concerned. In a recent survey by Bonner (1998) a lack of existing guidelines for contemporary interfaces was identified. The consequences of the development of microchips, embedded control systems, multi-function controls and multi-modal displays gives huge choice to the designer for the product/user interface. Many of these products are small, hand-held and require the hands to be used concurrently but the ergonomics literature provides little beyond the anthropometric data for the hands; data which is of minimal relevance for a product the size of a matchbox with ten different functions. The opportunities for getting design wrong are enormous.

There are also other emerging issues in ergonomics which are of huge interest to designers such as designing pleasure into products. In the work of Jordan (1997,1998) a theoretical framework for the different types of pleasure one may experience in product use is described and also some of the emotions associated with pleasure in product use are identified. However, there is no advice to the practitioner as to how this information may be best used as a part of the design process.

This paper describes three very different ways in which we have attempted to overcome some of these problems whilst working with design students. The first is an ergonomics 'checklist' devised to aid student designers in defining their product and the ergonomics issues associated with it, in the very early stages of the concept design process. The second method is an interactive 'package drawing' which allows a first pass at issues of fit and visibility and vehicles and the third is the way in which a student designer adapted ergonomics research into pleasure in products into a practical input to the design of an automotive product. In each case data and methods have been adapted to suit both the needs and the natures of the 'user population'.

3.2 Co-designing Example 1: The Checklist

Ergonomics is a part of all Product and Transport design students' project work, at Coventry School of Art and Design; in many cases carrying a separate mark. In the short timescales available for many projects there is a need for quick and easy tools. A review of the available tools and techniques to facilitate the identification of ergonomics issues was undertaken (see Stanton, 1998, for a review of some of the methods suitable for product evaluation); no existing tool appropriate to the needs of students working on projects as short as 5 weeks with little available ergonomics support was found. Most tools required either more time than the students had available, or the support of a trained ergonomist in order to be successfully used. It was also extremely difficult to identify a tool which could be used in the very early stages of concept design. The reality of many student design projects is that the outcome is based very much on the early designs with short timescales leaving little time for the process of iterative design. Tools for use in the early stages of concept design do exist; for example, the Automotive Designers Ergonomics Clarification Toolset (Woodcock and Galer Flyte, 1995) provides a structure for the systematic investigation and recording of ergonomics issues in the early stages of the automotive design process. This tool, and others like it, provide a comprehensive approach but are much too time consuming to suit the needs of designers working on short projects. The primary aim was to produce a tool which allows the early identification of ergonomics issues in product design, is easy and quick to use. It was also desirable that students could use the tool without staff intervention or support and that it provided a record of their thinking.

The checklist was developed, using the expertise of both authors. It covers approximately 4 sides of A4 and is divided into 3 sections. In section one the user is required to characterise their user population (e.g., sex, age, range of percentile values to be accommodated, *etc.*), in section two the user is required to characterise the intended product (e.g., how often is it used, are users forced to use it or do they choose, is it used in emergencies, is it a luxury product?), and in section three the user is asked to consider a broad range of ergonomics issues and how they are relevant to the particular design proposal (e.g., control and display type, allocation of function, maintenance, documentation, vandalism, *etc.*). There is also a space in each section for the users to make notes and respond to the questions and issues. See Figure 3.1, overleaf, for an example of sections of the Product Design checklist.

A pilot study of the checklist was carried out using two groups of students. The first group was of final year students (n=15) who were undertaking their major project over a period of five months, with a research and design phase of approximately 2.5 months. The second group was of first year students undertaking their first design project with only 5 weeks from start to completion (n=60). Both groups of students had completed basic ergonomics theory courses and had individual and large group tutorial time, respectively, with a qualified ergonomist. The final year students were issued with the checklist, completed it outside class and it was used as a basis for individual tutorial sessions at a later date. In the first year group the students were undertaking the design of one of five different products and the list was used as the basis for group 'brainstorming' sessions with an ergonomist.

The final year students were all able to complete the list in approximately 30 minutes and the interrogation of the information at a subsequent tutorial session was found to extremely valuable in ensuring student effort in the right areas. The ergonomist was able to direct the student to the appropriate literature and help them in the design of 'quick and dirty' empirical data collections where necessary. For the students, it meant that they considered all ergonomics aspects of their product as they began design work and not (as is sometimes the case) only towards completion of their design.

Ergonomics Checklist - Product

In all of your design projects there are ergonomics issues. This checklist is designed to help you identify the ergonomics issues, and the design implications, in your project in the early stages of the design process.This will allow you to incorporate the ergonomics requirements as you design the aesthetic of your product.

Name Project

Characterise your users. Their

	DesignConsiderations	
sex?		
age range?		
educational capabilities? e.g., varied, degree		
Are they a skilled or naive user group?		
What % of the user population do you intend to design for?		
Which body dimensions will be critical to the efficient usage of the product? eg popliteal height, hip width		
Are you designing to accomodate any population extremes, e.g., elderly or disabled? If so, who?		
Does your user population have a particular culture or lifestyle, e.g., 'club', 'youth', 'japanese'		

Below are listed a wide range of ergonomics topics/issues which may affect your design. Please tick those which do and state, which particular aspect is of importance and describe their relevance. We will discuss the best way to address your requirement (e.g., recommendations in the literature, user trip, task analysis or 'quick and dirty' user trials) during our appointment.

Topic	√	**Particular relevance and plans for information acquisition**
Allocation of function between product & user; will this be as is current convention, e.g., man controls car.		
Displays - location - type - aesthetic		
Controls - location - type - aesthetic		
Layout of controls and displays		
Labelling/icons		
Fit - user to product		

Figure 3.1. Product design checklist

The way in which the tool was used with the first year students gave them an effective introduction to the tool and expectations for ergonomics in design projects. The brainstorming sessions took approximately one hour and left each group with a completed checklist to use as a basis for their design. Some minor alterations were made to wording and content following the pilot study.

The checklist is now in place as a part of all student individual and group project work. It has resulted in the formalisation of ergonomics requirements of projects (regardless of whether there is input from a qualified ergonomist available) and is becoming widely accepted a starting point for all projects.

3.3 Co-designing Example 2: The Package Drawing

The 'package' drawing is the industry standard for presentation of the layout of mechanical components and occupants in a vehicle. Traditionally, these are a set of static 2-dimensional (2D) orthogonal views (side, plan and front/back) drawings in full size or to scale which describe the vehicle layout. The mannequins used are generally static in standard 5th and 95th percentile sizes and postures. In industry the issues of 'fit', visibility and layout are now more commonly studied in 3-dimensions (3D) using a CAD package. For transport design students, they are often the basis of (in some cases the totality) of ergonomics presented for assessment. Previously, the emphasis has been placed upon the package drawing as a piece of artwork, rather than on the content, with students being encouraged to use their presentation skills to improve readability (see Figure 3.2).

The aim was to extend the use of the package drawing to include more evaluation of ergonomics issues at a basic level such as ingress/egress, reach to grab handles, storage, maintenance and worst case fit scenarios. Most students do not have the opportunity, for reasons of time and resource, to explore ergonomics in 3D using either CAD or physical mock-ups.

Students are encouraged to make jointed movable mannequins which describe the population for whom their product is designed. These include worst case scenarios, e.g., long legs and short arms making comfortable legs and reach to the steering wheel a complex problem. The most recent accessible data are used (Peebles and Norris 1998) to maximise the usefulness of the activity. The students are then able to make a visual assessment of, for example, comfort in the vehicle, ingress and egress to the vehicle, reach into storage areas, reach into maintenance areas and reach to grab handles. This exercise can be particularly rewarding in vehicles other that cars, e.g., heavy goods vehicles, agricultural and other industrial vehicles, where there is often a much greater functionality to be explored. Once investigation is complete the student is able to draw in appropriate fixed mannequins which demonstrate fit and ranges of adjustment.

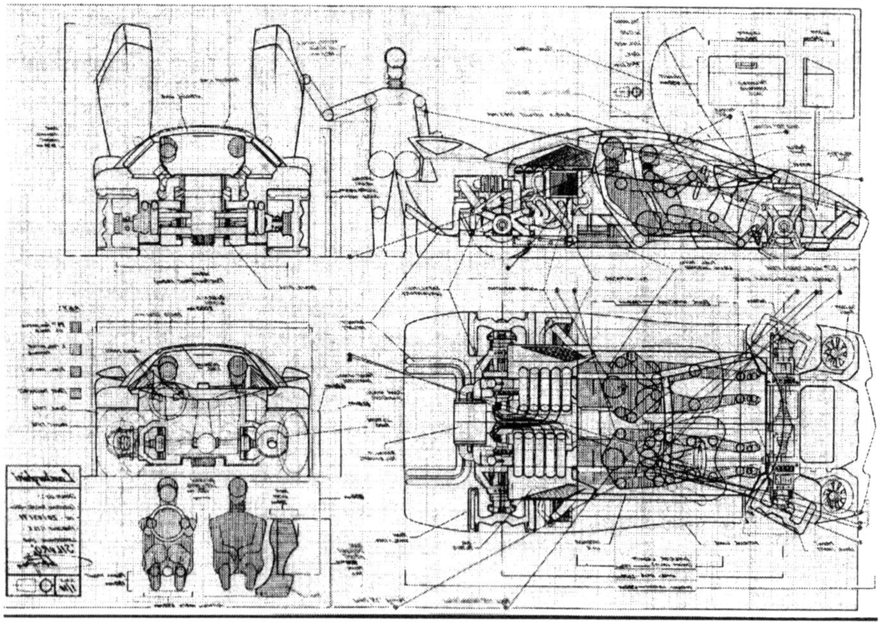

Figure 3.2. Example of a student package drawing

The technique is proving extremely successful in encouraging transport design students to think more broadly about how the occupant packaging issues in their vehicle design may be resolved. It is essentially a 'visual' method which appeals to 'visual' people and allows them to do a greater amount of basic ergonomics than would otherwise be possible. Accepting that the technique is fairly crude, highly subjective and that validation would only be possible with some form of 3D realisation, use of the package has greatly expanded the number of informed design decisions and compromises that the designers are making.

3.4 Co-designing Example 3: Designing Pleasure into Products

Jordan (1998) has demonstrated that pleasure in products involves more than simply usability. He identified feelings associated with both pleasurable products (security, confidence, pride, excitement and satisfaction) and displeasurable products (annoyance, anxiety, contempt and frustration). He also described salient product features which influenced feelings of pleasure/displeasure such as features, usability, aesthetics performance and reliability. In a later publication (Jordan, 1999) he adapts Maslow's hierarchy of human needs into a hierarchy of user needs to be satisfied in product design. At the base level is functionality rising up through usability to pleasure (something offering an emotional benefit). Pleasure is defined in four ways, based upon the work of the social anthropologist, Tiger:

- Physio-Pleasure - to do with the body. It includes pleasure associated with touch, taste and smell, e.g., the smell of a new car.
- Socio-Pleasure -pleasure derived from the company of other people, e.g., a product which is a 'talking point' and, therefore, facilitates social interaction
- Psycho-Pleasure - pleasure gained from accomplishing a task, e.g., a word processor which enables quick and easy formatting may provide psycho-pleasure
- Ideo-Pleasure - pleasure derived from theoretical entities, e.g., the aesthetics of a product or the values that the product embodies.

As Jordan states there is much more work to be done to develop this theoretical base into useful information that designers are able to incorporate directly into their designs. This case study describes an automotive design student's attempt to do just that.

The aim of the project was to design a Sports Utility Vehicle (SUV) with a first priority of pleasure for the user. Using the Jordan framework as a basis, a number of research activities were carried out. The methods and tools used were simple and where possible visually-based to suit the student and the time available.

Data were collected about the pleasure associated with 6 different products; cars, clothes, computer consoles, jet skis, mountain bikes and watches. Two owners of each product type were given a semi-structured interview concerning the aspects of the product which gave them pleasure. Subjects were between the ages of 18-40 and were predominantly male.

Concurrently, a questionnaire survey was conducted to investigate the reasons for purchase, the advantages, disadvantages, usage and pleasure aspects of SUVs. The survey was carried out in the car parks of supermarkets, DIY stores, electrical superstores and home stores in the Midlands and the South East. 18 owners of vehicles were interviewed, 9 female and 9 male, between the ages of 25 and 48.

An analysis was made of the pleasure adjectives associated with the 6 products and the SUVs. They were placed within the hierarchy of user needs described earlier. The SUV owner interviews were further analysed and revealed the motivations for purchasing an SUV and the advantages and disadvantages of ownership. This information was used as a part of the design specification and is described in detail in Rice (2000).

- <u>Pleasure</u> Stimulating, Absorbing, Luxurious, Fashionable, Characterful, Sexy, Glamorous, Rewarding, Importance, Living on the edge, Stylish, Modern, Timeless, Smart, To be noticed, Feel of materials, Use to limits, Risqué, Camaraderie, Raw.
- <u>Usability</u> Simplicity, Practical, Reliable, Tough, Refined, Ease of use, Comfortable, Technical, Control, Rugged, Quality, Capabilities
- <u>Functionality</u> Powerful, Educational, Economical, Precise, Sporty, Multi-purpose, Freedom

There are many adjectives associated with pleasure and further analysis was conducted to find the most popular adjectives for each product type. They did vary from product to product, e.g., for the watch the pleasure was associated with materials and aesthetics. The most common adjectives used to describe the owners pleasure relationship with an SUV were:

- Rewarding
- Use to the limits
- Living on the edge
- Being noticed
- Luxurious

Whilst the adjectives are interesting they give no direct information which helps the designer. A method was devised which would give the designer a better understanding of the adjectives and some visual clues as to how to express them in design. An informal focus group was organised. It included both design students (non- SUV owners) and SUV owners. Images of all types of SUV were collected and included in the activity. The group was directed to discuss the adjectives and associate images with them, as follows:

- Rewarding - Personal satisfaction from multi terrain travel, obtaining best performance from the vehicle, multi-tasking nature very satisfying. Image chosen: Cherokee Jeep.
- Using to the limits/ Living on the edge - Reaching the boundaries of the driver and the vehicle capabilities, being able to tackle the most difficult terrain, driving hard on open road. Image chosen: Landrover Defender.
- Being noticed - Status, high social rank, higher, bigger, ostensibly superior, pleasure from being safe. Image chosen: Range Rover.
- Luxurious–Quality - finish, interior more like a Jaguar or Rolls Royce. Image chosen: Mercedes M class.

This exercise provided some visual clues to the designer as well as giving more detail to the adjectives. The obvious limitation was the population used which contained only a small number of SUV owners and student designers so the results may not have been representative. However, adapting and extending the framework developed by Jordan (*ibid.*) facilitated a much greater understanding of the pleasure issues of the automotive product being designed. It also demonstrated one way in which a subject developed in a theoretical manner might be used in a practical way in the design process.

3.5 Conclusion

As stated earlier, in an ideal world designers and ergonomists would always work in teams together. However, if they are not able to codesign then the type of tools and data ergonomists provide need to be very carefully considered and specified. Designers have different needs to those of the ergonomist and they use ergonomics information and tools in different ways. To successfully design for them, the ergonomists need to fully understand their needs and requirements, in precisely the same way as they do for any other user group for whom they are designing.

3.6 References

Bonner J (1999) Personal communication

Burns CM, Vicente KJ, Christofferson K, Pawlak WS (1997) Towards viable, useful and usable human factors guidance. Applied Ergonomics 28, 5/6:311-322

Haslegrave CM, and Holmes K (1994) Integrating ergonomics and engineering in the technical design process. Applied Ergonomics 25, 4:211-220

Hasdogen G (1995) The nature and limitations of user models in the household product design process. Design Studies 17, 1:19-33

Jordan PW (1998) Human Factors for pleasure in product use. Applied Ergonomics 29, 1:25-34

Jordan PW (1999) In: Green WS and Jordan PW (eds.) Human factors and product design. Taylor & Francis, London, pp15-25

Peebles L, Norris B (1998) Adultdata. Department of Trade and Industry. Crown copyright

Porter CS and Porter JM (1997) In: Green, WS, Jordan PW (eds.) Human factors and product design. Taylor & Francis, London, pp15-25

Rice S (2000) Pleasure versus utility. Internal Report. Coventry School of Art and Design, Coventry University

Vicente KJ, Burns CM, Pawlak WS (1998) Better handbooks, better design. Ergonomics in Design :21-27

Woodcock A, Galer Flyte M (1995) The opaque interface - the development of an on-line database of ergonomics information for automotive designers. HCI '95 Adjunct proceedings, pp96-104

Chapter 4

Creating Enhanced User Experiences: The Designer is a 'Co-operator' by Facilitating Communication

R.A. Young, G.J. Van Der Veen, M.E. Illman and F.J.B. Rowley

4.1 Introduction

'We still currently think of the camera as a product dedicated to the task of simply capturing images, but it is not just a camera any more. It's a camera, an album and a way of editing and choosing. Somehow the design expression has to support all these things. For Kodak, the challenge is to create an interactive experience that supports the brand and the feeling of the right "Kodak moment".' (RSA Journal, 1999)

Product design, together with manufacturing industries, is moving towards what is referred to by Pine and Gilmore (1999) as the 'Experience Economy'. Goods and services have become commoditised and company success now depends more than ever before on meeting the improved quality expectations of the customer base.

Where designers have had some practice of the process of creating 'added' functionality in products, the designer's role has begun to move towards that of 'co-operator' by communication *i.e.*, a Co-Designer. This paper sets out to justify and expand upon this statement, based on the interim findings of a research programme. The feature underlying the statement is that the designer is someone who has to work alongside colleagues from other disciplines within industry. A collaborative working process of this nature requires a significant willingness to cede excessive personal aims to those of a mutually agreed agenda. Furthermore, the designer appears to play an increasingly influential role in facilitating this co-operative process, (Fujimoto, 1990). The chief characteristic that the design profession appears to possess, which is influential to the success of the process, is the designer's role as a generalist. This implies the capacity to see problems from both holistic and specialist perspectives and to empathise with the emotional needs of the user.

Alongside the essential creative design contribution to the operational activities of a project, the designer as 'co-designer', has to act as arbiter for all the other disciplines which are involved. In order to deliver these undertakings the 'co-designer' must communicate the most appropriate actions, not only with regard to the designer's own set of rules, but also according to those of the entire team.

The product creation process functions by continuously sending information to others, processing the responses it elicits and feeding back the outcomes to progressively tune and control the track of the emerging design, (Fawcett, 1990). Within this era of product creation, it has become clear that the participating disciplines have also to evolve their own modes of integrated practice in order to communicate their intentions and realise their design concepts as manufacturable products. The social dynamic of the designer's role is becoming more important, in effect through 'co-operation' a new breed of design practitioner is being born. What follows, are the first results from our continuing investigation into the designer's role within the process of creating products.

4.2 Form Follows Function Follows Sustainable Satisfaction

4.2.1 Today's Communication by Consumer Product Manufacturers

Today's management briefs for new products start by recording market demands, sales and strategic gap analysis. Similarly, consumer lifestyle and intelligence reflect the behavioural patterns and demands of potential new customers. The problem with this data is that it only shows a picture of the past. It gives no indication of the current situation or an anticipation of the future.

New emerging societal needs are forcing an enhanced product satisfaction.

'Now: U= f(m,P), in Future: U= f(M,p)'

(Where: u = product use, f = function, m = mental attributes, p = physical attributes)
(NOTE: Capital letter indicates point of emphasis)

Manufacturers see the 'functionality' of the majority of today's consumer goods as the key purchasing criteria for consumers. Such functionality can be seen as the interplay between physical attributes and emotional factors. When both are successfully applied, the expectation is that it will lead to a product which satisfies the user.

At this moment, much of a product's functionality has been reduced to purely materialistic concerns and therefore offers little in terms of emotional engagement and interaction. Many products are still developed along the 'Form Follows Function' principle, (Sullivan 1896), but is has become too focussed and short-sighted, as the principles address a materialistic viewpoint. Satisfaction in such products and services depends only on the pleasure derived from the interaction or direct physical contact with one or more of the senses. Ergo, much of today's effort is spent gaining optimal technological performance in only one of those senses, rather than basing the potential for derived pleasure on the user's broader notion of satisfaction. This is a satisfaction created by more or all of the senses including the creation of a total contextual experience, *i.e.*, a 'sustainable satisfaction'! The definition of sustainable satisfaction is one based on human needs and a better balance between the multiple contextual issues which generate it, *i.e.*, ranging from instinctive personal, physical and emotional factors to even those of societal change.

It is foreseen by many commentators (e.g., Harvel, 1994) that these contextual, rather than physical, issues will become increasingly relevant in future products and that physical issues will shift to have a more mental or emotional relevance. If this is accepted as a basis for change, it is possible to imagine that all functionality in the future will be derived from conceptual needs. Other new streams of thinking such as sustainable technologies, *i.e.*, 'less is more' (Schulze, 1995), will also support this shift.

The follow up to this development is that the next generation of products will offer added value in user interaction and satisfaction in relation to combinations of the senses. The interaction design will be based on a carefully considered contextual reasoning, which supports the fulfilment of fundamental human needs.

4.2.2 Enhanced Communication by Consumer Product Manufacturers

These new values demand a corresponding improvement in communication. In the main, these current methods of communication offer only simplistic visual indications about where a company's product range is positioned against its main competitors. Though these visuals often include other information, there is little evidence of tools being used to convey a common and universal understanding of the sensorial information, qualities or attributes, which lie at the heart of user expectation and experience. This implies that there is an emergent need for an enhanced level of communication and the authors believe this can be achieved by re-examining the senses and starting from a human-centred Product Creation Process (PCP), based on the aim of creating sustainable satisfaction. After the materialistic view of the 'Form Follows Function' principle, it is now time to re-create our co-design communication within the PCP.

4.2.3 Enhanced Design: Revived Interest in the Senses and Human Satisfaction

Special interest in the senses and design is not new. During the nineteen seventies, the senses were well researched. Debates were held as to what touch is and how the five senses are addressed in modern product design (Mendini, 1989). Regular studies emerged giving overviews of the impact of the five senses in contemporary design (Branzi, 1984). Still many of these developments were based on the exciting technological progress made at that time. Latterly, during the eighties (Manzini, 1986), new materials with new properties epitomised the change in (human) interaction, creating a new era in product functionality.

During the Nineteen Nineties there was a strong revival of the senses in product design but with an additional starting point; human needs as a driver for new product functionality and the emergence of an emotional value being quoted as a sixth sense (Marzano, 1992). While the reasons for a revived interest are conspicuous, the issue of brand and brand value are important. Presently this research has indicated a great need for the development of tools to measure and control sensorial perception of attributes across disciplines so that the PCP can be more holistic in its progress towards human-centred product values and attributes. The role of the designer in this process is therefore becoming that of an 'agent for change' (Mitchell, 1993) and the question posed by this evolved role is, how can the emotional value be conveyed? Historically, visual icons and text were used, e.g., mood boards. However, today's products have to comply with more constraints. As a result, more holistic thinking is needed. Therefore, the profession will have to develop new methods and means of communicating. The new modes of communication will need to reach and inform those involved in the PCP more effectively than existing modes of two and three-dimensions. In developing new methodologies and approaches the design profession must become effective in accommodating the full range of human senses. This requires the exploration of the new mental and physical contexts in which new concepts, artefacts and services will exist.

4.3 Co-operating Design: New Roles

Today's designers are already trained to work and collaborate in several stages of the PCP and they are equipped to visualise and communicate solutions to problems from the generation of the idea to its development as a product. Most of this work involves the realisation of the physical configuration and hardware of the product. As tomorrow's product creation strategies begin to cater for the emotional dimension, so must the framework of perception surrounding the designer's contextual thinking.

4.3.1 The Triple Change of Design

Armed with a greater knowledge of issues that lie at the forefront of the development process, tomorrow's designers will not just be involved as a visual creator in the PCP. Design is already stretching its input into future prediction and scenario writing; in the future, this role will extend further into developing conceptual models that can define strategic business and product planning paradigms. In order to be able to embed the consideration of emotional and physical attributes into the planning process, the discipline is, and will continue to change, in three key areas. The triple change involves:

4.3.1.1 Societal Change: Design Bridging Consumer and Product Manufacturer Domains

There is clear evidence supporting the fact that consumers are rapidly becoming more knowledgeable and demanding in their purchasing habits, (Markham, 1998; Underhill, 1999). This is now being seen through the established routes of commerce and is putting greater pressure on both retail and manufacturers. In contrast, manufacturers have been forced to create astigmatic products because of increased global competition. Such products are increasingly tailored to suit 'first fit', narrowing the consumer flexibility (Kotler and Armstrong, 1989). The ability to create a superior product is compromised when it is set against a strictly limited manufacturing budget. However, the designer is in the best position of any of the PCP disciplines to balance and 'communicate' both legitimate demands, (Fujimoto, 1990) and bridge these 'conflictive' interests.

4.3.1.2 Product Creation Process Change: Technologically Driven by Human Pull

The rapid changes occurring in consumer product markets mean that the traditional notion of the PCP will also need to change quickly. Where technology was once the predominant factor, this is no longer the key driver. Consumer preferences, human factors and satisfaction will re-shape the PCP and override today's efficiency focussed systems.

The question remaining is; how will manufacturers deal with creating products for this most demanding of markets? There are still very few companies equipped to know how consumer demands can be described in their product portfolio. The human element is the constant 'red thread of interest' which necessitates the future co-operation between the designer, engineer, ergonomist and businessman alike. Only through better co-ordination, communication and co-operation will tomorrow's companies be able to satisfy tomorrow's demands. It is therefore crucial to generate a mutual understanding, a 'creation' language perhaps.

4.3.1.3 Professional Change: Creation of Emotional Bonds

As explained previously, it is the contextual meaning of a product that creates a higher level of interaction between tomorrow's products and users. This meaning could even create a deeper emotional or bonding process that could be either more intense or longer lasting. It is this meaning which links the 'business' interest of many of today's companies to the design process of tomorrow. Our research has shown that companies now (start to) recognise the need to develop and change their strategies to meet future markets.

Through these triple changes and their increasing relevance to strategic issues, designers have to learn to communicate and translate the consumer language to the manufacturer and *vice-versa*.

4.3.2 The Enhanced Industrial Designer: the Designer as a Mental and Physical Linguist

> *'We often conceal new functions - lighting features are often "built in", for instance, but the new "invisible" functions may equally well be given visible form, not because it's a necessary consequence of their physical presence, but simply in order to announce that presence. Such indexical forms, originally pointing to the end function may eventually even re-define the aesthetic of the original object.'(Stefano Marzano, 1998)*

By improving the profession, the enhanced breed of designers will become mental and physical linguists. He/she will acquire the skill of 'translating' conceptual thinking beyond today's physical media of two and three-dimensional modes of communication. While these qualities will remain within the designer's core set of skills, the focus will shift towards the ability to convey mental or cognitive pictures of 'potential' product users and the contextual settings in which both physical and virtual objects will be applied or exposed. What will be key for the designer, is the ability to understand and interpret the emotional values required to fit a particular contextual setting, and have the methods and means of communicating that with other disciplines to ensure that the necessary values of the setting are fully met.

4.4 An Enhanced Product Creation Process: Co-designing User Experiences

4.4.1 Sensory Attributes and Contextual Design

The authors have been researching differences in the consideration given to the importance of Sensory Attributes (SA) across different fields of consumer product manufacture. Communication within the PCP team is researched, e.g., how the different disciplines participate and co-design during the PCP. The team visited several consumer product manufacturers and interviewed designers, product managers and other disciplines. Case study reviews (Yin, 1994) were held to reflect on successful product creation (UNN CfDR, 1998 – Sensible Products).

4.4.2 Manufacturing Industry (UK Based)

The aim of the research is to explore the co-design of a significant variety of consumer products across UK based manufacturing industry. To generate a more complete picture of today's industrial activity, it researches companies within and across specific industrial sectors, such as automotive, sports and DIY.

During each case study visit to a particular company, a dedicated and structured interview is held to generate research data. By examining and cross analysing (Yin, 1994) the material, it has been possible to gain specific knowledge and insight into:

- current evidence of co-operation or non-co-operation between disciplines in relation to the integration of sensorial product attributes;
- future possibilities for areas of collaboration and co-operative development for sensorial attributes in tomorrow's products.

The resultant data comprises a representation of the awareness versus the contextual application of sensorial attributes within companies. This is subsequently processed and converted into textual and visual material, selected examples of the results to date follow below:

4.4.2.1 Tools 1: Snowflakes, Single Senses

Constructed as an indication of interest and awareness in SA during the company visit, the snowflake is a simple graph consisting of five star axes. Each axis is representative of one of the five senses. Product Development Managers (PDM) and Product Development Teams (PDT) indicate their impressions of importance on each of the five axis. Although it is only used as a crude indicator of relative awareness and interest in sensorial product attributes, the information provides an interesting insight into the differences in cross-disciplinary vision and communication.

Figure 4.1 illustrates an extreme example of where data feedback from the PDM (Dark Overlay) conflicts with the response of his PDT (Lighter tone overlay) The PDM's extremely narrow viewpoint (in this particular situation) sharply conflicts with the broader sensory overview illustrated through the PDT's snowflake outline.

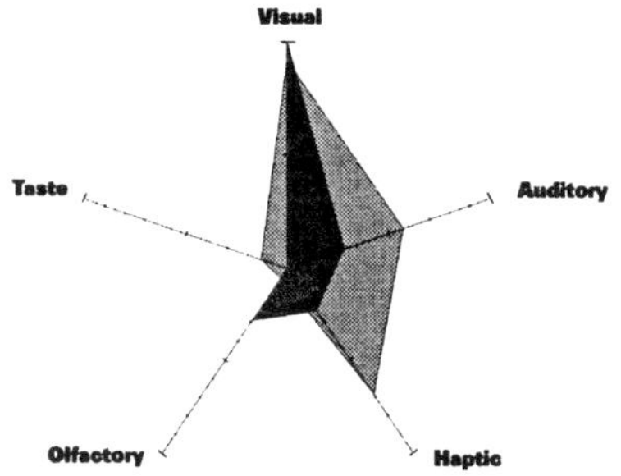

Figure 4.1. SA - Snowflake PDM vs. PDT

4.4.3 Measuring Sensory Attributes and Contextual Design

During different stages of the analysis it is possible to discriminate the physical Sensory Attributes (SA) and even the emergent emotional aspects of products which, as cited previously, are becoming more important to future products. The research is conducted in such a format that multi-sensorial information is also recorded and analysed to finally determine the overall product perception developed during the PCP.

4.4.3.1 Tools 2: Matrices

Visual matrices are later developed to reflect pulling 'fields' (Figure 4.2a) within the company culture, the different types of integration of the sensory attributes and how the companies are trying to create a total perception of their product(s). The four parameters measured to create a company profile, are the relation to the Senses, Implementational Tools, the Business Perspective and the Design Discipline.

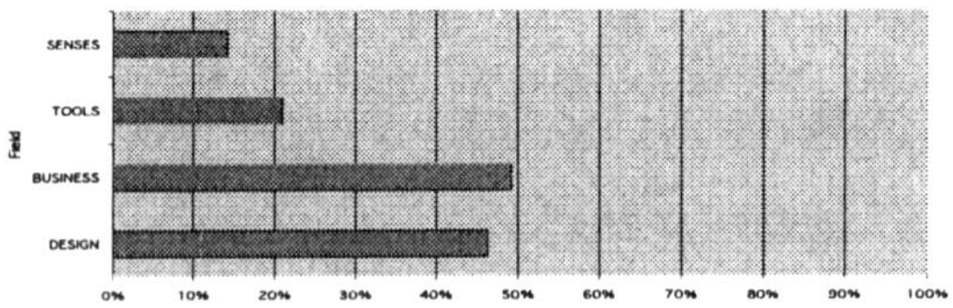

a

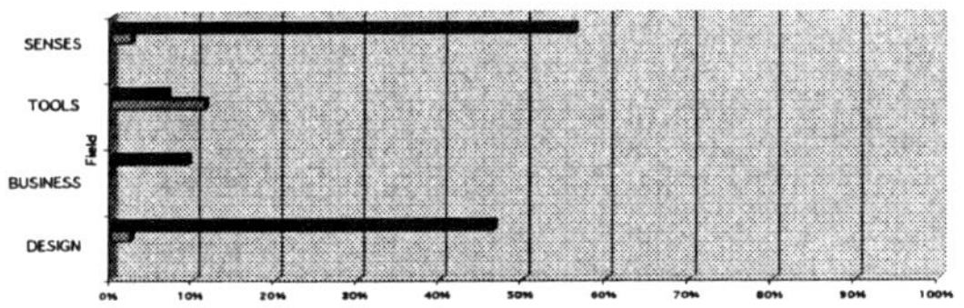

b

Figure 4.2. a. company ethos; b. description (dark) vs. practice (light)

Current findings from the research are producing good indications in respect to how SA are understood and finally applied in a product (see Figure 4.2b).

4.4.3.2 Tools 3: Lexicon

The research is especially interested in the 'product creation' language and the terminology used across disciplines involved with the PCP. During the research a methodology has been developed to monitor the vocabulary which might constitute the creation of a universal PCP language or a 'design' language.

It is important to explain the use of verbal expressions and meaning for a sensorial attribute to other disciplines. The 'company language' is analysed as part of the research and specialist remarks are recorded and consigned to a separate data collection file as the company lexicon. Collating these lexicons could lead to a more formalised protocol for describing the wide variety of physical and mental attributes across disciplines. This could result in a Design Lexicon or Dictionary.

4.5 First Research Conclusions

4.5.1 Understanding Creation

Understanding the PCP is a very broad field for investigation. However, first results from the research show that there are considerable differences across disciplines when discussing the very same sensorial topic, which indicates that our findings are at the start of a potentially enormous emergent field of study; the creation of enhanced product experiences and services.

By conducting this research the authors are convinced that the design discipline is creating its own field of knowledge concerning experience. This knowledge can become comparable with that supporting the other main PCP disciplines, which can refer to a longer involvement in product development, such as engineering, ergonomics and marketing.

4.5.2 Results

Results to-date are leading to remarkable overviews, e.g., how different disciplines truly collaborate, and give insight as to how SA are integrated during a company's creation of enhanced user experiences. Without mentioning specific companies, some of our first findings are:

4.5.2.1 Co-design & Communication

- In some PCPs there is no common communication structure on SA and contextual design. This is leading to unclear communication across the different disciplines.
- In some PCP's, borders between disciplines are unclear resulting in ambiguous input and responsibility for SA integration, e.g., design is addressing business issues and marketing is indicating relevant design issues.

4.5.2.2 Co-designing Sensorial Attributes

- There is awareness within companies of SA integration during product creation, however the overall awareness level is low. First findings show companies tend to have a strong focus on one particular sense and then specifically, a focus on the physical rather than emotional attributes.

4.5.2.3 Co-designing Sensorial Attributes & Contextual Design, the Experience

- There is little evidence of combinations between physical attributes or between physical and contextual integration.
- There is a considerable difference between companies on how experiences and emotional factors are integrated into the PCP.

- Most Case Studies do show that the disciplines in companies are capable of describing what they would like to achieve, but very little evidence was recorded in relation to applying them. Even in those that are good, there are considerable differences in the continuity shown between disciplines.

4.5.3 Final Conclusions

Current findings have led to the most interesting research question; 'are all disciplines developing the same product?' The research has identified the critical role that the designer plays in the process of facilitating co-operation between disciplines and also the need for an effective PCP vocabulary to support communication across disciplines. The value of the existing research has been proven in terms of these initial findings. Our intention is to pursue the work, including the aim; to explore the possibility of creating 'design lexicons', and to investigate the influence of an individual's capability to express SA's during the PCP.

4.6 References

Moggridge W, Brown T, (1999) Beauty with a smile: expressing experiences in design. RSA Journal, Frairy Press

Pine JP, Gilmore JH (1999) The experience economy. Harvard Business School Press

Fujimoto T (1990) 1.1 – 1.6 Product strategies for the nineties: The growth of international competition and the importance of effective product development management and the role of design. Financial Times Conferences Seminar Papers Division, Dudley

Fawcett D (1990) 6.1 – 6.4 Product strategies for the nineties: managing product development against time pressures. Financial Times Conferences Seminar Papers Division, Dudley.

Sullivan L (1896) A Philadelphia review: the tall office building artistically considered. Lippincotts. Re-Printed: Sullivan L (1947)

Havel V (1994) The need for transcendence in the postmodern worl. Conference Paper. World Futures Society, Bethesda USA

Schulze. (1995) F Mies Van Der Rohe. A critical biography Reprint Edition, University of Chicago Press

Mendini A (1989) Giancarlo Politi Editore

Branzi A (1984) Hot house: the Italian new wave design. Idea Books, Thames & Hudson

Manzini E (1986) Materials of invention. Arcadia srl Milan

Marzano. S (1992) Flying over Las Vegas. ICSID 1992. Philips Design. Re-printed 'Thoughts': Creating Value by Design - Philips Design (1998) V&K Publishing

Mitchell CT (1993) Redefining designing : from form to experience. Van Nostrand Reinhold, New York

Markham JE (1998) The future of shopping: traditional patterns and net effect. Macmillan

Underhill. P (1999) Why we buy: the science of shopping. Simon & Schuster

Kotler P, Armstrong. G (1998) Principles of marketing. 6th Edition Prentice–Hall International

Fujimoto T (1990) 1.1 – 1.6 The growth of international competition and the importance of effective product development management and the role of design. Product strategies for the nineties. Financial Times Conferences. Seminar Papers Division. Dudley.
Marzano S (1998) 'Thoughts' creating value by design - Philips design. V&K Publishing
Yin R.K (1994) Case study research: design and methods. Revised edition. Vol 5. SAGE Publications Inc, California
Sensible Products (1998-1999) Case study review(CSR): unpublished research reports. Centre for Design Research University of Northumbria at Newcastle. www.senses.org.uk

Chapter 5

Project-related Development of Building Products as a Co-operative Process

M.A.R. Oostra

5.1 Introduction

Manufacturers are generally regarded as the building industry's main product developers. In fact, however, the limited progress that is being made in the industry results from initiatives taken by a variety of different parties. Manufacturers are not the only ones who initiate the development of building products. Architects too play an important role, more particularly in initiating project-related product development. Since architects operate within the industry as generalists, combining different products and techniques to realise buildings, it is impossible for them to be fully informed on every technical detail and all the latest technological developments. The creation of new products, tailor-made to particular building projects, requires expertise from others; manufacturers and various kinds of consultants. The development process can therefore be seen as a co-operative process.

5.1.1 Project-related vs. Project-independent Products

In the first instance it is important to make a distinction between two product categories: project-independent products and project-related components (inspired by Eekhout, 1997). A project-independent or standard product can be manufactured independently without a client being involved. All products are identical and do not need adjustment to a specific building project, even though manufacturers can include a number of varieties in the product range, differing for instance in colour or measurements. They will know for certain precisely what requirements the product in question must comply with. In the case of a special component or project-related product, manufacturers may get a request from a client that cannot be resolved with a product from the standard product range. A completely new product must be developed for a specific building task.

This paper describes the role of architects in project-related product development through the frame of game theory, focussing on constraints imposed upon them and the strategies architects use to control such processes. These constraints and strategies have been investigated in four different cases, and in a number of supplementary interviews, as part of a PhD research project. Before the results are presented, essential parts of the research methodology and game-theory will be presented.

5.2 Methodology

In the first instance the aim of my doctoral research, of which this study is a part, was to catalogue and describe the product development processes in the building industry. First, case study research was carried out focussing mainly on manufacturers. Six initial cases revealed that architects play an important role in initiating product development by requesting manufacturers to supply non-existent project-related components to realize their projects. It also turned out that manufacturers use some of these ideas to generate new standard products. In contrast to the development processes of standard products in the building industry, product development initiated by architects is a virtually uncharted territory. I therefore decided to make this the central subject of my doctoral research.

5.2.1 Cases and Additional Research Data

Since very little research has been done into the development of project-related products in the building industry, the obvious choice was to use an exploratory research method. In principle, it would be possible to use exploratory, descriptive or explanatory research strategies. On the other hand, these strategies have different characteristics which make one more suitable for a particular study than the other. The decision to opt for case study research was based on Yin's (1989) considerations.

This paper considers the results of four exploratory case studies that inventorised how architects guide product development. The first case study investigated how Lignostone is used in the building industry. The second case study concerned the design and development process of a new crash barrier. In the third case an arched prefabricated concrete beam was developed. Finally, the fourth case study considered the design and development of a facade panel; the sound proofing properties that were required in this case had never before been met with a panel of a mere 8 cm.

Semi-structured interviews were held with employees of manufacturing companies and architectural practices. The sensitizing concepts underlying the interviews came from business administration literature on innovations, and product development in other branches of industry. The interviews produced tools to help get to grips with product development in the building industry. An overview is to be found in a state-of-the-art report (1996). The case studies investigated to what extent the sensitizing concepts could be applied to component design. This approach is based on a long tradition of the grounded theory methodology that originated in the social sciences (Glaser and Strauss, 1967).

The case study results will be supplemented with statements made by six architects who were not part of the case studies. The research results are linked together using the game theory. Before going into the theoretical framework underlying this study, I shall sketch one of the case studies.

5.2.2 One of the Cases in More Detail: The Lignostone Case

An example of a product innovation case used in this study is the first application of Lignostone, a wood-like material, in the building industry. Lignostone components were designed by the Dutch offices of Hubert-Jan Henket, for the extension to the Teylers Museum in Haarlem. The architect was impressed by the qualities of the material, which is normally used to produce golf clubs, hockey sticks and gymnastic rings, and used it in the structural mullions supporting the glass façade of the museum. Although the manufacturer was not very keen at first, Henket was insistent and persuaded him to help with the production of the Lignostone components. He and André Jorissen, the structural engineer, conducted research, solved a problem with the joints, and designed modifications to the manufacturer's machinery to enable it to produce the required four-metre long components.

5.3 The Building Process as a Set of Games

Historians date the origins of game theory as far back as 1654. Initially, game theory started as an abstract mathematical invention; the theory of probability. Since then it evolved in a model for all kinds of interdependent decision-making processes, and the social sciences started to use game theory as a framework for explaining and predicting behaviour (Colman ,1995).

Crozier and Friedberg (1980), among others, showed in their book *Actors and Systems* that games provide an excellent frame to study companies and industries. According to them, the creation and use of structured games is one of mankind's more important inventions, making possible the achievement of commercial goals, which require the cooperation of many different individuals. Contributing to the common goal allows individuals to advance their own interests at the same time. The way in which an industry and companies are organised has, however, an important influence on the behaviour of the individuals involved. People are not controlled by the rules of the game but by its structure. In other words there is scope within the rules for the players to determine their own strategies which will allow them to play in a way they personally find satisfactory. Usually individuals will only be able to achieve their own goals as long as they go on contributing to the common goal (Crozier and Friedberg, 1980).

5.3.1 Two Games in Which an Architect can be Involved

The group of individuals involved in a building project can be described as a collection of players, each with a different frame (Schön,1983; Reddy, 1978). These players will work towards their own goals by contributing to the common goal, the realisation of a building project. One of these individuals plays the role of the architect. For present purposes an architect is defined as someone operating alone or on behalf of a group of individuals, usually an architectural firm or department, who designs a plan for a building and looks after its realisation. An architect is involved in a number of different overlapping, cooperating groups or games. This paper considers only two such games: the building game and the product development game, with particular emphasis on the latter.

The main goal of architects is to realise their (re-)designs for buildings or other objects, for which they need cooperation with people outside their firms. The building process is the process in which the initiator tries to convert an idea into a physical building. The achievement of this goal requires many different types of expertise, generally represented by individuals from different companies. In what follows, this game is referred to as the building game or the building process game. Participation in building process games is a precondition for an architect to be able to initiate the development of a project-related product.

Within the building process game an architect is free to get involved in another game, the PD (product development) game, the goal of which is the development of project-related products. Whether architects will consider entering this second game or not is dependent on the way they frame their role (Schön, 1983). The contribution made by architects to this game is termed component design. The building game imposes certain limiting conditions on the PD game, and interactions with the building game will have consequences for the product development process.

5.4 The Building Game

Putting together a building involves different companies. These companies will be constantly moving from one building project to another and will often have several projects on the go at the same time. The external collaboration that is set up for the duration of a building project will change its composition as the project progresses. The people taking part can try to change the rules of these games in whatever way will be most advantageous to themselves.

The building game involves a whole range of different people, all from different companies usually; architects, clients, contractors, project developers, manufacturers, subcontractors, *etc*. Not all individuals are involved in every aspect of the building game, and different building games overlap even within a single company. In general a player will be involved in several building games at the same time, so that the limited amount of time available will mean that attention paid to one building game will sometimes be at the cost of other building games. It may happen that one individual combines different roles.

There are different ways of structuring the building process: the 'traditional' form in which an architect directs the other players on behalf of the client; the building team; and the turnkey concept (Bonebakker *et al.*, 1997). The chosen structure of the building process has consequences for the position of power that the different players within the building game are able to occupy. In the traditional building process the architect is in charge, whereas in other types of process the architect has much less chance of achieving a position of absolute power. In particular, if the architect has to work under the supervision of a building project manager his opportunities to influence others are reduced. This paper examines the situation in a 'traditional' building game. The dominant view amongst architects is that even in the traditional building process there is little possibility of designing components.

5.4.1 Constraints Imposed on the PD Game

The time frame, within which the project must be completed, lays an important constraint on the PD game. The more time available for a building project the more opportunity there is to invest time and energy in PD activities. Where the lower limit lies depends on the complexity of the product to be developed, the complexity of the building assignment and the situation within which everything takes place.

Secondly, the size of the budget available to complete the building project also plays an important role. It is generally accepted that an above average price per sq m or a higher fee for the architect is required if any PD is to be done. Research shows that this kind of decision should not be dependent on the budget. Various firms have shown that PD can be done within customary budgets (Oostra, 2000).

Finally, innovation goes hand-in-hand with uncertainty (Crozier *et al.*, 1980), particularly in the initial phases. In general the other players will try to keep such uncertainties under control and try to escape limitations which others try to impose on them (Crozier *et al.*, 1980). In a defensive reaction (Argyris and Schön, 1996) the constructional engineer will stick to the usual methods of calculation, the manufacturer to the usual methods of production, and the contractor to the products and constructional techniques with which he is familiar. Theoretically, the architect is able to design the component. An idea for a new technical solution or product, to be used in a building he is designing, forms an obvious extension to that task. This means, however, that the architect must be well informed about costs and feasibility. If the architect has a particular interest in building technology, this may well be possible. But most architects find it difficult to keep their technical knowledge up to date, because as generalists, they need to know about so many different fields. That is why cooperation from other players is needed. When an architect presents a product idea to someone who sees an opportunity to achieve his own goal, the idea will fall on fertile ground. In the language of the games theorist, opposition from other players is termed a constraint and their cooperation a resource.

5.4.2 Strategies

Architects use different strategies to improve preconditions for the PD game via steering the building process, for example:

5.4.2.1 Money

- to persuade the client to raise the budget while convincing him of the advantages created by the new products;

5.4.2.2 Organisation

- considering whether PD can or should be kept separate from the building process game;
- adjusting the set-up of the building game in favour of the PD game when changes occur in the building team.

In the next section the second example will be further examined.

5.4.3 The Position of the PD Game Inside the Building Game

If architects see opportunities for product development in upcoming building assignments, they deliberate over which position the PD game should have within the building process game.

The first case is where the architect tries to separate the PD game completely from the building game. In this case the architect will enter into a co-operative agreement with manufacturers, in advance of the building process, to work together on developing a new product, which might or might not be related to the particular project. The architect in fact takes on a different role, the role of industrial designer or building technologist.

The second case is where the development of the product is definitely linked to a particular building project, but the building game is separated from the PD game. The architect may agree with the client to exclude the PD process from tender. The effect of this is to limit the influence that the players in the building game, and the contractor in particular, are able to exercise on the PD game. The product is included in the specification as if there were no comparable alternative, so making its use an essential part of the building plan.

The third, most common case, is where product development and the building project remain linked to one another. This type of case is the one most frequently met in practice. As in the previous case, the building project will impose limiting conditions on the product development process in terms of duration and budget. But in contrast to the second case, the new product is part of the tender. The contractor has therefore more influence on the product development. In the previous case the contractor's only power was to refuse to allow the product to be installed. But in this case the contractor can try to replace the product which is to be developed by an equivalent substitute. He or she also has direct control over the budget.

5.5 The PD Game

Architects who want to play the PD game need other people to provide their cooperation and expertise. At the same time, constraints on the PD game can be formed by people involved in the building game. These factors can be seen as resources and constraints.

5.5.1 Resources

An architect depends on the support of other parties, those who have knowledge and facilities needed to realize a new product. For this, an architect needs at least a manufacturer who is willing to make a new product. But the Lignostone case proves that this is not always as obvious as it seems. The manufacturer, who specialized in sports goods, was not interested in delivering building products. It took much power of persuasion to move this monopolist to collaborate.

Because the parties involved in a building process do not see the development of new products as an essential part of it, an architect will have to motivate the parties whose collaboration he needs to develop a product. Money can play an important role in this respect. But other reasons could also motivate a person or an organization to contribute to product development. The following motives for co-operative behaviour towards architects have been found in the case studies:

- a client organization - because product development is appropriate to its declared aims;
- a researcher at a university of technology - because he is involved in research into the material in question;
- an institution concerned with the development of steel structures - because it is prepared to subsidize a project where it sees the possibility of encouraging the use of steel in buildings;
- a structural engineer - because he is very interested in applying the material in buildings, or because he enjoys thinking about something new;
- a supplier - because he would enjoy being involved, while he does not see the project as an immediate financial success for his company ('hobbyism') ;
- a supplier - because he sees opportunities to increase his turnover.

The sandwich panel case illustrated that the development of a special product does not always require direct contact between architect and manufacturer. In this particular case, the latter maintained contacts with the contractor. The architect specified that the sandwich panel should be no thicker than 8 cm, but the sound proofing requirement had never before been met by a panel of a mere 8 cm. When drawing up the specifications the architect had probably assumed that it was technically feasible to meet this requirement. The manufacturer had taken the order on the assumption that he would be able to manage the job. He took such a big risk because his aim was to penetrate a new market.

Architects whose aim it is to develop special products need other parties to collaborate with them and to offer their expertise. However, the same parties could also obstruct the component design process.

5.5.2 Constraints

A variety of different people are in a position to obstruct a project-related product development process. For example, the insurer may hinder the application of existing products in an unconventional way, let alone the application of new products. There may also be problems if suppliers have set up a cartel with the aim of dividing the market between them. Things can be difficult if there turns out to be only one supplier who can supply what the architect requires, but that supplier is barred from doing so by the terms of an agreement with his fellow suppliers. The required products will not be forthcoming. Attempts to cut out subcontractors will discourage innovation, since a low price will not guarantee the quality of the subcontractor's work, especially if the subcontractor is to contribute to the new component this can have unfavourable effects. This situation can also result in a worsening of mutual relationships, which in turn means communication between the various parties will suffer resulting in coordination problems between the various different activities and products.

5.5.3 Strategies

If an architect thinks about the players and the obstacles that can be put in the way of the development process, he may be able to anticipate those obstacles and do everything he can to avoid them or reduce their effect.

Architects use strategies to improve preconditions within the PD game to make the realisation of new components possible and to ensure that the process goes forward as smoothly as possible. The following strategies were found during the PD process:

5.5.3.1 Time

- starting the product development process as early as possible in the building process, so increasing the amount of time available

5.5.3.2 Money

- looking at different kinds of manufacturers when determining the price to ensure that the price is in line with the market. This means looking at the cost of alternative solutions. An architect who goes about this cleverly, by, for example, by thinking carefully about the product description or the way in which he invites tenders, may achieve a more favourable price

5.5.3.3 Quality

- calling on the know-how of third parties - competing manufacturers or product developers - to overcome a problem

5.5.3.4 Communication

- ensuring that the client remains enthusiastic not just about the building project but also about the component to be developed. This is important. If there is any uncertainty about possible risks the client may reconsider his support for the new component.

5.5.3.5 Atmosphere

- improving motivation in the pre-tendering phase by paying the manufacturer for any development work.

5.6 Conclusions

Architects generally cannot avoid working with other players in the building game if they want to realise new project-related components. This may involve facing two types of obstacles: restrictions resulting from the building game, and restrictions within the PD game.

In the light of the number of possible obstacles they might encounter, it might seem that the development of project-related products is almost impossible, and indeed this is the way that architects generally view the situation. Despite this, a select group of architects regularly take initiatives leading to project-related products, using a variety of tactics to control the process. Other architects can take advantage of the same strategies and so achieve components that are better adapted to their own and their clients' requirements and demands. In other words, by playing the PD game, architects can increase their chances of being called on to play in future building games.

2.7 References

Argyris C, Schön DA (1996) Organizational learning II; theory, method and practice. Addison-Weseley, Amsterdam

Bonebakker M, Flapper HAJ, Hofkes K, van Tol A (1997) Jellema hogere bouwkunde; 1 Bouwnijverheid. Spruyt. van Mantgem & de Does, bv Leiden, the Netherlands.

Colman AM (1995) Game theory an its applications: in the social and biological sciences, 2nd edn. Butterworth-Heinemann Ltd, Oxford

Crozier M, Friedberg E (1980) Actors and systems: the politics of collective action, 2nd edn. The University of Chicago Press, Chicago London

Eekhout M (1997) POPO; Of ontwerpmethoden voor bouwproducten en bouwcomponenten. Delft University Press

Glaser BG, Strauss AL (1967) The discovery of grounded theory; strategies for qualitative research. Aldine Publishing Company, Chicago

Oostra M (1996) Productontwikkeling van bouwcomponenten. unpublished

Oostra M (2000) Architecten als component ontwerpers: de rol van de architect in product ontwikkeling, unpublished concept version 1 PhD thesis, Delft University of Technology

Reddy MJ (1978) The conduit metaphor; a case of frame conflict in our language about language. In: Ortony A (ed.) Metaphor and thought. Cambridge University Press, Cambridge UK

Schön DA (1983) The reflective practitioner; how profesionals think in action. Basic Books, Inc

Yin RK (1989) Case study research; design and methods. Sage Publications, Newbury Park

Chapter 6

Symmetry in Design Participation in the Built Environment: Experiences and Insights from Education and Practice

R.A. Fowles

6.1 Introduction

Those who propose the use of design participation in the built environment recognise the positive contributions that users and communities can make to the quality of decision-making. They are also becoming aware of the beneficial changes that can take place within the participating group as a whole, and within individuals, as a result of participating. These changes embrace a number of issues including the raising of environmental awareness and the development in people of a real sense of ownership of the buildings they use, both important ingredients of sustainability.

In the UK, following a fairly dormant period for participation in the 1970's a number of pioneering projects were completed during the 1980's principally under the umbrella term of community architecture. These addressed social issues: housing for the homeless; rehabilitation of neglected estates; provision of community facilities, *etc*. In the early 1990's the social emphasis was beginning to be juxtaposed with environmental and ecological concerns. This emerging synthesis was acknowledged in the Declaration of Interdependence for a Sustainable Future by the UIA/AIA World Congress of Architects in 1993, 'We are ecologically interdependent with the whole of the natural environment; we are socially, culturally, and economically interdependent with all of humanity; sustainability in the context of this interdependence, requires partnership, equity, and balance among all parties'. Significantly, for the development of participation, emphasis is placed on a fundamental component of successful participation: *i.e.*, the relationship between players in the process in the form of 'partnership, equity and balance'.

6.2 Participation

Models of participation in the design of buildings are often simplified to embrace two parties, the architect and the user. However, many give hints of the differential status of the two parties, with the simplest noting that design may be carried out either AGAINST people, or FOR people, or WITH people, or BY people.

In the USA, where community participation preceded that in Europe, the move towards defining a geography of power based on participation was helped by Arnstein (1960) in her model, which she named, a 'ladder of participation'. Her message that 'citizen participation is citizen power' has broad application today. 'It is the redistribution of power that enables the have-not citizens, presently excluded from the political and economic processes, to be deliberately included in the future'. Arnstein's ladder has eight levels: 'manipulation - therapy - informing - consultation - placation - partnership - delegated power - citizen control'.

For architectural educators and practitioners the most useful of the models is the 'scale of participation' by Wulz (1986), which has seven distinct participatory relationships between the architect and the user. He places these between the poles of 'expert autonomous' architecture and 'user autonomous' architecture. They are: Representation (the most passive form, in which the architect 'has consideration' for the wishes and personal needs of the user-client); Questionnary (statistical gathering of the population's requirements); Regionalism (sensitivity to the specific and cultural heritage within a geographically limited area); Dialogue (often based on informal conversations between architect and users, with the architect reserving the right to make the final decisions); Alternative (users are given alternative solutions from which to choose); Co-decision (user-clients are in balance with the architect with regard to the degree of influence); and Self-decision (the architect is called in as required; people's independence from all forms of authoritative intervention is seen as the purest form of democracy). Notable is the suggestion that there does not have to be a direct or first-hand contribution by the user for a form of participation to be said to be taking place.

6.2.1 The Idea of a Symmetrical Relationship

If one accepts that the user can be brought into the process on a contributing basis, and in a face-to-face situation with the professional, then a number of questions arise. Can the parties actually reach a 'partnership, equity and balance' in their relationship? What information or expertise can the user contribute which equates with that of the professional? In other words, is it possible to find a 'symmetry' in the relationship between the parties? The possibility of the two parties meeting mid-way is in fact recognised by both Arnstein (*i.e.*, partnership), and Wulz (*i.e.*, co-decision,) but what does this mean in practice?

Arnstein may have indicated the possibility from a political viewpoint, but it was Horst Rittel, working in the more theoretical context of the world of design methods in the early 1970's, who suggested that users should be party to the process, as it is the 'knowledge' held by a wide range of people affected by a problem which should be utilised in seeking solutions to that problem. Rittel (1972) claimed 'expertise does not reside solely in the professional, but in all those whose interests are affected by a design or planning problem', recognising that there are gaps in the knowledge and expertise of the professionals, *i.e.*, a degree of ignorance on their part, which can only be filled by other people. He succinctly emphasised this point by introducing the concept of a 'symmetry of ignorance'.

6.2.2 Participation and Sustainability

It took two decades for this concept of symmetry to be more widely valued although during this period it was being pursued by a small but significant number of architects and planners in different countries. Although the wording of the Declaration by the World Congress of Architects recognised that 'sustainability requires partnership, equity, and balance among all parties', were the implications of this declaration fully realised by its authors? Were they really advocating 'equity' between themselves and building users? Did they fully understand the implications when participation is identified to be a key mechanism in working towards sustainability?

Chronologically, this declaration followed from Agenda 21 and the 1992 Earth Summit in Rio de Janeiro, which emphasised the importance of involving whole populations in broad processes to achieve change. The rationale for this is simple: in a society with no common visions, only short-term decisions can be made. Visions, to be of value and to be realisable, have to be shared by all, *i.e.*, by experts and lay people. A non-participatory community, one that cannot join together to develop and share visions and their implementation, is inherently unsustainable.

6.2.3 Symmetry: Some Implications

In a conventional planning and building process the decision-makers, *i.e.*, financiers, professionals and developers, often work in isolation from the communities they serve. When participatory processes are initiated from the outset, with respect for contributions from all participants, frameworks for physical, social, economic and ecological change can be agreed by consensus. A common sense of ownership is established and action plans can be implemented in true partnership. Environmental change, which is more than superficial, depends upon such partnerships, which tap the core values shared by all and releases energy and creativity at all levels.

For many this approach remains a utopian vision and there are those who still argue against opening up the design process to all those affected by, or who have an interest in, the problem. Many architects hold the fear that they will abdicate their responsibilities and expertise. However, there are an increasing number who are realising that so much is to be gained. Here is one such expression of confidence, from Peter Blundell Jones (1987), when referring to the work of architect Peter Hübner who joined with the Youth Club at Stuttgart-Wangen,

Germany, to design and build their own building, 'We need have no fear that the architect's role will be lost in such a participative process - in some ways it becomes more vital. For without Hübner nothing would have happened at Wangen, and his is the most obvious shaping hand behind the project. Nonetheless he has taken the role of interpreter and co-ordinator rather than dictatorial designer, and has allowed the whole process to express itself without imposing too many of his own prejudices.

User participation in the built environment design process is well documented by Hatch (1984), Woolley (1989), Towers (1995) and Fowles (1997). In these compilations we see influential roles for tenants, employees and neighbourhoods, in a range of building types at different scales of the built environment, including banks, company headquarters, libraries, care facilities, community halls, new housing and estate and urban regeneration programmes. It would be a valuable exercise to examine the distribution of participatory projects along Wulz's scale of participation. Probably many projects can be located at each level. Wilkinson (1999) has found that a project may start at one level and move its location on the scale during the course of the design process. In my own work, whether designing and tutoring participation procedures for live student-community projects, or in practice, I have tended to start with the view that if a symmetry does not exist, or is not understood, at the outset then one can be reached over time. Behind this approach is the basic belief that the design process should be open and transparent and follows democratic principles.

6.3 The Projects

As a Registered Architect and member of staff at The Welsh School of Architecture I carry out an integrated programme of teaching, research and consultancy work in participation, community architecture and ecological design. The following project descriptions indicate the type of work undertaken. All projects are monitored, material generated by participants is retained, and feedback interviews are held at the end of each design process. Learning is passed to the next project.

6.3.1 Live Student-community Participation Projects

A network comprising the School of Architecture, community technical aid centres (TAC) and community organisations, generated from 1985, a programme of live student-community participation projects, for a wide range of building types. A description of the procedures and techniques of these projects has been published (Fowles, 1994). Typically a community group will approach a TAC to seek help and advice on funding and design. If appropriate a student involvement is agreed. A series of participatory design workshops are held from which alternative proposals emerge. Public meetings and fund-raising activities may then follow. Projects vary in duration but are typically five weeks with four design workshops.

Originally focussing on participation issues, the programme later embraced the concept of Eco-participation (Fowles and Hines, 1994), 'in which building users and communities participate, with professionals, in the design of their buildings according to ecological principles'.

6.3.2 New Headquarters for SGI-UK

SGI-UK is part of Soka Gakkai International, a worldwide lay Buddhist organisation. Taplow Court in Buckinghamshire was purchased as a new national headquarters. A strategic framework for design participation was sought which would be clear to understand, simple to implement, not greatly extend design time and cost, and which would take the process through from vision to detailed design. It would aim to maximise the involvement of the membership who, through their everyday professional lives, hold expertise to complement that of the architects and their consultants, and who, through their work with SGI, hold knowledge about the needs of the different activities of the organisation. Members from throughout the UK were provided with a range of opportunities to contribute, including regional and subject based Focus Groups, two design weekends, and a concluding Public Exhibition. Architype Ltd are the architects, and the author designed the six-month participation process which took place in 1997 (Fowles and Hines, 1999).
Burry Port Activity Centre and town study

6.3.3 Burry Port Activity Centre and Town Study

Research funding allowed the design, facilitation and monitoring of a participatory design process in Burry Port, Dyfed, based on the needs of a core community group of Scouts and Guides who wished to build a new HQ and activity centre. The research report (Carter and Fowles, 1992) documented the process, and concluded with a set of guidelines designed to assist future enablers and community groups. It relates the activities of the community group and the involvement of the author and his colleague, John Carter, over a two year period, firstly as tutors in conducting two student-community design participation projects, and secondly as enablers and researches of a broader participation process in the town.

6. 4 Symmetry and the Related Concepts

6.4.1 Symmetry

6.4.1.1 Symmetry of Ignorance or Symmetry of Knowledge?

Dudley (1996) refers to development projects in third world countries where there is a need for field workers, when drawn from the community itself, to understand the ignorance of professionals, who may come in from outside and who are 'sometimes foreigners and more often middle-class intellectuals'. He observes that many of the day-to-day motivations, concerns, and assumptions of impoverished villagers and shantytown dwellers are totally alien concepts to the professionals. He concludes: 'only when there is a mutual understanding of each other's ignorance is there likely to be productive communication'. I have found it very useful, when discussing my approach to design with potential participants, to refer to this concept of 'ignorance'. However, in the early stages of a participatory process when trust and friendship being nurtured, to appear to indirectly accuse a participant of having a degree of ignorance is not a very good starting point. So I have substituted it with the concept of 'symmetry of knowledge', as this represents a more positive approach by placing value on what people actually know: *i.e.,* their particular expertise.

6.4.1.2 Pre-requisites to Symmetry

Most processes start with an initiating group which should be clear about who they are and who they represent. A group will only be representative of its community if members of that community are supportive of it. The aim should then be to inform everyone likely to have an interest that the process is happening, to reach beyond the few 'outright volunteers' to seek active, willing participants, to allow time to identify the knowledge and skills which participants bring to the process and then to make every effort to utilise the knowledge and skills.

6.4.1.3 Enabling Symmetry

In democratic participation, the role of the enabler should be to provide opportunities for every individual to contribute. Although there may be procedures and techniques which are common to every situation, each participatory process is different, and therefore any strategy should be sufficiently flexible to respond to new situations as they arise. In particular, the enabler must ensure that instances of creative suggestions are followed up. This places value on the participant who had the idea, thereby motivating further involvement. Furthermore, the idea itself will then be seen as a 'user or community initiative'. In workshops and discussions the enabler should ensure that all points made by participants are recorded and brought into the process. In short, the enabling skill is getting people to reveal and utilise what they have. They may not perceive that their knowledge or skill is relevant or useful at the start, but once each individual's knowledge or skill is put to use, then a degree of ownership, control and power follows. Enabling people to realise what they know is confidence-building for the participants, and consequently the nature of enabling must change with the level of knowledge and skills of the group.

6.4.1.4 Symmetry of participation

McLagan and Nel (1995) reveal how discriminating our normal frameworks for human activity can be. Creating conditions for equal opportunity to contribute requires an awareness of these powerful constraints. Non-discrimination means that everyone should have opportunities to take part in ways that are appropriate to their personal circumstances. At a very practical level some may not be able to get to meetings due to employment, disability and looking after children. Others may not be able to participate due to their language. Other practical areas, which the enabler has, control over are the workshop seating patterns and the degree of conviviality of the participation materials particularly in 3D modelling. Avoiding profession-centred and discipline-oriented jargon is also most important. Finally, participation will often require a dialogue between people who may rarely communicate in their everyday work and lives. The early stages of a project should facilitate this.

6.4.1.5 Symmetry of Learning

Participation should always be a two way learning process in which all participants benefit. In the educational situation, bringing students into direct contact with people and their design problems ensures that projects become broad based learning vehicles embracing a diversity of issues that are keenly debated. Students develop an awareness that design has a socio-political dimension, and can involve a wide range of people and organisations often with conflicting objectives that have to be resolved. They make contact with real people, their needs, values, perceptions and expectations. They experience the importance of developing interpersonal skills and good social relations as the basis for group design work. They become aware of the different roles of an architect, particularly in situations in which designing is made a shared experience in which users have the opportunity to contribute their knowledge to the design process. If possible design workshops should be located in the locality of the participating group and in the School of Architecture. Bringing the public into the School helps to de-mystify the profession, and taking students into the community furthers their understanding of the problem and its context.

My approach to design is founded on the belief that with the participation of users in the design process a building will be more responsive to their needs. With the participation of members of a community, the community itself can be strengthened. As with others who work in the field of community design I hold the belief that the environment should work better if the people, who live, work and play in it are actively involved in its creation and management.

6.4.2 Role Interchanging

As previously noted, references to the act of participation invariably simplify it to a two party activity of user and architect. In my experience this is rarely the case. Often there are a number of parties involved and roles are frequently interchanged during the process. The outcome of this overlap and complexity is of extreme benefit to all concerned and to the process as a whole: people learn from each other, view things from a different perspective, and establish a sense of a 'design community' or 'team'. For example in the SGI-UK project, who were the people who directly participated in the design process? There were the eight members of the Building Committee, who had initiated the project and became the client team. They had frequent meetings with the architects but they also facilitated at the workshops. There was the architectural practice, comprising a core design team, plus other members of the practice who joined in as facilitators at workshops. There were consultants and specialists from a range of disciplines all of whom also facilitated at workshops. There was a participation observer/reporter who helped summarise workshop outputs. Finally, a large number of the SGI-UK membership of 6,000, *i.e.,* the users, contributed ideas and information to Focus Groups, whilst members who attended the workshops became designers at both the strategic and detailed levels. This is a simplified grouping of participants with a brief indication of the different roles they undertook. On-going analysis is examining the extent of influence of these parties on the design itself, but at the time of writing this is not yet complete. My belief is that individuals and the process benefited from the fluidity of role swapping as, for example, designer and client became participation facilitators, and as users became designers and experts.

In student-community projects a further benefit arises during the various sessions within the overall process. This arises from who participates with whom. Formal dialogue takes place between: students with students; students with user-client group; user-client group with wider community; students and community group with architectural practice or technical aid centre; and as tutor I have also taken on the roles of enabler or a member of a student research group.

6.4.3 Transparency

6.4.3.1 Recording the Process

For participation to achieve its goal the contributions of the participants have to be recorded at all stages, so that the output from one stage may be seen to be taken forward to become the input to the next stage. 'Transparent' procedures which utilise generative graphic techniques and create a documentary record should aim to capture the visions, concerns, desires and ideas of all members of the community, and form a record of all proposals and decisions, not only for use within the process itself but for future reference, research and dissemination.

6.4.3.2 An Output Should Become an Input

A seminal model for a participatory process, of: '...workshop event - output - development by architect - input to workshop event...' is represented in the process followed in the design of the community library at Boulder Creek, Santa Cruz, California (Oberdorfer, 1988). In this project 100 community members participated in four design workshops to create, firstly the library's site plan, then the floor plan, then the building form, and finally many of the architectural features. The architect designed and facilitated the community participation process that incorporated this well-defined strategy of design workshops. The complementary roles of architect and users are clearly articulated in this design process, with the output generated by the users being taken by the architect, worked upon and then fed into the next workshop.

I have developed and frequently utilised this model in live student-community projects and in consultancy work. It is simple to comprehend and works well. A typical project timetable of four evening workshop events on a student-community project would be: Week 1, The Brief Workshop, in which 12 members of the user-client committee work with 12 students to develop the building brief, after which students research the brief requirements and prepare input to next workshop; Week 2, The Community Ideas Workshop, in which 120 community members participate in a 'planning-for-real' exercise, after which students produce a number of strategic approaches to the problem; Week 3, The Building Design Workshop, in which the committee and students develop in greater detail a range of ideas from the previous workshop, after which students refine designs and prepare presentation drawings; Week 4, The Public Presentation Workshop, in which a set of proposals is exhibited to gain wider public response and further publicity, after which the Committee then selects, final design and funding is sought.

6.4.3.3 Communication

It is essential that all participants irrespective of the size of the group be equally and regularly informed about progress, events and decisions. With SGI-UK, reports on each workshop were distributed to all participants. Everyone should be kept up to date on major events and issues. In this way, all participants are made to feel equally valued and the spreading of unproductive rumours and disinformation is avoided. If possible build on established methods of community-based information gathering and dissemination. Effective use of the media, through accurate notices, press-releases and statements, and carefully considered publicity opportunities, are essential to encouraging participation and keeping the wider community informed of progress and events. In most projects we publish a regular newsletter. It is not only those most directly involved in a project that need to be kept informed. Consider also those who are resident in the community and who may not have a direct interest in or be directly affected by the proposal.

6.4.4 Ownership

Transparency can be assured if information and all documentation relevant to the process are made freely available to participants. One consequence of transparency is participants being able to develop a sense of ownership of the process. This will also be heightened if the creativity of people is realised right at the beginning, their influence on the process identified, and the benefit of their contribution acknowledged. It is important to participants to have a sense of achievement at regular intervals, so the aim must be to realise objectives in a tangible form and ensure that all events have tangible outputs. The keeping of minutes of meetings and recording events by the participants themselves will help achieve this. I believe that the sense of ownership of the process is likely to be translated into a sense of ownership of the building when it is occupied, which may for example translate into greater care and lower maintenance costs. This issue has not yet been followed up in the projects referred to in this paper. However there are indications of pride being translated into greater care of buildings in a number of participation projects but well researched empirical evidence to support this is needed.

6.4.5 Personal Transformation by Engaging with Environmental Issues

The way to fully comprehend the subtleties and complexities of democratic participation in design is through the 'learning-by-doing' procedure of joining with people to discuss issues, define problems and work out solutions. It has become very clear that when people do participate in the creation of the environment that they themselves will later inhabit, they begin to critically examine broader environmental and ecological aspects of building design. Through observation, and the comments of participants, I am able to claim that the synthesis of a participatory approach with an environmental and ecological agenda results in a significant personal level of change within people. This personal transformation may be one of the most beneficial outcomes of the participation process and an important ingredient in contributing to sustainability. It is a spiritual outcome, strongly related to the creation of a socially responsible and ecologically sound architecture.

For my students, extending the studio into the community re-integrates experience into education and increases their awareness of social and ecological issues. It aids their own personal transformation and moral development, and nurtures a caring attitude to people, place and planet. At the end of one project a student group concluded: 'We had become aware of three fundamental dimensions to our involvement as human beings: personal transformation, co-operative endeavour and of our responsibilities as designers'. With SGI-UK the situation was different. The opportunity to open up the design process to allow members to express their ideas, and actually influence the design, was for many, simply a variation on the Buddhist philosophy of co-operation and collaboration, yet one that was welcomed with some surprise 'I didn't think architects listened to what people had to say'. Once aware of this, the process is underway.

6.5 Conclusion

6.5.1 Creativity and Expertise

The trend towards recognising the value and benefits of 'symmetry' in the built environment design process does suggest that we are beginning to move beyond the confines of expert culture into extending our resource base of design information and ideas by recognising the expertise that resides in all people. This approach not only provides architects with new insights into the needs of people but it releases for the common good our greatest resource, the creativity of people. We begin to contest the former privilege of experts, but at the same time we are extending the privileges of professionals by creating opportunities for them to design with the people who will work, rest or play in 'their' buildings, and we are giving people the privilege, if not the right, to influence decisions about their own environment.

6.5.2 Participation

A participatory or community architect, will start each process with a clearly articulated vision of a participation process designed to create a cycle of awareness, involvement and ownership amongst the whole membership of the user-client group. This vision will need to be maintained for it to become an integral part of the development of the brief, in agreeing priorities, functional content, practical design issues and the spiritual and philosophical ideas to be expressed through the architecture. However, participation is not easy, and the 'symmetrical' participation equation holds a complex set of interrelationships between, amongst many other things, motivation and self-interest, threat and goal, leadership and collective co-operation, and personal and community benefit. It must be recognised that this approach to design, even when participants want it to happen, is set against a background of paternalism in society.

6.6 Acknowledgements

The author thanks the many user-client groups, students and professional colleagues who directly or indirectly through their open attitude to sharing the design process, have contributed to the concepts described in this paper.

6.7 References

Arnstein S (1969) A Ladder of Citizen Participation. AIP Journal July:216-224

Blundell Jones P (1987) Three kinds of participation. Architectural Review 1081:60-76

Carter J, Fowles RA (1992) From small beginnings.....: community participation in Burry Port. Report on the project 'Community Participation in Building Design' ARCUK

Dudley E (1996) Educating field workers. In: Hamdi N (ed.) Education For Real: The Training of Professionals for Development Practice. Intermediate Technology Publications, pp 183-194

Fowles RA (1994) Design as Process, Context as People. In: Proceedings ACSA European Conference, The Urban Scene and the History of the Future, London pp 232-235

Fowles RA (1997) Participation Bibliography, Social Aspects Module, Welsh School of Architecture, Cardiff University

Fowles RA, Hines J (1994) Eco-participation. In: Proceedings of the First International Conference 'Buildings and the Environment' The International Council for Building Research Studies and Documentation, BRE, Garston, Watford, UK

Fowles RA, Hines J (1999) Transformative Architecture: a synthesis of ecological and participatory design: New buildings for SGI-UK at Taplow Court, Bucks., UK. EcoDesign Vol VII, No.1:14-17

Hatch RC (ed.) (1984) The Scope of Social Architecture. Van Nostrand Rheinhold

McLagan P, Nel C (1995) The Age of Participation. Berret-Koehler

Oberdorfer J (1988) Community participation in the design of the Boulder Creek Branch Library. Design Studies, Vol 9, No 1:4-13

Rittel H (1972) DMG 5th Anniversary Report. Design Methods Group Paper No 1:5-10
Towers G (1995) Building Democracy. UCL Press
Wilkinson N (1999) Issues raised by Community Participation in Urban Housing. BArch Special Subject Dissertation, Welsh School of Architecture, Cardiff University
Woolley T (1989) Design Participation Today: A Report on the State of the Art. ARCUK
Wulz F (1986) The Concept of Participation. Design Studies Vol 7, No 3:153-16

Chapter 7

Does 'inclusive design' Require an Inclusive Design Process?

R. Luck

7.1 Introduction

This discussion paper has been prepared to stimulate debate on design for inclusivity, particularly whether it is an achievable ideal and whether there is an approach to design that is truly inclusive of user preferences. To set the scene, the meaning of inclusive design is explained. The methods currently used to include user representation within the design process are described, as well as criticisms and deficiencies of these processes (e.g., for not representing disability needs). The social model of disability is described, presenting the case for the greater empowerment and inclusion of disabled people in decisions that affect their lives. The paper concludes with the description of an approach being used at the University of Reading to actively involve mobility, sensory and cognitively impaired people in the design of a building they will occupy.

7.2 What is Meant by Inclusive Design? - The Philosophy of Universal Design

The evolution of building accessibility in America is traced from the Architectural Barriers Act of 1968, via The Americans with Disabilities Act (ADA) in 1991, to current thinking in the twenty-first century on universal design. Universal design and accessibility are considered to be different concepts; accessibility focuses on modifications, whilst universal design focuses on developing new solutions that anticipate diversity of users and offer multiple options for use.

The Americans with Disabilities Act (ADA) was considered to be an opportunity for designers to move away from a dimension-driven approach to design towards a performance focused approach. This opportunity was perceived by designers and owners as legal quicksand and lobbied for increased specificity of requirement. The Dept. of Justice responded with a regulatory framework. The

Americans with Disabilities Act Advisory Guide (ADAAG). A process for certifying state building codes was introduced along with attempts to harmonise the many regulations on access and make it easier for designers to comply with the new law.

In the UK inclusive design has followed the American model and has similarly been legislation-driven, where changes in thinking are reflected in the refinement of successive Acts. In the UK the legislation has gone through the following stages:

1970 The Chronically Sick and Disabled Persons Act.
1981 The Disabled Persons Act.
1992 The Building (Disabled People) Regulations: Part M Access for Disabled People.
1995 The Disability Discrimination Act.

In both countries impairment is viewed from a deficit model and designers are conscious of increased liability because of the introduction of accessibility legislation. Jones (1999) has observed the following barriers to universal design that are present in the current approach to design:

- designers lack personal familiarity with people who have disabilities
- they have no training in critical thinking, to consider who is privileged by design norms
- inclusivity should be included in design discourse as a design value.

Jones considered that if these issues are addressed universal design has the potential to be emancipatory because it moves the discourse from compensating a disenfranchised group, to framing a consumer market. In this way universal design can be emancipated from its single-issue origins to a more compelling and collaborative position-one of changing the process and artifacts of design to be more equitable.

To illustrate the changes in legislation and thinking with respect to design for accessibility Jones (1999) has studied the change in step and stair design within a building. Over time there has been a changed approach to step and stair design; classically a formal grand entrance had steps to a central front door, making unaided entrance by mobility-impaired people impossible. The first response to making the building more accessible was to either retro-fit a rear entrance, or ensure that there was a route through the building from an existing back door to other parts of a building. Ramped access to the main entrance was the next response. It was an attempt to create separate but equal access for mobility-impaired people. The Americans with Disabilities Act (ADA) introduced in 1991 was followed soon after in the UK by the Building Regulations in 1992 and the Disabilities Discrimination Act (DDA) in 1995, which required ramped access to all new public buildings. An American civil rights case ruled that an optional ramped route was not separate but equal; it was unequal, as the ramped route was secondary. It wasn't until a 'universal design' approach was taken that examples can be found of designs that incorporate both stairs and ramps in an integrated manner, without preference for either route.

7.3 Criticisms of the Design Process for not Representing Disability Needs

This section of the paper is concerned with user participation and the representation of user needs in the design process. The broader user perspective is taken, as disability needs are viewed as one type of user need that should be understood and represented in an inclusive design process. The rationale for including user needs in the design processes is presented together with the user models used in theory and practice, the methods used to represent user needs and design practitioners views on the suitability of these methods.

User participation in the design process has received more critical attention since the 1980's (Darke, 1984) when the lack of inclusion of the user perspective in the design process was considered an oversight. It was considered that the inclusion of the end user's views in the design process would result in greater user satisfaction with the building. An inclusive approach acknowledges that giving users a voice reflects an organisation's consideration for its employees' working conditions. Lipman and Harris (1998) have explained this in terms of power within organisations. They made the observation that participation in design has become inextricably linked with the empowerment of those whose views are sought. User representation can be sought for different stages and levels of involvement with a project.

In construction user needs have been given much consideration in the field of facilities management, where the area of post occupancy evaluation has evolved as a quantitative measure of the performance of a completed building. Obtaining a measure of the users' perception of the building in use is part of the evaluation process. The work of Preiser (1988) and Kernohan (1992) into user needs evaluation methods has resulted in the design of different post-occupancy evaluation techniques which have varying levels of building user involvement. The methods range from checklists administered by experienced evaluators to more involved discussions with users led by facilitators measuring the effectiveness of the design and briefing process, from a building users' perspective.

A participatory building design and management approach was developed by Kernohan *et al.* (1992) to create good relationships amongst a project team. This approach generates a group-enabling power to complete a project together, rather than the power of one party over another. In this way user participation can be included in building design and management. The approach is of particular interest to this study as it is concerned with the representation of views from different parties and is also concerned with gaining experiential knowledge through dialogue. It is considered that knowledge is socially constructed through dialogue. Knowledge is generated and confirmed in a dialectic between concrete knowing and abstract knowing. This is a key observation, which coincides with the approach of this project that knowledge and new ideas are generated through the process of discussion.

Hasdogan's (1996) work differs from the previous references, which make observations on user involvement, as the models of users used in the design process to assess user needs is the subject of his analysis. He has classified the user models and developed a matrix of these models. The models he observed were; empirical models, experimental models, scenario-based models and theoretical models. It is of interest to note that ergonomists were critical of empirical anthropometric models, finding the data unreliable. This is especially true when designing for disabilities, as generalised anthropometric data is rarely useful. The representative user model Hasdogen observed was considered to be useful, but was rarely used because of cost and time constraints of a project.

Sharrock's work (1994) is also of interest when considering the attributes of the design process, which inhibit the representation of disability needs. Sharrock has identified ways that 'the user' is presented in the design process. The user is introduced through the use of typificatory structures, types developed by the designer. According to Schön (1988) it is the organisation of design knowledge into types and rules which underpins design creativity. Sharrock uses data from his own studies of design to standardize modes of talk about 'the user'. A 'typified' approach allows designers to manage their tacit knowledge and 'the users' cultural experience as part of design reasoning. In this way 'the user' becomes a scenic feature of the space. In this sense 'the user' is considered in the reasoning of designers and is also a feature of the design space. This work gives insight into the way designers use 'user' information.

7.4 Social Model of Disability-representation of Disability Needs

Linked with a need for user participation in the design process is the issue of hegemony and the representation of disabled building users. Hegemony concerns the location of power and decision making by a majority on behalf of others. There is a feeling amongst disabled people that many decisions which have a strong influence on their lives are made by others on their behalf (Darke, 1984). This includes the design of environments, even those with specific access concerns, where design decisions are made with little or no consultation with people with disabilities. There has been much written about design for disabilities which is uncomplimentary about the process. The work of Hall and Imrie (1999) investigated architects' understanding of disabled people's building needs, based on their relationships with building users and found that many designers were assuming a knowledge and understanding of environmental issues for disabled people which they don't have. This presumptive designing has led to dissatisfaction. Mason's (1992) work found that when disabled people are able to choose for themselves their decisions differ from those made on their behalf. This observation reflects that their interests have not accurately been represented in the past.

The issue of user participation and representation of views is particularly sensitive when designing for disabilities as activity in the field of disability theory and the politics of disability has increased since the 1970's. There has been a rejection of the hegemonic labelling of a person as 'handicapped' or 'disabled' amongst disability theorists and an introduction of disability as a social construct

(Barton, 1998). Disability studies look at impairment from a sociological perspective, where a social model of disability considers that societal values determine what is considered abnormal and this defines normality (Davis, 1995). The politics of disability and the social model have brought about a movement amongst disabled people for greater involvement and empowerment rather than representation of their needs and values. Although there are mechanisms for the inclusion of various disability perspectives to influence the design of environments, e.g. access committees to influence local authority access policies, there is criticism of the effectiveness of these methods (Imrie, 1999). It has been observed that the advice given to a local authority by an access committee can be ignored, but the fact that an access committee exists meets the authority's consultative duties. This illustrates the main thrust of this paper that although design exercises may involve a user consultation process, ultimately there is an autocratic decision made on behalf of impaired people.

Taking a social model of disability perspective, to address this imbalance disabled people should be emancipated and their views should be considered from an empowered, rather than a consultative position. This would be possible if disabled people were designing environments for themselves, but few architects and planners are disabled (Hall and Imrie, 1999). There are several issues that make this emancipatory approach difficult. Design is a specialist activity, with liabilities, knowledge and expertise that cannot arbitrarily be passed on to someone else. Another issue is representation: how representative is one disabled person of a larger group? Even if a person is able to represent someone with a similar disability, a consultation exercise is flawed if a person with one disability has to make presumptions about the needs of a person with a different disability. The next section describes an accessible approach to design that is still being refined, but attempts to address some of these issues.

7.5 An Accessible Approach to Design

The approach being used to design the concept building at the University of Reading specifically addresses the issue of representation of user needs, so for this reason time has been apportioned to develop a method to better understand user requirements. Two mechanisms have been used to collect data and knowledge from the user and commissioning client about future building requirements. The mechanisms are: semi-structured interviews and Steering Group meetings.

These mechanisms formalise the social processes within the project for gathering user and commissioning client information to develop a design brief. The mechanisms are discrete events to gather information about building user preferences and to elicit the commissioning client's requirements for the building. The approach is systematic without being a checklist. There is a process to the collection of this information, but each person presents their information and their requirements in their own terms.

7.5.1 Semi-structured Interviews

The method used to gather user preference data was to separately interview four people with different disabilities: mobility impairment (a wheelchair user), a registered blind person, a profoundly deaf person lip-reading the interviewer and a profoundly deaf person using British Sign Language to communicate through an interpreter with the person conducting the interview. There were several methodological issues to consider when gathering data, such as the ability to generalise the findings from these people for a larger group of building users and how to structure the interview so comparable information is gathered from each person interviewed.

A key concern was selecting who should be interviewed. Individuals with different disabilities and experience of the built environment were interviewed, enabling them to have direct influence on the design of the project. The researcher had the opportunity to interview the actual people who will use the building in order better to understand their personal requirements. This selection included representation from all disability groups that will use the building. This method of inclusion of views empowered those involved, as their unsolicited opinions were included in the exercise.

One of the interviews was conducted with an interpreter using British Sign Language to enable the discussion between the researcher and a deaf person. This changed the dynamic of the interview and introduced another level of interpretation. The interview became more of a three-way discussion of ideas rather than a prompt and response conversation. Although the words and terms referred to in the *aide memoire* were easy to understand, they were all relevant to construction and their use needed to be clarified to the interpreter so they weren't misrepresented to the person being interviewed. The response to the prompt was also influenced by the person's knowledge of construction and a keenness to respond in an informed manner. A key factor to get across to the respondent was that there was no correct response. Each response was an expression of their personal wish list for the building based on their own experience. In this way the response was individual.

The ability to generalise the findings from one person to a larger population is a representation issue to consider. The concern is that the needs and views expressed by one person with one type of disability may not represent the needs and views of others with the same disability. It is also that case that one disabled person may not be able to represent the views of a person with a different disability. Although the time constraints of the project meant that a small range of commonly occurring disabilities were represented by the structured-interviews this was considered adequate because of the structure of the interview and way the data was interpreted.

The interviews were conducted by the project researcher using a semi-structured approach and an *aide memoire* to steer the discussion. The *aide memoire* included 120 prompts such as 'access and escape', to encourage a person to comment on their access needs within a building and how they would escape from a building in the event of a fire. The *aide memoire* was developed through experience in architectural practice, to gather information about different aspects of a building to prompt the respondent to generate ideas for the building as part of the briefing process. The same *aide memoire* was used for each person interviewed. This allowed consistency across the sample of people interviewed, but didn't impose a structure as to how an individual should respond. The responses were the unbounded ideas of the person interviewed, expressed in their own terms. The prompts encouraged people to reflect on their experience of other buildings and environments, making qualitative judgements on the relative merit of these environments. This approach generated rich data of personal perceptions of their experience of buildings and suggestions for improvements to the built environment. Interviewing people individually had the advantage that their ideas were personal and not affected by group pressures and influences.

The rich verbal data gathered was transcribed, so it was available as text to be interpreted. The interview data was treated as direct access to 'experience' rather than actively constructed 'narratives'; both approaches are considered valid by Silverman (2000). Using each *aide memoire* prompt as a heading, the responses were combined within a single document, so the response from one person to a prompt was followed by those of another with a different disability. The suggestions made by those interviewed were given equal value, all ideas and suggestions were included in the briefing document. Any conflicting user needs were resolved using the second management device, the Steering Group, rather than the judgement of a single person.

7.5.2 Steering Group

The Steering Group is a management device within the project organisation introduced to formalise decision making. The Steering Group is a useful mechanism for trade-off and negotiation of problems. It is a forum for problems to be raised and resolved. A key to successful trade-off is the composition of the Steering Group and the understanding each member of the Group has of his or her role within the Group. Kelly, MacPhearson and Male (1991) wanted to ensure that the representatives on the Group were of sufficient status, part of the decision-making unit, for views to be expressed without reference back to others. In this instance each person was representing themselves as users of the final building. If there were different views amongst members of the Steering Group, in order to resolve disputes, and move forward and make a decision, the members would be reminded of their role as a representative of the wider objectives of the project, *i.e.*, to build an accessible building. This shift in perspective and reminder of the broader responsibilities is a useful mechanism of the Steering Group structure for group problem solving. The Steering Group should include representation of the building user group as well as the commissioning client. For this project this includes the involvement and representation of various disability groups. The minutes of the Steering Group meeting are a paper trail of decisions made on the project.

The issues to be addressed by the Steering Group are to aid the resolution of the differences of opinion expressed during the semi-structured interviews. Each stakeholder on the Steering Group examines the options prepared by the team, which includes the development of the brief as well as the direction the design is to take.

Careful examination of the language used and the emergence of ideas is the main thrust of this way of working. It is recognised that metaphors illustrate views and subtly encapsulate and communicate complex ideas. The synergy of the group is a good platform for generating and developing ideas. An advantage of this method is that ideas are discussed and more fully understood by all concerned, which may lead to a more valid and thoroughly negotiated result. The process is transactional and relies on the shared understanding of language to arrive at a valid result.

7.6 Conclusion

This paper has attempted to address the question whether inclusive design requires an inclusive design process by firstly considering the current socio-political pressures to include consultation with users, especially those with disabilities, as part of a design process. This political stance considers that disabled user involvement is essential to avoid assumptions being made on their behalf. To answer the question of this paper and to draw conclusions for others who are co-designing, an inclusive design process is absolutely necessary to emancipate disabled users and to meet social models of disability pressures.

The allied question of whether there is an approach to design that is truly inclusive of user preferences was addressed through investigation of the methods currently being used to represent the user within a design process. It was revealed that there are deficiencies within these processes that result in dissatisfaction with the built environment. To address some of the user representation imbalances a method was prototyped that involves user consultation in a structured manner to resolve conflicts of user need. The validation of this approach will be through user satisfaction and post occupancy evaluations of the scheme once the project is complete. This review of design processes has shown that no truly inclusive method exists, but the approach suggested in this paper is a positive step.

7.7 References

Barton L (1996) Disability and Society: emerging issues and insights. Addison Wesley Longman Ltd, Essex

Darke J (1984) Architects and user requirements in public sector housing. Environment and Planning B: Planning and Design 11:389-404

Davis L (1995) Enforcing normalcy; disability, deafness and body. Verso, London.

Hall P, Imrie R (1999) Architectural practices and disabling design in the built environment. Environment and Planning B-Planning and Design 26(3):409-425

Hasdogan G (1996) The role of user models in product design for assessment of user needs. Design Studies 17:19-33

Imrie R (1999) The role of access groups in facilitating accessible environments for disabled people. Disability and Society 14 (4):463-482

Jones,S, Welch P (1999) Evolving Visions: segregation, integration and inclusion in the design of built places. In: Proceedings of Evolving Design Research Association Conference. Florida, pp106-116.

Kelly JR (1991) A method for determining a user's value system and its application to briefing. Management, Quality and Economics in Building.E&FN, Spon

Kernohan D, Gray J, Daish J (1992) User participation in building design and management. Butterworth Architecture, Oxford

Kohlert C. (1998) Briefing through visualisation. Conference at DEGW, 19 Feb. 1998

Lipman A, Harris H (1998) User participation- a closed dialogue. Environment and Planning B-Planning and Design .25(1):1-10

Mason P (1992) The representation of disabled people: a Hampshire centre for independent living discussion paper. Disabled, Handicap & Society 7:79-84

Preiser WFE, Rabinowitz HZ, White ET (1988) Post-occupancy evaluation. Van Nostrand Reinhold, New York

Schön D (1988) Designing rules, types and worlds. Design Studies. 9(3):181-90

Sharrock W, Anderson B (1994) The user as scenic feature of design space. Design Studies. 15(1):5-18

Silverman D (2000) Doing qualitative research.Sage, London

Chapter 8

For the Love of the People: Participatory Design in a Community Context

A. Noble and C. Robinson

"Design is a personal activity and springs from the creative impulse of an individual. Group design or design by committee, although occasionally useful, deprives the designer of the distinct pleasure of personal accomplishment and self-realization. It may even hinder his or her thought processes, because work is not practiced under natural, tension-free conditions..." (Rand, 1993)

"Documentary [photography], as we know it, carries (old) information about a group of powerless people to another group addressed as socially powerful...[it] implores us to look in the face of deprivation and to weep..." (Rosler, 1989)

"In the past the local community where Hinetemoa lives, has only been described negatively...people living in Housing Corporation flats - poverty issues, bad health issues. Something comes on TV about benefits or low income housing and there's a five second clip where they sort of flash along the outside of these places...I mean they've got a really tight-knit, proud community here but those aren't the images that are ever portrayed" (Muriel Tunoho, Community worker, 2000)

8.1 Introduction

The Massey University School of Design takes on 'live projects' for 'real clients' as part of the curriculum in its Bachelor of Design degree. This gives students experience of the sort of issues, work and clients they might encounter in the 'real world', while still receiving the support, guidance and constructive critique of tutors and classmates.

In terms of its broader research and public service objectives, the charter of Massey University calls for links to be established with the wider community in order to ensure the relevance of the University's teaching, research and scholarship.

This paper describes a 'live project' that attempts to reconcile traditionally individual activities like photography and design with a union campaign concerned with challenging public perceptions about the lives and work of a group of contract cleaners. For the Massey University School of Design this project explored new ways of teaching and research; new ways of thinking about the roles of photography and design; new ways of dealing with a client group; and new outcomes from those normally expected from a student 'live project.'

'Live projects' are initiated when a client, usually a non-profit community organisation, seeks the assistance of the School of Design with a project. Provided the project meets the learning outcomes associated with a particular course, a representative from the organisation might come to the classroom, brief the students and leave information material. Individually the students then research the client and the issue, come up with a solution to the client's brief and present the work to the client; the client will select the piece of work that meets their needs and take it away to get published, printed or manufactured.

8.2 Sole-practitioner Model

The pedagogical model usually followed is what we term the sole-practitioner model, based on professional practice in the disciplines of graphic design and photography. This approach has its advantages. It is efficient in operation, can be completed within a short time period and is relatively easy to assess. Control over the creative process is in the hands of individual students. Although they may receive briefings from the client/affected groups, they are largely placed in a position outside and separate from a relationship with those whom the work is about or for. The model assumes that armed with some research knowledge (usually found in a library) and assistance from a tutor, a student will communicate the appropriate meaning to the appropriate audience, will receive personal satisfaction, a portfolio piece (and a good grade.) The 'essence' of this model is described by modernist designer and educator Paul Rand in the opening quote of this paper.

To each project the students are also expected to bring knowledge they have gathered throughout the course of their degree: knowledge of modernism and post-modernism; the issue of meaning - its creation, interpretation, analysis, deconstruction and reconstruction; the relationship of design to an audience, 'fragmented', 'passive' or 'active'; issues of subjectivity and objectivity; context; visual rhetoric; appropriation and representation of symbols, designs, images and people from other cultures or minority groups.

8.3 Participatory Design

The sole-practitioner model was not the model that we decided to use for this project, choosing instead to approach it as an exercise in participatory design education in a community context and to research the process used to generate a collaborative social documentary photography and design project.

Our decision to use a participatory design model had its origins in post-modern analysis of documentary photography. This analysis exposed the lack of objectivity in the genre of documentary photography; how the meaning and context of supposedly 'truthful' photographs was constructed by those who made them and the people who viewed them; how the photographer was supposed to sit outside and separate from the subjects of photographs and encouraged to filter images and information through a personal perspective while the subjects of the photography remained powerless as a silent 'other'. Rosler (1989), in her influential essay on documentary photography, summarised the cynicism surrounding a practice which located itself in galleries or environments that showed personal stories and social problems as grand or dramatic. "Documentary testifies, finally, to the bravery or (dare we name it?) the manipulativeness and savvy of the photographer, who entered a situation of physical danger, social restrictedness, human decay, or combinations of these and saved us the trouble" (p.308).

We sought to develop new ways of working collaboratively, developing strategies that placed a greater emphasis on the uses of photography, enabling participants to contribute to the ways they are photographed and the uses to which images of them are put. A priority was to honour the knowledge and experience of the participants in our project. It seemed to us that the most appropriate way forward would be through Participatory Action Research (PAR), described in Reason (1994) as a "methodology for an alternate system of knowledge production based on the people's role in setting the agendas, participating in the data gathering and analysis, and controlling the use of the outcomes" (p.390).

Our project thus began as a social documentary project for a final year photography paper entitled ***Advanced Social Documentary*** - a paper recently established to consider collaboration as essential to social documentary practice[1].

Aware of Massey University's Archive of Contemporary Culture (and the School of Design's community-based social documentary initiatives), the Projecta Foundation - a philanthropic foundation that supports projects for social change with an interest in photography, brought together the School of Design with a community arts group (HVCA)[2]. HVCA had been approached by a group of cleaners and a trade union (the Service and Food Workers Union), who wanted to create a photographic exhibition about the life and work of seven contract cleaners.

---

[1] Photographic essays produced in this paper are housed in the Massey University Archive of Contemporary Culture at the Alexander Turnbull Photographic Archive at the National Library of New Zealand. The Massey archive was established to ensure ongoing access to the photography by the participants/or subjects and to establish the possibility for other applications of the work for research and publication by the groups involved.

[2] Hutt Valley Community Arts (HVCA) is an organisation dedicated to encouraging community groups, often minority groups, to express their culture in creative ways within a community setting. They have a community art gallery which they use to facilitate projects that create increased understanding of a group by the community at large.

The objective of the exhibition would be to honour the contribution to society of this group of people, challenge the lack of awareness and negative public perceptions about the work of cleaning, and make visible the problems facing contract workers in New Zealand, in particular their concerns as members of a large group disadvantaged by the Employment Contracts Act (1991)[3].

8.4 The Photography

The project was developed in two stages. The first stage involved a group of photography students from the social documentary paper working alongside seven contract cleaners for four months. This was a small but representative group of Maori and Pacific Island workers, all members of the Service and Food Workers Union, in low paid work with little job security. Working at night, they felt their contribution to the economy went unrecognised. They were concerned that employers were cutting their hours and increasing their workload as cleaning companies competed to provide services based on price. These concerns informed a brief prepared for the students with input from the participant groups. Together the students and cleaners made images of home and of work, of participation in community life and of dedication to improving conditions for fellow workers. A telephone network kept students informed of the cleaners' involvement with pickets and industrial action.

Figure 8.1. Mafoe Eric at home with her family (L) and at a Service and Food Workers' Union picket (R). Photographer: Vicky Birkinshaw, student.

[3] Contract workers are employed throughout New Zealand in cleaning and kitchen work, and in many other areas of the service industry. Their jobs depend on a contract between their employer and a client company for the provision of a service. The Employment Contracts Act provides no protection for contract workers if their employer loses the contract to provide a service. New Zealand has no Transfer of Undertakings legislation to enable workers to transfer from one contractor/employer to another without reduction of their wages or conditions. To change this was one of the key objectives of the contract workers campaign.

Traditional social documentary photography might have portrayed these workers as poor victims, oppressed by the capitalist system. Instead, in response to a brief that had been devised to facilitate participation, the photographic essays produced by the students showed people who were proud of their contribution to the community, enjoyed and valued their work, were committed to their families, their wider community and fellow union members. The students presented the photographic essays to the cleaners as a record of their lives and work and their participation in the project. A second set of essays, along with the negatives, is on public record, gifted to the Massey University Archive of Contemporary Culture.

8.5 Exhibition and Publication Design

The photography completed, the participant groups met regularly to design a photographic exhibition. These meetings occurred over a three month period. Text to contextualise readings of the photographs was designed collaboratively. A union worker interviewed the cleaners and the transcripts were edited during readings at the meetings. Much negotiation took place. The union argued for more polemic. The student photographers stressed the need to discern those images that by their strength and relationship to other images made a point that did not need restating in words. The cleaners kept the focus on the stories and images that were key to their participation in the project.

Figure 8.2. Photography student John McCormack with cleaner Olive Harding at a meeting of all participants at the HVCA Far Site Gallery to discuss the exhibition layout (L). SFWU campaign poster designed by graphic design student Grant Williams (R). Photographer: John McCormack.

The opening of the exhibition, named by the cleaners *For the Love of The People*, was planned to coincide with the national launch of the union's contract workers campaign on 25 October 1999, Labour Day. Uses of the exhibition beyond the venue (such as resources for schools) and the value of publishing a record of the project had been discussed at the regular meetings, and pointed to the need for a publication. Funding was found to publish a brochure containing the cleaners' stories and images from the exhibition, as well as a poster publicising the union's campaign to raise the profile of the rights of contract workers. A class of third year graphic design students was given the printed material to design as a project. Although these students had not been involved with the project to this point, they were briefed by, and collaborated closely with, the union, the community arts group and the photography students. Twenty students each designed a brochure and a poster, and the participant group selected the final printed pieces.

8.6 Review of Process

Interviews were conducted during and after the project. The purpose of the interviews was to establish the degree and quality of ownership of the project. The following is a selection of responses from the interviews:

> *"...Because people were involved in a lot of the decision-making, it broke down the difference between - this is someone else's project and this is our project. And so people like Mafoe and Pua and Hine and Bill and even Olive, who is actually a very quiet person – by the time the project was up on the wall they were saying to people 'come and see our project – this is about us – this is about you'." (John Ryall, Union Secretary, 2000)*

> *"...In terms of the process, it was good to have space and time to discuss things as a group, but also to give the cleaners time to go away, reflect on things without the pressure of knowing that a decision made at that gathering was the final one. Having that number of meetings paid off in the end. It was really valuable." (John Ryall, Union Secretary, 2000)*

> *"...We met to discuss the layout of the exhibition. They (the cleaners) had started from a not subservient, but a very passive place. They didn't want to rock the boat and were just happy to kind of leave everything in other people's hands. There were comments like 'Just put me down the back of the room in a corner by a cupboard' – that kind of thing. They were absolutely trusting really, which was good, because they had built up a lot of trust through the project, but I talked to them and planted the seed of the significance of their role as tangata whenua within the project. This gave them the strength to feel confident about seeing the relevance of their role in deciding where their photographs ended up." (Muriel Tunoho, Community worker, Wellington 2000)*

> *"...I think the photography project is portraying us as we are. I see my grand-children have got something to look at and my great-grandchildren. It's a good thing to leave behind...it's history." (Hinetemoa Kahu, Cleaner, 1999)*

"...I'm a shy person, but I put my shyness behind my back...to stand up for my future and for my people...they are not my blood — but they are the Samoan people." (Mafoe Eric, Cleaner, 1999)

" ...This project has given me the chance to work with two warm and wonderful women, their families and workmates. The environments that I have entered have been alien to me, but both women have embraced me and the project. This has given me a feeling of honour and gratitude that I only hope I can repay through these images of their families and their work." (*Vicky Birkinshaw, Photography Student, Wellington 1999*)

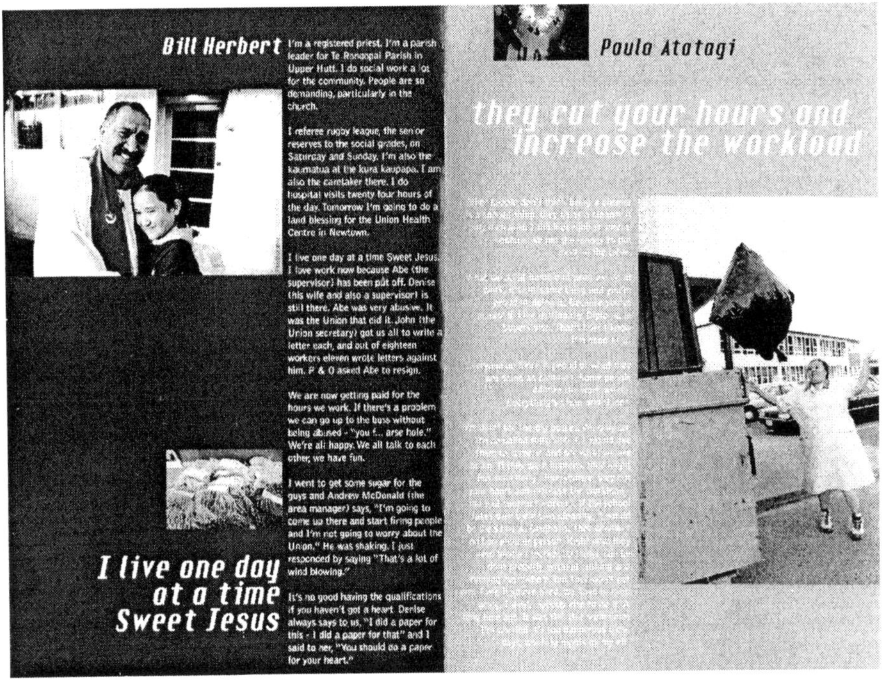

Figure 8.3. Excerpt from exhibition brochure, designed by graphic design student, Sarah Delany

8.7 Observations

From our interviews with the participants we made the following observations:

1. *The participatory design model enabled participants to speak for themselves.* The cleaners felt that their ability to define their own lives and issues of concern, to empower their families and speak on behalf of their wider community through the images and their stories, were of greatest significance.
2. *The traditional power relationships in the practice of documentary photography were overturned.* The opening of the exhibition created sufficient interest to attract national television and the local newspapers. Whereas the

subjects of photographs are often silent in a gallery context, the voices of this previously invisible group were broadcast loudly and clearly. The cleaners' initial objective of wanting their concerns to be heard was met.

3. *Authorship was shared and quality was not compromised.* With the sharing of creative decisions no one person could lay sole claim to be photographer, author, designer, editor or curator. Contrary to expectations, when authorship was no longer privileged, the collaborative work produced was still of a consistently high quality.
4. *Process and experience were of great significance for all concerned.* The participants valued the process, as much or even more than the outcomes. The physical outcomes of the project - the exhibition, brochure, posters, photographic essays, image resource and oral histories, were the visible but not the only significant outcomes. For the photography students the first hand experience of the contribution photography could make to community development was of greatest significance. This enabled their skills, both personal and photographic to grow. Contrary to Paul Rand's comment that self realisation only occurs when design is a personal activity, the students gained an enormous sense of personal accomplishment and self-realisation from having contributed so much to the community.
5. *Collaborative design proved to be a valuable learning technique.* As well as benefiting from the experiences highlighted above, having to collaborate on the design of the brochure and posters made students confront their own assumptions about the primacy of the typeset word and the photographic image in a printed piece of work. The photography students felt that the graphic design students did not give enough significance to the power of the image; the graphic design students felt that the type and its layout said as much, if not more, about the lives of the cleaners as the image. The resulting pieces, all graphically strong, are compromises. But it was through the experience of collaboration in the classroom, rather than through being lectured at, researching, or individually exploring the relationships between text and image, that the student's addressed the issues and were able to meet the learning outcomes of their paper.
6. *Massey University was able to promote research and serve the community.* For Massey University, the outcomes were mostly positive. To be associated with an exhibition that received much positive publicity on television and in the newspapers, and positive feedback from within the community, was extremely desirable from the perspective of meeting its Charter objectives.

However, the University also got caught up in some of the unforeseen difficulties associated with such a complex participatory project. In a case of the right arm not knowing what the left arm was doing, the same week that the exhibition opened, the University's own cleaning contract with a particular firm expired and many contract cleaners lost their jobs. The irony of this was spotted by a local community newspaper which ran the headline 'UNIVERSITY SPONSORS PHOTOS THEN DUMPS CLEANERS'. Not only did this result in some negative publicity for the University, but it incurred the wrath of the very union with which we were trying to collaborate!

Figure 8.4. SFWU Newspaper, Front Page, November 1999

Further conflict occurred over the issue of co-ownership of the images. Until the images were lodged with the Archive of Contemporary Culture they were still technically the property of Massey University and not available for public use. The Service and Food Workers' Union however, saw a more immediate use for the images in the lead-up to the 1999 General Election in New Zealand and without consultation published them on the front page of the pre-election issue of their newspaper alongside a message urging union members to vote the then government out of office. This placed the university in a potentially awkward position as owner of the images.

Issues of ownership and copyright were complex to resolve and might have been helped by clearer contractual agreements at the start of the project. However, given the experimental nature of the process, at the out-set no-one was prepared to predict the exact nature of the outcomes. It might be seen as an aspect of participatory design practice that where no one individual or organisation drives the process, some of the outcomes will be unpredictable, while others will be not entirely to the satisfaction of all concerned. Another aspect is that objectives will be formed and modified in the process of developing the work. It would be potentially limiting to set all the boundaries at the beginning.

8.8 Conclusion

The project supports the view that there is an important role for design educators in establishing inventive, experimental models of practice and research – models that question traditional values, challenge assumptions about ways of learning and working in the field of visual communications design, allow for engagement in the work and reflection on the process.

Research in design and photography is often theoretically driven and its outcomes unrelated to community-based uses and applications. In the teaching context a greater focus is placed on the sole-practitioner model of design practice, privileging the author and the designer/client business relationship. Other models of design practice and research, like participatory design, may need to be offered as alternatives. In this case, the PAR model offered a rich and rewarding experience for all concerned.

"While working with Olive and Hagavavae my appreciation for the cleaning professional grew. I gained an insight into these people's lives and feel privileged to have been able to work with them. Their positive outlook and pride in their work was inspiring. I always looked forward to seeing them." (John McCormack, Photography student, 1999)

Figure 8.5. Paula Atatagi, at a commercial cleaners' picket, May 1999. Photographer: Lynette Shum.

8.9 References

For the love of the people exhibition brochure (1999) Far Site Gallery, Wellington

Freire P (1982) Creating alternative research methods: learning to do it by doing it. In: Gillette A, Hall B, Tandon R (eds.) Creating knowledge: a monopoly? International Council for Adult Education, Toronto, pp 29-37

Rand P (1993) Design form and chaos. Yale University Press, New Haven

Reason P (1994) Three approaches to participative inquiry. In: Denzin NK, Lincoln YS (eds.) Handbook of qualitative research. Sage Publications, London, pp 324-339

Rosler M (1989) In, around, and afterthoughts (on documentary photography) In: Bolton R (ed.) The contest of meaning; critical histories of photography. The MIT Press, Cambridge Massachusetts, pp 302-340

Tandon R (1982) A critique of monopolistic research. In: Gillette A, Hall B, Tandon R (eds.) Creating knowledge: a monopoly? International Council for Adult Education, Toronto, pp 79-84

Ryall, J, Tunoho M (2000) interviewed at Wellington, 26 January 2000

Chapter 9

Games and Movies: Towards Innovative Co-design with Users

J. Pedersen and J. Buur

9.1 Introduction

User-Centred Design in industry is moving towards a common understanding that involving users in design orientated collaboration is much more beneficial than classical usability testing, (Madsen and Borgholm, 1999). The collaborative workshop formats developed in university settings in particular in Scandinavia, (e.g., Greenbaum and Kyng, 1991), 10-15 years ago are becoming increasingly popular now, but they need further development, both to accommodate new understandings of user collaboration and to fit into industrial organisations.

When establishing close and continuous collaboration with users, we extend their role into the design process from being a source of information to being a design partner. Actually we ask a lot of the users. We want them quickly to adapt to the new social setting they are invited into, to establish fruitful collaborative teamwork with people they do not know, either personally or professionally, and to accept a design process towards a goal posed by us. All without losing their contextual foothold.

The User-Centred Design Group at the Danish component manufacturer Danfoss is grounded in the Scandinavian tradition of participatory design. We have a long tradition of close collaboration with users throughout the design process (Buur and Bagger, 1999).

A major challenge in User-Centred Design is to organise and facilitate user workshops that are engaging and productive for all participants – and to make them work in industrial practice. When organising events involving users not just once or twice in experimental settings, but repeatedly in ongoing projects, one has to build ways of wôrking that ensure a positive outcome every time.

One track to take could be to standardise workshop procedures and, as facilitators, to administer the process strictly. But this direction tends not to leave room for questioning the initial problem and also to inhibit further development of the competence. Instead we have chosen to work with metaphors, which allow participants to improvise social interaction and design collaboration within a framework.

In this paper we will discuss how we combined the two metaphors of game playing and film making to engage the creativity of participants with different competencies both in a fun and efficient way. The two metaphors offer easy, adaptable frameworks for social interaction, and help establish a design vocabulary for fruitful idea generation and design discussions. The combination provides an opportunity to gradually move the focus of the design discussions from goals and intentions to desirable product properties, continuously developing and enriching the initial ideas.

9.2 The User as Design Partner

Establishing design collaboration between users and the development team is the capstan in User-Centred Design. Since users are seldom there as a constant resource, it is common to plan collaboration to happen in concentrated sessions, e.g., invite participants to user workshops or 'design events' (Buur and Bødker, 2000).

Brandt and Binder (1997) point out the importance of establishing a common ground, because the participants are each firmly rooted in their worlds of competency. There is a gap between the competencies of users and designers, and there are different approaches to span the gap.

One can select users, who have already proven that they are able to think in design terms. The lead-user approach of Von Hippel (1988) is to find users, who themselves experiment with modifying products to suit their needs: '..in fast moving fields such as high technology, the experience of ordinary users is often rendered obsolete by the time the product is developed or during its projected commercial lifetime. For such industries I propose that "lead users" who do have actual experience with novel product or process concepts are essential to accurate marketing research.'

Or one can support 'ordinary users' in seeing their work in the light of new technological opportunities. Lanzara (1983) states: 'It is crucial to understand that the ability to see a situation as another is not at all an analytical ability.' And he continues: 'a large part of the design process, especially in large-scale projects and organizations involving several actors, is not dedicated to analytical work to achieve a solution but mostly to efforts at reconciling conflicting frames or at translating one frame into another. Much work of the designer is less concerned with finding a solution to a specific problem than with defining collectively what is the relevant problem, how to see it.'

In the second approach, supporting design collaboration is about providing resources that help the participants to see one situation as another; making room for the individual participants and the team to frame and reframe problematic situations (Schön, 1983); exploring and experimenting with the situation they are inquiring into; and dreaming but still maintaining a clear picture of reality.

The environment seems to have a significant influence on the flow and outcome of user collaboration. At Danfoss, we wanted to combine some of the advantages achieved in user workshops organised in the designers' environment with those gained when collaborating with users in their own work context.

When collaborating with users *in our design environment* (e.g., a meeting space at the company), we can invite a number of users from different plants and learn from hearing them exchange work experiences. We can include a fair number of internal people, which is important because user collaboration is about influencing attitudes. Being in a foreign environment (and with other users), users will tend to take a more general view of things.

When collaborating with users *in their work context*, users tend to feel more at ease as they are on home ground – we are the visitors. Tools and environment are physically present and easy to refer to. This makes for a conversation grounded in concrete and specific work experiences. The idea was born to create a type of design event with activities in both environments and with two sets of resources to support design collaboration.

9.2.1 Board Games: Constructing Virtual Worlds

Two concepts took a central role in shaping the 'design environment' part of the event: That of 'language games' and that of 'constructing a virtual world'. Based on a Wittgensteinian approach, Ehn and Sjögren (1991) suggest that 'by understanding design as a process of creating new language games that have a family resemblance with the language games of both users and designers, we have an orientation for doing design as skill-based participation.' So, we need to support participants in creating a design language.

Lanzara (1983) points out that the designer tries to reduce ambiguity by constructing a virtual world: 'While working toward setting a problem a designer continuously produces examples, picks up instances, draws analogies, shapes virtual worlds.' We should provide resources not simply for designing a product, but for shaping a virtual world.

Horgen *et al.* (1999) use board games to involve employees of R&D departments in transforming their own work environment. In a very concrete sense participants use a game board and game pieces to 'construct a virtual world', *i.e.*, to design the physical layout of the workspace. But the interesting thing here is that though Horgen's game board is a scale print of the building layout, the game pieces do *not* depict tables, shelves and cupboards. The pieces have arbitrary shapes and do not represent particular entities until the players decide what they mean to them. The pieces turn into reification of concepts during the design discussion. They become vocabulary in a design language that the players develop as they go along.

Inspired by Horgen, our approach was to develop a board game, which would help participants create a common design language and construct a shared virtual world.

9.2.2 Movie Making: Acting the Future

For the collaboration with users in their work context, we have been greatly inspired by Binder (1999) who has achieved amazing results with single users improvising and acting out future use scenarios in their work place. Binder found that users needed 'props' to act with in the form of very crude foam blocks: 'things' that would trigger the users' imagination and to which they themselves could associate meaning. Binder videotaped his scenario sessions continuously and observed how users can jump between acting in front of the camera and directing the video movie – 'Did you get this?'

Minneman (1991) suggests an explicit use of video 'letters' for communicating accounts of design problems and ideas, because video can cope with the ambiguities in early design work.

The challenge to us was to take Binder's way of working from individual to collaborative work (*i.e.*, several users co-acting an improvised scenario) and then to make the role of the video camera more explicit by putting the users in control of when to record and when not to.

9.3 The Co-design Event

In a recent vision project at Danfoss, we planned a full day workshop in which 30 designers, developers, marketing people and usability experts collaborated with users to explore new ways of improving wastewater treatment processes and to create ideas for future products within this industry. It was the second out of four workshops during a six-month period. Some of the participants had been involved in the project previously; others were participating for the first time. In parallel, the workshop also had the purpose of making the design project transparent to in-house circles of stakeholders, instigating learning on wastewater treatment, moving attitudes towards user sympathy, and shaping consensus across business units.

The participants were divided into three teams with two users in each. The teams were asked to focus on their own subprocess of wastewater treatment: mechanical cleaning, the biological process, and sludge processing.

9.3.1 Part One: The Design Game

The teams were seated around tables and after a presentation round we introduced them to the design game and its purpose. A map of a plant layout served as the game board, Figure 9.1. Foam pieces in different colours and shapes worked as game pieces for the team to attach meaning to. A team member from the User-Centred Design group served as game master explaining the rules, and contributed to the game with his own knowledge and his understanding of the project teams' findings so far.

Figure 9.1. Playing the board game: the workshop participants use game pieces to discuss the instrumentation on the present and future wastewater plants

In the first round the objective was to re-build the existing process, each participant, in turn, equipping the map by placing foam pieces representing a specific type of component (sensor, pump, display) on it, Figure 9.2.

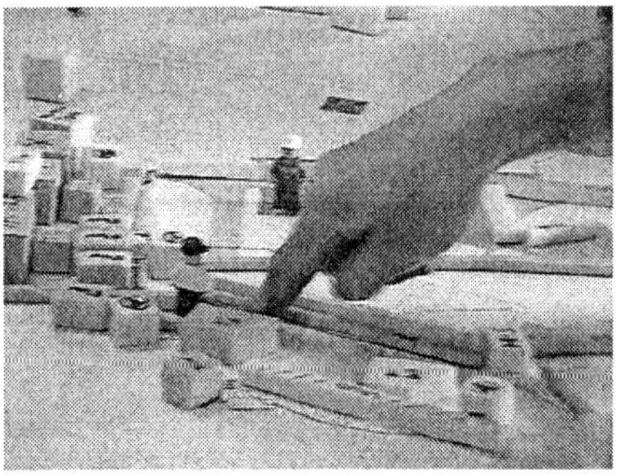

Figure 9.2. Game pieces: participants build a vocabulary of varying pieces in varying shapes. The Lego character serves as a springboard for talking about operator roles in he future plant.

By naming, shaping and placing game pieces, the participants explore the plant step by step. With each piece they offer a part of their understanding for negotiation in the group. Primarily the user from the plant commented on the placement, but it also challenged the other group members' theoretical understanding of what 'ought' to be there, and a lively dialogue was initiated. The aim was to make the group gradually build a common understanding of the process to be re-designed. Often, in the beginning of the game, the placement of the piece was only accepted when touched by almost everybody. The turn-taking, allowed all participants to engage in the process, and to stay in an open-minded 'game mode'.

In the second round, improvements were suggested by modifying the pieces on the board, e.g., introducing new sensors, changing control loops or adding interaction resources. The participants were forced to justify the placement, which fostered a fruitful dialogue about goals, intentions, benefits and effects. People were asking each other such things as, for example: 'what if we change this?', 'on our plant we do this, because...', 'would you benefit from this?'.

9.3.2 Part Two: The On-site Movie

Figure 9.3. Movie making: Two users act operator and apprentice in a scenario where a handheld device is used during the daily round at the plant

In the second part of the workshop the teams were asked to choose the most promising ideas and produce a movie showing a future scenario at a nearby wastewater treatment plant. To make sure the focus would be on how humans interact with technology, we suggested the teams should stage a scenario, in which an experienced operator would introduce a newcomer to how to do his job in the future plant.

Figure 9.4. Premiére: the presentation of game board and movie initiates a broader discussion among the teams about future plants and how work will change for operators

During a quick walk-through at the plant, participants tried to relate the game pieces to reality. Then the team started planning the video production. The team worked as scriptwriters, instructors, actors and camera crew. The plant worked as the scene and a box containing foam blocks was available for props. The aim was to let the groups act out their board solution in the scene of reality. They came to use the foam blocks much like they used the game pieces. Playing with new artefacts in a real context forced the group to enrich the original idea by discussing embodiment and user interaction. One of the teams produced a handheld device for distributed control, the other two teams introduced new types of sensors and interfaces, which they placed at well-considered spots in the plant.

The teams continuously discussed, negotiated, agreed and recorded the future scenario. At the end of the day the three teams presented their game board ideas with their on-site movie as a basis for a final discussion about the new concepts generated and visions of the future wastewater treatment plant.

9.4 What Did We Learn?

We believe that the event at Danfoss illustrates how the game and movie metaphors can be used to engage the creativity of users in the co-design process. People who were not used to collaborate, collaborated, and people who were not used to design, designed, in an open-minded, explorative atmosphere – in a setting suitable for industrial practice.

Not only did the participants quickly adopt the social framework we offered to them, they also established a design vocabulary, which fostered lively dialogues, made the participants express their ideas, negotiate their understandings and carry out rapid experiments. They developed a common understanding by building a shared virtual world, represented by physical artefacts.

The two metaphors helped the users administer their shifting roles as user and designer, keeping a present and visible path between the virtual world and the concrete reality. The game board and the context grounded the design process, and the negotiated placement of the game pieces and asking the cameraman to record, served as a clear boundary between today's reality and ideas for the future. As a result of the metaphor combination, it was possible for the teams to take the design discussions from goals and intentions during game playing, towards embodiment and user interaction of the ideas during the movie making process. The shared virtual world created on the board game helped to smooth the shift of focus from abstractness to concreteness, allowing the overall idea survive when new questions were raised about fitting it into context. In two of the teams the design discussion drifted outside the game board, leaving our original idea of what to redesign behind. But in the game there was room for reframing the problem - the design process was in the hands of the participants. There was room for the users to articulate the ideas they viewed as important.

From an industrial viewpoint the result of such events is a large number of 'rich' ideas, collaboratively developed by internal experts and users. The ideas were enriched throughout the day, drawing on process control experience, trends in the wastewater industry, 'horror stories', and supporting arguments from the other competencies present. This is very important for establishing a clear picture of the validity, vitality and feasibility of the proposed ideas.

When evaluating the process, all the participants noted that the event had been rewarding. The designers had gained new insights into the users' work practice, and the users had exchanged experiences with colleagues and technical experts, and pictured what their plant instrumentation and work practice might look like in the future. The positive experience with this form of design collaboration is an important precondition for a continuous involvement of users throughout the design process.

9.5 References

Binder T (1999) Setting the Stage for Improvised Video Scenarios. In: CHI'99 Extended Abstracts. ACM Press, pp230-231

Brandt E, Binder T (1997) Customer/User Workshops in Product Development – Creating Fora for Collaboration Between Different Compentencies, ICED97, Tampere

Buur J, Bødker S (2000) From Usability Lab to "Design Collaboratorium": Reframing Usability Practice. DIS 2000, New York

Buur J, Bagger K (1999) Replacing Usability Testing with User Dialogue. Communications of the ACM, 42, 5

Ehn P, Sjögren D (1991) From System Descriptions to Scripts for Action, Design at Work: Cooperative design of computer systems, LEA publishers, Hillsdale, New Jersey

Greenbaum J, Kyng M (1991) Design at Work: Cooperative Design of Computer Systems, LEA Publishers, Hillsdale, New Jersey

Minneman SL (1991) The Social Construction of a Technical Reality: Empirical Studies of Group Engineering Design Practice. Dissertation, Stanford University

Horgen T, Joroff ML, Porter WL, Schön DA (1998) Excellence by Design – Transforming Workplace and Work Practice. Wiley & Sons, Inc

Lanzara GF (1983) The Design Process: Frames Metaphors, and Games. In: Briefs U, Ciborra C, Schneider L (eds.) Systems design for, with, and by the users. North-Holland, Amsterdam, pp29-40

Madsen KH, Borgholm T (1999) Cooperative Usability Practices. Communications of the ACM: 42, 5

Schön DA (1983) The Reflective Practitioner. Basic Books, New York

Von Hippel E (1988) The Sources of Innovation. Oxford University Press, New York

Chapter 10

Working Anywhere: Co-design Through Participation

U. Dandavate, D. Steiner and C. William

10.1 Background of the Project

This paper outlines a participatory research project in which multidisciplinary teams from SonicRim and ACCO Brands worked to understand people's experiences of working away from their primary workspaces. The paper will use this project to illustrate the processes of collaboration (between SonicRim, ACCO and end-users), cultivation (of customer intuition) and ideation (through continual empathy with end user experience).

We believe that only those companies that can internalise customer experiences will be able to resonate with their customers' aspirations. We propose that this activity calls for designers to embrace a new mindset. We call this approach Postdesign.

10.1.1 Industry Overview Summary

The office products industry (*i.e.*, office supplies, excluding computers, other technology-based products and office furniture) is a traditional industry in terms of its product types and categories. Its evolution in the last 50 years and its adaptation to changes in the work environment has been slow. Very few traditional office products have evolved beyond mechanical improvements since their conception. The biggest influence in the office products area, specifically in the United States, has been the introduction in the 1980s of office super stores, causing increased commoditisation of existing products rather than functional or usage improvements.

10.1.2 State of Design Research in the Office Product Industry

The office products area can be organised into several categories, including paper organisation products, consumable supplies, presentation products, personal organisers and computer accessories. This last category has evolved as a result of the widespread use of computers for personal and work related purposes.

All the other categories contain products invented in the first half of the twentieth century, such as the paper clip, staplers, binders, and punches. Many of these devices support clerical work processes and continue to provide sufficient functionality for specific tasks. However, these existing products do not adequately address the tacit needs of the Knowledge Worker. New needs associated with knowledge-driven project work, computer oriented work styles, and collaborative processes have drawn the attention of the industry towards the urgency of understanding people's emerging work styles and associated needs/aspirations.

10.1.3 Design and Design Research

A new approach to product development was introduced, focusing on user-centred design. New to the company and possibly the industry, the main objective was to achieve strategic differentiation from the competition in new product development through intimate user understanding.

Initial research approaches centred on the belief that new functionality combined with new approaches to appearance should appeal to the end-user and create acceptance of new lines of products. This was the intention of the newly founded new product development group (NPD). Its first task was to introduce to the company the function of design in general and design research specifically. The challenge was to evolve and integrate a new process into an industry that was guided by a market driven primarily by the channel customers.

Ethnographic research was the first methodology applied to learn more about end-users. Over a period of three years, the NPD team conducted research projects in collaboration with E-Lab (Robinson, 1994; Cain, 1998) and Design Research to observe user activities in the office environment as they relate to objects/products, activities and interactions. The process of structured planning (Owen, 1993) was used to organise and analyse research materials and develop a systemic understanding of the work environment.

The ethnographic research did yield insights about user behaviours and needs. ACCO gained a better understanding of the usage of products and generated ideas for improving existing products by redesigning them to new requirements. Some of the learning provided the basis for innovative concepts in existing categories.

The communication approach chosen by the NPD team to mediate learning and insights to all participants was, however, found ineffective. Although representatives of different groups were involved in the projects from the beginning, presentations of insights discovered along the project development did not result in meaningful learning experiences that could be translated into successful concepts. Continuous guidance was needed to apply and implement insights and frameworks.

10.1.4 Search for New Methodologies and Approaches

The NPD team was faced with the question, 'Are there other ways to communicate these insights, user understanding, and the implications for marketing and channel strategies?' On further introspection the group also realised that it might be missing an understanding of a component of user experience that cannot be accessed through mere observation. The ACCO team wondered, 'Are we possibly missing the opportunity for developing products that serve less explicit needs?' and, 'Could a new approach give us these insights and at the same time improve internal and external communication and understanding?'

ACCO created an interdisciplinary team to explore potential opportunities that could become the focus of new research and development projects. The diversity of the backgrounds and departmental ties of the team members was a first step to assure company-wide participation in, and ownership of, this project.

The first third of the project was dedicated to research and analysis with three focus areas: users, technology, and industry. The team was limited in conducting primary research and instead reviewed the previous three years of research to identify new overall patterns and possible trends. Also, the team reviewed research conducted in similar industries, such as the office furniture industry.

The team identified patterns of change in work habits and environments, electronic work and communication tools, and the use of information, that will have long-term implications on the use and development of office products. The ACCO team discovered, among other things, a need for knowledge workers to work away from their primary workspaces. There are approximately 38 million mobile professionals in North America. Mobile Professionals are employed full time and spend 20% or more of their total work time away from their primary work environment (Meyers, 1999).

The team proposed a new user research project focused on working away from the primary workspace. The responsibilities for design and user research were given back to the NPD team to apply different methodologies and approaches for this project. The primary objective was defined as a search for opportunities for new product ideas for working anywhere.

The search for state-of-the-art in design research led ACCO to SonicRim, a recent start-up that has developed design research capabilities based on participatory research principles. The need for continual discovery and assimilation of insights was a critical consideration in selecting the participatory approach.

10.2 The Approach

Drawing from the domains of marketing, anthropology, psychology and design, SonicRim has developed different types of tools that allow participants to articulate their experiences. Rather than taking a functional/behavioural view of work practices of people, the focus is broadened to include emotional and cognitive experiences of working anywhere. Besides seeking an understanding of the more explicit facets of people's experiences, which they could articulate through conversations and journals, we needed them to reveal what they knew, how they felt and what they aspired to – things they could not easily articulate, nor reveal in their observable behaviours. The SonicRim team gave the participants cameras and workbooks to document what they **do** over a one-week period. They were visited in their homes and in their workspaces to hear what they had to **say** about their experience of working away from their primary workspace. And they were given exercises, such as collage kits and cognitive mapping kits, which allowed them to **make** images/artefacts that embodied their concepts of ideal experiences of working anywhere. SonicRim believes that by providing people with these tools that afford different modes of expression, researchers can gain access to deeper meanings associated with people's everyday experiences (Sanders, 1999).

10.2.1 What Did We Do?

The project began with a workshop with the NPD team. Prior to the workshop each member of the team was given a disposable camera and a workbook to document their personal experience of working anywhere. Besides setting goals and responsibilities for team members, the workshop provided the team members first-hand experience with participatory tools, and it set up the relationship between SonicRim and ACCO as collaborative learners, as opposed to client and consultants.

10.2.2 Generative Search

Figure 10.1 depicts different types of research relevant to different phases of the design process (Sanders, 1999). Much of traditional design research has depended on information derived from evaluation (e.g., usability research). Opportunities for breakthrough will come from generative search at the front end of the design process, designed to integrate aspirations of stakeholders (including end users) with design ideation. Generative search requires its own set of tools, and a unique mindset that allows design researchers to assimilate the understanding of collective aspirations of stakeholders into their inspiration process.

Figure 10.1. Types of Design Research

We conducted generative search in two geographic markets: Chicago and San Francisco. We conducted one-on-one participatory work sessions with individuals within their work environments. In each market we organised five contextual visits with people in a mix of small, midsize and large offices, along with home offices.

Three participatory group sessions (six people per group) were conducted in each market. We used a methodology similar to the contextual visits, yet customised for the facility research setting. In selecting the participants we recruited for high mobility within the office buildings; high mobility outside the building for client meetings; and those involved in free-lance work. We also selected a few participants who work from their car.

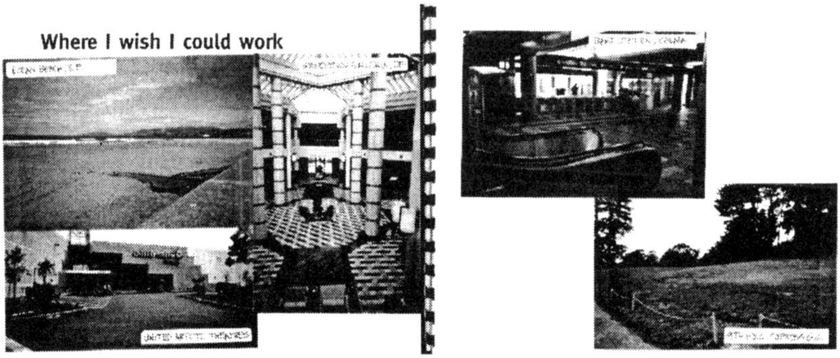

Figure 10.2. A page from the workbook

Two weeks prior to the visit, participants were sent a workbook, in which they were asked to take pictures of the things that help them work anywhere, the places they work and the places where they wish they could work. It also contained open-ended questions and seven point rating scales for participants to rank the importance of different activities (e.g., think, remember, create, *etc.*) when working away from their workspaces. Figure 10.2 depicts a page from one of the workbooks completed by the participants.

We planned the participatory activities in a sequence to allow people to reflect over their past week's documentation of their experiences, and then project their current and ideal experience. Figure 10.3 depicts a collage created by a participant. It contains four zones representing different personal experiences of the participant. The area inside the circle represents experiences around hi-tech products; the area outside the circle represents experiences with low-tech (or no-tech) products; the area below the line represents current experiences and the area above the line, the person's aspiration for work experiences away from the primary workspace.

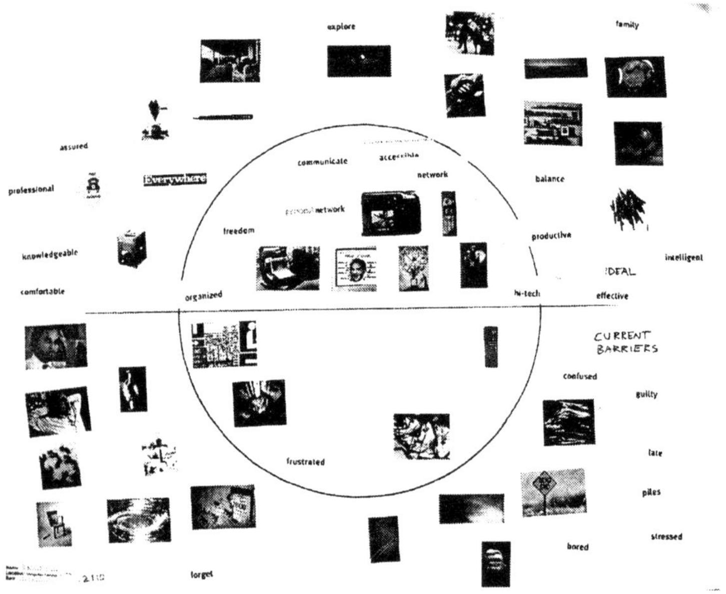

Figure 10.3. Collage - the experience of working anywhere

Finally, we gave them another set of tools consisting of words that allowed them to create their ideal system of low-tech products for working away from their primary workspace. The toolkit consisted of a white board, three sets of words representing products, functions, and contexts of use, and a set of coloured dots and sketch pens. They used dots to represent those activities that they rated very important in the workbooks as nodes of their diagram, and created clusters of words around these nodes to describe components of their ideal low-tech system of products.

10.2.3 Analysis

The participatory activities with end users were successful in getting the ACCO team to start thinking about the emotional and cognitive issues surrounding the experience of working away from the primary workspace. We used different ways of structuring, analysing and interpreting the data. For example, we used a multi-relational database to store, structure and compare textual information from workbooks and transcripts of narrations/conversations; and multidimensional scaling techniques such as Pathfinder networks and cluster analysis to find patterns in the rating scales, collages, and cognitive maps.

The core team communicated regularly to draw inferences from visual organisation of data and to identify experiential themes. We discovered three user segments: team workers, creative planners and information catchers. We also identified distinct activity segments that were central to the experience of working anywhere: thinking, creating, writing, managing paper, planning, remembering, communicating/being accessible, storing, organising. Together, the user types and activity types helped us gain insights into the experience of people working away from their primary workspaces. We developed an experiential model that visually depicted the emotional experience of working away from the primary workspace. An experiential model allows the design ideation team to stay focused on core experiences people have/seek when working away from their primary workspaces.

10.2.4 Ideation Workshop

The findings were embodied in 25 inspiration boards that allowed the larger ACCO team immersion in the insights. Smaller teams used information about the three user segments and activity segments as a framework for developing scenarios. They created a narrative of a day-in-the-life of each user type, then collaboratively generated new product and marketing ideas for the persons portrayed in the scenarios. Subsequently the team reduced the number of inspiration boards to seven, and displayed them in the ACCO product development area. The framework was put to use in a series of internal brainstorming sessions, and based on the input received from these sessions, SonicRim conducted a final ideation workshop.

The goal of the final workshop was to create user scenarios as a basis for ideation. In the first exercise each team member had to identify him/herself as one of the three user types identified from the research. Then, we provided them with a poster board with a circle drawn in the centre. The circle represented a typical workday in the life of this team member. We gave them a set of time icons (Dandavate and Welker, 1998) to visually depict their typical day. Next, we gave them four sets of stickers that contained words and images of products and activities, both high-tech and low-tech.

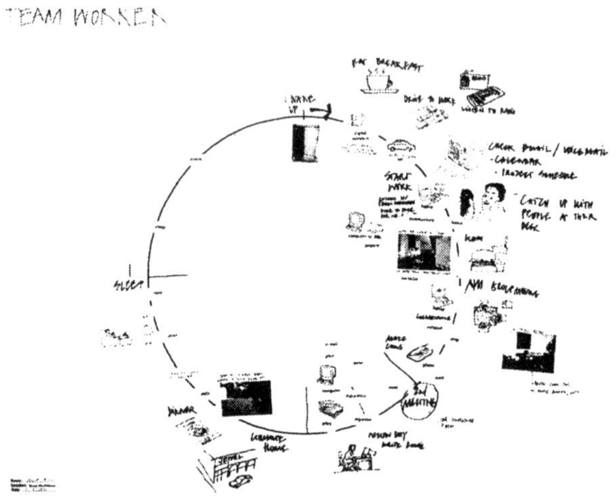

Figure 10.4. A day-in-the-life scenario

They arranged these stickers along their map to depict a scenario: an ideal day-in-the-life of the user type they selected to be. The new visual format for developing a scenario (see Figure 10.4) proved to be a more effective embodiment of scenario development than other formats we experimented with earlier in this project.

After developing the scenarios we asked the participants to draw or write product/service/branding ideas on index cards and place them along the path of the ideal scenario. Finally the workshop culminated in a sharing of individual ideas and their relevance to the day-in-the-life of the user type. Ideas continued to emerge from this framework. We had to stop the ideation for lack of time.

10.3 Reflections from the ACCO Perspective

Below the reflections of the ACCO team are presented for the purpose of understanding the value derived from the participatory approach undertaken during this project.

10.3.1 Internal Workshops

After a first presentation of the results of the ideation phase the ACCO team conducted a number of brainstorming sessions to further internalise the materials. The participants were members of the NPD team who had been part of the project from the beginning. The goal of these sessions was to gain a deeper understanding of the issues, activities and frameworks, and to think about them in terms of ideas, concepts and general implications.

In the first session the group discussed initial insights, ideas, and concepts of each participant. The discussion guided the brainstorming to create potential themes for new product development, specific concepts, and general conclusions. In the second session the group developed a number of scenarios for each of the user types. These scenarios described the personal profile, activities, aspirations, and a day-in-the-life of the person. The results of the both sessions were compared in order to refine concepts, identify additional ones, and target the concepts to individual personal profiles.

All the sessions were documented in a summary given to all participants. This document was designed so as to allow the readers to add notes, make sketches or illustrate other ideas. This report and all the materials generated in other workshops will continue to build the basis for future ideation sessions. All these materials will be analysed, filtered and integrated in the marketing planning process.

The final workshop facilitated by SonicRim provided the NPD team with additional materials and tools that facilitated better integration of multi-disciplinary team ideation. The visual format for scenario development, using the day-in-the-life format proved to be a breakthrough for developing a tool for future internal ideation. Overall the SonicRim/ACCO collaboration was successful in generating a framework for design ideation that could be used in the future, even without SonicRim's direct participation. The shared experience of this project allowed the ACCO team to experience and accept the concept of resonance and the power of empathy (with user experience/aspiration) in inspiring infinite design ideas.

10.3.2 Participatory Research vs. Observational Research

It was clear to the NPD team that, of the two different research approaches, the participatory approach provided additional means not only to learn about the end-user, but also to communicate the learning internally. In terms of user understanding the two approaches complement each other very well: observational research revealed activities or observable facts that the users are not aware of and are not able to articulate if asked. The participatory approach delivered higher-level learning about user aspirations that are not accessible through observation alone. The objective of the group is now to combine prior learning from observational research with the learning from the participatory research where it is appropriate.

10.3.3 Client Participation and Commitment

The inclusion of all participants throughout the process - not only through participation in research activities, but also going through the same generative exercises - was clearly a successful way to enhance understanding and eventually ownership of the project and its result. The inclusion of members of the executive management, sales groups, and other company areas helped to promote the project in the beginning and was a start in having them understand the topic and its implications. Continuous inclusion of significant company groups and individuals in the process are crucial for the success of such endeavours, because this approach does require a new way to look at end-users and how products are marketed and sold to them. If this approach is not embraced by a significant number of people in the company the whole initiative is at risk of not being implemented.

Support on all levels will not only insure success for project implementation, but it also will allow for a continuous and concerted effort to follow through with the project. This will require commitment from the beginning of financial, human and internal resources. It also will require a clear long-term strategy that supports commitment of resources and is flexible in supporting the implications of the project, and determination to introduce meaningful results in the market.

10.4 References

Cain J (1998) Experience-based design: toward a science of artful business innovation. Design management journal 9, 4

Dandavate U, Welker K (1998) Making meaningful changes in user experience: emerging role for the design community. Electronic Symposium of the Arizona State University. Department of Industrial Design

Meyers K (1999) The device user constituency: Sherwood Research. The Sherwood Report

Owen CL (1993) A critical role for design technology. Design management journal 4 , 2

Owen CL (1991) Product integrity by design. Chicago IL: Institute of Design, Illinois Institute of Technology

Robinson R (Winter 1994) Making sense of making sense: frameworks and organisational perception. Design management journal.5, 1

Sanders EB.-N (1999) Postdesign and participatory culture. Paper presented at the Useful and critical: the position of research in design conference. University of Art and Design in Helsinki

Sanders EB-N, Dandavate U, Stuart S (1996) User empathy in the product development process. Proceedings of IDSA conference on alternate realities (CD-ROM) IDSA

Sanders EB-N, Dandavate U (1999) Design for experiencing: new tools. Design and emotion conference Department of Industrial Design, Delft University of Technology Delft, The Netherlands

Chapter 11

Structured Methods and Creativity: A Happy Dutch Marriage

M. van Welie and G.C. van der Veer

11.1 Introduction

DUTCH is a design method for complex interactive systems. Such design projects require multidisciplinary teams, where the various members each have the responsibility for some aspect of design. This poses constraints on the use of documentation, representation and communication within the team and in relation to clients and users. The required representations vary in character, from formalisms to scenarios and highly informal sketches.

From our experience in applying DUTCH in teams in educational settings and industry, we will show that the combination of representations does not limit creativity but, on the contrary, stimulates creative input from designers, clients and prospective users. Informal and open scenarios are built on the basis of formal representations of task models, and evaluation is done by systematic heuristic walkthroughs as well as by acting out the scenarios.

Designing complex interactive systems is rarely an individual activity. As soon as different user roles can be distinguished (technological innovations affect roles, procedures, and other organisational aspects) various disciplines have to be involved, and these have to work in a structured design team. In industry, such large systems are typically designed by groups of 10 to 30 persons where subgroups are formed that handle particular activities. DUTCH is intended for the development of such systems. We suggest allocating complementary activities to small, specialised groups of around 3 to 5 persons. The groups have to work together to complete the design. Breaking up the work is necessary when such large projects are undertaken. When groups work together it puts constraints on the way communication is handled. We found that a combination of formal and informal representations is well suited to this kind of communication.

In the following sections we will highlight the kind of representations that we use in the various design activities. First we will give a short overview of DUTCH in order to explain the activities it comprises and their purposes.

11.2 DUTCH Design

Over the past years we have taken useful bits of theories and combined them into a coherent practical method for designing complex interactive systems. The resulting method is called DUTCH (Designing for Users and Tasks from Concepts to Handles). The method has been used successfully in both industry and education proving the practical value of the method. From experiences such as those outlined in the example given below, we learned that a practical method requires one to: a) define a clear process; b) define the models and representations including their semantics; and c) support the method and models with tools.

Our design process is task-based which means that it uses the tasks of users as a driving force in the design process. The goals are to design both usable and useful systems. We think it is important to base the design on the work that has to be done by the users. Therefore, the users play an important role in acquiring knowledge about their work as well as for usability testing.

Our process consists of four main activities: (a) analysing a 'current' task situation, (b) envisioning a future task situation for which information technology is to be designed, and (c) specifying the information technology to be designed. In parallel to these activities, (d) evaluation activities make the process cyclic. Figure 11.1 gives an overview of the whole design process with all activities and sources of information.

When using DUTCH it is required that the design team works in groups, because of the different types of expertise that are needed. Typically teams could consist of a variety of disciplines including computer scientists, psychologists, ethnographers and industrial designers.

11.3 Groupware Task Analysis (GTA)

The design process starts by an extensive task analysis using our method GTA (Van der Veer *et al.*, 1996). A task analysis of this type includes a description of the work, of the work situation and of users and other stakeholders of the system to be designed. We distinguish two task models. The first task model we make is a descriptive task model and is used for analysing the current task situation. The second task model is a prescriptive task model for the system that is to be designed. Both task models are usually done by task modelling specialists, e.g., psychologists and ethnographers. However, for the second task model all design disciplines should participate by contributing ideas, which are then described by the task modelling group.

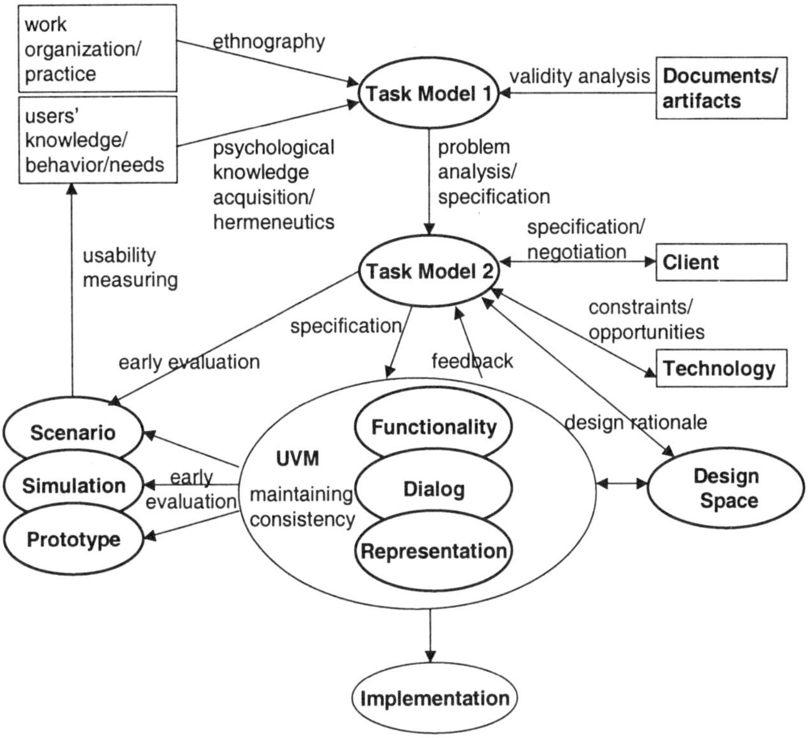

Figure 11.1. The DUTCH design process

11.3.1 Analysing the Current Task Situation (Task Model 1)

In many cases the design of a new system is triggered by an existing task situation. Either the current way of performing tasks is not considered optimal, or the availability of new technology is expected to allow improvement over current methods. A systematic analysis of the current situation may help formulate design requirements, and at the same time may, later on, allow evaluation of the design. In all cases where a 'current' version of the task situation exists, it pays off to model this. We use a combination of classical HCI techniques, such as structured interviews (Sebillotte, 1988) and CSCW techniques, such as ethnographic studies and interaction analysis (Jordan , 1996).

11.3.2 Envisioning the Future Task Situation (Task Model 2)

The actual 'step' from Task Model 1 to Task Model 2 is probably the most important one and at the same time the most open transition in DUTCH design.

Based on the requests from the design client and the problems and inconsistencies found in the current situation (Task Model 1), and wishes and ideas of the users and stakeholders questions for redesign are formulated. All available design disciplines can now be invited to create options for improvements or changes, without discarding each other's ideas but possibly reacting to ideas from other disciplines. After exhaustive generation of options, all disciplines should contribute to generate criteria to evaluate all option against affordability, usability, technical feasibility, and any other type of assessment that may seem valid. Task Model 2 will, in general, be formulated and structured in the same way as the previous model. However, in this case it is not considered a descriptive model of users' knowledge, although in some cases it might be applied as a prescriptive model for the knowledge an expert user of the new technology should possess. Initially, development of Task Model 2 is driven by the results of the analysis of Task Model 1.

11.4 Detailed Design

The detail design phase consists of the specification of the technology to be designed. This phase is started by Task Model 2 that contains specification on the task level but does not contain any details concerning the technology. During detailed design decisions have to be made on the look and feel of the technology, including the dialogue design and hardware ergonomics/design. The task specification is used as a basis for the functionality of the technology as far as relevant to users. Using the task structures an initial grouping of functions and the main navigation structure is described. In a continuous process choices are made that affect other design activities. For instance, hardware design can pose restrictions on the number of buttons or screen estate that is available. In commercial design there is often a tension between purely usability-directed design and commercial issues such as product marketing.

11.5 Designing in Teams

In DUTCH, team based design is a requirement. Teams need to work together in order to make the right decisions and to obtain a high quality end product. For example, in a typical project we found the following teams:

- Task Analysis Group;
- Detailed Design Group;
- Evaluation Group;
- Scenario and Prototyping Group;
- Management Group.

We found that it is important to have one independent group that manages all the other groups. The management group directs the time schedules, deals with delays and group conflicts, and the overall project documentation. The management's main purpose is to direct the dialogue between disciplines as well as between design phases. The management group monitors the use of representation to communicate between design sub-disciplines, as well as between the design team and the users, stakeholders and client.

Designing in teams can work well and potentially has several advantages. First of all, working in groups on separate issues creates a competition effect. Each group depends on each other and is being judged by other groups. If one group delivers poor work the other groups immediately alert them to this fact. On the other hand if a group has developed really new ideas, it is very disappointing if the other groups just discard their creativity or fail to see its merits. Not surprisingly, it is usual that certain groups get really upset by the work of others. However, this even leads to qualitative improvements of the final work as it forces the management, as well as the whole team, to reconsider and argue about misunderstandings and rejected proposals. Another advantage is that in this way the expertise people have is used optimally. In educational settings, the management team forms the groups and designers have to apply for a position. This way everyone gets to determine what he or she does, within some limits of course. In industry, a comparable 'marketing' situation and resource management is in fact common practise.

11.6 Design Representations

Each activity in the DUTCH method requires special representations. Moreover, for a single activity different representations are used depending on the purpose. Some are for internal use and other are intended for communication between disciplines or with clients and domain experts. Representations of a formal or 'informal' graphical type can be created with our tool Euterpe (van Welie *et al.*, 1998) while other representations are created 'by hand', e.g., sketches, scenarios and role-playing.

11.6.1 Representations for Task Modelling

Task analysis has been a field of mostly formal methods. Methods such as GOMS (Card, Moran, and Newell, 1983) are rather formal and require considerable training to understand. Such representations are certainly unsuitable for communication with clients or domain experts. In our experience, 'formal' graphical representations, like task trees, are reasonably understandable by client or domain experts, as long as they are not too big. Usually parts of a tree containing 10 to 20 tasks can be understood. For communication within design groups the formal task models work well, especially if they are fully linked, *i.e.,* when in fact actual representations are views on the same elements of a coherent structure embedded in a database. When task models are constructed using our tool Euterpe, the whole model can be exported to a set of linked web pages.

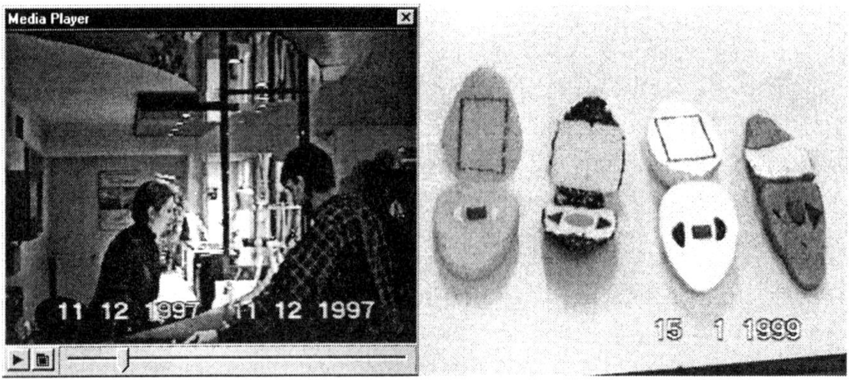

Figure 11.2. A video fragment and some hardware mock-ups

Besides a linked-model representation, the original material from which the model was built can also be included. We use images (e.g., scanned sketches), recorded interviews and video clips of ethnographic observations and scenario play, digitally stored and attached to the information they belong to, e.g., a certain elements of a task structure. Figure 11.2 shows a video fragment and some hardware prototypes for a handheld device.

11.6.2 Representations for Detailed Design

In the detailed design phase, the design is worked out in such detail that a prototype can be made out of the specifications and finally a full-blown implementation. Representations for this phase include sketches, screenshots, interactive prototypes, hardware mock-ups and NUAN (the DUTCH version of a complete formal representation). Especially in this phase the combination of formal and informal representations are working out well. The creative ideas are born in the informal representations and then become more detailed with iteration.

At the end of the process a detailed specification is handed to the implementers and this specification needs to be unambiguous. If not, the final product may not have the intended look and behaviour. Figure 11.2 shows some hardware mock-ups used to determine the basic shape and texture of a handheld device. The prototypes were made of sponge, clay and polystyrene foam. Figure 11.3 shows an example of a NUAN interaction specification.

11.6.3 Scenarios

Scenario analysis is a technique that can be used during the development of Task Model 2 as well as for the evaluation of detail specifications, both early and late. Scenarios are informal descriptions of tasks that are likely to occur together in a specified situation, including well-described user involvement. Some scenarios can be heuristically evaluated by a group of designers, possibly together with future users. On the other hand, in some cases scenarios show their value when actually acted out. When scenarios are acted out, the actors are instructed to try as many variations as possibly within their task description.

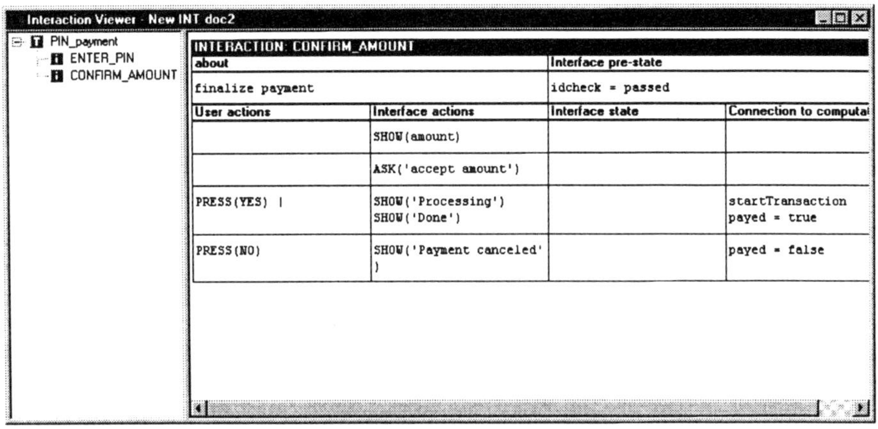

Figure 11.3. A NUAN interaction specification

It is useful to have some of the scenarios additionally to be acted out by intended users who in fact are sceptical or afraid of using the future technology. These actors will both show how the design could fail and how to manoeuvre around the supposed problems. To make sure nothing is overseen the whole scenario is videotaped. After several scenarios have been played a claims analysis is done. Together with each scenario, claims have been identified and during this analysis it is being evaluated whether or not the claims really hold.

11.7 Formal Methods and Creativity

Formal methods allow precision and informal methods allow vagueness and creative interpretation. At some point in the design process both are needed. In our experience we found that formal and informal methods can work complementarily. At some point details need to be worked out and by using formal methods hidden faults and unnoticed issues may be found. Another issue is communication. When design groups need another group's results as input, the work needs to be 100% clear and unambiguous.

Using a formal method is never a guarantee for unambiguous descriptions but it makes it easier to understand as description and finds imperfections. This is facilitated because formal methods have an explicitly defined syntax and semantics.

11.8 Does It Work?

So far, DUTCH design has been adopted into regular practise in University design education in the Netherlands (4 Universities) and Romania (2 Universities), as well as in Dutch post-academic education (3 institutes). In Industry it has been applied at several large design enterprises in Europe, e.g., it is regular practice in a design company in Austria for the design of safety-critical systems. In all these cases, we learned that the design teams claimed to benefit strongly from the facilities for combining formal representations with 'informal' representations (scenario play, user evaluation of sketches, and video clips) guided by the structure of design activities and their interdependencies. We developed special design cases for education, focusing on envisioning design. The illustrations in this paper are taken from the elaboration of a student exercise to design new procedures and related technology for future regular hotel guests of hotel chains, aimed at accommodating simple administrative procedures and providing optimal information and travel assistance.

11.9 Conclusions

DUTCH is a design method that uses both formal and informal representations. We found the combination of these representations to be useful and desirable. The formal representations give precision when needed, while informal representations work well to communicate ideas both within design groups and to clients.

11.10 References

Card SK, Moran TP, Newell A (1983) The psychology of human-computer interaction. Lawrence Erlbaum Ass, Hillsdale

Jordan B (1996) Ethnographic workplace studies and CSCW. In: Shapiro D, Tauber MJ, Traunmueller R (eds.) The design of computer supported cooperative work and groupware systems. North-Holland, Amsterdam

Sebillotte S (1988) Hierarchical planning as a method for task-analysis: the example of office task analysis. Behaviour and information technology 7(3):275-293

Van der Veer GC, Lenting BF, Bergevoet BAJ (1996) GTA: groupware task analysis - modeling complexity. Acta Psychologica 91:297-322

van Welie M, van der Veer GC, Eliëns A (1998) Euterpe - tool support for analyzing cooperative environments. The ninth European conference on cognitive ergonomics. Limerick, Ireland

Chapter 12

24hr Design and Development: An Engine Design Project

U. Lindemann, R. Anderl, H. Gierhardt and G. Fadel

12.1 Introduction

Globalisation, characterised by the worldwide merging, acquiring and collaborating activities, enforces and enables enterprises to distribute their processes. '24hr Follow the Sun Design' is an example of a purposeful distribution across the borders of time, companies and continents. Through a case study the BMW Group wanted to investigate the potential of '24hr Follow the Sun Design' to reduce costs and time as well as to improve quality. The basis for planning and realisation of the project is the Distribution Methodology - a holistic approach for identifying typical problems as well as developing and implementing solution activities. The present paper gives a basic introduction to the Distribution Methodology by describing its main elements and its application in the '24hr Follow the Sun Design Project'. The paper also presents the structure of the entire '24hr Follow the Sun Design Project' as well as the concept for the realisation of the project. The initial results of the project are also presented.

12.2 Distribution Methodology

The problems and challenges in distributed product design are networked, complex and mostly not transparent. From this it follows that there is a need for methods and tools in order to acquire and integrate information on the typical problems of distribution, as well as to develop actions solving these problems. In this regard the Distribution Methodology is a holistic approach to analyse, improve and enable distributed design projects. The terminology of distribution includes not only the local aspect but also distribution has a more diverse sense. Based on the theory of Systems Engineering, this means that distribution describes the combination of process elements and the manner of their interconnection (Gierhardt *et al.*, 1999a).

12.2.1 The Elements of the Distribution Methodology

As shown in Figure 12.1, the Distribution Methodology consists of 6 elements. These elements represent the integrated and interrelated areas of action required to succeed in distributed product design. They are:

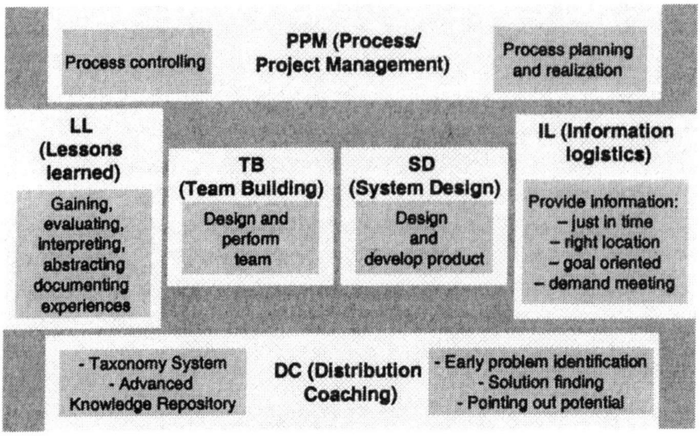

Figure 12.1. Distribution methodology

12.2.1.1 System Design

In the sense of Systematic Design (Pahl and Beitz, 1997) and Integrated Product Development (Ehrlenspiel, 1995) design methods are used and adapted to the special requirements of distributed design. Special requirements include: clear and current information at each location; obvious proceeding and tasks for all partners; common understanding; and consistent documentation. They are the basis for every kind of collaboration in engineering design, but gain much more importance within a distributed design environment.

12.2.1.2 Process and Project Management

Within a distributed environment, methods and tools of project and process management are used and adapted to the special requirements of distributed design. According to the idea of process orientation (Gaitanides, 1994; Davenport, 1993), process definition modelling and continuous process improvement are the basis for project realisation and organisation. Therefore the process is the framework for realising projects. Especially in a distributed environment process, orientation enables transparency and clarity of action and decision. The main tasks of project management are organisational structuring, forming the team and scheduling as well as setting and controlling goals. This also enables transparency and clarity of action and decision.

12.2.1.3 Distribution Coaching

In accordance with Gierhardt *et al.*, (1999a; 1999b) Distribution Coaching is used as a method and tool to support project planning and realisation. In the planning phase the Taxonomy System (Gierhardt *et al.*, 1999c) was used to characterise the state of distribution within the project. This characterisation identifies typical distribution problems before the beginning of the project. Based on the problems identified, the Taxonomy System generates a solution concept to be implemented by the Distribution Coaches.

12.2.1.4 Information Logistics

According to Fraunhofer (2000) information logistics describes the supply of information to individuals and groups within a distributed design environment. The goal-oriented placing and demand-meeting distribution ensures that information is delivered just in time to the location where it is needed. The backbone for this is a set of tools and methods that include a telecommunication infrastructure, CSCW, groupware, reasonable and structured meetings, ooRM (Object Oriented Requirement Model), Management techniques (e.g., Affinity Diagramming) and web-based information management systems.

12.2.1.5 Lessons Learned

In order to learn from distributed projects, methods are needed to gain, evaluate, abstract, document and interpret information about the experience. Therefore a learning kit was elaborated (Figure 12.2).

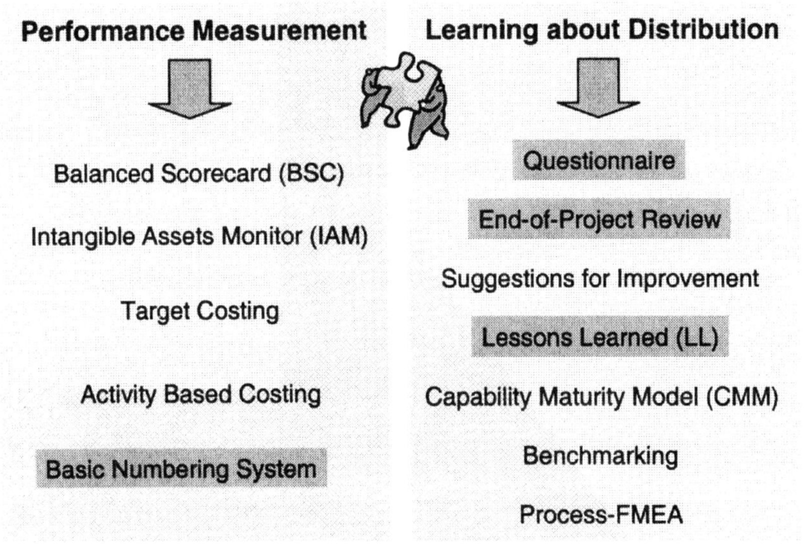

Figure 12.2. Learning kit

12.2.1.6 Team Building

In addition to the other elements of the Distribution Methodology, the process of building the dispersed team is a significant challenge. The primary requisites of efficient and effective teamwork are trust and cohesiveness among team members (Carmel, 1999). It is a particular challenge to build and maintain trust and cohesiveness in dispersed teams. Sample methods, techniques and activities addressing the team aspect are shown in Section 12.3.1.

12.3 The '24hr Follow the Sun Design Project'

In collaboration with Clemson University (Department of Mechanical Engineering), University of Technology Darmstadt (Department for Computer Integrated Design) and Technische Universität München (Department for Product Development) the BMW Group has initiated, planned and realised a '24hr Follow the Sun Design Project'. Three student teams (one at each University) planned and designed a new concept for a 1-cyl. test bed engine (see Figure 12.3). The Preliminary Drive Train Department of the BMW Group was the customer for the new engine concept.

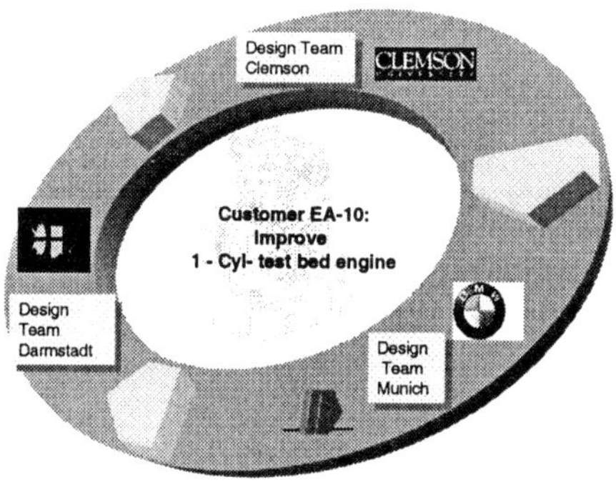

Figure 12.3.

The goal of the project was to answer the following questions:

- Is it possible to realise '24hr Follow the Sun Design' for a typical engineering design task?
- Does the concept for the '24hr Follow the Sun Design Project' work?
- Is the Distribution Methodology a useful and effective approach?
- What is the potential contribution to improved quality and reduced cost and time?
- What is the additional effort?
- How does distribution impact on team work?
- What can be learned for future projects?
- What methods and tools supporting design work are necessary and useful?
- What methods and tools supporting information logistics are necessary and useful?

12.3.1 Product

For basic research in the thermodynamics of combustion the BMW Group uses 1-cyl. engines (Ricardo P 8800, FEV Systemmotor E4). These engines are expensive and do not satisfy all the research requirements. Within the (24hr) design process an improved concept for a 1-cyl. test bed engine was to be elaborated. The focus of design was in three different areas:

- Variable stroke - Changing the stroke for the FEV and Ricardo engines requires designing and producing a completely new crankshaft. This is costly and imposes a delay of almost 3 months. The goal was to design an engine with a variable stroke to reduce cost and minimise delay.
- Mass balance - In order to guarantee functionality and to minimise vibrations an optimised mass balance has to be designed.
- Design for Ease of Assembly - The engines used result in a complex process for test bed set-up. The new concept was to simplify assembly in order to minimise integration and modification time at the test bed.

12.3.2 Distribution Scenarios

In order to realise the ‘follow the sun’ cycle the three teams worked in a special shift system (see Figure 12.4). Each shift was 5 hours. Every shift had two exchange periods of 30 minutes. Within this time period the actual task was handed over to (and received by) the continuing design team. All information (work results such as documents, models and descriptions of work progress) was given to the continuing location. Also, the plan for carrying on the work was discussed, determined and documented.

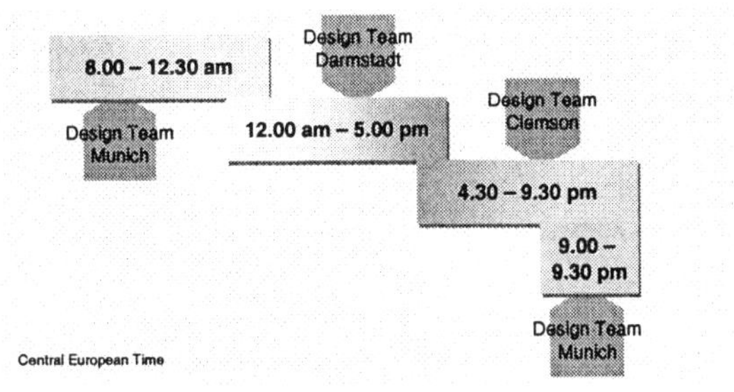

Figure 12.4. Shift system in the project

The project as a case study of distributed design provides only absolute results because there was no reference project to compare to. This has to be considered when evaluating the results of the work. In order to enable some comparisons to be made, three modes of distribution were employed in three different intervals within the design process.

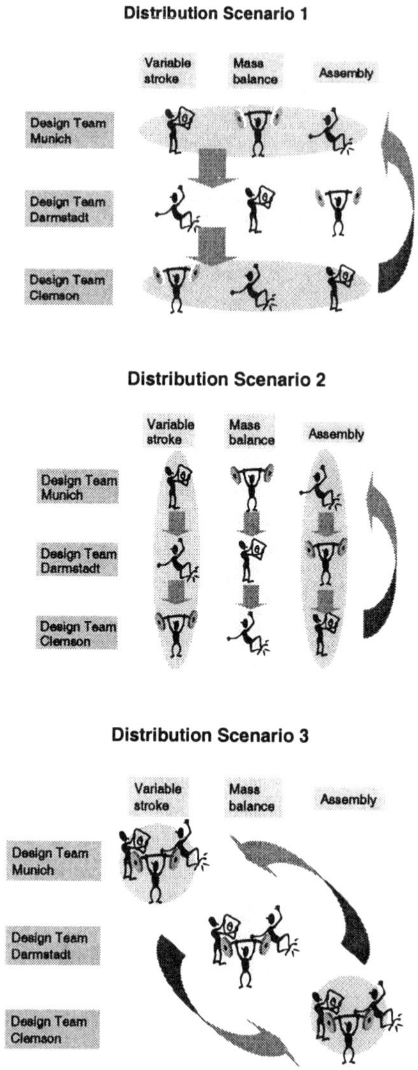

Figure 12.5. Distribution scenarios

Thus, three different Distribution Scenarios were employed within the three phases of the design process (see Section 12.3.3. Process and Project Management). During the analysis phase the teams at each location collaborated sequentially with each other and addressed all three of the design problems (Scenario 1 in Figure 12.5). In the solution-finding phase each design problem was assigned to one member of each design team. They still co-operated sequentially (Scenario 2 in Figure 12.5). During the concept phase, the teams collaborated in the classical manner of Concurrent Engineering (Prasad, 1996), (Scenario 3 in Figure 12.5).

12.3.3 Use of the Distribution Methodology

In all phases of the project the Distribution Methodology described in Section 12.2.1 was used.

12.3.3.1 System Design

One situation where the use of design methods in a distributed environment is appropriate, is the modelling of requirements. The object oriented Requirement Model (ooRM) (Steuer, 1999), which contained all requirements concerning the test bed engine, was used. Furthermore intuitive methods have to be adapted. So the Distributed Creation Method was elaborated. It is a combination of Brainstorming, Method 635 (Pahl and Beitz, 1997) and the Gallery Method based on the Solution Form. The process of solution-finding circulated among the three design teams. In the beginning, one team starts brainstorming in order to generate solutions for a problem.

The results are documented and explained in the Solution Forms. The forms are handed on to the following design team. It starts solution-finding with a brainstorming session and continues by commenting on, questioning and varying the elaborated solutions of the first team. This work is also documented in the Solution Forms. The next team proceeds in the same way. The solution finding process circulates as many times as is productive. The process is finished when the teams decide that all ambiguities concerning the solutions are resolved and no more ingenious solutions can be found. The forms are stored in an idea repository.

12.3.3.2 Distribution Coaching

The Taxonomy System was used to characterise the three Distribution Scenarios and to elaborate solution concepts. The scenarios differed mainly in the relative importance of the various problems. The solution concept was represented and detailed within the areas of Information Logistics, Team Binding, Project and Process Management. The Distribution Coaches implemented the solution concept and supported the project continuously (e.g., maintaining team performance).

12.3.3.3 Project and Process Management

Project managers, design teams and the coaching team (Distribution Coaches) were the main elements in the project team structure. The project was running 3 months beginning January 10th, 2000. The schedule and goal definition were represented in a gateway plan, which was the basis for goal controlling (Figure 12.6). Based on the design process theory of Pahl and Beitz (1997) three main process phases were defined and modelled: Analysis; Solution-finding; and Concept. The use of design methods and the integration of information and communication technology were also implemented in the process model.

The schedule and goal definition were represented in a gateway plan, which was the basis for goal controlling (Figure 12.6). Based on design the process theory of Pahl and Beitz (1997) three main process phases were defined and modelled: Analysis, Solution Finding and Concept. The use of design methods and the integration of information and communication technology were also implemented in the process model.

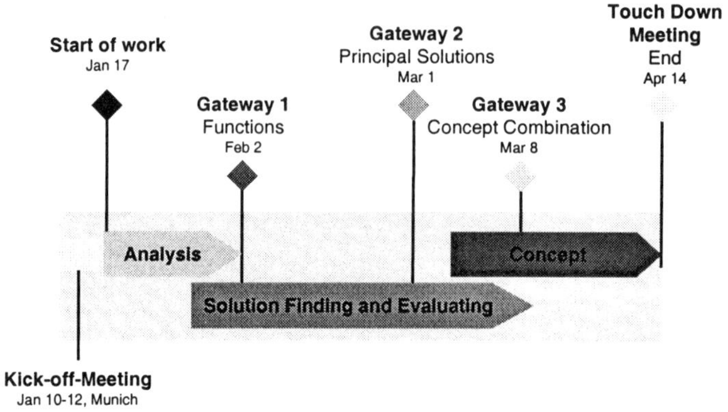

Figure 12.6. Gateway plan

12.3.3.4 Information Logistics

A core element of the solution concept generated by the Taxonomy System was the Project Centre, which consists of:

- *Information platform:* Hyperwave Information Server as WWW-based system for information management (documents, personal web pages, contact and general project information, exchange containers).
- *Communication tools:* Microsoft NetMeeting, e-mail, telephone, video conferencing.
- *Communication methods:* daily information transfer meeting between each shift, regular phone conferences among coaches and project leader, reasonable and situation-based video conferences.

12.3.3.5 Team Building

Within the '24 hr Follow the Sun Design Project' several coherent methods, techniques and activities concerning the team aspect were applied:

- Kick-off meeting to enable face-to-face contact and to build personal bridges between the team members.
- Milestone and team issue meetings (by video conference) in order to maintain a sufficient level of personal contact to work together effectively.
- More formalised and deliberated communication in order to establish networked information flow (rules in exchange periods, exchange form, more detailed documentation of work results).
- Encouragement of informal communication across and between team members and team leaders during meetings.
- Professional support for analysing and improving team processes by a team trainer.

- Intelligent application of communication tools (e.g., NetMeeting during design phases, video conferencing system during milestone meetings).
- The application was executed by the Distribution Coaches.

12.3.3.6 Lessons Learned

In order to learn from the '24hr Follow the Sun Design Project' a basic numbering system for measuring project performance was chosen out of the learning kit. Distribution issues such as communication or the work of the Distribution Coaches are captured by questionnaires, an End-of-Project-Review and individual lessons-learned stories.

12.3.4 Initial Results

At this stage, evaluation of the project results is not complete. The following list presents findings from an initial analysis of experience and data. The project ran, worked and produced design results. The project revealed no new problems not already addressed by the Distribution Methodology. The holistic Distribution Methodology proved to be very practical and beneficial for planning, realising and improving distributed design projects. Tthe main problems were identified early, the Solution concept worked and it served as the main support to be used by Distribution Coaches in the planning phase.

The primary management activities required during the project were: transparent and interactive project management, clear and well-founded decision making, and team maintenance

The technological implementation of the Project Centre has worked. The intercultural impact on work process was not distinguishable from the individual impact of personalities.

12.4 Summary and Outlook

This appears to be the first time that a '24hr Follow the Sun Project' has been realised in classical engineering design. Also a complex product has been developed going through all the design phases.

The described project represents basic research on collaboration across cultures and continents. The focus was on the general feasibility and basic problems of distributed design in all process phases. The results of the case study demonstrate the feasibility of sequential 24hr work in engineering design across time zones. But in order to answer the main research questions total evaluation has to be finished. At this stage, the time advantage posed by some researchers (e.g., Carmel, 1999) is questionable.

The basis for the feasibility of the project was the holistic Distribution Methodology. This indicates that the methodology is a general tool to support distributed design projects.

Goals for future research are improvement and continued implementation of the Distribution Methodology. The results from the 24hr project can also be integrated into less distributed design projects. In this regard more realistic case studies are the next step, e.g., using an extremely distributed project not with students, but with engineers in an industrial environment. Moreover, the integrated and intelligent use of advanced information and communication technology within distributed projects is a focus of continuing work. The combination of virtual reality tools, groupware and CSCW as an integrated communication and information space appears to offer great potential along these lines.

12.5 References

Carmel E (1999) Global Software Teams: Collaborating Across Borders and Time Zones. Prentice-Hall, Upper Saddle River (NJ)

Davenport TH (1993) Process Innovation. Harvard Business School Press, Boston (MA)

Ehrlenspiel K (1995) Integrierte Produktentwicklung. Carl Hanser Verlag, München, Wien

Fraunhofer Institut für Software und Systemtechnik (ISST) (2000): `http://www.informationslogistik.org/lexikon/index.html`

Gaitanides M (1994) Prozeßmanagement. Carl Hanser Verlag, München, Wien

Gierhardt H, Ott T, Gaul H-D (1999a) Distribution in Product Design and Development Processes. Proceedings of 1999 ASME Design Engineering Conferences, 11th International Conference on Design Theory and Methodology, Las Vegas, September 12th – 16th, 1999

Gierhardt H, Ott T, Gaul H-D, Lindemann U, Anderl R (1999b) Planning and Improvement of Distributed Product Development. Proceedings of 6th European Concurrent Engineering Conference (6th ECEC), Nürnberg, April 21st – 23rd, 1999

Gierhardt H, Ott T, Gaul H-D, Lindemann U, Anderl R, Thomson B (1999c) Investigation of Distributed Product Design and Development. Processes Proceedings of 12th International Conference on Engineering Design (12th ICED), Munich, August 20th - 24th, 1999

Pahl B, Beitz W (1997) Konstruktionslehre. Springer Verlag, Berlin, New York

Prasad B (1996) Concurrent Engineering Fundamentals, Volume I. Prentice-Hall, Upper Saddle River (NJ)

Steuer C (1999) Objekt-Orientierte Anforderungsanalyse eines Einzylinder Motors. In: Hinblick auf Verteilte Entwicklung Semetserarbeit am Lehrstuhl für Produktentwicklung der TU München

Part I Collaborative Design Methods

Summary

Co-designing is essentially design with, for and about users. Working in this manner stretches the designer in terms of the conceptualisation of the design brief, requires a reconceptualisation of what it means to design (and be a designer), and the adoption of new working methods to allow the development of a symmetrical partnership, to be established based on mutual trust, understanding and respect.

The contributions in this Part have addressed all these issues at both local and global levels. They have promoted different ways of engaging the user and establishing user and product requirements throughout the design life cycle, for example, through participatory design, steering groups, ergonomics checklists, the discovery of new product attributes and the identification of new user groups. Also, this Part has introduced a second theme, that of the role of the designer in the team, and how team activity can be supported and enhanced.

In conclusion, Part 1 emphasises the progress within the profession towards greater user involvement, the development of methods of feeding the results of this involvement effectively into the final design, and a growing sensitivity to users and their needs.

Part II

Issues in Developing Integrated Collaborative Design Approaches

Part II Issues in Developing Integrated Collaborative Design Approaches

Introduction

To compete successfully in highly volatile global markets, many design companies are continually having to explore ways to reduce product-development times whilst simultaneously enhancing the quality and saleability of product ranges. In recent years there has been a growing realisation that integrated collaborative design approaches, such as concurrent engineering, can deliver significant benefits to companies and consumers (e.g., in terms of product quality and cost, and design efficiency). The eight contributions in this Part address key issues surrounding the development of integrated collaborative design approaches that have also been touched upon more tangentially in the contributions in Part I concerning collaborative design methods. These issues are broached primarily through the presentation of detailed case studies of well-established, inter-company and inter-designer networks.

The initial contribution, by Bhamra, Heeley and Tyler, outlines a research project concerned with innovative approaches to product-design activity such as Concurrent Product Development (CPD). The CPD strategy seeks to ensure the delivery to market of better and cheaper products by investing in early design activities through the effective deployment of multi-disciplinary teams of designers, manufacturers and marketing experts, who pursue parallel-processing activities. Bhamra *et al.* focus on the role of CPD within the UK textile and clothing industry, which has traditionally approached product development as a linear process involving sequential contributions from team members, which can result in costly, late-stage product changes. Their research has concentrated on facilitating the adoption by companies of CPD procedures by instigating workshop sessions aimed at securing senior management support, the setting of company standards, and the building of effective project teams. Bhamra *et al* overview their preliminary results relating to the effectiveness of CPD-based design practices within their target companies.

The contribution by To and Harwood continues to focus on the global fashion industry in its discussion of issues relating to the development of integrated collaborative design processes. Like Bhamra *et al.*, To and Harwood also emphasise the need for companies to implement more flexible and agile collaborative processes based around multi-functional teams that can rapidly respond to changing market demands. They present case studies of eight international fashion-buying companies, and focus on the 'integration mechanisms' that the companies adopt to support collaborative design communication and collaborative design activity.

Filson and Lewis explore the process of 'networking' between companies, which involves small-to-medium enterprises (SMEs) working in close collaboration in order to offer products or services in competition with larger organisations. Filson and Lewis note that several countries have already attempted to mimic the success of the first networking model - developed in Italy – which is generally considered to represent best practice in terms of the close collaboration between SMEs and the innovative and flexible management of products and workforces. The focus of the authors' research is an ongoing attempt to improve the performance of SMEs within the manufacturing sector in South Wales, through the provision of mechanisms to encourage networking. In order to secure such improvements, their intervention attempt in South Wales has been informed by a benchmarking study of the successful practices employed in established networks in Italy, Denmark and the UK. They review the main principles of effective inter-company networking gleaned from their benchmarking research.

Lofthouse and Bhamra's contribution addresses the topic of 'ecodesign' (*i.e.*, the development of design solutions that are sensitive to the environmental impact of the whole life-cycle of a product). They note that ecodesign has come of age such that it is now widely recognised as an activity that must be included at the earliest stages of product development. The paper reports the initial findings from a three-year project which aims to provide information for the development of approaches to improve ecodesign in companies. Like Filson and Lewis, Lofthouse and Bhamra's research also involves a benchmarking study of industrial designers to determine information about current best practice and contemporary opinions about ecodesign. Lofthouse and Bhamra additionally discuss the findings from a one-week participant-observation exercise at Electrolux Industrial Design, which facilitated the acquisition of grounded descriptions of ongoing industrial-design work on 'live' design projects within a business context.

Erol, Cooper, Press and Thomas also focus on a novel and important application of collaborative design practice, this time in the area of crime reduction. Their contribution explores how collaborative design involving all stakeholders associated with the product-development process can be used as a means to introduce crime-reduction mechanisms (such as security devices) into products and environments in the early stages of the design process. The reported research examines the extent to which crime-reduction measures are incorporated within current design practice, and presents the results of semi-structured interviews conducted with a range of stakeholders within design companies and related industries and professions (e.g., manufacturing and design education). The data were analysed to identify key design issues in designing against crime, and to determine case-study exemplars of contemporary 'design-against-crime' practices.

Cooper, Bruce, Tierney, Daly, Harun and Svengren explore the concepts of the 'designer as innovator' and the 'designer as integrator' and discuss the importance of designers taking on such roles in contemporary design practice so as to ensure the mapping between rapidly changing technology and the demands of the global economy. They emphasise how increased competition in the market-place means that companies can no longer simply rely on the cheapest price-point for products or the shortest lead time from design to market. Instead, more and more consumers nowadays seek fashionable and effective products that place high demands on designers' abilities to generate *innovative* design ideas. One key element of such innovation can stem from the designer capitalising on the knowledge of other experts, including technology specialists, managers and marketers. Cooper *et al.* argue that since designers are often the initiators of the interaction between different specialists, then they have a unique and vital role to play as *integrators* throughout the product-development process.

The arguments of Cooper *et al.* about the need to move beyond simply designing affordable, high-quality and reliable products toward designing innovative ones is also resonant in the contribution by Burns and Evans. They propose that since many product sectors are characterised by large numbers of competing products with similar specifications and features, it is no longer enough for companies to match the best products available since these will nether capture new customers nor encourage loyalty from existing customers. Instead, the key to success, they argue, is for companies to differentiate their product, to innovate, and to engage the customer on an emotional level. Their paper describes research into the role of 'customer delight' during product evaluation that forms part of an approach to improve product development at Nissan Technology Centre Europe. The customer-oriented design approach that they describe is based on a combination of observation, interview and self-report techniques. For example, the naturalistic observation of customers using products can reveal delight responses to distinct vehicle features.

The final contribution, by Demirbilek and Demirkan, continues the theme of end-user involvement in the design process, this time with a focus on issues relating to engagement by the elderly in product development. Demirbilek and Demirkan adopt participatory-design methods (discussed extensively in Part I) as a way of exploring the comments and ideas of elderly end-users concerning how best to design domestic doors and door handles. Two sets of participatory-design sessions were undertaken with small groups of three to six elderly people. In the first sessions, the participants were asked to assist in the design of doors and door handles by considering all possible requirements, needs, wishes and ideas. The sessions involved unstructured interviews, scenario building and brain storming facilitated by the designer. In the second set of participatory sessions the designer presented the visual design concepts generated in the first sessions to the same groups of individuals for correction and critical assessment.

Chapter 13

Developing an Integrated Approach to Supplier Design Activity Within the UK Textile and Clothing Industry

T.A. Bhamra, J. Heeley and D. Tyler

13.1 Introduction

This paper describes some of the results of a 2-year research project entitled 'E-Co Challenge', which is a collaboration between Manchester Metropolitan University and Cranfield University as well as six companies from the UK textile and clothing sector. This work is funded as part of the DTI Sector Challenge Programme. This project benefits from having the collaboration of a number of the UK's leading companies in the Textiles and Clothing Sector, as well as two universities with different and complementary expertise.

This research project is concerned with novel approaches to the product development activity, particularly Concurrent Product Development (CPD). The CPD strategy involves the delivery of better, cheaper and faster products to market by using multi-disciplinary teams (involving design, marketing and manufacturing) from the beginning of a project, planning and investment in early design activities, and parallel processing activities. Extra time and effort spent in the early phases of design reduces the need for costly and time-consuming design changes, lessens the chance of product recall, as well as improving communication throughout the product development process and the supply chain.

CPD has been used in other sectors for a number of years and has delivered measurable benefits, and there is a strong case for some sector-specific technology transfer. A survey of 19 companies which implemented concurrent engineering in their design processes, conducted by Cranfield University, indicated on average a 90% improvement in quality, a 25% improvement in cost and a 70% improvement in time.

Throughout the world for the past two decades, many industries have faced a number of challenges. To compete successfully, companies are continuously looking to reduce development times. The need for better quality and shorter product development lead times is now widely acknowledged. There is a growing realisation that the CPD approach offers an effective way to achieve these objectives, and it is now becoming a recognised company strategy.

13.2 UK Textile and Clothing Industry

The textile and clothing sector is the sixth largest industrial grouping in the UK and provides over 250,000 jobs, 9% of the total UK manufacturing work force (Taylor, 1990). Over the last decade there has been a growing trend in the UK Textile and Clothing sector has been increasingly losing ground to cheaper foreign imports; import penetration in the clothing sector was 65% in 1991, and is set to reach 80% by 2001 (Traish, 1994). These statistics are an indication of the market pressures faced by the sector, and for the urgent need to seek greater competitiveness. Some manufacturers have started to respond with innovative design input improving the quality and style of products. This quest for higher value-added consumer products has resulted in a variety of technological and organisational changes (Rhodes and Carter, 1995).

Failure to plan, predict and pre-empt fashion change has been the downfall of many companies, particularly independent manufacturers and retailers without the resources to change direction and pace every time fashion does. In an environment, particularly in clothing manufacturing, where the UK industry is dominated by very large companies, producing good quality products at the right price, with on-time delivery, the only differentiating factor is the design of the product.

13.2.1 Product Development in the UK Textile and Clothing Sector

The build-up of knowledge about product development in the textile and clothing industry appears limited, although there seems to be an increasing move within companies to make their product development activities explicit in an attempt to identify the characteristics which differentiate between success and failure (Davies-Cooper and Jones, 1994). Conventional approaches to product development in the textile and clothing industry have been characterised by functional independence. Each participant contributes to the process sequentially, which results in excessive costs and rework in production associated with late stage design changes. A recent Apparel Supply Chain benchmarking study (Anon, 1996/7) highlighted product development lead times as a weakness in the UK's international competitive position. Efficiency of the product development process was also found to be low. On average, only 30% of products developed actually find their way into the store, with designs often being changed more than four times from first sample to final approval (Watson, 1997).

In traditional organisations, design teams believe they know what their customers want. They produce sample garments and ranges - only to find that they have not got things quite right. The result is a series of late-stage product changes that are time-consuming and costly.

The adoption of formal methods within the framework of Concurrent Product Development (CPD) has customer orientation as the target. This entails establishing better communications with customers, reducing unproductive debate, and generally providing a sharper focus for all activities. Extra time and effort spent during the early phases of design reduces the need for costly and time-consuming design changes, and lessens the chance of product recall. Teamworking within organisations can also unify the work of people with very different backgrounds, technical discipline, experience, and organisational responsibilities, although it is widely recognised that this diversity of team members also represents significant challenges in maintaining effective communication.

13.3 Research Approach

The industrial collaborators involved in this research project are involved in different stages of the production chain: from fibre production through to retailing of the final product. We have also been able to explore relationships and communication exchanges throughout the supply chain. We have compared the product development activity with other sectors where Concurrent Product Development (CPD) has been successfully implemented. Using pilot studies involving three case study companies, we have adapted and modified these techniques to suit the individual companies and the sector as a whole.

The research has concentrated specifically on measuring the effectiveness of current practices within the three case study companies: a fibre producer and two clothing manufacturers. It was important that we were involved in activities that allowed us to get close to informants and to gain access to the data. This involved a combination of site visits, semi-structured interviews (to gain insight into their particular roles and responsibilities and their relationship with other functions), and observing and participating in project review meetings. Secondary evidence has also been obtained from publications, conference papers and reports.

The research has followed one product in each of the three 'case study' companies in order to build up a more in-depth picture of their product development procedures. In the fibre company we followed a particular R&D project; in the clothing manufacturers we examined the development of a knitwear range and a hosiery product. We felt it would be more meaningful and valid to analyse three of the companies in detail. Realistically, the industry we are working in is fast-moving, and constantly changing so that it has only been possible to examine a small number in detail.

13.4 Tool Adaptation

As one of the main aims of the E-Co Challenge project is to assist the UK textile and clothing sector in adopting Concurrent Product Development (CPD) procedures the research examined techniques which have been used successfully within a variety of sectors (including automotive and electronics). The CPD approach allows companies to improve product development by delivering better, cheaper products to market faster and more cost effectively. The tools involve multi-discipline teams, right-first-time methods and parallel processing activities.

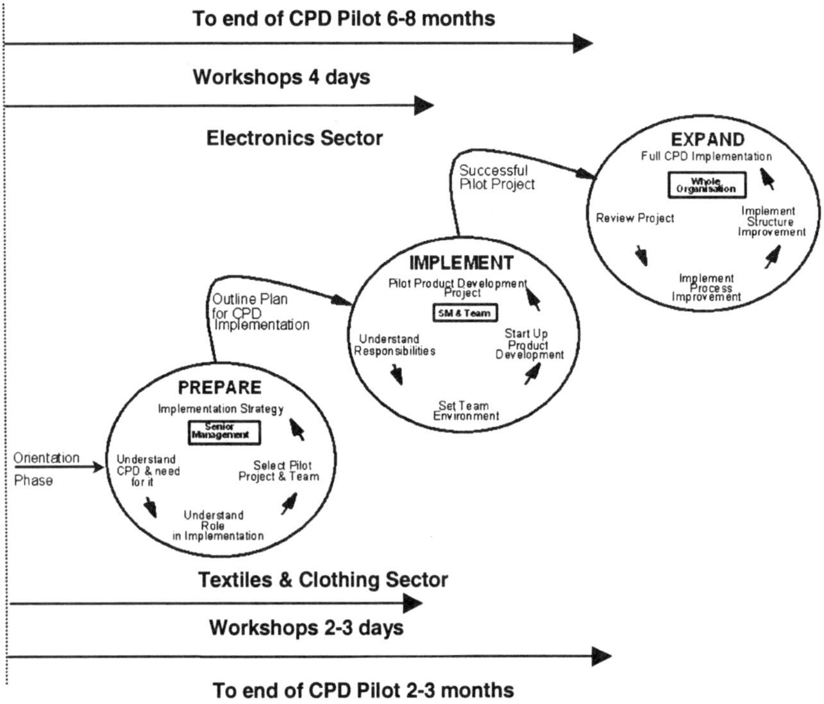

Figure 13.1. CPD approach

Previously Cranfield University developed the Fast approach as part of an EPSRC funded project for CPD implementation in the automotive and electronic sectors. This formed the starting point for our project. The Fast CE workbook (Lettice *et al.*, 1999) introduces and provides the support for three workshops. The first covers the understanding of CPD and the gaining of senior managers' support. The second, which is linked to the launch meeting of the product development team, gives management the opportunity to explain what they expect to gain from the CPD approach. The third workshop explores ways of extending and applying the principles of concurrency more widely across the business. It is recommended that the initial two workshops take two days each to ensure that a full understanding is built up with the company. This is illustrated in Figure 13.1 below. Figure 13.1 also highlights the differences between the industry sectors and the adaptations that have been made to the process to ensure its successful implementation within the textile and clothing sector.

The benefits of these workshops are that they facilitate effective communication. They also provide a format for encouraging good practice and setting company standards. A fundamental benefit is that when problems do arise they get solved more rapidly rather than being forwarded to the next stage of the product development.

The negative aspects of the workshops are that the success factor is dependent upon the skills of the facilitator. The attendance of team members also impacts upon the ability to deliver solutions. The tracking of the team is also a potential area of concern and it may be necessary to establish administrative support to record the outcomes of meetings.

13.4.1 Adaptation for the Textile and Clothing Sector

One of the first observations we made when adapting this approach to suit the textile and clothing sector was the speed at which everything happened. It became clear when we started our first Case Study (Heeley *et al.*, 1998) that we would be unable to start off the project by running the usual two sets of two day workshops; people could not justify spending two days out of their work for an unproven process. We therefore reluctantly agreed to squeeze both two-day workshops into a single one-day event. This meant combining the senior management and the project team workshops. Although the activities are very similar, it proved to be very difficult to achieve our aim of building up senior management support for the process prior to its launch. However, we did achieve our aim of establishing an effective project team who would carry out the project using the CPD approach.

Following on from this we started our second case study (Bhamra *et al.*, 2000). Once again we had to take into account the speed of events within the sector but we also wanted to ensure that we had senior management commitment prior to the launch of the project. Therefore we ran a one-day Senior Management event followed by a two-day project team event. In addition, we introduced a new workshop activity involving process modelling to try and get understanding from the team on how product development is currently carried out within their company. This is an activity that can be extremely useful in highlighting both the strengths and weaknesses in the current product activity, outlining duplication of effort and complex decision making processes.

The tools illustrated in Figure 13.2 are all workshop based and designed to ensure that a good understanding of the CPD approach is built up at the same time as developing an active team and selecting a profitable project to work on. These workshop tools have been tested and adapted through use in our collaborating companies. One of the key features of this approach is that it can easily be adapted to suit the textile and clothing industry. As it is workshop-based it is possible to mix and match activities as well as reducing the time spent on activities, missing them out or substituting them. We have shown that it can be run in 1 to 4 days depending on the level of effectiveness required and as long as the project team established are well supported.

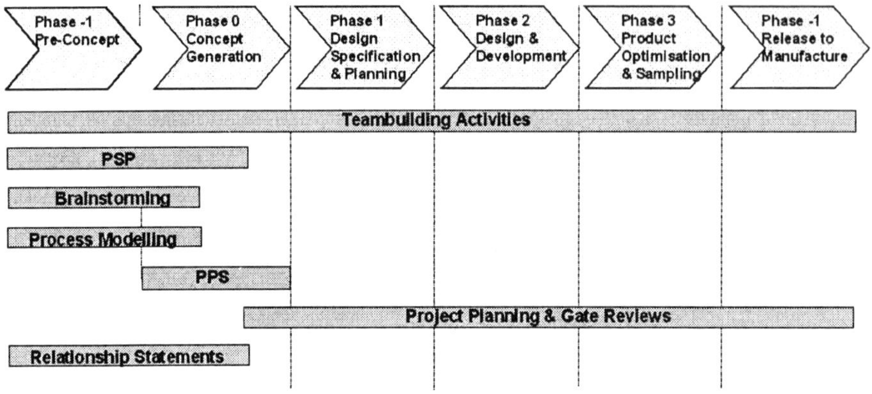

Figure 13.2. Tools used during NPD

13.5 Results

The CPD project was recognised throughout all the case study companies as an important change to the way in which product development was carried out. Some of the aspects of the approach that the teams identified as significant were the fact that using CPD brought together representatives from each of the companies involved in the working group often using co-location. Holding meetings off site, away from distraction to assist concentration, also helped to maintain focus. The creation of a steering committee (management committee) enabled the team to feel supported. This also led to the adoption of a policy of complete honesty and openness about all aspects of the project, which was something that had not always happened in the past, particularly in cross-company projects.

The use of facilitators (members of the E-Co Challenge Project Team) to advise and prompt, helped to maintain the focus of the team. A brainstorming session at the first meeting helped the team members understand the different processes and businesses within the company, and to address the problems of a blame culture. By understanding the roles and responsibilities of each person and of the team as a whole the team worked together well. In all companies it was the first time that a critical path and gate review for a project had been used successfully.

Processes were put in place to enable continuous communication about the project using meetings. Email and telephone enabled all team members and steering committee members to be kept informed. Finally, one of the most significant results was that from the start and throughout the whole project there was buy-in from all members of the team. This ensured that the project was a success.

All of the teams have continued their activities together. Interestingly two of the case study companies have approached their continuation of the CPD approach in very different ways. One of the clothing companies initiated a number of projects, almost immediately after the first pilot project had started, using a similar approach but all with the same team leader (an internal consultant). These projects all had limited success as the team leader was used to using an authoritarian approach and found it difficult to operate in the team environment. The expansion of the CPD approach stopped within the company when the internal consultant left the company.

One of the other companies is now slowly expanding the CPD approach throughout their business, following closely the way in which the pilot project had been run. In fact this company has taken some of the tools and further adapted them to suit their business.

13.6 Conclusions

A fundamental benefit which was found through adopting the CPD approach by the teams that we established in the companies, was that the problems which do arise get solved early on (involving a multi-functional team), rather than being forwarded to the next stage of product development. This includes rectification of any problems or changes to designs or processes. Additionally, communication within the team has also been greatly enhanced.

It was felt that the success of the projects was aided by involvement from the E-Co Challenge team. By taking the role of external facilitators, we were not viewed as a threat by the company. Interestingly, the different and complimentary expertise of the E-Co Challenge team (made up of design, engineering and manufacturing experts) proved to be of particular benefit to this research project.

The evidence is clear from our use of this tool within our collaborations that the CPD approach enables the company to build a good understanding of the need for effective product development. It also forges people from many different functional backgrounds and even companies into a project team with common aims and clear roles and responsibilities.

The key benefit of this approach for the textile and clothing sector is that it facilitates the development of a multi-functional team approach to product development, something that until now has not been practised within the sector. By taking this approach companies can expect significant improvements in costs, quality and time to market. The CPD approach equips the company with all the tools and techniques needed to successfully establish concurrent product development.

References

Anon (1996/7) Clothing World, Dec/Jan, 7

Bhamra TA Heeley J, Tyler D (2000) Case Study #2 Report. Internal E-Co Challenge Report

Davies-Cooper R, Jones T (1994) Successful Design Interfaces in New Product Development: Using Diagnostic tools to identify weaknesses in the functional relationships in new product development in manufacturing industries. 6th International Forum on Design Management and Education. Paris, France

Heeley J, Bhamra TA, Evans S (1998) Case Study #1 Report. Internal E-Co Challenge Report

Lettice F, Smart P, Evans S (1999) Using concurrent engineering for better product development: a resource for implementers. Cranfield University.

Rhodes E , Carter R (1995) Emerging Corporate Strategies, Teams And The Changing Role Of Design. Co-Design: Interdisciplinary Journal of Design and Contextual Studies 3:6-13

Taylor MA (1993) Technology of Textile Properties: An Introduction. Forbes Publications Ltd., London

Traish. R (1994) Retail Challenge. Clothing, Textile and Footwear Challenge Speech Pack. British Retail Consortium, BATC, London

Watson. K (1997) Benefits Supply Chain Partnerships. Putting Energy into the Supply Chain Textile Conference. Birmingham

Chapter 14

Integrating Collaborative Design Processes: Case Studies for Global Fashion Marketplaces

K.M.C. To and R.J. Harwood

14.1 Introduction

Pragmatically, global-oriented operations concern cross-functional and cross-enterprise process integration and optimization. Operations at several location facilities can be integrated to serve a given marketplace's requirements and a single upstream facility can be agilely designed to serve for several downstream operations. Global operation can therefore improve the process and cost performance on a network system basis (Flaherty, 1996; Singletary and Winchester, 1996). However, when designing and developing such global-oriented products, especially for today's fast-paced competition, no single functional entity or enterprise is able to perceive all the perspectives throughout all phases of a product life cycle.

To appreciate the significance of this in global fashion businesses, we can observe that marketing teams capriciously ask design teams to revise the designs in upstream processes in anticipation of changes in market demand or affordability. Manufacturing teams correspondingly adjust the design of production setup and re-schedule raw material distribution and financial arrangements. Such changes would in turn requirre the marketing teams to re-consolidate the end-market operations in view of the global image, pricing and quality consistency. Inefficiency and ineffectiveness of design and integration of all these interlaced design processes will give rise to various degrees of organisational conflict, loss of customer focus, product quality variation, and undesirable design iterations and failures.

How to implement more agile and collaborative design processes is therefore recognised as a pertinent research interest in the discipline of global product development. Indeed, a lot of contemporary research has contributed significantly to identifying the characteristics of, and formulating management approaches for, effective product design teamwork (see Reid *et al.*, 1999; Cross, 1995; Leonard-Barton *et al.*, 1994; Slusher *et al.*, 1989). In this paper, the authors address the issues of integrating collaborative design processes for global fashion products using case studies in leading Asian-based international fashion buying companies.

The findings and experience of the case studies provide insights in enabling and monitoring collaborative design processes.

Using the analyses of the cases, a strategic framework for choice of integration mechanisms and the underlying interface management issues are discussed. In the following sections, the authors discuss the research approach and the empirical issues that affect the performance of collaborative design processes in global-oriented operations.

Through the case studies, we explore the contextual aspects related to a variety of integration mechanisms. To explore the methodological possibilities of design process integration, a strategic choice of integration mechanisms will be discussed. Finally, the authors conclude with a discussion of the concerns related to the integration implementation.

14.2 The Research Approach

This paper examines the experience and implication of integrating collaborative design teamwork in the international fashion businesses. The research presented here is based on the contingency perspective of organizational integration, wherein the appropriate forms of collaborative design processes are subject to the kinds of design task and market environments with which the buying companies deal (Morgan, 1986). In the cases studied, eight well-known Asian-based fashion-buying companies that supported existing top-listed fashion retail institutions in Australia, Europe and the US were qualitatively researched. To allow the generalization of empirical findings, a conceptual model of designing new global-oriented products has been established, as shown in the Figure 14. 1.

The model reflects the fashion buying companies' dynamic business processes and the nature of process interdependency that should be considered in the course of new product development. The cases studied represent today's globally oriented virtual enterprising concepts, whereby fashion products in global-scaled volume are developed and distributed. While maintaining a high level of product exclusivity and supply certainty, the companies design and orchestrate the operations of global supply pipelines as virtual consolidating manufacturers, and therefore co-ordinate a large number of interdependent enterprise entities across countries.

The aims of the case studies are to investigate empirically the pertinent issues of integrating collaborative design processes, and the scope for strategic choice of integration mechanisms available in such complex contexts.

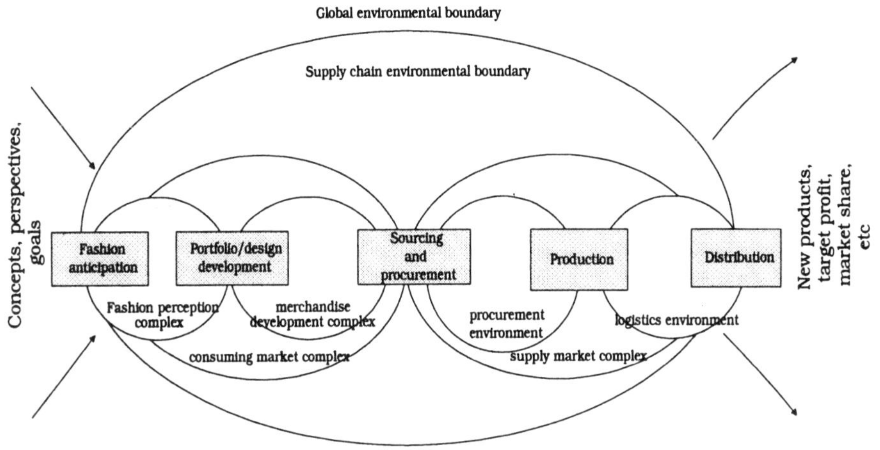

Figure 14.1. Contextual process elements for new product development in global fashion business

14.3 Discussion of Findings

14.3.1 The Empirical Issues of Collaborative Design Processes

Collaborative design refers to the activities and communication processes that enable multi-functional and technical teams, informally and cooperatively, to define and embody new product attributes. In the global context, it inevitably involves a great number of functional and technical teams in countries that bring their own perspectives and information to best during the whole cycle time of product development and production. In this context, many companies have attempted to resolve this complexity using well-defined contact schedules or structured information flow and technologies. In most empirical evidence, such dedicated information systems do ensure the adequacy of communication and interaction. However, it is still difficult to enable a high degree of collaboration supporting the development of a shared vision and understanding in the course of new product design processes (Jassawalla and Sashittal, 1998; Fleischer and Liker, 1997). In this section, we discuss the inherent characteristics of collaborative design processes within the framework of the research findings. The empirical evidence reveals:

- Intractable changes of end-market requirements that cause indecisiveness - this requires last-minute decision-making and action by individual teams. Consequently definitive and valuable information exchange amongst all teams within the supply pipeline cannot be encouraged in the early phases of a product design cycle.

- A large number of closely interdependent communication and activity processes amongst functional and technical teams requires substantial resources and time to ensure consistent coordination and work progress. Such kinds of interaction result in unfavourably long cycle times and greater coordination effort.
- The inter-dependent nature of these processes increases the possibility of progress changes and failure. Problems in individual processes are likely to trigger repercussive problems toward interdependent processes. Inevitably, this results in undesirable reciprocal work, and delay in new product delivery into marketplaces. Re-working and re-scheduling of preceding processes result in inefficient allocation of personnel and facilities.
- Collaboration is the centrepiece of design processes, especially in the context of new product conceptualization and embodiment. Integration of collaborative design processes should ensure sufficient co-existence of different perspectives, which are often domain-specific and characterized by expert judgement and professional intuition. How they interface becomes a crucial factor for ensuring proper exchange of opinion and documentation. Most problems in integrating collaborative team works indeed lie in improper contact, conflicting opinion and poor documentation amongst these well-partitioned design teams and functional groups in different countries. This is especially evident where high personnel turnover occurs, as found in most sophisticated economies, where the mobility of professionals is relatively high.

Facing these collaboration issues, the buying companies in these cases, according to their individual needs, resource constraints and business environment, adopt different forms of integration mechanisms to support communication and co-ordination in design processes. The next section will summarize the types of integration mechanism adopted in the cases.

14.3.2 Types of Integration Mechanism Adopted

Integration mechanisms are types of managerial or technological tools for effectively coordinating communication and activities across both organizational teams and supporting groups (Mills, 1998; Paashuis, 1998). They allow forms of control over process consistency, and overcome co-operation barriers stemming from different sources of geographical, organizational and even cultural differences. In the cases studied, all fashion buying companies had been endeavouring to invest in collaboration infrastructure, both organizationally and technologically. They also unanimously agreed that this was an alternative form of organizational learning, aimed essentailly at augmenting the teams' capability in group brainstorming and collaborative design decisions, based on concensus.

Integration mechanisms can be categorized in accordance with two polar dimensions: integration enablers to integration keepers, and technology-centric to people-centric, as listed in Table 14.1. Integration enablers act as initiation tools to promote or establish an effective infrastructure for teams to regulate the exchange of design information and trade-off in different design perspectives. Integration keepers are those that keep track of consistency in the act of design processes and diagnose potential design process incongruence. The effectiveness of integration keepers depends on the ways in which the teams concerned interface.

Table14.1. The categorization of integration mechanisms adopted in the companies studied

	Technology-centric	People-centric
Integration enablers	System analysis and engineering; Combinatorial process design; Standardization-for-information exchange; Virtual prototyping (e.g., CAD/CAM); Rapid real sample proto-making.	Team development; Team co-location; Standardization-for-skill; Management- and engineering-for-change.
Integration keepers	Knowledge-based document management support systems; Real-time information sharing, mark-up & automation systems; Integrated messaging (e.g., Computer-telephony integration)	Group thinking; Inter-team meeting; Telephone and videoconferencing.

Owing to space constraints, a detailed discussion of the theoretical aspects of individual integration mechanism will not be given in this paper (for further information, see Fleischer and Liker, 1997); Park and Cutkosky,1999; Paashuis,1998; Ramesh and Tiwana,1999).

14.4 Strategic Choice of Integration Mechanisms

To conceive and implement appropriate integration mechanisms in a diverse market environment, we often ask a few basic questions: what contextual issues should be considered for integrating collaborative design processes? What kind(s) of integration mechanism should be adopted in a specific design task? Can we generalize them as strategic concepts? These questions are discussed below.

To recapitulate, the research was conducted on the premise of a contingency view, which emphasized that there was no best and universal mechanism for integrating collaborative design processes. This view regards integration as the response of an intelligent organism to interacting contextual factors. Therefore, to ensure proper adoption of integration mechanisms, the relevant context of design tasks should first be identified. In Table 14.2, the empirical findings concerning the pertinent contextual aspects of integrating design processes in the eight fashion buying companies are summarized.

Table 14.2. The generalized contextual aspects of design tasks

	Category 1 High fashion	Category 2 Mass fashion
End-market distribution modes	Highly operation-standardized specialty chains or counters that distribute small quantity but wide range of product lines	Sale teams that consolidate allsorts of retailer requirements to optimize the quantity and delivery ordered
Product characteristics	Image and trend oriented; Competition in the sense of uniqueness and life style individualism	Items in each product line allow mix-and-match; Stable variety of classic and casual items with strong social group conformity; Strong manufacturers' or stores' identity; functionally or value-based innovating
Collaboration approaches	Cross-functional contact at all operational levels; Informal, horizontal and opinion-rich discussion	Procedural communication flow; Technical problem-type discussion
Interface methods	Periodical teleconferencing; Team meeting; Sketch works	CAD, MRP2 systems; Centralization of planning and decision making; GroupWare type systems
Supplier management mechanisms	Co-operative supplier development; "Paternalistic" liaison; Preferred sourcing; Long term contracting	Competition-based supplier selection; "Opportunistic" alignment; Multiple sourcing; Transaction-based contracting
Perceived novelties in design task	Wide varieties of product lines and features; Product based design task seeking for perceived fashion superiority and trendiness; Very seasonal, volatile and high rate of perishability; Short distribution life cycle	Process based design task seeking for innovating make-up processing methods; Relatively less seasonal; Long distribution life cycle

Fashion design tasks can be categorised broadly into two types:

- The high fashion category that focuses on trendy but volatile fashion novelties. These companies optimise their performance through a range of exclusive chain and counter operations in order to target designated market segments.

- The mass fashion category that focuses on the well-known manufacturers' or stores' brand concepts and values. These companies serve their markets through resourceful sale teams interacting with dispersed retailers. The product novelties are relatively stable and value-oriented.

From the findings, we can see that the design task for the high fashion business is to endeavour to pursue product attribute innovation in a market environment that is complex and turbulent. The design processes should be supported by mechanisms that entail sufficient *ad hoc* and market-oriented collaboration in the team structure. On the other hand, the design task for the mass fashion category favours integration mechanisms that effectively share, discuss and record information in a complex, but relatively stable, business environment. As such, the team structure is characterised by a strong expert-oriented outlook and the communication flow is inclined to be sequential and structured.

Figure 14.2 conclusively illustrates our interpretation of the contextual relationship between rate of change in perceived product novelty (in the inverse sense, the design task certainty) and the scope of team interdependence in global fashion marketplaces. It also illustrates the strategic model of choice of integration mechanisms for effective collaborative design processes.

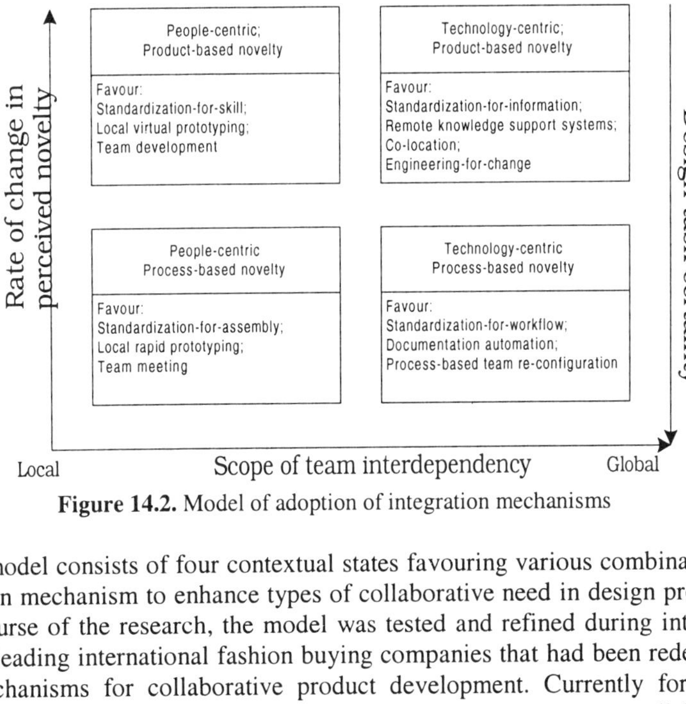

Figure 14.2. Model of adoption of integration mechanisms

The model consists of four contextual states favouring various combinations of integration mechanism to enhance types of collaborative need in design processes. In the course of the research, the model was tested and refined during interviews with the leading international fashion buying companies that had been redesigning their mechanisms for collaborative product development. Currently for design activities within a local territory, the companies tend to improve collaboration using traditional people-centric mechanisms. For instance, the matrix-based team structure is used to build up the skills of interlaced functional and technical teams, as well as their co-ordination and collaboration. However, when the design task involves a large number of global-based technical and functional teams, the companies manage to facilitate inter-team collaboration using remote technology (data)-centric interface systems. These systems mainly entail standardized plans and schedules (e.g., MRP/ERP systems), consistent operational and communication workflow (i.*e*., process optimisation and automation engineering), and mutually accessible design data and opinions (*i.e*., document management systems). In the cases, some companies explicitly stated that they are considering even more advanced collaborative document management and interface systems that allow the creation, review and revision of design information by designated teams on a network basis. Their existing videoconferencing and local station design systems will soon be replaced by more economical technologies, such as Java-based telephony integration and shared virtual CAD/CAM applications.

Meanwhile, for high fashion product design that emphasizes capricious product attributes, integration mechanisms should enhance rapid idea representation, retrieval and evaluation on a large amounts of potential design concepts. Currently, one company in the cases studied is studying the feasibility of implementing applications in shared and networked virtual environments. As such, a designated number of remote users can view simultaneously the common virtual images of potential design on their monitors, manipulate images at any time, and pass back and forth between participating users. The main technical constraints in this scenario are the complexity of design information transmitted each time, the available bandwidth and the network traffic.

For teams handling relatively stable mass fashion design, the extent of product novelty is restrictively determined by the 'cost-to-market', competitiveness and the requirements of quality and delivery consistency for mass volume production. Design is regarded as an evolving process through continual improvement in planning and production processes. New product development and realization needs to be carried out within certain limits of process modification and re-design. In this context, integration in process workflow, inter-team interaction, supplier co-operation and consistency of teams' skills across the product life cycle can reduce the burden stemming from design and environmental uncertainty. From this viewpoint, process-based design requirement standardization, rapid proto-sample making systems and document automation are widely adopted in today's mass fashion business practices.

14.5 Conclusion

We have described the findings of a qualitative investigation of collaborative design process integration in global fashion companies. It is concluded the integration problems in global-oriented design processes are pragmatic issues of interdependency. These issues induce concern about the appropriateness of different integration mechanisms. Using the cases studied, a strategic model for integrating design processes is suggested, to provide insight to assist the collaboration of a large number of multi-disciplinary teams across countries. The model categorizes four contextual states requiring different types of collaborative needs, and identifying various combinations of integration mechanism in global fashion buying companies. Above all, the empirical evidence has proved that team integration, once appropriately handled and addressed, would entail collaboration and process consistency, which would, in turn, result in more advantageous cost and overall performance in today's turbulent global marketplaces.

14.6 References

Cross N, Cross AC (1995) Observations of teamwork and social processes in design. Design Studies 16,3:143-170

Flaherty MT (1996) Global operation management, McGraw-Hill, New York

Fleisher M, Liker JK (1997) Concurrent engineering effectiveness: integrating product development across organizations, Hanser Gardner Publications, Cincinnarti

Jassawalla AR, Sashittal HC (1998) An examination of collaboration in high technology new product development process. J prod Innov Mgmt 15, 3:237-254

Leonard-Barton D, Bowen HK, Clark KB, Holloway CA, Wheelwright SC (1994) How to integrate work and deepen expertise. Harvard Business Rev 72(5): 121-130. Sep/Oct

Mills A (1998) Collaborative engineering and the internet. Society of Manufacturing Engineers, Dearborn, Michigan.

Morgan G (1989) Creative organization theory: a resource book. Sage, Newbury Park, CA, USA

Paashuis V (1998) The organization of integration product development. Springer Verlag, Berlin Heidelberg New York.

Park H, Cutkosky MR (1999) Framework for modelling dependencies in collaborative engineering processes. 11:84-102

Ramesh B, Tiwana A (1999) Supporting collaborative process knowledge management in new product development teams. Dec Supp Sys 27: 213-235

Reid FJM, Culverhouse PV, Jagodzinski AP, Burningham C (2000) The management of electronics engineering design teams: linking tactics to changing conditions. Design Studies 21: 75-97

Singletary EP, Winchester, JrSC (1998) Beyond mass production strategic management model for competitive manufacturing transformation in the US textile industry. J Tex Inst 89, part 2:5-15

Slusher EA, Ebert RJ, Rasdell KM (1989) Contingency management of engineering design. Proc Inst Mech Eng 1: 65-75

Chapter 15

An Evaluation of Inter-company Networks for New Product Development

A. Filson and A. Lewis

15.1 Introduction

Networking is the process by which small to medium enterprises (SMEs) work in close collaboration in order to offer products or services to compete with larger organisations. However, research has demonstrated that, through the wide range of terms used, it is difficult to find a common definition of the many forms of organisation of labour and the many different structures that exist in the often lightly used term 'network'. To further compound the issue this term is easily confused with concepts of industrial districts, co-operations, strategic alliances and collaborations (Pannicia, 1998). So much so, that some researchers would not consider differentiating between these different functional forms (Seremetis, 1994).

Several countries have already attempted to mimic the success of the first networking model, originally developed in Italy, and internationally respected in regard to its use of design, flexible specialisation and networking systems (Cooke, 1996c). These include Denmark, America and Portugal (Cooke, 1996a ; Seremetis, 1994 ; Lyberaki and Pesmazoglou, 1994 ; Newhouse, 1997). These models are considered to represent best practice because of their close collaboration between the SMEs and their innovative utilisation and management of flexible specialisation. This has enabled them to compete successfully with larger companies through flexibility in products and workforce. In view of the close geographical proximity of these SMEs, the Italian networks are often referred to as industrial districts.

A major problem encountered when determining best practice from all the examples is the definition of the typical network. The pioneer in industrial district research, Becattini (1990), provides the following definition of the Italian industrial district: 'a social-territorial entity which is characterised by the active presence of both a community of people and a population of firms in one naturally and historically bounded area'. Whilst GHK (1998) define a network more simply as:

'...a formal alliance between companies, as more or less equal partners in the collaboration – beyond the scope of normal contractual arrangements for procuring products and services, but stopping short of mergers and acquisitions.'

15.2 Research Methodology

This paper is based on research carried out as part of a European-funded project to improve the performance on SMEs within the manufacturing sector in South Wales, through providing mechanisms to encourage networking. In order to define the practices that contribute directly to the success of networks it was important to benchmark the best of current practice for the strategic use of design and manufacturing flexibility.

In particular this has involved significant research of networks in Italy, including visits and interviews with a number of key individuals. These included Giuseppe Calabrese, Secondo Rolfo and Giampaolo Vitali from the Institute of Research on Firm and Development (CERIS), part of the Italian National Research Council; John Edgar, the Chief Executive of Business Net Ltd (Part of the Danish Technological Institute); and Dr Luciano Consolati, the Director of Lumetel, the technological service centre for Lumezzane, located near Brescia.

15.3 Italy

Firstly, when identifying the best practice of the industrial districts and networks in Italy it is important to consider the significant networking features of the so-called 'canonical' industrial districts. This is the name given to the original model of industrial districts found in the Emilia-Romagna region of Italy and first widely reported in the 1970s (Brusco, 1982 ; Paniccia, 1998).

The small firm is a significant characteristic of the districts, enabling them to respond rapidly to changing product demands. The districts relied on a multi-skilled and flexible workforce and built up production networks of inter-firm co-operation (Lyberaki and Pesmazoglou, 1994). In the 'canonical' industrial districts, there were no particular industrial policies aimed at strategically encouraging inter-firm co-operation. These spontaneous networks emerged locally through a bottom-up approach (Seremetis, 1994). The SMEs competed with each other and outsiders on productivity, design, product quality and innovation, rather than relying on price cutting to increase market share (Economist, 1991).

Traditionally, the owners and employees of the SMEs had a culture of close knit personal and informal working relations, as the owners knew each other and their families. This resulted in the development of a trusting relationship between the companies (Seremetis, 1994 ; Brusco, 1995). The co-operation involved no formal contracts, was generally informal, and therefore relied heavily on high levels of trust (Piore and Sabel, 1983). Conversely, Paniccia (1998) does not consider this trust to be a crucial element in the development of the industrial districts, as this can not happen until after intitial co-operation. However, visits to

Italy to corroborate these findings, undertaken as part of current research, found the districts to be informal in character and working on a basis of trust and personal relationships. This universally recognised characteristic of Italian businesses was supported by research carried out at CERIS and Lumetel, which considers that trust is still one of the most important features of Italian industrial districts.

More recently, co-operatives, service centres and local government have encouraged SMEs to continue to work together to compete with recent competition from the rest of Europe and the Far East (Cooke, 1996a). These services, offering technical and market information to the SMEs, play an important role in the continuing evolution of the districts. Small firms are at a disadvantage because their size means that they can seldom employ managers to handle marketing, accounting, or even technical services. The co-operative service organisations aim to offset this disadvantage by pooling resources (Seremetis, 1994).

The Lumetel service centre is attempting to spread innovation in terms of technology, product processes, organisation, management and finance. This centre was started up initially by funding from the local Lumezzane authority. Dr Consolati considers it important that the SMEs pay for the services they are given, so they are taken seriously and enthusiasm maintained. Subscriptions from the members now fund the service. In addition, Consolati emphasised the importance of gaining the trust of the members.

Calabrese (during interviews) highlighted the need for the industrial districts to continue to innovate to retain competitiveness. However, CERIS research has uncovered that companies in the districts would like to innovate, but do not have the resource or knowledge to do so (Rolfo and Vitali 1999). The service centres aim to address this.

The key features of the Italian networks are therefore:

- the close relationships between the entrepreneurs and their employees;
- spontaneous inter-firm co-operation;
- service centres that are present to support the SMEs;
- informal communication mechanisms;
- high levels of trust.

15.4 The Danish Network Programme

In spring 1989, the Danish Minister of Industry announced a £13 million programme to encourage network co-operation among SMEs, thus enhancing the international competitiveness of Danish industry. This utilised a top-down methodology rather than the bottom-up spontaneous development witnessed in the case of the Italian districts. The programme was based on an attempt to adapt successful network experiences from countries such as the US, Italy and Germany to Danish conditions. The Danish Technological Institute (DTI) formalised the system into a policy where trained network brokers created networks of firms (Cooke, 1996a).

The programme was reported to be a success with Jacob and Martinussen (1991) estimating that more than 1,000 companies entered networks between 1989 and 1991 without programme grants. Huggins (1996) reports that the networks at both a local and national level in Denmark have had a significant benefit for the firms involved, as detailed below:

- small firms demonstrated an ability to learn from each other;
- increased awareness of product development issues from manufacturers;
- knowledge of access to sources of finance.

However, the programme additionally highlighted a secondary issue, not previously considered. The actual time required to establish a network was found to be greater than expected and there was an increase of resources required for the network (Huggins, 1996). Yet, Henriksen (1995) found that due to the small business culture in Denmark, in most of the networks, the members already knew each other as they had collaborated before and already trusted each other. In cases where the companies did not know each other the development of the network process was much more difficult as they needed to build some trust between them.

There was also evidence of the networking system being very vulnerable to the take-overs. Henriksen (1995) assumes that this is because the firms are so successful that large companies see an investment opportunity in the network, and will benefit from the network's marketing relations and technologies.

The key features of the Danish networks are:

- that they emerged from a programme utilising a top-down methodology;
- that they were supported through a centrally funded Government initiative;
- formal communication mechanisms.

15.5 UK

Although there are few naturally occurring product development networks in the UK that may be comparable to those in Italy, networks have recently been established from networking programmes run by Training and Enterprise Councils (TECs) and Business Links (Edgar, 1999). However, there are several significant business differences between the business culture of the UK and Italy (Zeitlin, 1995). Zeitlin expresses the opinion that in the UK the domination of major corporate enterprises limits the development of independent small firms and collaborative supplier networks by restricting access to final markets. This suggests a reason why networks have not emerged spontaneously. Additionally, the UK does not have the culture of close working relations between SMEs in the same area: so close, informal relationships of trust have to be formed between the companies before a network can be successful (GHK, 1998).

Nevertheless, there has been some success in establishing networks by Business Net Ltd. In order to form networks they have written a number of analysis tools (together with the Danish DTI) to provide product development and performance indicators to give direction during the growth process. In the UK they have been working with the Department of Trade and Industry and with Business Links to establish regional programmes and provide training for network brokers, who are paid through the Business Links or TECs (Edgar, 1999). Edgar considers that the most successful networks comprise of companies of between 20 and 30 employees and target growth markets.

There are two basic types of network promoted in the UK (Business Link Hampshire, 1999). Informal networks are business groups that run in an informal basis, sharing ideas and focusing on issues that are of common interest. These can be business clubs and sector groups. Formal networks are those that trade as an individual organisation and have written contracts. It is mostly the formal networks that Business Net offers advice on, and that the Business Links promote and facilitate.

The key features of the UK networks are:

- that they are largely bound by formal contracts;
- they use formal communication mechanisms;
- they are supported through funded initiatives.

15.6 Problems and Sustainability

Despite the benefits promoted about networks and districts over the years, they may suffer from problems with sustainability. Although the Italian districts experienced growth and success in the 1970s and early 1980s, in the late 1980s they suffered from competition from the rest of Europe and the Far East (Harrison, 1994 ; Cooke, 1996a). Therefore, there was an increased need for continuous innovation in the products and processes the districts offered. In the 1990s, these problems of the industrial districts were discussed in depth (Harrison, 1994 ; Cooke, 1996a ; Triglia, 1995 ; Seremetis, 1996). Likewise, although papers on the continuing success of the districts and innovation centres have been enthusiastic (Economist, 1991 ; Rossant, 1996), there are still many problems (Calabrese *et al.*, 1999). The following evolutionary tendencies were found (Harrison, 1994a; Cooke, 1996a):

- the partial de-localisation of activities towards low labour cost countries;
- the growth of a firm's 'hierarchy' with the emergence of leaders;
- family businesses reaching maturity;
- reduction in the extent to which the districts innovate;
- flexibility can now be achieved in large organisations.

There was evidence of this happening to the district of Lumezzane, where Consolati (1999) believes that this was all part of the evolutionary nature of the districts, allowing continual growth and improvement. Conversely, it was felt by researchers from CERIS that although it is advisable to accept this evolution, an effort should be made to maintain the most important aspect of the district (for example, product design) in Northern Italy (Calabrese *et al.*, 1999).

Sforzi (1989) wrote that to be an industrial district one of the main criteria is to contain geographical clusters of small firms specialising in the different production phases of a unified overall process. This is no longer the case where many districts co-operate with SMEs in Southern Italy and abroad, in order to reduce costs (Calabrese *et al.*, 1999) with the districts evolving to become networks.

There is limited evidence to support the theory of the evolutionary tendencies of the organised districts in Denmark. It is reported that some SMEs have moved on to new networks after successes in others (Huggins, 1996). Henriksen (1995) reports that the success of the Danish networks could be affected by the allocation of authority and responsibility altering the relationships within the network. By nature, entrepreneurs make their own decisions and they could be affected by giving up their independence in the network. Furthermore, with all networks, problems may emerge when companies need to work with people from different social cultures with different languages.

As the attempts to build inter-firm networks of SMEs in the UK have been recent, there is little feedback on the long-term sustainability of these formal networks. However, Edgar's recommendation (during private communication) that any SMEs that may instigate issues concerning take-over bids are eliminated during the network set-up phase, suggests that current UK networks aim to maintain their initial structure.

15.7 Summary

This paper has identified the many advantages to be achieved for SMEs working in industrial districts or networks. The network of companies can benefit from a competitive advantage because of their combined size and sharing of knowledge, innovation and best practices. Next, this paper has sought to define the main principles behind inter-company networking and has identified major differences in business culture between Italy, Denmark and the UK that affect how networks are formed in these countries. The key difference between the Italian networks and those in the UK and Denmark is that they are supported by a culture of informal communication resulting in trust and close relationships. In Denmark and the UK consultants or brokers have been required to develop this relationship between the SMEs. It was this difference in business cultures that allowed the networks to form spontaneously in Italy, whilst in the UK and Denmark they were established in a more formal manner.

The predominant finding was that commitment, mutual respect, understanding and trust from member companies lead to successful attempts to start new networks. Although no country can be identified as representing best practice in networking, it is suggested that organised, formal networks are recommended for the UK whose SMEs lack the trust, co-operation and local relations found in Italy.

This paper is a result of ongoing research by the DERC to create a network in South Wales. Once established the network will be continually monitored to provide new research material concerning the evolutionary tendencies for organised formal networks. It remains to be seen if the evolution of the networking model should be allowed to continue spontaneously (as in Italy and Denmark), or whether efforts should be made to maintain the very organised state of the network.

15.8 References

Brusco S (1982) The Emilian Model: Productive Decentralisation and Social Integration. Cambridge Journal of Economics 6:167-184

Brusco S (1995) Local Productive Systems and New Industrial Policy in Italy. In: Bagnasco A and Sabel C (eds.) Small and Medium Enterprises, Pinter, London.

Business Link Hampshire (1999) Business Networks (Online). Hampshire: Business Link. Available from: http://www.hampshire.businesslink.co.uk/ [Accessed 29 Nov 1999]

Cooke P (1996a) Building a twenty-first century regional economy in Emilia-Romagna., European Planning Studies, February

Cooke P (1996b) Reinventing the Region: Firms, clusters and networks in economic development. In: Daniels PW, Lever WF (eds.) The Global Economy in Transition. Addison Wesley Longman Limited, Harlow, England.

Economist (1991) All together now, September

Edgar J (JohnE@bnluk.co.uk), 1999. RE: Danish Technological Institute (DK DTI). Email to author

GHK Economics and Management (1998) A Practitioners Guide to Business Networking, Brokering SME Alliances

Harrison B (1994) The Italian industrial districts and the crisis of the co-operative form: Part I. European Planning Studies, Jan 01, 1994

Henriksen L (1995) Formal co-operation among firms in networks: The case of Danish joint ventures and strategic alliances. European Planning Studies 2:254-260

Huggins R (1996) Technology policy, networks and small firms in Denmark. Regional Studies 30, 5:523-526.

Lyberaki A, Pesmazoglou V (1994) Mirages and Miracles of European Small and Medium Enterprise Development. European Planning Studies 2:499-521

Newhouse J (1997) Europe's rising regionalism. Foreign Affairs; January 01

Paniccia, I (1998) One, a hundred, thousands of industrial districts: Organizational variety in local networks of small and medium-sized enterprises, Organization Studies19, 4:667-699

Piore M, Sabel C (1983) Italian Small Business Development: Lessons for U.S. Policy. In: Zysman J, Tyson L (eds.) American Industry in International Competition: Government Policies and Corporate Strategie. Cornell University Press, NY, pp391-421

Rolfo S, Vitali G (1999) Industrial Districts and Innovation: The Limits of the Local Technical Systems. Ceris-CNR, Turin

Rossant J (1996) Aprilia's Leader of the Pack. Business Week, June 10

Seremetis P (1994) SMEs in technological networks: Italy, Denmark and the UK. European Planning Studies, September

Sforzi F (1989) The geography of industrial districts in Italy. London. Routledge.

Zeitlin, J (1995) Why are there no industrial districts in the United Kingdom? In: Bagnasco A, Sabel C (eds.) Small and Medium Enterprises. Pinter, London.

Chapter 16

Ecodesign Integration: Putting the Co into Ecodesign

V.A. Lofthouse and T.A. Bhamra

16.1 Introduction

Since the 1960's industrial-design focused books, by authors such as Packard (1963) and Papanek (1971) have pioneered the need for a new agenda from the design profession by openly questioning the morality of affluent lifestyles and designing for obsolescence. Despite these efforts, the first response to environmental needs by industry was reactive and driven by legislation rather than design. Legislative requirements initially only addressed end-of-pipe problems, such as capping smoke stacks, then later called for 'cleaner manufacturing' from factories. Further research then recognised that for ecodesign to be successful, it needed to be considered earlier in the product development process (PDP). Attention shifted to design engineering. It was recognised that to be effective, ecodesign must be an integral part of the PDP and focus on the environmental impact of the whole life cycle of a product. In response to this, the most proactive companies started to develop solutions to aid design for disassembly, recycling and remanufacture.

Ecodesign has now come of age. Though it has taken time to work back through production and design engineering, ecodesign is now recognised as a design activity which must be included at the earliest stages of development, during discussions with marketing and industrial design (Bhamra *et al.*, 1999). The most proactive companies are now beginning to recognise that industrial design skills are key to the realisation of well eco-designed products (Philips Electronics N.V., 1997). Finally, Papanek's concerns are being recognised as it is accepted that it is now '...time to address the most fundamental aspect of industrial production: the product itself' (Brezet and Hemel, 1997).

This paper reports the initial findings from a three year collaborative research project between Electrolux Industrial Design and Cranfield University, which aims to provide new information and potentially new approaches to improve the way in which industrial designers tackle ecodesign. This paper outlines why industrial design is now considered to have the potential to make a valuable contribution to ecodesign. It then goes on to discuss a number of observed barriers that appear to be hindering the wide-ranging acceptance of ecodesign as an industrial design activity. Finally the paper discusses the need for multidisciplinary teamwork if ecodesign is to be achieved in 'live' product development projects.

16.2 Research Approach

The ideas discussed in this paper have arisen from a combination of different research activities with different groups of designers. Ideas and opinions were also collected during a workshop held by the Eco Innovations group at Cranfield University, which aimed to investigate which industrial design 'attributes' might be most beneficial to the development of ecodesign products. A number of semi-structured interviews were carried out with Electrolux industrial designers and industrial designers working in consultancies to establish current opinions on ecodesign. Information from a benchmarking study, which aimed to identify 'which' companies were achieving 'what' in the field of ecodesign, have been used to provide information about current best practice and state of the art thinking. Finally a one-week participant-observation exercise at Electrolux facilitated the collection of well-grounded, rich descriptions of the industrial designers working on a 'live' design project within the context of the business

16.3 Industrial Design and Ecodesign

16.3.1 Industrial Design's Value to Ecodesign

The main reason why industrial design is considered to have a valuable part to play in the development of ecodesigned products is to do with the context in which it is positioned. Industrial design sits at the early stages of the PDP, where the design brief is at its most flexible, and the most critical decisions are made with respect to cost, appearance, materials selection, innovation, performance, environmental impact, and perceptions of quality (longevity, durability, reparability). This means that industrial design plays a key role in determining the nature of the product. Observations at Electrolux have shown that though design during these early stages is a team activity, it tends to be led by industrial design, and fed by the specialist information provided by production and engineering experts. Thus industrial design is in a good position to implement changes and influence the environmental impact of the product.

In addition to the contextual position of industrial design, other attributes have been identified as being beneficial to the development of ecodesigned products. Industrial design provides a pivotal link between the consumer and the product. This position at the interface between supply and demand means that, with an understanding of need versus desire, industrial design could be heavily influential on reducing consumption levels, a major issue for sustainable development.

Designers by their very nature enjoy a challenge. Adding ecodesign into the remit of industrial design has the potential to create a more interesting and challenging design process, by encouraging greater creativity within individuals and design teams (Dewberry, 1996). Industrial designers have the ability to respond to this challenge by generating new ideas and scenarios that can lead to creative and innovative sustainable solutions.

It can also be argued that involving industrial design in the development of ecodesign products should lead to improved ecodesign. Research has shown that '...if ecodesign is only used at [the] post-specification stage, [only] relatively minor environmental changes can be affected' (Bhamra *et al.*, 1999). The fact that production and design engineering are specialist subjects with expertise in mechanical design, materials and electronics (see Figure 16.1a) means that the scope of environmental improvement is limited by the areas that they can tackle. Industrial design on the other hand, is a more wide ranging subject which covers a diverse range of issues from mechanical design to marketing, psychology and artistry (see Figure 16.1b). The added scope provided by industrial design means that new dimensions are added to the ecodesign solution. This may be likened to the effect that the introduction of industrial design had on the overall quality of the products developed during the Industrial Revolution.

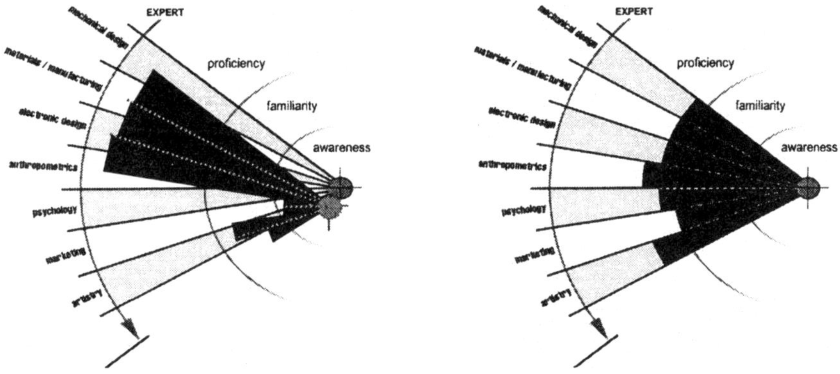

a b

Figure 16.1. Adaptation of the 'Cone of Industrial Design' (Bates and Pledgley, 1998). Illustration of the skills of: a. the design engineer; b. the industrial engineer.

The participants in the 'Attributes' workshop believed that it is the very essence of industrial design that makes it so appropriate to the development of eco-designed products. The essence of industrial designers can be seen as their ability to: think in abstract ways; be creative; be problem solvers; rise to a challenge; and bring great scope to their work, as illustrated in Figure 16.1.

16.3.2 How Industrial Design can Improve Ecodesign

A product developed in collaboration with aneco-redesign programme at RMIT clearly demonstrates a 'best case' example of how the involvement of industrial design in projects, where business and environment are joint drivers, can lead to the development of products which are more innovative than their predecessors (Sweatman and Gertsakis, 1996).

The Kambrook Axis kettle project aimed to reduce the energy consumption of the kettle during its use stage. Having found no suitable technological answer to the problem the design team studied the users' behaviour to see whether this provided any answers to the efficiency problems. The study concluded that users tend to fill the kettle far fuller than required, turn the kettle on, walk off to do something, then come back 5 minutes later and reboil the kettle before using the hot water.

These findings enabled the designers to identify three new ways of reducing the energy consumption of the kettle. The water gauge was made clearer and moved to the top of the jug, the jug cavity was insulated with a double wall, to keep the water hotter for longer and reduce the energy needed to reboil the kettle, and a temperature indicator was fitted so the user could tell whether it was necessary to reboil the kettle (Sweatman and Gertsakis, 1996).

By considering the psychology of the user the designers managed to vastly improve the environmental performance of the kettle, and using basic ergonomics and well considered materials, to create a much more innovative solution. In today's highly competitive markets where there is a need to retain a commercial edge, the benefits of developing ecodesigned products with industrial designers can be clearly seen. This edge can be simultaneously expressed through both product and environmental quality.

16.4 Potential Barriers to Ecodesign

Section 16.3 considered the idea that the very things which constitute the 'essence of industrial design' are what makes it an ideal discipline for dealing with ecodesign, and that its involvement can lead to the development of environmentally superior products. However, the benchmarking study identified that there are still relatively few examples of ecodesign projects involving industrial design in the UK. This section recognises that the abilities outlined in Section 16.3.1 represent only what industrial designers have the potential to do. The next section will consider a number of barriers that may be hindering the realisation of this potential.

16.4.1 Confidence

Recent interviews with a number of design consultants in the UK, indicated that many industrial designers feel they have a limited understanding of ecodesign and the types of issues that it involves. They also indicate that although they are enthusiastic about the potential of ecodesign they seem unsure as to where their role might lie beyond specifying suitable and minimal materials, and low energy strategies, which may or may not be within their design remit (Lofthouse, 2000).

It is possible that the technological roots of ecodesign (discussed in Section 16.1) may have alienated industrial designers and limited their understanding of ecodesign, allowing them to believe that ecodesign requires the specialist, technical knowledge of design engineers. If this is the case, then there is a need to demystify the subject by showing that ecodesign is merely an extension of good design, and educate designers so that they understand what and how they can contribute to ecodesign.

16.4.2 Tools

In 1996, Dewberry identified that industrial designers felt they would need new information if they were to carry out ecodesign-orientated projects, and the DEEDS project identified a lack of ecodesign tools available for these early stages of product development (Bhamra *et al.*, 1999). Recent interviews suggest that this situation does not appear to have changed. To combat this, a number of the more proactive companies responded by giving industrial designers the 'tools' which they developed to help design engineers with ecodesign. However, although checklists and environmental guidebooks had previously been considered to be useful design tools for design engineers, observations and interviews have shown that industrial designers find them inappropriate and difficult to use. They have even been found to have the negative effect of alienating industrial designers from ecodesign. The reason for this appears to lie in the fact that the two design disciplines have very different roles and as such industrial designers approach problems in very different ways to engineers (see Figure 16.2).

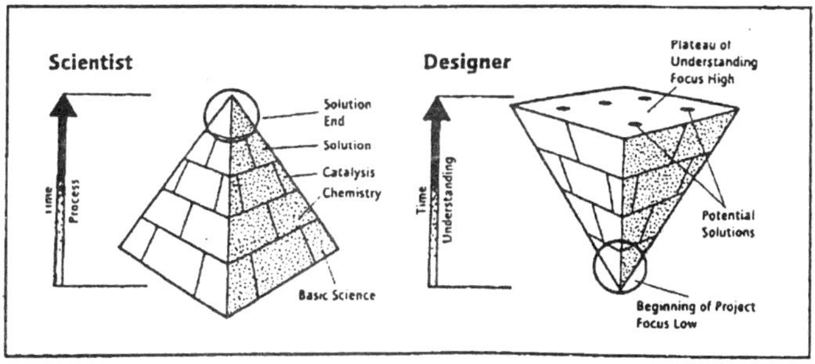

Figure 16.2. Divergent and convergent thinking (Joyce *et al.*, 1998)

Joyce *et al.* (1998) explain that designers are divergent thinkers. 'Divergent thinking is characterised by ideation and a fluency with unusually associated ideas... Any one of the ideas generated may be acceptable' (Durling *et al.*, 1996). In contrast to this, engineers are convergent thinkers (Joyce *et al.*, 1998).

'Convergent thinking progresses toward the production of a single, right answer to a problem... characterised by a logical, analytical approach to problem-solving' (Durling *et al.*, 1996). So although both disciplines deal in problem solving through 'design', they each approach it in very different ways; industrial designers are trained to communicate through design using sensory language and visualisation skills, whereas design engineers focus on integrating methods to achieve a final goal.

These differences suggest that design tools which are appropriate to one discipline, can not automatically be transferred to another. This suggests that there is a real need for appropriate methods to be identified and developed to help industrial designers integrate ecodesign into their every-day design practice.

16.4.3 Project Type

The type of design project attempted also appears to be critical to the success of ecodesign projects. A number of good examples, such as Herman Miller's Aeron chair (Herman Miller, 1999), and Philips' Typhoon colour monitor (which uses less materials and packaging and has fewer components and printed circuit boards, Philips Electronics N.V., 2000), demonstrate how successful ecodesign projects can be when there is a flexible and open brief. The situation seems to be quite different for 'continuous improvement' type projects.

Though 'continuous improvement' projects often provide the companies' 'bread and butter', and may result in tens of thousands of products entering the market, they often have massive restrictions and limitations. Time restrictions can limit the opportunity for formal training and the sourcing of new environmental information. Coupled with budget restrictions, this can mean that there is neither time nor money available to develop new, superior components, so older stock parts, which are cheap and easily available, are used. Added to this is the lack of flexibility associated with this type of project, which can often lead to the project being 'railroaded' down a predetermined development path, allowing little opportunity for reflection. So although industrial design is considered to be one of the most flexible stages of the PDP, observations and the work of other authors (Dewberry, 1996) indicate that the 'brief' which is given to design has often had all the main decisions made. The lack of flexibility at this stage may act as a barrier to ecodesign, as it restricts the input that industrial design can have on the project. The DEEDS project identified that, without commitment from marketing, at the very earliest stages of the PDP, it is 'very difficult to ensure that ecodesign [is given] priority' (Bhamra *et al.*,1999). It therefore follows that unless industrial designers, as the link between the product and the customer, are given the opportunity to be involved in drafting up or amending the brief, their potential influence over the environmental impact of the product will be dramatically reduced.

16.4.4 Team

Although Section 16.3.1 considered in detail the value of industrial design skills to the development of ecodesigned products, it is important to understand that this paper recognises that the industrial design team cannot work in isolation to achieve successful ecodesign. Instead, it is believed that multidisciplinary teams with high levels of communication are needed for successful eco-product development. Section 16.4.3 has already highlighted the importance of marketing involvement if ecodesign is to be successful. In addition to this it is also important that industrial design, design engineering and production are working towards the same product goal. Section 16.1 described how, in the 1980's, ecodesign was considered initially during the production stage. In the 1990's it was involved both at the production and the design engineering stages. It now needs to be considered at all three stages of the product development process, as illustrated in Figure 16.3. It is naïve to believe that the industrial design team can develop an improved ecodesigned product in isolation from the rest of the product development team. If, for example, the situation arose where production and industrial design were motivated towards ecodesign but the design engineers were not, it would be likely that ecodesign attempts would fail, as the engineers would have a different set of requirements (or brief) to the rest of the team. So in much the same way as quality, ecodesign must be recognised as an important element of good design which can not be achieved by one discipline in isolation, but must be a common goal at every stage of the PDP.

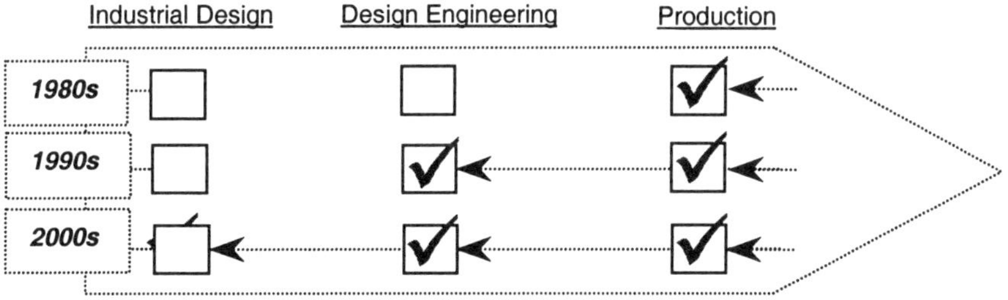

Figure 16.3. Consideration for ecodesign in the product development process

To compound this idea, interviews with Electrolux employees indicated that unless ecodesign is a clear part of the structured PDP it can not be considered at all, as it is 'almost impossible' to consider things which fall outside 'the system'. This suggests that management support is required to ensure that the structure is there to allow for continual emphasis and opportunity for reflection, throughout the project. This is a sentiment that echoes the DEEDS findings which also recognised the importance of management support (Bhamra *et al.*, 1999).

In companies that have traditionally been environmentally proactive, there may be a need to change the mind-sets of employees who still view ecodesign as a technical activity dealt with by engineering and production. Interviews with designers revealed that they often feel that the rest of the team does not expect 'eco' comments from them.

So it appears that if ecodesign products are to make their way into the market place, there is a real need for all members of the product development team to be aware of the different range of skills that each has to offer, and to recognise that each discipline will make a different contribution to the product. The most successful companies in the benchmarking study were those who took a holistic approach to ecodesign, rather than thinking it was something that could be dealt with by one discipline, be it production, design engineering or industrial design (Lofthouse and Bhamra, 2000). The aim must be to recognise and use the most appropriate discipline at each stage of the development process towards a common aim, as is the normal approach with product development.

16.5 Conclusion

This paper has discussed the potential that industrial design has to improve ecodesign and has shown that its very nature and the context in which it sits appears to make it suitable for realising well eco-designed products (Philips Electronics N.V., 1997). The key must be to tap into these skills.

A number of potential barriers that appear to be hindering industrial design's ability to be involved in ecodesign projects have been identified. It is hoped that these provide the first steps towards a solution. The main problems occur when industrial design is marginalized and not able to be an active part of the development team. To combat this, the whole design and development team needs to be educated so that everyone has a clear understanding of the skills and abilities that each discipline can contribute to ecodesign. To achieve this, it is important that industrial designers have a clear understanding of what their contribution to ecodesign can be, within the context of the business in which they work. There is also a real need for appropriate tools that can help industrial designers integrate ecodesign into the PDP.

Ecodesign clearly needs to be integrated into the PDP as easily as possible, so that it can become part of the teams' everyday practice, as there are often time restrictions that limit the usefulness of structured training sessions. It is also important to recognise that the restrictions and limitations, such as time, cost and flexibility, inherent in some types of project may lead to a situation where it is impossible to improve environmental performance.

It should also be recognised that there is a real need for ecodesign achievements to be measurable. By setting and meeting objectives, companies can recognise successes, confirm the business benefits of ecodesign and provide the designers with the opportunity to learn from their achievements, thus perpetuating the learning experience.

Finally it is clear that without a combination of management and team commitment at all the stages of the PDP - production, design engineering, industrial design and marketing - ecodesign products will not be achieved. Good communication, relevant tools and common goals within empowered multidisciplinary design teams can bring us a step closer to developing innovative, environmentally responsible and profitable products with an improved competitive edge.

16.6 References

Bhamra T, Evans S, Simon M, McAloone T, Poole S, Sweatman A (1999) Integrating Environmental Decisions into the Product Development Process: Part 1 The Early Stages. In: EcoDesign '99: First Symposium on Environmentally Conscious Design and Inverse Manufacturing. IEEE, Tokyo, Japan, February 1-3, 1999

Brezet H, van Hemel C (1997b) Ecodesign: a promising approach to sustainable production and consumption. In: Industry and environment. UNEP, Paris, 20,52

Dewberry E . (1996) EcoDesign - Present Attitudes and Future Directions. In: The Design Discipline. Technology Faculty, Open University, Milton Keynes

Durling D, Cross N, Johnson J (1996) Personality and learning preferences of students in design and design related disciplines. In: IDATER 96. Loughborough University, UK, 2-4 Sept 1996

Miller H (1999) http://www.hermanmiller.com/product/index.bbk?id=21. Herman Miller Inc., 9th February 2000

Joyce M, Franklin K, Neale P, Kyffin S., Veronesi B. (1998) What stimulates the creative process? The Journal of Design and Technology Education, 3:113-116

Lofthouse VA (2000) Interviews with design consultants: ecodesign. Cranfield University, Cranfield, 25th May 2000

Lofthouse VA and Bhamra TA (2000) Benchmarking to Understand Appropriate Communication of Ecodesign - A Collaborative Project. In: Design Research International Symposium on the dimensions of industrial design research. Milan, May 18-20, 2000

Packard V (1963) The Waste Makers. Penguin, Middlesex

Papanek V (1971) Design for the Real World

Philips Electronics NV (1997) From necessity to opportunity - corporate environmental review. Philips Electronics NV (CEEO), Eindhoven

Philips Electronics N.V. (2000) http://www.news.philips.com/archief/199805290.html., 25th April 2000

Sweatman A, Gertsakis J (1996). Eco-kettle: keep the kettle boiling. Co-Design

Chapter 17

Co-designing Against Crime

R. Erol , R. Cooper, M. Press and M. Thomas

17.1 Introduction

Crime and crime reduction currently receive little attention in both design practice and education. However, informed design can be used effectively as a tool for reducing crime through designing in crime-reduction measures during initial stages of the development process ,as described by Ekblom (1995; 1997; 1999a&b) and Ekblom and Tilley (1998). This paper focuses on the current extent of inclusion of crime reduction measures within design practice.

Design can be used as a tool to aid crime prevention by incorporating features into potential targets which make criminal activity less appealing to the criminal. This can be done through introducing mechanisms of deterrence into a design (preventing crime through influencing the offenders' perceived risk of negative outcomes), such as reducing the potential reward from the crime or increasing the risk of detection. Alternatively, crimes can be discouraged through increasing objective difficulty of criminal activity (physically blocking the offence), for example the use of physical barriers such as locks, thus requiring more skill or equipment on the part of the offender.

Existing work addressing the application of design against crime has concentrated on environmental design (Clarke and Mayhew, 1980; Clarke, 1997 and Newman, 1972) and some areas of product design. Urban design can reduce the perceived risk and actual incidence of crime though improved lighting systems, building design and security. Travel and public transport is another area which has received considerable attention to improving crime reduction through design (Atkins, 1989). The automotive industry has had to address the problem of high incidence of car crime (Webb and Laycock, 1992), and has adopted crime reduction features into car design as standard, which are now used as a selling point for vehicles. Further reduction in retail crime, domestic burglary and IT related crime could be achieved with greater consideration to these issues within the design process.

Crime prevention features are often considered to be a retrospective add-on extra, and generally implemented once a crime has been committed rather than as an initial consideration in most product development. Taking account of crime resistant features in design requires extending the designers' thinking from normal product use to product misappropriation, mistreatment and misuse. Considering how a product, system or service could be misused in the hands of a criminal, and what aspects of design should be addressed to reduce the possibility of criminal misuse should be an integral part of the design process. Neglecting to do this results in the failure of designers to anticipate the vulnerability of their product to crime.

Reports produced by the Home Office looking at various aspects of crime reduction have pointed out the significant impact design could have in reducing crime. For example, badly designed features of the built environment were seen to contribute to causes of vandalism (Baker and Bridgeman, 1994). The suggestion was made that manufacturers should receive feedback on the effectiveness of current designs in order to be able to meet the required specifications for future alternative designs. Manufacturers appear to be reluctant to incorporate crime reduction features into their products if they perceive it to have no direct benefit for themselves (Kock *et al.*, 1996), so a possible compromise could be for manufacturers to design for security upgrades by the consumer (Clarke, 1999). Whichever direction is taken, the resulting action, whether or not designers drive it, needs the co-operation of all the stakeholders in the design process. Co-designing needs to occur, with the client, customer, manufacturer and crime prevention experts contributing to collaboration throughout the design process.

17.2 Design Against Crime Programme

This paper results from research undertaken for the Design Against Crime programme, initiated by the Home Office and DTI through the Design Council, as part of the governments Crime Reduction Programme. The aim of Design Against Crime is to influence those involved in the development of new products to design in crime resistant features before products are launched on the market. The objectives of the research are twofold:

1. to increase the crime resistant aspects of products through raising awareness and practical knowledge amongst designers;
2. to contribute to effective incentives of all those involved in product development to take crime reduction into account.

Initial work concentrated on determining the current levels of awareness of the role of design in reducing crime amongst all those involved in the design process. The specific 'goal' for designers is to incorporate good quality design against crime, in ways which makes it likely that the products will be appropriately produced, widely purchased and effectively used. They should do this by anticipating, or taking account of prior experience of the susceptibility of the product as a target of crime or a tool for the commission of crime. Part of the research programme involved identifying design disciplines where crime reduction is already considered in design, and undertaking case study research to illustrate how the designer collaborates with others to design against crime.

17.3 Methodology

Semi-structured qualitative interviews were conducted amongst key stakeholders within design and related industries, to ascertain the current awareness of crime reduction amongst designers. People interviewed included key figures in all the design industry bodies, designers from a wide range of design disciplines, design educators, including respondents to a postal survey of design course leaders, and manufacturers. Following analysis of the interviews, case study exemplars of current inclusion of design against crime were identified and relevant design issues highlighted.

17.4 Crime Reduction Awareness

In general, people do not concern themselves with crime reduction until they have had personal experience of being the victim of crime. Crime itself provides a certain degree of fascination, and the media frequently cover programmes and articles on crime related subjects, although these tend to concentrate on the commission of crime rather than on crime reduction.

The responsibility of improving crime prevention awareness amongst the general public in each police district rests with the Crime Prevention Officers (CPO) whose main role is to raise awareness amongst the public of how to reduce crime. This is often targeted at specific groups, for example small businesses, teenagers, and the elderly, (Crime Prevention College Digest, 1998a; 1998b) with schemes such as Neighbourhood Watch being effective in promoting crime prevention awareness amongst the general public.

The effect of design in reducing crime was clearly apparent to the CPOs interviewed given their experience in practical measures to reduce crime. Within pubs and clubs, design changes have been implemented to prevent drug taking, which include producing towel holders and windowsills in toilets with rough or sloping surfaces to deter cocaine use, and using blue light bulbs in toilet cubicles which make it impossible to locate veins for drug injection. These design features were reactive to problems identified in particular areas, with one CPO commenting that, 'no thought is given to crime reduction in any sort of design.'

Schemes have been set up to increase awareness amongst specific groups of professionals. The Secured By Design initiative promotes design of the environment to reduce opportunities for crime affecting new and refurbished homes, commercial properties, and car parks. It also encourages manufacturers of security products to design to meet the minimum police recommended specifications.

Certain industry bodies provide information on crime specific to that industry. For example, the British Retail Consortium (BRC) produces a Retail Crime Survey annually, which reports on trends in retail crime and the associated costs, generating numerous newspaper and trade press reports (Campbell, 1998). The BRC is proactive in promoting Retail Crime Partnerships across the country, which involves collaboration between local authorities, major retailers and the police in reducing crime in the city centre environment.

The majority of initiatives promote crime reduction awareness within businesses themselves, to reduce the possibility of becoming a target of crime. For manufacturers, little is done here to promote the idea of designing crime reduction measures into their own products. The focus is on using security products rather than producing secure products.

17.5 Product Design

Product design as a discipline encompasses a vast array of issues, such as, for example: aesthetics, functionality, user needs, environmental issues, market requirements and possibilities, technical issues, ergonomic concerns, volume and cost of production, and production/ manufacturing capabilities. With all these issues to consider, crime issues are rarely at the forefront of the designer's mind during product development.

Often manufacturers do not address the issue of product security, as it does not directly affect them, since crime occurs further down the supply chain after they have sold their products. Clarke (1999) suggests that 'add on' products can be created to make original products more secure. The consumer is therefore involved in two stages, buying the original products and buying the additional product. Designing in security features in the initial design removes the need for this, and is an issue which product designers need to be aware of, along with the possibility of products being misused for criminal activity.

17.5.1 Case Study: The Adshel I+

Adshel is a leading world brand in street furniture and operates over 3,000 street furniture agreements with municipalities in 20 countries. Having an awareness of the potential of both malicious and accidental damage occurring to their products, and having a good understanding of material and manufacturing technology available, the company can actively manufacture against crime.

The Adshel I+ information terminal is an electronic service that provides free, up-to-date, interactive information on towns and cities, in a visual and attractive format, with touch screen technology and attractive graphics, combining editorial information with video, music, voice-over and maps. Whereas traditional Adshel products were prone to vandalism, the I+ would also be vulnerable to theft due to it housing a PC and other items of electronic equipment.

The design consultancy PSD Associates were appointed to design the I+ so as to be suitable for external use, needing to be both weather and vandal resistant. The brief for Adshel I+ was written jointly by PSD and Adshel, the essence of which was to extend and market an existing network of information terminals that Adshel had already established in hotel foyers.

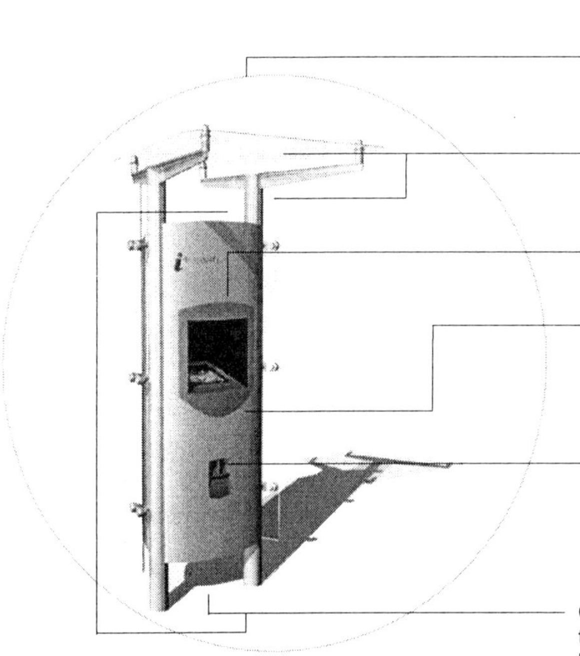

Figure 17.1. Focus on specific crime resistant aspects of the Adshel I+

The designers worked to get the internal arrangement of all the equipment to deliver maximum efficiency, whilst minimising the potential of product misuse. In addition to this, different orientations of the body were considered: whether it might be obstructive on the pavement through being sited parallel or perpendicular to the curb; how it might be positioned in relation to a bus shelter; and the effect of possible queue formation if people wanted to browse the internet for long periods.

The decision was taken to develop the unit as a self-standing product rather than as an extension to bus shelters due to vandalism and product life cycle considerations. The design team focused on the individual areas of the product, taking crime reduction into account in a number of areas. This included strengthening the area around the monitor, removing spaces which could be used to conceal bombs or allow the build up of litter, and standing the unit on two legs, giving clear visibility around the product and allowing for difficult site conditions such as uneven ground. The door was positioned on the back, with the lock interface designed inside the shell to prevent it being prised open. The opening for dispensing the tickets, maps and receipts was designed to prevent it from becoming blocked and to discourage users trying to access the interior of the machine. Figure

17.1 details the crime resistant features that were considered during the development of the I+.

It was important for the product to convey a message of strength and security to reduce vandalism attempts. According to Barry Jenkins, Design Director at PSD, during a research interview, 'Much of the psychology of strength is that it looks like it's strong ... normal people with a slight criminal tendency wouldn't want to try and vandalise it because it is too much hassle. So by not making this flimsy in its appearance or fine in its detailing it does look more resistant.' Product location was also an issue in achieving maximum product use with minimum risk.

A prototype was built to test product use and suitability of materials, and to monitor vandalism. It was found not to be subject to criminal damage. Following detailed design approval Adshel went into manufacturing. Adshel and their customers will continue to monitor crime resistance throughout the product life.

17.6 Interior Design

Crime reduction in environmental design is an area that has received considerable attention over the last 20 years. This covers urban design and city planning, or design of the 'built environment', building design and architecture, and interior design. These all play a role in situational crime prevention (Clarke, 1997).

Interior design has received less attention than that of the wider built environment. The Portman Group report (1993), entitled 'Keeping the Peace' suggests various design features which should be taken into account in the design of licenced premises. For example, lighting and colour can reduce levels of frustration; layout can be designed specifically to minimise potential for jostling, and to allow adequate monitoring and surveillance. The design of retail outlets is another related area that could offer improved crime reduction. Retail design is generally based on increasing the number of customers to a store, and increasing the volume of customer purchase, rather than concentrating on reducing shrinkage. There is a definite conflict between open, attractiveness of retail environment and requirements of security, as explained at the Retail Innovations Conference, reported in Design Week (1997): 'although designers take security into consideration, modern design and merchandising practice could hamper efforts to stop crime, as making products more accessible to shoppers generally increases sales.'

Retail design played a significant part in the work of the interior designers interviewed, with the retail industry as a whole attracting a large amount of interest from various sectors within the design industry. Design Week publishes an annual supplement about the retail industry detailing many diverse factors about retail design (Dowdy, 1999). However, crime reduction is never mentioned as a factor within this. The following case study illustrates how the interior designer, as a member of a development team, contributed to effective co-designing against crime.

17.6.1 Case Study: Blazer

The Bluewater shopping centre near Dartford opened in March 1999. It is the largest retail and leisure complex in Europe housing over 320 shops and restaurants, based on an 'integrated, innovative design policy' (Cole, 1999). The Moss Bros. chain of menswear stores, opened three outlets within Bluewater. The interior design for its Blazer store was assigned to the Tugman Partnership, a small multi disciplinary architecture and design consultancy whose clients include the fashion retailers Kookai, Jigsaw and Cerruti.

The design brief for the new Blazer store was set by the Visual Display Department at Moss Bros., after discussion with the directors. This required the designers to create a design concept appropriate to the overall image of Blazer. The brief referred to security features such as CCTV and a tagging system which were to be installed.

The initial design concepts were presented to the client at an on-site meeting. It was at this stage that Moss Bros security department became involved. Paul Bentley, Head of Security at Moss Bros accepted that at the initial concept stage 'the designers want something pretty, [and] the security side is often not at the forefront of design.' The main concerns were about the CCTV system, the position of the monitors, and the tagging system to be used, resulting in 'detailed discussions' between themselves and the designers.

Whilst a compromise was reached regarding the positioning of the monitors, the Security Department came to an agreement with the designers about the tagging system to be installed. All Moss Bros. stores had previously adopted a tagging system that required the use of large pedestal detectors at the entrance to the store. As this was a narrow retail unit, it was important to the design to have as wide an entrance as possible, allowing uninhibited access to the store. The extra width meant there was a need for multiple detectors across the entrance. The designers argued that this was not appropriate in this case, as the 'repulsively ugly' pedestals were not in keeping with the image of the store. They suggested an alternative system that could be incorporated easily into the aesthetic design. Whilst the designers and visual display department wanted to adopt a system which was as unobtrusive as possible, the security department were mindful of a visible deterrent, as an effective way of preventing shoplifters entering the store, and thus avoiding the possibility of confrontation with staff. After investigating the system further, Moss Bros. agreed, and installed the new system in all three outlets in Bluewater. Figure 17.2 shows the completed Blazer store design.

Figure 17.2. Blazer at Bluewater shopping centre

Since the completion of Blazer at Bluewater, Moss Bros. have continued to work with the Tugman Partnership. Having found that 'the designers were initially not conversant with the retail crime aspects,' they embarked on a 'steep learning curve' and have worked successfully together since then, with new projects including many security features that needed to be amended in the Bluewater project.

17.7 Issues

The case studies described in this paper are representative of areas within the design industry where crime reduction issues have been considered as an integral part of the design process. A number of issues have arisen from the research relating to co-designing against crime:

- the designer must relate crime prevention to the total product environment;
- a well-trained designer should be incorporating crime issues as basic common sense;
- products must be designed as 'fit for purpose'. To fulfil commercial and manufacturing briefs, there will always be a trade-off between different design considerations;
- the designer can act as an integrator and visualiser, integrating stakeholder needs and requirements;

- designers currently build up awareness in crime reduction through experience – this needs to be addressed earlier through further training;
- designers can be the catalysts in informing manufacturers about the types of issues influencing product design, including crime and crime reduction;
- the designer's role is often in finding a creative solution to the potential conflict between the aesthetic needs of the original design brief, and the requirements of security and crime prevention. This is best done when crime prevention requirements are considered simultaneously with others, and not as a subsequent add-on;
- understanding the security requirements at the start of the process allows the design and development process to progress much more quickly;
- few incentives exist to incorporate more crime reduction spending.

There is a role for the designer to act as an integrator through the co-designing against crime process: however, this must be explicit and considered at the front end of the design and development process. All stakeholders must collaborate to address design against crime.

17.8 Acknowledgements

This paper presents results from research conducted as part of a project for the Design Council, funded through the Government's Crime Reduction Programme on behalf of the DTI and Home Office.

17.9 References

Atkins ST (1989) Critical paths: Designing for secure travel. Design Council, Guildford

Baker M, Bridgeman C (1994) Preventing vandalism: what works? (Crime Detection and Prevention Series Paper 56) Home Office Police Research Group, London

Campbell D (1998) Retail staff suffer more violence but arrests fall. The Guardian 19 Feb 1998: 21

Clarke R, Mayhew P (1980) Designing out crime. Home Office Research Unit Publications, Oxford

Clarke RV (ed) (1997) Situational crime prevention - successful case studies. Harrow and Heston, Albany, NY

Clarke RV (1999) Hot products: understanding, anticipating and reducing demand for stolen goods. (Police Research Series Paper 112) Home Office Policing and Reducing Crime Unit, London

Cole C (1999) The Big Blue. Design Week 14: 22-23

Crime Prevention College Digest (1998a) Business Crime CD ROM. Crime Prevention College Digest April 1998: 1

Crime Prevention College Digest (1998b) Teenage personal safety. Crime Prevention College Digest April 1998: 3

Design Week (1997) Thieves say design helps them Design Week 18 Apr 1997: 4

Dowdy C The Big Picture - Retail. Design Week (Supplement), June 1999

Ekblom P (1995) Less crime by design. Annals, AAPSS 539: 114-129

Ekblom P (1997) Gearing up against crime: a dynamic framework to help designers keep up with the adaptive criminal in a changing world. International Journal of Risk, Security and Crime Prevention 2(4): 249-265

Ekblom P (1999a) Can we make crime prevention adaptive by learning from other evolutionary struggles? Studies on Crime and Crime Prevention 8(1): 27-51

Ekblom P (1999b) The conjunction of criminal opportunity - a tool for clear, joined up thinking about community safety and crime reduction. Home Office Research, Development and Statistics Directorate, London

Ekblom P, Tilley N (1998) What works database for community safety/crime reduction practitioners - towards a specification for an ideal template. Policing and Reducing Crime Unit, Home Office RDS & Nottingham Trent University

Kock E, Kemp T, Rix B (1996) Disrupting the distribution of stolen electrical goods. (Crime Detection and Prevention Series Paper 69) Home Office Police Research Group, London

Newman O (1973) Defensible Space: people and design in the violent city. Architectural Press, Guildford

The Portman Group (1993) Keeping the peace: a guide to the prevention of alcohol related disorder The Portman Group

Webb B, Laycock G (1992) Tackling car crime: the nature and extent of the problem. (Crime Prevention Unit Paper 32) Home Office, London

Chapter 18

The Designer as Innovator

R. Cooper, M. Bruce, L. Tierney, L. Daly, R. Harun and L. Svengren

18.1 Introduction

A new millennium meets the challenges of rapidly changing technology and the demands of the global economy. Design and technology are catalysts for innovation and this is becoming increasingly important. As such this will place increasing demands on the role of the designer as an integrator between design and technology. In order to establish the context we must first clarify our understanding of the issues surrounding the 'Designer as Integrator'. The first part of this paper gives brief definitions of Innovation, Design and Technology. It then goes on to discuss the role of the designer as integrator, within single organisations and the entire supply chain.

18.2 Innovation

Innovation has been defined as, 'The introduction of novelties, the alteration of what is established by the introduction of new elements or form. A change made in the nature or fashion of anything, something newly introduced, a novel practice or method, *etc.*' (The New Shorter Oxford English Dictionary, 1993).

Innovation essentially refers to change and this can be applied to products and services offered by an organisation and the ways in which these are created and delivered (Tidd *et al.*, 1997). A new car design and a new car insurance package are examples of product innovation. A change in the manufacturing methods and equipment used to make the car, and the new office procedures and sequencing used to develop the insurance package are examples of process innovation. Incremental innovation often involves the re-styling of a car model, whereas radical innovations can transform the way we think and use a product or service.

Increased competition in the market place means that companies can no longer rely on the cheapest price point or the shortest lead times to sell products and services. Non-price factors, namely value, quality and design, must be utilised instead to provide the consumer with fashionable and effective products. Rodney Fitch came to the conclusion that, 'Only one company can be the cheapest - all the others must use design' (Fitch, 1989).

18.3 Design

Design is many different things to many different people and it can be viewed as a discrete activity, as a total process, or in terms of its tangible outcome (Cooper and Press, 1995).

There have been numerous definitions of design. Holt (1990) defines it as follows: 'The process of seeking to optimise consumer satisfaction and company profitability through the creative use of major design elements (performance, quality, durability, appearance and cost) in connection with products, environments, information and corporate identity'. For the purposes of this paper this represents a fairly comprehensive definition that clarifies design as a creative activity utilising market and company information to produce a two or three-dimensional product which satisfies the consumer and aids company profitability. This illustrates the holistic nature of the process, together with the that designing, as an activity, is part of a multifunctional design process.

18.4 Technology

'Ongoing technological development by thousands of organisations located around the globe creates a broad technological force that contributes to rapid change in other environmental forces as well as in markets for new products' (Thomas, 1993). The global community as a whole is geared towards technological advance. The aims for this advance can be generalised as producing cheaper products, using materials and production methods more effectively, and reducing assembly and labour costs. The current climate for technology has been described as 'The second industrial revolution –the digital age – offers a rare opportunity to rethink the tools of mankind (Buchanan *et al.*, 1995). Buchanan *et al.* discuss how there is an opportunity, as rapid change is taking place to create a new structure for the use of technology. When considering designers and their interaction with technology there is a shift towards making designers more influential in the use of Technology. Technology encompasses four core areas: knowledge, materials, processes and techniques.

Knowledge links pieces of information, which allows for understanding. It is not until the knowledge is encountered that the information can be used effectively. 'As this century comes to an end, the defining characteristic of the current wave of technology is the role of information...The newer technologies have widened the potential for economic well-being. Government should seek to foster that potential' (Greenspan, 2000). Greenspan refers to the need to treat technology as information. There is a growing necessity to widen the audience for information on technology.

Materials are the physical matter, which allow an actual product to be realised, without which the design, knowledge, processes and techniques are merely notional. 'Materials account for 55% of the cost of an automobile and a larger portion in other manufactured products' (Marino, 1997).

'*A process* is a specification of how to proceed in order to achieve some practical end. In this sense they resemble algorithms, and like algorithms they are manifested in some textual or pictorial language, that is, they are abstract; furthermore, processes have objective reality in that they can be communicated between people, examined, criticised, and modified. In all these respects, processes are abstract artefacts' (Dasgupta, 1996). Processes are selected due to their known attributes and performances. Edwards and Endean discuss the fact that there is no one individual process which suits all applications. Each application has specific requirements and it is the combination of these requirements which will inform the selection of the best process (Edwards and Endean, 1990).

Techniques are methods and skills in which manufacturing and production are completed using materials and processes and applying them in a series of stages of recommended best practice. This is also a phase of evaluation of the chosen material and process and considering the alternatives. A technique in manufacturing could be a number of considerations, to name some examples: a series of stages to treat a material before processing, an adhesive, which is compatible with a specific material, an application of a wax to a mould, which enhances the finish on the final product.

18.5 Designers as Integrators

Designers, technology specialists, marketers and managers all have different roles, objectives and responsibilities within the infrastructure of an organisation. These activities often interact in cases where problems occur. The interactions between the activities are often unstructured and occur sporadically. Integration between activities if implemented effectively can encourage communication, the pooling of resources, the extraction of information from new sources and the construction of new ideas. Designers are often the initiators of the interaction between activities, as they are involved in creating new concepts and therefore become instigators of change. Developments in technology can create new design directions for new product development. Designers and technology specialists will both benefit from the collaboration by expanding their knowledge of each other's activities. 'In fact, it's precisely when there's a fundamental upheaval in technology that design proves itself most valuable. New technologies enable new behaviours and bring new market opportunities that must be discovered and understood' (Campbell and Ryan, 1999). Using new technology as a tool to generate and inspire innovation is a technique commonly used. As technology is an evolving field it is a constant source of change and will always therefore have an effect on design and innovation. It is often the role of the designer to link organisational activities together. Designers, working in teams with other specialists (for example, marketing or manufacturing), can create more effective work practices and reduce time to market. If designers are going to be effective integrators they need to be flexible; Fujimoto discusses designers integrating within teams. 'It also requires a balance among structure and flexibility, focus, and freedom, leadership, and mutual respect' (Fujimoto, 1991). Harnessing and integrating design and technology together requires sensitivity, as end users are often afraid of new technologies, therefore design must package the technology into an easily acceptable form. 'A lot of our innovation is aimed at making sure we can lever current learning and experience' (Campbell and Ryan, 1999). Designers as visualisers and creators are able to harness rapidly changing technology and market knowledge into ideas to be realised through their design process.

18.6 Information Integration

Integrating the activities of an organisation can reduce the research time within New Product Development and assist innovation in offering new perspectives on design. The main benefit of the designer becoming an integrator and collaborator is the increased knowledge and information exchange between technology specialists, marketers, managers and designers. Within New Product Development, knowledge and information on technology and marketing can provide missing links for a designer in creating a concept. However knowledge and information is often dispersed not only throughout an organisation but also throughout the entire supply chain. There is a need for better access to and integration of such information, to enable faster innovation.

18.7 Interaction Between Designers and Technology Specialists

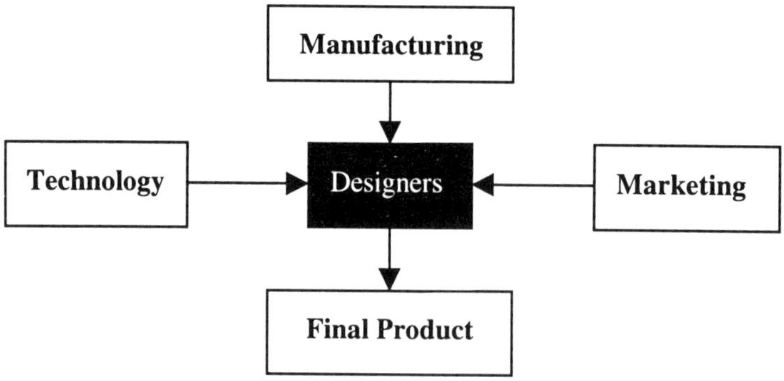

Figure 18.1. Model of current interaction within the organisation

Ford (1995 writes that: 'Now is a good time for designers to become involved in shaping technology – we need to make technology more responsive to human values'. The model in Figure 18.1 shows how the designer are at the centre of the interaction between marketing, management, manufacturing, production and technology specialists. Designers need to embrace this role and use it to its full potential. Support and integration will provide a framework for effective innovation.

Jones describes how the British have become renowned for producing inventions which they often fail to carry through to the production stage. Engineers at EMI in London invented the CAT scanner in 1972, for example. However, organisations such as Toshiba in Japan harnessed the technology, developed it and incorporated the original invention into innovative scanning and imaging equipment now found in most major hospitals (Jones, 1997). The CAT Scanner is a real example of an opportunity, which was not exploited to its full potential, although the technology was available it was not carried through into a successful commercial application.

Designers and technology specialists would benefit from adapting and integrating their fields of interest. For instance, designers need to be aware that technological advances can develop unpredictably. 'The essence of technological progress is its unpredictability' (Mokyr, 1990). Designers must learn how to identify and exploit technology to work for them. 'Technology will work better when it solves real problems instead of manufacturing them retroactively' (Ford, 1995). Designers therefore need to be in close contact with the sources of technological development - the biochemists, the genetic engineers, the material scientists.

18.8 Collaboration

Generally, designers interact with clients and with marketing, manufacturing, production, and management departments. Collaboration enhances these relationships, improves communication and provides a higher level of satisfaction for the client. 'The gap between designers and engineers is narrowing, due to collaboration...The key is to integrate design and engineering in the most effective way possible. Whether that means that the engineer and the industrial designer learns to use a CAD tool, or that traditional CIAD tools are designed for the artist are becoming more closely tied to engineering the industry is moving in the right direction. And that can only mean one thing innovation is alive and well in industrial design today' (Rowell, 1997). The future for collaboration lies not only with interaction in terms of communication and discussion but in terms of creating tools for collaboration that are used across the organisation. Much has been written about the value of collaboration between marketing and design (Bruce and Cooper, 1997). It is known that closer collaboration between marketing both as a functional operation, and as a link to the end user is essential to successful new-product development. We can now see therefore that such integration must be orchestrated within organisations.

Figure 18.2, illustrates a proposed model of collaboration within the New Product Development Process could take place. There are three main phases of collaboration: brainstorming, incubation and evaluation. Technology, design and marketing representatives should all be integrated in these phases. Management needs to oversee these activities, and create a structure and ensure understanding of the aims and objectives of the project.

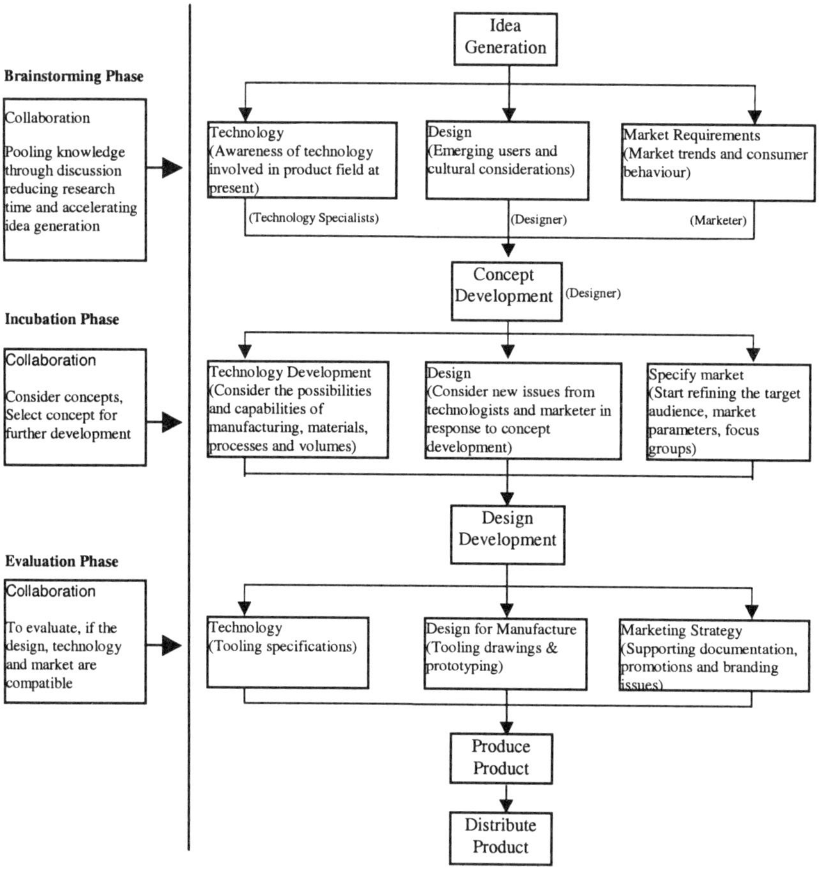

Figure 18.2. Proposed model of collaboration within the New Product Development process

18.9 Drivers

'In the future, creativity in industry, founded on the interaction between design and technology, will be the driver of national and international growth' (Lord Stone, 1999).

It is creativity that will drive the interaction and communication between design and technology. Creativity involves initiative and if the designer does not initiate the connection between design and technology it will not occur. Often designers use technology when necessitated by need, for example, when the market requires a new material for a product with properties not found in the existing materials used in production. However technology needs should preferably be considered at an early stage in the design process alongside concept development.

'Firms that lead in product innovation strive for marketing knowledge which is as extensive as their understanding of the technology itself' (Johne and Axel, 1987). The technology itself is not enough to substantiate a successful new product; the market must also be proved before the product is produced. The integration of technology is most successful when the products are designed in collaboration with the technology rather than being led by it. With technology being integrated into the design process the consumer/user needs are at the forefront of the design process. Technology-led design produces technology-driven products, which are adequate, if the main purpose of a product is functional. If other factors are important, such as, for example, aesthetics or markets, technology-led design is not advisable As Ulrich and Eppinger state: 'The primary characteristic of a technology-driven product is that its core benefit is based on its technology, or its ability to accomplish a specific technical task. While a product may still have important aesthetic or ergonomic requirements, consumers will most likely purchase the product primarily for its technical performance' (1995).

'In most British and American businesses product development is still predominately either technology-led or marketing-led'(Johne and Axel, 1987), this statement depicted the New Product Development climate of the late eighties. Since then, we have seen progress in the management of New Product Development in the form of integrating these two functions. Yet frequently integration is pursued by a product manager, leaving the creative skills of the designer outside the triangle.

The potential for introducing new technology needs to be nurtured for radical innovation to take place. There is still a lack of structure and systematic approach to managing New Product Development in many companies. Organisations would benefit from mechanisms to encourage New Product Development. The integration of *new* technology would seem to be an obvious role for designers.

18.10 The Designer Within the Supply Chain

Today organisations are rarely self-contained in terms of creating and delivering a product; most rely on suppliers and customers as part of the total value chain. Designers need to harness not only their own organisational knowledge and technology but also that which exists in the supply chain. The supply chain provides links between information, knowledge and requirements. How does the designer gains access to this information? The way in which the information is managed can be fundamental to the role of the designer and innovation success.

New management philosophies must be introduced to deal with these new demands. One such philosophy to be applied is 'supply chain integration' and 'partnering'. There has always been a plethora of collaborative strategies (such as co-makership, co-design relationships, strategic alliance, network, hybrid organisation, virtual organisation, concurrent engineering, parallel product development and others) which facilitate co-development.

Formal partnerships vary from full joint alliances to collaborative R&D. Relationship Management Theory proposes four types of partnerships.

Supplier Partnerships	**Lateral Partnerships**	**Internal Partnerships**	**Buyer Partnerships**
Goods Suppliers Services Suppliers	Competitors Non-profit Organisations	Business Units Employees Functional Departments	Intermediate Customers Ultimate Customers

Figure 18.3. Partnerships

Indeed whether the relationships are formal or informal, didactic or multiple, the information and knowledge held throughout the entire supply chain is extremely beneficial to the innovation and product development process.

Partnerships are considered as a means of harnessing more effectively the knowledge and skills in the supply chain in order to compete in a global marketplace. The DTI (Learning from Japan, DTI/ Anderson Consulting, No. 19957) proposes a three-stage strategic development programme consisting of survival, bootstrapping and expansion. The survival stage is based on the prerequisite that the company must keep hold of its present position by meeting customer basic needs. Bootstrapping involves gaining knowledge and experience from customers and partners and utilising this to improve internal performance. The expansion stage represents the application of new skills into product development and geographic markets. All stages demand the development of expertise in accessing information and data for New Product Development. Establishing a good procedure for designers to access knowledge, will enable the capture of information throughout the entire supply chain from third- and fourth-tier suppliers through to end-users and is, therefore, vitally important. Also, more integrated and better-managed supplier interfaces through relationship management and partnering will facilitate more accurate and easier access for the designer to knowledge and information from the entire supply chain.

18.11 The Changing Role of the Designer

As we have discussed, design is often not considered by managers as integral, and decisions involving New Product Development have a greater emphasis on marketing and technical issues than design. 'Design plays a significant role in 67% of large firms in the UK, but in only 47% of smaller firms, and yet design as a management tool is regarded by only 24% of companies is integral' (Summers, 2000).

However in a climate of increasingly complex products, the holistic integrated process is important and will change the role of the designer. 'With industrial design and its synthesis of elements belonging to the engineer, the artist and the scientist, a special language will at the same time evolve to enable the three to co-operate, and new ways will be found to explain their purposes; perhaps also a new higher creative activity will come into being; designing new material and spiritual values, a kind of synthesis of all tasks and methods involved in pursuing all the aims of science, technology and art' (Soloviev, 1968). Technology must be considered a source of inspiration for designers.

'Good design is about looking at everyday things with new eyes and working out how they can be made better. It's about challenging existing technology' (Dyson, 2000). 'The first major impact of technology on the development process is its ability to accelerate information flows in the design process...By providing information early we can increase the speed of the design process and thereby reduce its expense' (Smith and Reinersten, 1997).

Designers need to be open-minded to seek out new directions. Researching areas of technology for a design application may require working in a scientific, methodical way. Designers must be able to rationalise the most important issues; they must be able to absorb the new technology and consider what would be achievable within their company culture and what would be acceptable in terms of the market.

Ellis describes the changing role of the designer as 'The role of the designer is evolving toward being an effective liaison between marketing and engineering, an integrator, a product creator, and a visionary' (Ellis, 1994). Effective integration requires structure and also understanding from the individuals involved of the aims and objectives of the integration activity.

The interaction required is not just within the organisation, therefore the designer will need to develop means of accessing and pulling knowledge and technology from the supply chain. Therefore the skills required would involve understanding how to map supply relationships, identifying sources of technology and tools to access knowledge and information.

One such means of accessing the supply chain knowledge and managing the information which comes from it, is through intranet technology. Cooper *et al.* (1999) cite the development of legacy archives and information warehouses as a means by which distributed organisations can share knowledge throughout the New Product Development lifecycle. This has been developed for the construction industry. A virtual product team is established and collaboration throughout the development of a building is maintained through such systems as Data Stream. This not only provides the collaborators throughout the supply chain to understand the whole product development but also to contribute their knowledge through the Intranet to the designer. For instance, a supplier may be aware of a new technique or material which would enhance an aspect of the design, adding value and innovation to the whole product. The legacy archive enables continued learning from one project to another and the maintenance and transfer of knowledge.

In future there will be a need for designers to use such systems to draw knowledge and technology into their design process. The authors are currently working on a funded research project to develop such a tool and methodology for companies and their supply chains

18.12 Conclusion

New Product Development (NPD) relies on innovation, creativity and design to produce new well-designed products. There is an opportunity for design to be a driving force in harnessing technology for innovation. This means the role of the designer must change from being mainly involved in creative activities within New Product Development to becoming an integrator and collaborator in the fields of technology and marketing. Organisations need to be structured so as to encourage designers to be involved in activities in other areas of the organisation. It is argued that designers could have wider sources of knowledge and information by integrating with other areas of the organisation involved in marketing and technology. If the organisation is integrated in this way designers, marketers and technology specialists will all have a greater understanding of each other's activities, a fact which will assist them to support one another.

18.13 References

Bruce M, Cooper R, (1997) Marketing and Design Management. International Thomson Business Press, UK

Buchanan R, Ford S, Humphreys A, Thomas W, Swinehart R, McClusk, J, Ballay J (1995) Seeking Tools for Conviviality: Trends in Design and Technology. Design Management Journal

Campbell A, Ryan D (1999) Design Management in a changing world: Creating the Human Experience with New Technologies. The Design Management Journal, Summer 1999

Cooper R, Press M (1995) The Design Agenda: A Guide to Successful Design Management. John Wiley & Sons Inc, UK

Cooper, R, Aouad G, Kagioglou M (1999) IT in Construction, A driver or an Enabler. Journal of Logistics and Information Management 12 (1/2)

Coyne R (1997) Creativity as Commonplace. Design Studies 18 (2)

Dasgupta S (1996) Technology and Creativity. Oxford University Press, New York

Dyson J (2000) Facts, Figures and Quotable Quotes, Design in Britain 2000. Design Council

Ellis S (1994) Towards the Design Era, the Evolution of the Designer as Functional Interface with Marketing and Engineering. The Design Management Journal, Summer

Edwards L, Endean M (1990) Manufacturing with Materials. The Open University, London

Fitch R (1990) Fitch on retail design. Phaidon, Oxford

Ford S (1995) Designing Technology. Design Management Journal, Fall

Fujimoto T (1991) Product Integrity and the Role of Designer-as-Integrator. The Design Management Journal 2 (2)

Greenspan A (2000) Facts, Figures and Quotable. Quotes Design in Britain 2000. Design Council

Holt S (1999) New talent design annual: the international annual of design and communication Graphis. New York, NY

Johne A , Snelson P (1987) Follow the Leaders. The Engineer, 26th November

Jones T (1997) New product development: an introduction to a multifunctional process. Butterworth-Heinemann, Oxford

Mokyr J (1990) Lever in Riches. Oxford University Press, New York

Marino D (1997) Materials Data in Design and Manufacturing

Rowell A (1997) Tools of the Trade. Computer Graphics World 20 (3)

Smith P, Reinersten, D (1998) Developing Products in Half the Time. Van Rostrand, USA

Soloviev Y (1968) Quoted by d'Archy Hayman in his 'UNESCO Symposium on Technology and Artistic Creation in the Contemporary World'. Leonardo 11 (4). John Libbey and Company Limited, London

Lord Stone at Blackheath (1999) All our futures DFEE

Summers A (2000) Facts, Figures and Quotable Quotes. Design in Britain 2000. Design Council

Thomas R (1993) New Product Development- Managing and forecasting for strategic success. The Portable MBA Series 1993. John Wiley & Sons Inc., Canada

The New Shorter Oxford English Dictionary, 1993

Learning from Japan Anderson Consulting DTI No 19957

Tidd, Bessant, Pavitt (1997) Managing Innovation: Integrating Technological, Market and Organisational Change. John Wiley and Sons Ltd, Canada

Ulrich K, Eppinger S (1995) Product Design and Development. McGraw-Hill Book Company, Singapore

Chapter 19

Insights into Customer Delight

A.D. Burns and S. Evans

19.1 Introduction

Literature from various sources highlights the need to move beyond simply satisfying customers to delighting them (including Von Hippel *et al.*, 1993 and Ealey and Troyano-Bermudez, 1996). Many product sectors are characterised by large numbers of competing products with very similar specifications and features. Matching the best in class is no longer enough to capture new customers, and products that merely satisfy fail to encourage loyalty in existing customers (Schneider and Bowen, 1999; Ludvigsen, 1996). Differentiating your product, innovating and engaging the customer on an emotional level are now the keys to success (Ealy and Troyano-Bermudez, 1996). This paper describes research into the nature of customer delight during product evaluation, and forms part of an approach to improve product development at Nissan. Increasing the designer's understanding of the customer at the earliest stages of product development should facilitate the design of cars that stand out from the competition, that are wanted rather than needed by customers and that delight their users. The research described here forms part of a live vehicle development programme at Nissan Technology Centre Europe, and is carried out in collaboration with the Motor Industry Research Association (MIRA).

19.1.1 Market Research, QFD, and the Merely Satisfied Customer

Over the last two decades most major European and American vehicle manufacturers have adopted the working practices of their Japanese competitors in their quest for increased market share (see Clark *et al.*, 1991). The continual search for advantage within an almost stagnant market has led to a focus on the competitive dimensions of cost, quality, flexibility and time-to-market. However, adoption of lean philosophies and a reliance on economies of scale has left the industry in a state of gross over-production and with design departments fixated on cost-cutting and product reliability rather than product appeal (Ludvigsen, 1996). The result is airfields full of cars that nobody wants to buy and that customers can barely distinguish from one another. Within the typical concurrent engineering approach to product development, car designers tend to gain their understanding of customers through communication with a separate marketing or market research department. Customer information has to cross the boundary between the marketing department and the design department, this transfer usually taking place in the form of written summary reports, survey results, and target customer profiles (Macht, 1998). In short, product designers work with customer information that tends to be either negative (what is wrong with previous products), or translated and summarised, providing them with a somewhat diluted understanding of the customer. Even much touted customer focus processes such as QFD (Quality Function Deployment) cannot make up for this shortfall in the quality of customer information. This situation represents the last obstacle to the concurrent approach to product development and manifests itself in large numbers of expensive, late, design changes, products that fail to differentiate themselves from the competition, and customers who are satisfied but essentially fail to be engaged by the product (Rosenberg and Thompson, 1993).

19.1.2 'Go Forth and Delight your Customers!'

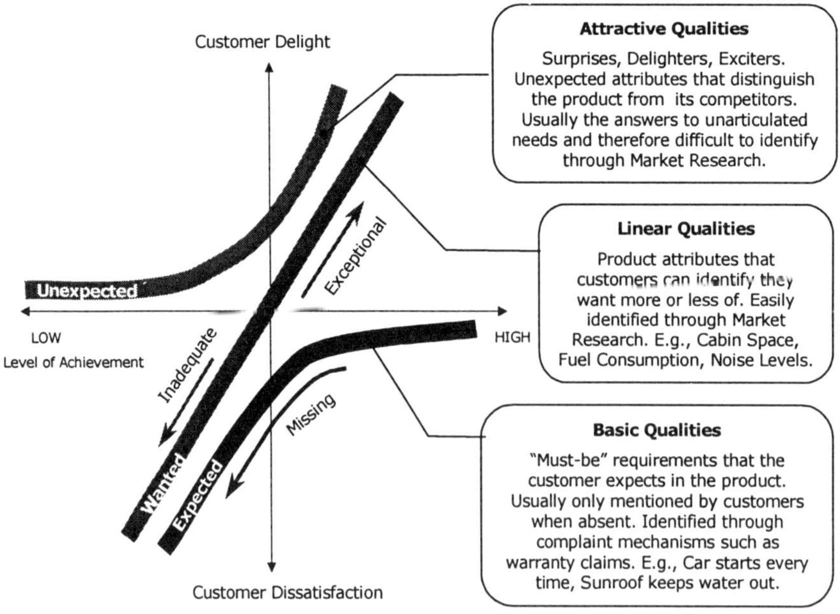

Figure 19.1. Kano's model of product quality

'Delight your customers' seems to be the current rallying cry of the business and consultancy literatures. Indeed, companies are being advised that the adoption of this goal is their only means of survival. Several definitions of delight are used, generally associated with exceeding customers' expectations and delivering the pleasant surprise that results when a product or service answers an unvoiced need. Various models are used to help companies evaluate their products in terms of customer needs, and these introduce the idea of delightful or differentiating product attributes. The most widely used of these is Noriaki Kano's Model of Product Quality (see Figure 19.1, and Kano, 1995). This suggests there are two ways to delight customers through product design - by providing exceptional levels of those qualities customers identify they want, and by identifying features that answer the unspoken customers needs.

19.1.3 Empathic Design

Empathic Design is a term originally coined by the American authors Dorothy Leonard and Jeffery Rayport (1997). They essentially advocate the use of observational research in the quest for the unarticulated customer information that can sow the seeds of innovation. Observing real world product use allows designers to uncover the information that customers cannot, or do not report:

- customer frustrations and confusions with the product;
- wasted time or time-consuming activities during product use;
- doing things wrong, misuse, abuse or unexpected use of the product;
- customer modifications of the product or use of aftermarket products;
- observation of fringe customers;
- the context of product use (including mood, environment and social setting).

The principles of Empathic Design are routinely used in the computer software industry, and several leading design consultancies tout the approach as their guiding philosophy (e.g., Segal and Fulton Suri, 1997). However, examples of the use of Empathic Design tend to occur in small design teams and often within freethinking design-centric or innovation-centric environments. Examples of the approach's use in large cost and quality focused design teams, like those in the automotive industry, are just that, isolated examples.

19.2 CUPID = Empathic Design Plus

Under the banner CUPID (Customer Understanding Processes In Design) this research builds on the concept of user observation, broadening the scope of Empathic Design to include any process that aids the product design team's in-depth understanding of the customer. It is hypothesised that by enabling designers to get inside the heads of their customers, and to feel a certain amount of ownership of the customer, the design of product attributes that delight and excite will be facilitated.

CUPID approaches the 'delighter' question from two directions. Firstly, due to a relative lack of research in the area (Keiningham *et al.*, 1999, but for exceptions see Oliver *et al.*, 1997), CUPID attempts to define delight from a consumer's rather than a consultant's perspective. Specifically, what is delight; how do products delight during purchase and ownership; and is there such a thing as a 'delighter' feature? Secondly the research evaluates methods of identifying 'delighters' within the product development process. This paper describes the former approach, whilst a summary of the latter can be found in Burns *et al.* (1999). CUPID has taken an exploratory approach to studying the nature of customer delight during product evaluation using a combination of observation, interview and self-report techniques.

19.2.1 The Murmur of the Customer

A hidden camera was used to observe customers evaluating a Nissan vehicle at a major UK Motorshow. Over 750 vehicle evaluations were recorded over a three-day period. As well as revealing many examples of what a delight reaction looks like (and examples of 'delighter' features), the technique provided more general insights into vehicle evaluation, including sequences of behaviour, important areas of the vehicle and customer comments.

19.2.2 Delighter Clinics

Four car clinics were conducted. In each, four subjects evaluated four cars statically and on test drive. Whilst evaluating the vehicles (a sports coupe, a family hatchback, a mini-MPV and a city car) each customer was accompanied by an interviewer and asked to concentrate on the things they liked about the cars, the output essentially forming sixteen in-depth interviews. Each vehicle attribute identified was scored on a 1 to 5 scale from Like to Delight, and customers explained their reactions in terms of eight factors: 1. Operation (the way the item works); 2. Function (the function the item performs); 3. Novelty (the item is new or different); 4. Surprise (the item is surprising in this vehicle); 5. Emotional Appeal (the item appeals on an emotional level, e.g., 'sexy', 'cosy', 'clever'); 6. Look; 7. Feel; and 8. Sound (the item appeals because either it looks, feels, or sounds good). The 64 vehicle evaluations during these Delighter Clinics produced 277 occurrences of subjects scoring vehicle attributes over 4 on the 1 to 5 Like-Delight scale. These were made up of 108 separate vehicle attributes, 22 of which were identified by 4 or more subjects and would seem to be reliable 'delighters' from a customer's perspective.

19.2.3 Summary Findings

Firstly, as expected in the face of overt research, customers find it difficult to identify aspects of vehicles that delight them and in particular find it hard to explain their reaction to attractive qualities. The subjects in our Delighter Clinics found it much easier to explain why they disliked something, than to describe why they liked something. This effect was reduced as interviews progressed and subjects became used to what they were being asked. Despite the concept of the 'delighter' feature appearing to be unnatural for some subjects, others (notably those with subjectively higher product involvement) found the concept meaningful, and footage from the Murmur of the Customer exercise revealed many overt and instantly recognisable delight responses to distinct vehicle features. These delight reactions have several defining features. These include positive facial expressions, exclamations or positive comments in reaction to a feature (but often generalised to the vehicle as a whole, such as 'Wow, look at that gearstick, SOLD!' or 'I've got to have one'), and prolonged or repeated attention to the feature. These findings provide support for the existence of product 'delighter' features. The Delighter Clinics were used to validate the use of the eight factors, Operation, Function, Novelty, Surprise, Emotional Appeal, Look, Feel and Sound for the description and explanation of delight reactions. Interviewers found it easy to code subjects' reactions in terms of these factors. The difficulties subjects encountered differentiating between Operation and Function, and Surprise and Novelty, and what distinguishing classified as Emotional Appeal, guided the writing of instructions for the Self-Report exercise. The only factor subjects felt could be included in future was 'Whole', referring to the experience of product appeal that cannot be split down into separate items or features. While the findings support the existence of 'delighters', but it is difficult to determine their effects. Some customer reactions and comments suggest the reactions are generalised to the overall evaluation of the product, whilst discussion after the Delighter Clinics revealed that a single 'enraging' feature was much more powerful than many delighting features.

19.2.4 Delighter Self-report Exercise

Twenty-nine volunteers, representing a variety of customer segments, were recruited to take part in a self-report version of the Delighter Clinic. Volunteers were provided with recording books and comprehensive instructions asking them to identify vehicle features and attributes that appealed to them. Volunteers were given a free ticket to a major UK Motorshow and encouraged to view as many vehicles as possible, scoring their reaction to features on a 1-5 Like-Delight scale. Volunteers were asked to explain why vehicle attributes appealed to them in their own words and in terms of the eight factors - Operation, Function, Novelty, Surprise, Emotional Appeal, Look, Feel and Sound. The volunteers reported 413 features and attributes that they considered to be 'delighters', rated 1 to 5 on the Like-Delight scale. 397 were given scores of 3 and above and were considered further. Figure 19.2 below shows the percentage of 'delighters' appealing to each of the eight factors. Respondents tended to explain the appeal of vehicle attributes according to combinations of these factors but some clear trends can be seen. Firstly, the majority of 'delighters' (up to 85% of those scoring 5) appeal, at least in some part, due to styling or the way they look. Not surprisingly, considering the importance of styling in cars and the motorshow situation, over 100 of the 'delighter' items identified by customers were either the vehicle's styling as a whole, or specific styling features (e.g., headlamps, roof lines, *etc*.). However, the results also suggest that the styling and look of everyday features can elevate them to the level of the 'delighter' (gearsticks, handbrakes and switches were all identified as 'delighters').

Secondly, Kano's suggestion that 'delighters' will be novel items is borne out by these findings, with roughly 63% of all 'delighters' appealing due to Novelty. Likewise, the assertion that delight is surprising pleasure is supported by the fact that 69% of the 'delighters' scoring 5 on the Like-Delight scale surprised subjects. These findings suggest that products can delight by surprisingly exceeding expectations

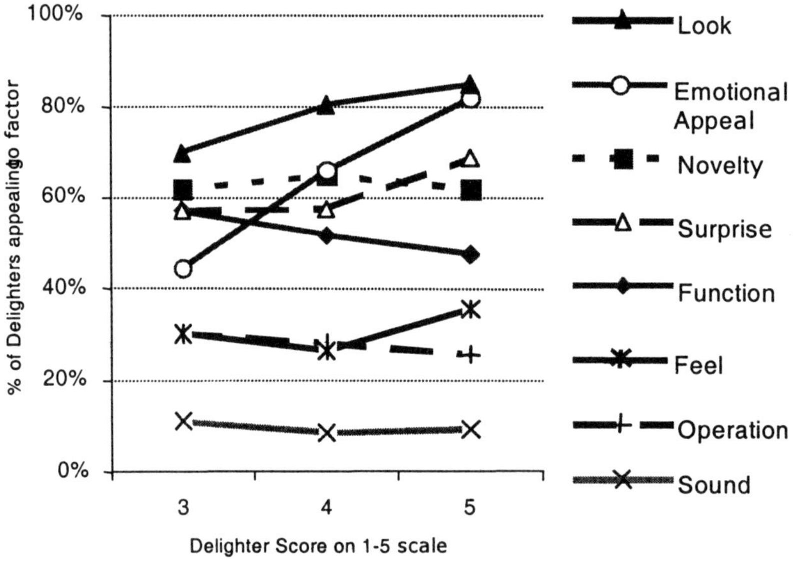

Figure 19.2. The percentage of 'delighters' appealing to eight factors

An interesting finding is that as the scores given to items increase, a smaller proportion of them appeal due to their function. The observed trend may be marginal but the fact remains that only 44% of the delighters given scores of 5 by subjects appealed because of the function they performed. This finding is somewhat contrary to the Kano Model that implies 'delighters' will be the functional answers to customers' latent needs, and lends only limited support to the automotive and other industries' 'more features' approach to product advantage.

The strongest trend observed in these data is that as the scores given to product attributes increase, more of them are appealing at an emotional level. In Figure 19.2 this effect can be seen clearly; only 44% of 'delighters' that scored 3 appealed at an emotional level, compared to 82% of those receiving a score of 5. Examples of this emotional appeal were gleaned from the respondents' qualitative reports. These ranged from comments like 'purposeful and stylish', 'looks very masculine and sexy', to 'it just says warm, clean, well made' and 'it makes me feel like a movie star'. In addition to the importance of the Look factor described above, the vehicle attributes that delighted respondents the most seem to be those that evoked emotional reactions in the form of feelings and cognitions. Many responses included 'the whole car' as a 'delighter' item. This finding was mirrored in the Delighter Clinics where some subjects felt that they could identify 'delighter' items but that their reaction in terms of satisfaction or delight was to the product as a whole. This finding is in line with the theories of the Gestalt School of psychology that suggest that objects, including products, are perceived as wholes or *gestalts* and that the *gestalt* is greater than the sum of its parts (see Baxter, 1995; Bruce and Green, 1992).

19.3 Conclusions

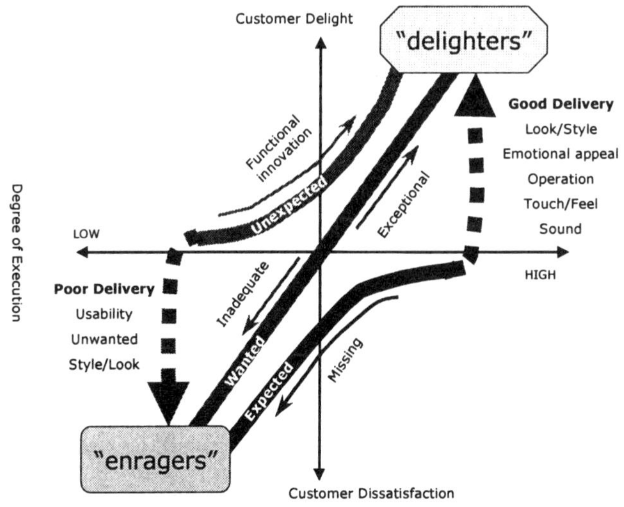

Figure 19.3. An improved Kano model of product quality highlighting the importance of *delivery* as well as *feature*

The findings described above seem to suggest that the customer's view of the 'delighter' is somewhat different to the industrialist's. As Kano's model suggests, there are indeed many 'delighters' that appeal because they have novel functions. However these functional innovations make up only a limited proportion of the 'delighters' identified by this research. Styling, feeling, delivery and emotional appeal were the important factors in the attraction of a large number of the items identified by respondents here.

The authors therefore propose a re-assessment of Kano's Model of Product Quality in light of these findings (see Figure 19.3). 'Delighters' should not be thought of simply as unexpected functional innovations that answer unvoiced customer needs. Instead of the focus on *features* that Kano's model tends to instil in its users, a consideration of the *delivery* of those and other features should be included. The results of this exploratory research seem to suggest that delivery, meaning, and emotional appeal can act as drivers for delight, elevating unexpected, wanted and even expected product attributes to the level of the customer 'delighter'. Moreover, despite not being within the aims of this research, it is clear that getting these drivers wrong can result in a much more powerful type of product attribute, the 'enrager'.

19.3.1 Co-design Implications

It is proposed that these drivers for delight can be identified at the earliest stages of product design by allowing designers to empathise with the customer through partial adoption of the marketing role. The authors take the view that any process that provides the designer with a more intimate and direct knowledge of the customer than they receive from a market research report will facilitate the design of both *feature* and *delivery* 'delighters'. To this end CUPID has developed and deployed the following tools (for a more complete summary see Burns *et al.*, 1999):

- Product-in-Use - the naturalistic observation of customers using products with a focus on identifying how products are actually used rather than how they are meant to be used, and including the consideration of mood, environment and context;
- Scenario-of-Use workshops - informal customer focus groups that employ role-plays and acting-out techniques as cues for recall in the identification of latent customer needs and 'delighter' ideas;
- Engineer/Customer workshops - a 'design-with-customers' exercise adopting role-plays to uncover latent needs from customers. Design engineers then work with customers to generate the best solution ideas.

These techniques designed and deployed with Nissan have been successful in three respects. Firstly, the tools have proved effective at uncovering unarticulated customer information. For example the Product-in-Use tool produces video footage that can be used in Engineer/Customer workshops to identify latent customer needs and to stimulate the design of 'delighters'. Secondly, there is limited evidence to suggest that these tools have the ability to change the approach of designers by increasing their level of insight into the lives of customers. Direct contact with customers in workshops seems to be highly effective, whilst Product-in-Use video footage proved to be both stimulating and beneficial when used in design workshops. Finally, the tools have succeeded in identifying many possible 'delighter' features generated from customer insight rather than intuition.

However, the success of these features cannot be accurately measured until the launch of the Nissan vehicle concerned.

In light of the delight research described in this paper, the tools used so far remain limited due to their focus on *feature*. Working with, and observing customers seems to be an effective means of identifying novel and innovative feature ideas that answer user needs, even previously unarticulated ones. However, the approaches used fail to identify the physical and emotional dimensions that we have shown are so important in the delightful product. It is therefore necessary to develop a means by which designers can consider the *delivery* of product attributes within the product development process, and this is the subject of further CUPID research.

19.4 References

Baxter M (1995) Product Design: Practical methods for the systematic development of new products. Chapman and Hall

Bruce V, Green PR (1992) Visual Perception: Physiology Psychology and Ecology. Lawrence Erlbaum Associates Hove E Sussex UK

Burns AD, Barrett R, Evans S, Johansson C (1999) Delighting Customers through Empathic Design. In:Proceedings of the 6th International Product Development Management Conference. Churchill College, Cambridge, pp157-171

Clark KB, Chew WB, Fujimoto T (1991) Product Development Performance; strategy organisation and management in the world automotive industry. Harvard Business School Press, Boston

Ealey L, Troyano-Bermudez L (1996) Are Automobiles the Next Commodity? The McKinsey Quarterly 4:62-75

Kano N (1995) Upsizing the Organisation by Attractive Quality Creation In: Kanji G H Total Quality Management: Proceedings of the First World Congress. Chapman Hall

Kenningham TL, Goddard MKM, Vavra TG, Iaci AJ (1999) Customer Delight and the Bottom Line. Marketing Management Fall:57-63

Leonard D, Rayport FF (1997) Spark Innovation through Empathic Design. Harvard Business Review Nov-Dec:102-113

Ludvigsen KE (1996) Creating the Customer-Driven Car Company. International Thompson Business Press, London

Macht J (1998) The New Market Research. Inc. Magazine July:86-94

Oliver RL, Rust RT, Varki S (1997) Customer Delight: Foundations Findings and Managerial Insight. Journal of Retailing 73 (3):311-335

Rosenberg MS, Thompson BM (1993) Rooting out the Causes of Inefficient Product Creation. Prism 2nd Qtr:97-111

Schneider B, Bowen DE (1999) Understanding Customer Delight and Outrage. Sloan Management Review Fall 41(1):35-45

Segal LD, Fulton Suri J (1997) The Empathic Practitioner: Measurement and Interpretation of User Experience. In: Proceedings of the Human Factors and Ergonomics Society 41st Annual Meeting, pp451-454

Von Hippel, E Thomke S, Sonnack M (1999) Creating Breakthroughs at 3M. Harvard Business Review Sep-Oct:47-57

Chapter 20

Collaborating with Elderly End-users in the Design Process

O. Demirbilek and H. Demirkan

20.1 Introduction

The fact that end-users can participate and contribute to the design process, was pointed out in previous studies on user participation to the design process conducted by Cavanagh (1996), Ciccantelli and Magidson (1993), Mitchell (1995), Morini and Pomposini (1996), and Reich *et al.* (1996). For Howes, *et al.* (1998), participatory design is a design methodology, European in origin, giving an important contributory role to the end-user in the development of products they would eventually use. This paper presents a study (Demirbilek, 1999) in which elderly end-users were invovled in the design process by means of participatory design sessions. In these sessions, the expertise of designers and the comments and ideas of elderly end-users were applied to how doors and door handles for domestic use should be designed. Two different design sessions were run for each group of elderly end-users.

20.2 Participatory Design Sessions

20.2.1 Characteristics of the Sample of Elderly End-users

The participatory design sessions were held with potential end-users, each consisting of 3-6 elderly people, mostly widowed, male and female over 65, from the city of Ankara. Random sampling was used among a group of volunteers. A pilot session was conducted with 4 elderly participants (one male and three females between 68 and 75 years old), at the end of which the participatory design session was revised. A sample of 13 potential elderly end-users forming 3 different groups (see Figure 20.1) took part, each completing both participatory design sessions.

Figure 20.1. Views from the first participatory sessions for each group

The sample of volunteers, consisting of 10 females and 3 males, had a mean age of 75. Nine of the elderly participants were living in a residential home, with two of them having their own homes but preferring to stay in this sheltered accommodation where they have a one-room studio with a bathroom and a small kitchenette. The other four participants were living in their own homes, two as a couple, and two with their children. None of them had any severe disability limiting their physical activities. Four of the women used canes and three of them had problems when sitting down and standing up. Only one of the women had difficulty walking due to her age (93) and weight.

20.2.2 General Conditions of Participatory Design Sessions

Small groups consisting of 6 people are said to successfully produce up to 150 ideas in half an hour at their first attempt (Jones, 1992). Therefore, this study, on participation in design by the elderly users was conducted in small groups of 3-6 elderly people each. The participants were not prepared before the participatory design session.

20.2.2.1 The First Participatory Design Session

In the first participatory design sessions, the participants were asked contribute to in the 'design' of doors and door handles for the house that they want to age in, considering all their own possible requirements, needs, particular wishes and ideas. In other words, the end-users were asked to talk about, and to 'design' the doors and door handles according to their own requirements, an in their own words. The setting was a room, with chairs around a table. The designer and the participants sat around the table, having papers and pens available for everybody.

This session involved a combination of unstructured interviews, scenario building, and brain storming facilitated by the designer. To start the session, a group of questions prepared in advance was used to stimulate conversation and set out the problem areas. When the participants started to respond directly to the questions, the designer introduced scenario building in order to help the participants to express themselves more independently, without being limited by the questions. Once the participants started to create various scenarios, the designer encouraged them to start brain storming in order to propose any kind of solutions to the problems they have depicted while building their scenarios.

The first participatory design sessions for each group of participants began with a blank page. A pre-existent solution was not proposed because the aim was that the elderly participants should propose their own solutions. The outputs of these design sessions were not intended to deliver finished products or designs, but to be a point of departure for the remainder of the design process.

Different scenarios in which the elderly could imagine themselves were used during these sessions. Examples of scenarios were:

- You are coming back from shopping, hands full. Nobody is at home and you have to open the door. The keys are somewhere deep in your bag (or pocket). What happens?
- You are in the kitchen, the door is closed, and your hands are all greasy and dirty because of preparing a meal. The telephone is ringing in the other room. How do you open the kitchen door? What do you do?
- You are alone at home. Somebody rings at the door. You look through the 'eyehole' but you cannot see the visitor well. What do you do?

The designer asked the participants to share their ideas, whether positive, negative, or neutral regarding these scenarios. From the negative reactions put forward by the participants, the designer asked them to describe how things could have been better, and how the relevant parts could have been designed. The sessions were conducted with blank papers in front of the participants for them to draw and write when necessary, and they were highly encouraged to do so. This was believed to give them control over the situation, and the feeling that they are really designing (see Figure 20.2).

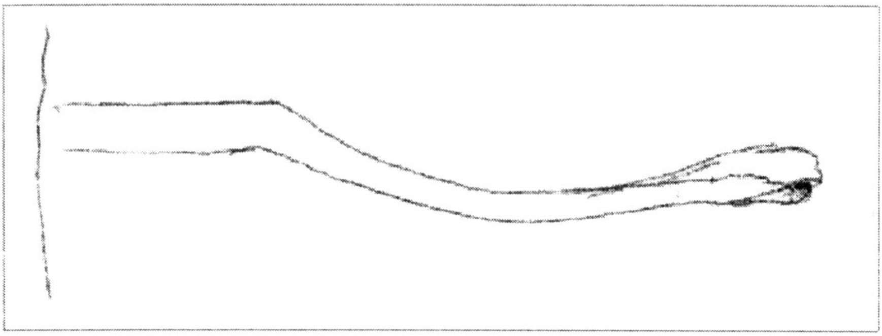

Figure 20.2. Proposition of an elbow operated door handle (drawn by a 74 years old participant)

The session facilitator collected the outputs of the sessions (in written, oral, sketch and gesture format). This information, which was recorded on video for later evaluation (to recall all the details), provided a basis and starting point for the author to prepare the preliminary design proposals for the products. This technique is similar to the video ethnography technique (Mannoy, 1995). The analysis of those videotapes allowed a full range of behavioral traits to be observed; and a sequence of events to be analysed.

After the completion of the first participatory design sessions, the outputs (responses to the questions, drawings and additional comments) were grouped under the topic they were related to (such as: physical aspect of a door; door operation; accidents related to doors, *etc*.). They were then listed in matrixes to assess their existing correlation with technical design considerations.

20.2.2.2 Results of the First Participatory Design Sessions

As none of the participants have experienced a participatory design session in their lives, almost all were very shy at the beginning. The designer had to make them imagine themselves in some related scenarios to be able to really start the session. The elderly users participated in their own ways and in their own words. At some stages, where they could not adequately explain something in words, they mimed the scenes that they wanted to describe (see Figures 20.3, 20.4, and 20.5), using hand gestures and body language.

20.2.2.2.1 Questions and Answers of the First Participatory Design Sessions

Figure 20.3. A participant showing door-handle operation at the door

Figure 20.4. A participant drawing a sketch for the handle design in his mind

Figure 20.5. Two participants explaining their ideas with hand gestures

The elderly were presented with eight groups of questions relating to the subtasks of domestic door use, starting with the entrance door. The questions were grouped under the following headings: problems faced with main entrance door; problems with keys while opening or closing doors; door safety while opening and closing doors; reasons for closing interior doors; problems and recommendations related to door handles and knobs; problems and recommendations related to the glazed parts on doors, and the materials chosen; and different kinds of doors available.

After having grouped the answers, the most important points identified by the elderly users were carried into matrixes to see whether there was any relationships between the elderly users' requirements, design limitations and technical design requirements. The matrixes were prepared using the answers and comments elicited by the pre-set questions and the product type specifications from the literature survey. Both the elderly user's requirements and the design specifications were rated according to their importance levels. The importance of elder users' requirements was rated according to the number of participants having (or approving of) the same opinion. The importance of design specifications relating to the design of a door was rated according to the professional judgement of the designer.

While designing the main entrance door the most important features for consideration were ranked in order of importance as follows: the visibility range of eyehole; the placement of keyhole on the door; the lighting level outside the door; the mode of operation of the key; the lighting direction outside the door; and the placement of door handle on the door.

While designing the main entrance door the most important features for consideration were ranked in order of importance as follows: shape of the grip part of the handle / knob; the door material(s); the level of transparency of the glazed parts; the surface finish of the handle/knob; the handle/knob material; the way of opening of the door; the placement of the door handle; the colour of the door; the dimensions of the door handle/knob; and the appearance of the door handle.

20.2.2.2.2 Conceptual Designs

Three design concepts were generated after having completed the first participatory design sessions. Among them, an elbow operated door handle concept and a conceptual design for a door screen that will allow elderly users to see a visitor without having to go near the door were proposed. Another conceptual design created was a structure to be erected outside, near the main entrance door, serving as a shelf for shopping bags and a seating unit while searching for keys.

20.2.3 The Second Design Participatory Design Session

In the second participatory sessions the designer presented the concepts reached as the outputs of the first session to the same groups of elderly participants (see Figure 20.6). The presentations were in the form of free-hand and computer drawings, not too perfectly drawn, in order that the participants should not feel that every thing was already decided, and that nothing was left for them. Each participant received A4 copies of the drawings on which he/she could draw as well.

This time, the participants were asked to criticize the presented drawings. Every comment, and the copies of the design concepts which were 'corrected' by the participants, were collected at the end of the session.

20.2.3.1 Results of Second Participatory Design Sessions

During the second participatory design sessions, the elderly participants seemed more comfortable, probably because they were familiar with the process. They have listened to the presentations of the design concepts carefully and made their comments. Some of them did draw corrections on their copies of the drawings.

Among the three conceptual designs presented to the elderly, the lever handle received the most correction. The second conceptual design, the foldable shelf/stool to be hung on the wall of the main entrance door also underwent some modification. The third conceptual design, the door screen was accepted as it was, and the idea was very much appreciated by all of the participants. No corrections were made on the drawings of the door screen.

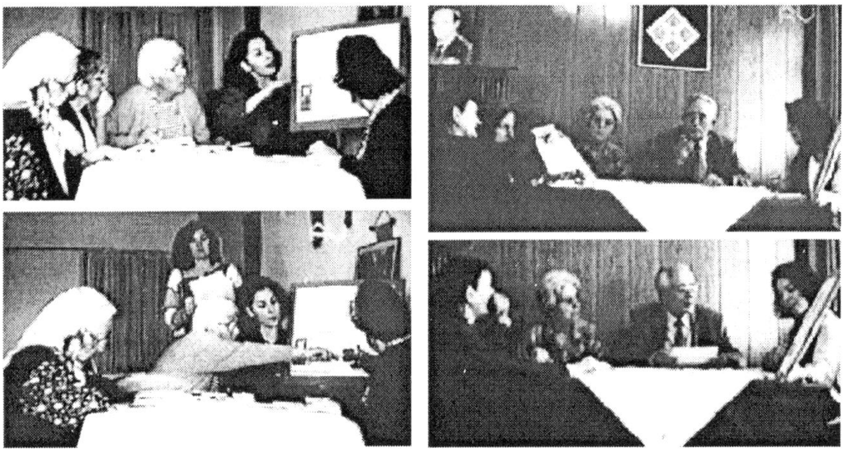

Figure 20.6. Views from the second participatory design sessions

20.3 Conclusion

All three concepts were generated entirely with the help of the responses given and proposals made by the elderly participants during the first participatory design sessions. This result reinforces the statements of Howes, *et al.* (1998) who claim that design knowledge is present in all the people potentially affected by a design, and that these people can contribute to improve that design. Druin (1997) adds that in this way, researchers can directly identify new product possibilities that they might not have been able to think of on their own. This latter issue is reinforced in the following statement by Haigh: 'Design for the young and you exclude the old; design for the old and you include the young' (1993:p14). This point becomes more valid as the number of elderly people increases. Furthermore, Coleman (1997) says that older people are, to a greater and greater extent, commanding substantial wealth and are prepared to spend it on things that will really improve their lives. This means that they are an important potential consumer market which needs to be explored more deeply. This study has tried to illustrate a concrete example of the involvement of elderly end-users early in the design process. Prototypes of the three design concepts that have been developed will be produced in order to be tested in suse, again by elderly users.

20.4 References

Cavanagh S (1996) The Space we Need: Principles of Housing Design for Older Women, Women with Children, and Parents with Disabilities. In E. Komut (Ed.) Housing Questions of the Others. Ankara: Chamber of Architects of Turkey, 67-76

Ciccantelli S, Magidson J (1993) Consumer Idealized Design: Involving Consumers in the Product Development Process. Journal of Product Innovation Management. 10:341-347

Coleman R. (1997) Breaking the Age Barrier. Speech at the Royal Society of Arts, June '97 as part of the Design Council - Design in Education Week, London: DesignAge, Royal College of Art, March, http://valley.interact.nl/DAN/NEWSLETTER/NEWS97/home.html, 6, 6, 1998

Demirbilek O (1999) Involving the Elderly in the Design Process: a Participatory Design Model for Usability, Safety and Attractiveness. Unpublished PhD Thesis. Ankara: Bilkent University

Druin A (1997) Participatory Design: Children Participation in Computer Design". http://mtsnmc.unm.edu/intel97/html/participatory_design.htm, 10, 3, 1998

Haigh R (1993) The Ageing Process: A Challenge for Design. Applied Ergonomics. 24. 1:9-14

Howes G, Gordon P, and Masellis B (1998) Perspectives on the Role of Participatory Design in Computer Mediated Communication. CMNS 353. http://oscar.cprost.sfu.ca/~cmns353/Projects/GroupJ/index.htm, 10(3)

Jones CJ (1992) Design Methods: Seeds of Human Future, Second Ed. Van Nostrand Reinhold, New York

Mannoy R (1995) The Scholarly Practitioner. Co-Design, Milton Keynes

Mitchell CT (1995) Action, Perception, and the Realization of Design. Design Studies. 16:4-28

Morini A, Pomposini R (1996) New Designs for Aged People Housing. In Ural O, Altinbilek D, Birgönül T (Eds.) XXIVth IAHS World Housing Congress. Middle East Technical University, pp791-800

Reich Y, Konda SL, Monarch IA, Levy SN, Subrahamanian E (1996) Varieties and Issues of Participation

Part II Issues in Developing Integrated Collaborative Design Approaches

Summary

The eight papers in this Part have illustrated the challenges that face contemporary designers in their attempt to compete successfully in fast-paced global markets. Challenges which have been discussed include not only the need for products to be high quality, functional and cheap but also for them to be innovative, stylish, fashionable, secure and emotionally engaging. All of the contributions reflect various attempts to meet such challenges head on through the development of integrated, collaborative design approaches such as concurrent engineering, inter-company networking and participatory design. However, these contributions not only describe the kinds of approaches that are available to designers but also present in-depth case studies of their application, with a detailed focus being maintained on the full exploration of issues associated with the actual success or failure of integrated collaborative design practices.

One theme that dominates the contributions is the importance of companies establishing effective *multidisciplinary* and *multifunctional* team-working methods if design is to be achieved that meets contemporary market challenges. For example, Lofthouse and Bhamra note that whereas industrial designers appear to be better at fluent, divergent thinking and idea generation, design engineers tend to display convergent reasoning that serves to progress design problems toward single solutions. Lofthouse and Bhamra further emphasise that it is only by capitalising upon the strengths of multidisciplinary teams (e.g., ones that include industrial designers, design engineers and production engineers) that successful ecodesign of industrial products can be achieved. They note that the most successful companies in their benchmarking study were those taking an holistic approach to ecodesign, rather than attempting to approach it from a single discipline. Similarly, Filson and Lewis, in discussing the development of inter-company networks for new product development, emphasise the role of shared knowledge and practices between different network partners as being a key ingredient for success. In their contribution, Erol, Cooper, Press and Thomas argue that the need to take account of crime-resistant features in products requires the extension of the designers' thinking from normal product use to product misappropriation, maltreatment and misuse. The surest way to achieve such a focus is to enlist the collaborative involvement of all stakeholders in the product-development process, including clients, customers, manufacturers and crime-prevention experts.

Continuing the theme of multidisciplinary team-working methods, Cooper, Bruce, Tierney, Daly, Harun and Svengren, in exploring the concepts of the 'designer as innovator' and the 'designer as integrator' also adopt a contemporary definition of design that emphasises its holistic and multifunctional nature. However, despite the importance of designers in driving integration that is noted by Cooper *et al.*, interesting difficulties in the attainment of integrated design practices are noted by Filson and Lewis in their research on inter-company networks. They identify a key difference between Italian networks and those in the UK and Denmark, which centres around the informal nature of communication in the former (which tends to forge close relationships between partners and high levels of trust), whereas in the latter, consultants and brokers have often been required to develop the relationship between SME partners before networking is successful. A particularly good summary of the range of issues and complexities surrounding the development of integrated, multidisciplinary, collaborative design approaches within a global context is provided by To and Harwood, who present case studies that explore the 'integration mechanisms' that companies adopt to support collaborative design. They note that most problems in integrating collaboration across design teams arise from groups having impoverished modes of communication, conflicting opinions, and poor documentation processes. These problems can be exacerbated when companies have a high rate of personnel turnover. To and Harwood argue that to choose the right integration mechanism for a particular design network requires a careful analysis of the 'context' of the design work being undertaken. In the case of fashion design, for example, this context is typically polarised in terms of whether mass fashion or high fashion is the driving factor behind design activity.

A second recurrent theme throughout the contributions in this Part concerns the vital importance for companies of applying extra time and effort in collaboration during *early* design phases in order to ensure the effective design of products to meet contemporary design challenges (such as fast-moving, global-market requirements in the clothing industry, and emerging political requirements for the ecodesign of industrial products). Early collaboration in multifunctional design teams can reduce the need for costly and time consuming late design changes, thereby lessening the chances of product recall and improving communication throughout the product development process and the supply chain. Bhamra, Heeley and Tyler's results from their case studies of the use of a Concurrent Product-Development approach within three UK textile and clothing companies attest to the improvements that can be made to design practice. One key benefit that they observed was that the approach facilitated the teams' early resolution of problems. Additionally, communication within the teams was seen to be greatly enhanced. Lofthouse and Bhamra's research on ecodesign integration also emphasises how ecological and environmental considerations (e.g., concerning the life-cycle of industrial products) need to be dealt with at the very earliest stages of product development. Similarly, Erol, Cooper, Press and Thomas note that crime-prevention features in products are often considered to be a retrospective add-on that are only implemented once a crime has been committed – rather than being viewed as an initial consideration that is integral to product development.

A third theme that is prevalent in this Part concerns the importance of *customer-involvement* or *user-orientation* in design to ensure that products meet market needs. Such a user perspective during product development can also serve to reduce unproductive debate among designers, and can provide a sharper focus for all design activities. A user-oriented approach is well illustrated in examples of ecodesign presented by Lofthouse and Bhamra, where it is shown that the innovative design of industrial products (e.g., kettles) with improved environmental performance (e.g., in terms of energy consumption) can best be achieved through careful consideration of the behaviour and attitudes of end-users. Likewise, in the process of co-designing against crime, Erol *et al.* emphasise the importance of the designer relating crime prevention to the total environment associated with the user or consumer of a product - whether this is an information terminal or a shopping complex. In the context of car manufacturing, Burns and Evans argue that increasing the designer's understanding of the customer at the earliest stages of product development can facilitate the design of cars that stand out from their competition, that are wanted rather than needed by customers, and that 'delight' their users. Burns and Evans point out, however, that many contemporary concurrent-engineering approaches tend to gain an understanding of customers by communicating with separate market-research departments. As such, customer information (e.g., in the form of reports) has to cross a boundary between the marketing department and the design department, which means that designers tend to work with customer information that is summarised and diluted. Burns and Evans argue that what is needed is to integrate *effective* customer-oriented approaches into the early stages of the design process, such as ones based on observational research methods like 'empathic design' that can fully address the question of customer delight. Demirbilek and Demirkan extend the importance of end-user involvement in design to the elderly, who are increasingly commanding substantial wealth and who are prepared to spend on innovative products that improve the quality of their lives. Demirbilek and Demirkan demonstrate, through a case study, how ideas and critical feedback derived from participatory-design sessions with elderly end-users can successfully enhance product development

It is evident from the papers in this Part that the implementation of integrated, collaborative design approaches within organisations has great potential to bring together the activities of people with very different domain expertise, technical backgrounds, experience and responsibilities. This can lead to demonstrable successes in the quality and efficiency of design productivity. All these papers, however, also attest to the fact that the sheer diversity of team members can present significant challenges to the maintenance of consistent and effective communication. Issues of communication across design teams also form a central focus for research discussed in subsequent Parts of the present volume such as Part IV which looks at work on mediated communication in design.

Part III

Studies of Collaborative Design

Part III Studies of Collaborative Design

Introduction

The empirical investigation of *collaborative* design activity is emerging as a vitally important element of contemporary design research. The recent burgeoning of studies of collaborative design contrasts with research undertaken prior to the 1990s, which tended to focus on 'de-contextualised' activity where individuals tackled only small-scale simulations of real design problems in laboratory-like conditions. Such research is now generally acknowledged to provide only limited insights into the design activity that actually takes place in the socially-dominated contexts associated with modern design practice (e.g., where colleagues, managers and clients all impact regularly on an individual designer who is working to satisfy both personal and organisational goals). The ten papers in this Part each describe studies motivated by the aim of advancing a theoretical understanding of the cognitive, social and cultural processes underpinning real-world, collaborative design activity. Together, the papers cover a representative cross-section of the types of empirical research that are starting to flourish in the drive to broaden our knowledge of the nature and determinants of collaborative design phenomena.

Gill, Deshmukh and Athavankar begin this Part by discussing a study which examines how a designer's imagery may be influenced by the verbalisations and sketches of co-designers. The study involved two professional architects collaborating across a video-link. Analyses focus on independent, retrospective reports taken at intervals from the designers, with reports being coded to examine patterns of cross-cueing between one designer's activity and the other's imagery.

Investigating patterns of social interaction in design teams is a theme also taken up by Reid and Reed, who aim to understand what determines the periodicity of the rhythms observed in the discussions of design groups - where bursts of heightened conversational activity are interspersed with phases of reduced conversation and turn-taking. Their hypothesis is that conversational cycles are moderated by oscillations between periods of visually-dominated figural reasoning (linked to solution-generation processes) and verbally-dominated conceptual reasoning (linked to evaluative processes). The study recorded discussions of multi-disciplinary teams working on engineering-design problems. Data analyses involve a range of sophisticated mathematical and statistical-modelling techniques.

The next three contributions also focus on small-group design meetings. Ball and Ormerod's analysis of a design-review meeting seeks to illuminate the social dynamics of solution-exploration activity in collaborative design, with an emphasis on the potential benefits of collaboration over individual design. Turner and Cross's aims are broader, being directed toward understanding both the processes that arise in design meetings and the technological requirements needed to support virtual design groups. Their data derive from meetings of a team pursuing a software-engineering project. The analyses assess proportions of time spent using various workspace resources and engaged in different design activities. Van der Lugt's study, like Reid and Reed's, is also concerned with cyclical aspects of design. It explores the hypothesis that graphic idea generation involves a cyclical process of slow idea production, reflection and building on earlier ideas, whereas sentential idea generation involves rapid, associative processes leading to high quantities of less well-integrated solution concepts. Design groups were required to generate ideas using either a sentential technique (brainstorming) or a graphic technique (braindrawing). Generated ideas are analysed in terms of their connectivity to earlier ideas.

Lloyd's contribution departs from the previous empirical studies in this Part by examining not so much the design process itself but rather the way in which it is represented on television. Lloyd presents a qualitative content-analysis of three programmes about the design process, and focuses, for example, on the dramatic characterisation of those involved in design work (including the design chiefs in the client organisations) and the rhetorical devices that individuals employ to justify their design ideas and preferences.

Mazijoglou, Scrivener and Valkenburg's study is methodological in emphasis. They focus on examining the validity of their 'Rich Picture' scheme which structures verbal data, derived from design teams, in an essentially uninterpreted form that is based around discontinuities in the designers' attentional focus. Their motivation is to test the alignment between their atheoretical Rich Picture description of a data-set and an independently-derived, theory-based description of the same data-set that is informed by Schön's concept of reflective practice.

The final three contributions to this Part turn to an examination of the design activity associated with *innovative* design projects that are characterised by the need for adaptability and flexibility in the interpretation and exchange of ideas. Milne's paper pursues a comparative assessment of the organisation of activity within two design groups, and applies coding schemes and analysis methods similar to those used by Turner and Cross and van der Lugt. Milne's particular interest is in the creative, idea-generation processes that arise in the early, conceptual stages of product design. The study by Reid, Culverhouse, Jagodzinski, Burningham and Parsons foregrounds the role of technical and social challenges in shaping the trajectory of team design in terms of the progress and morale of team members. Interview diaries form the data for an assessment of the interplay between the various challenges confronting the design team being studied as it struggled to translate customer requirements into a detailed functional design specification. Finally, the research reported by Steele, Macmillan, Austin, Kirby and Spence returns to the theme of cyclical processes in design with its focus on the patterns of collaborative activity in the conceptual phase of interdisciplinary building projects. The paper outlines the genesis of a framework for understanding conceptual design which is derived from an extensive review of the design literature and from the authors' own empirical research. This framework comprises five high-level design 'phases' which map on to 12 low-level 'activities' that are specific to certain phases. Steele *et al.* outline the patterns of design progression evident in two of their studies of building-design teams. They demonstrate how the results from their empirical work have enabled the development of this conceptual design model in ways that enable it to capture the iterations that are observed across activities and phases of the conceptual design process.

Chapter 21

Imagery as a Private Experience and Architectural Teamwork

N. Gill, H. Deshmukh and U. Athavankar

21.1 Imagery and its Role in Early Creative Phase

Recent accounts of the role of imagery in creative processes have gone beyond the anecdotes dealing with creative literature, artistic achievements and discoveries in science. These studies have shown how creative processes can benefit from imagery. Reasons are not difficult to seek. Typical attributes of imagery – its spontaneity, non-linear nature, depictive qualities and the possibilities of voluntary control - point to the enormous potential of imagery in the development of ideas. It encourages private fantasies and yet it allows simulation, permitting evaluation of ideas. The fact that images directly depict the layout of the represented material and preserve between-part relationships is significant to designers, who creatively manipulate these relationships. So, its no wonder that most designers, in casual observations, refer to the rich imagery that they often experience during designing. Yet there is little published work available in this area. The reasons for this lack of literature could be attributed to two unique properties of mental imagery. First, the experiencing of mental images is so well integrated with the thinking process that most people are not even aware of it. Second, it remains a totally private experience inaccessible to outsiders.

This inaccessibility raises some new questions relevant to this paper. If a private experience like mental imagery is so significant in problem solving, can a design team ever succeed in solving design problems, particularly when the team members are geographically apart?

Simulating the design process where two architects work at distant locations but are connected by a video-link, the pilot experiment explores 'issues' that such private experience would raise in a team design effort. We will turn to the team issues after a brief review of work on mental imagery.

21.1.1 Studies of Mental imagery

Explorations of the relationship between imagery and creativity are not new to research literature in cognitive science and design. We have excluded a review of literature on 'mind's eye' due to space constraints, but readers may want to refer to an earlier publication (Athavankar, 1997) (see also a compilation by Roskos-Ewoldsen *et. al.*, 1993). It is not surprising that some of the design research should take its inspiration from this work by cognitive scientists. Two specific areas from this work seem to have influenced design research. The first deals with the mental synthesis of component parts, where subjects are asked to mentally combine given parts to create useful objects (Finke, 1990). Finke's findings suggest that people are able to use mental imagery to evolve creative and interesting object ideas. The second, refers to the sketch-like depictive qualities of images, and explores its potential to inspire discoveries. This has prompted researchers to compare the effectiveness of sketches as displays that could be restructured and reinterpreted. In general, imagery seems to perform poorly as a depictive display, at least in experimental situations when it comes to restructuring and reinterpreting displayed material (Anderson and Helstrup, 1993; Kaufmann and Helstrup, 1993).

It is not clear why the idea of 'mental synthesis' merited so much attention in design literature. This task differs from typical design problems in several ways, particularly when compared to Goel's well-documented criteria for design problems (Goel, 1995). These studies fail to conceptually separate invention and discovery from typical design problem-solving. Similarly, the kind of image reinterpretation tasks used in these experiments is only occasionally experienced in design problem-solving. It is difficult to agree with this conclusion, particularly in the light of findings presented by others (Athavankar, 1996; Singh, 1998). If mental images were so poor as displays, designers should find it impossible to develop creative solutions, if they were blindfolded. These papers show that designers and architects used their mind's eye as a virtual design studio. They almost lived in their mental space, and created or manipulated ideas of objects with amazing ease.

21.1.2 Collaborative Design

While this literature does reiterate the contribution that mental imagery makes in individual creative efforts, it also poses new problems to the current trends in the practice of design. Modern design is increasingly seen as a team effort, often with expert team members interacting across the globe to develop new models of teamwork. Almost no material is available on the role of imagery in team problem-solving. Seen in the context of the discussion above, it is not surprising that the literature on collaborative design should also neglect the role of imagery. In fact most of the literature in collaborative efforts appears to be motivated by creation of computational representations and tools, obviously external to human mind. In any case, we plan to briefly review the representative literature before we proceed.

Communication has been a major focus of a number of these studies. For instance, Sommerwald gives a brief overview of work in software and engineering designs and neatly categorizes the research approaches (Sommerwald, 1996). Others, like Badke-Schaub, investigate the factors that influence teamwork in engineering design practices (Badke-Schaub, *et al.,* 1999), and, going further, also proposes a model of collaborative work. Taking a different view point, Cross looks at team design activity as a social process and explores the individual's role and relationships within the team (Cross and Cross, 1995)

Focus on teamwork in architectural design is equally common in the research literature. Peng explores the properties of team sustaining frameworks for supporting the complexities and spontaneities observed in multi-disciplinary interaction. He also proposes a pilot collaborative modeling environment (Peng, 1999). Giving a new dimension to the collaborative modeling environment, Scrivener and his colleagues have examined the potential problems that arise when team are geographically separated and have to also cope with different time zones. ROCOCO project proposed and evaluated 'Sketchpad' in which a common drawing space is shared by distant teams (Scrivener *et al.*, 1993).

Concerned more with interaction and communication, the literature appears to have neglected the key role that private mental images play in thinking and creative work. This paper attempts to fill this gap.

21.2 Experiment Set-up and Procedure

Exploration of the role of mental imagery in team effort is unfamiliar territory. So the experimental set-up and procedure were evolved through experiments, and accommodated to the limited equipment at our disposal. The useful results that the procedure yielded do give guidance for future studies in this area.

The subjects for the experiment were two architects, SF and YR, with several years of professional experience. They did not know each other, nor was their identity revealed. SF is a successful practising senior architect in his forties. YR is a budding architect in his late twenties, mainly involved in interior projects. The subjects were asked to design an information centre for a historical monument, *vi21.*, Elephanta caves, eleven kms. from Mumbai. They were sculpted on an island in the Oman Sea, built in the 6th Century by Hindu Kings, and contain huge images of Hindu deities. A printed design brief was given simultaneously to both the partners, containing area specifications for an exhibition hall, six guest rooms, a reception area, recreation space, kitchen and library, *etc*. This remained with them throughout the experiment.

They were ushered into two different rooms and were given assumed names. Simulation of distant interaction was implemented as follows: each room had a worktop with normal stationary required for sketching; a video-link ensured that the sketches, as well as conversations, were available to each other through monitors on their tabletops; camera angles ensured that the designers' identities were not disclosed. They were told that the study dealt with team process, but the study's focus on imagery was not revealed to them.

21.2.3 Accessing Mental Imagery

As mentioned earlier, a private experience such as mental imagery remains inaccessible to outsiders and can be only accessed through immediate retrospective reports given by the subjects. In earlier studies on an individual designer, Singh (1998) has elicited retrospective reports by replaying the videotapes to subjects at the end of the session, as a strategy to access imagery experiences. To avoid the subjects having to rely too much on memory of the entire 75 minutes of the design session, pauses (of collectively 120 minutes of reporting) were used for retrospection, mostly at the point of completion of important arguments. There were three pauses during the session and a fourth at the end of the session. To ensure structured comments, the two experimenters noted clues, such as key words and phrases from the conversation between the two subjects, and used them for structured discussion. On the face of it, use of intermediate pauses may appear as unusual interruptions, but earlier exploratory studies showed that such brief pauses were not a major hindrance to continuity. The subjects remained separated and their narration during the pauses was also separately video recorded.

The reports during the intermediate pauses were rich with information due to the recency of memory. There was another important advantage. Because the subjects had no idea of what they would do subsequently, they had no opportunity to use 'after-the-fact-relevance' of their imagery experience as a filter in their reports during pauses. The entire data was logged onto a four-column table. The first two columns contained the transcripts of the conversation between SF and YR respectively. Columns three and four contained their after-the-fact reports on the imagery experiences during the pauses. An attempt was made to classify these experiences, and search for probable clues in the transcripts that could have acted as triggers. Some of the important observations are given in the following sections.

21.3 Word-image Relationship

Conversation was not only central to team interaction, but as it appears now, it was also a source that prompted imagery experiences in the teammates. Are there patterns that can be found in the relationship of words and the images that they prompted? The following observations reveal some of these patterns.

There were several instances, when SF's words prompted YR to quickly conjure up images, thus broadening his imagery canvas and enabling him to experience new images. Surprisingly, the very same words did not generate any images in SF's mind, who had actually spoken the words.

Take for example, words like staircase, display and activities [00:09:25] which were used in the discussion in the first session. When SF talked about a staircase, YR had a clear stereotypical image of a staircase with a dead wall on one side. The wall had a stone mural. The staircase had a balustrade and was semicircular in plan. He saw himself walking up the stairs along with other people when he visualised the staircase. Similarly, the word 'activities' evoked in YR images of people in the lounge, waiting, reading, sitting, and looking outside. He described them as tourists, wearing casual clothes. Much of this seems to be on the border of fantasy.

Imagery prompted by someone else's words can throw up interesting surprises. For example, midway during the design, SF says [00:59:40]: '... Also I am looking for one "Aha" factor.' The word 'Aha' referred to the 'eureka' feeling in the process. Though the word was used as a motivational tool and was not intended to prompt any images, YR visualised himself standing in the road at night, seeing the building lit by floodlights. Considering that it was very early in the session and no spatial issues had even been discussed, it appears even more surprising.

21.3.1 It Works... Sometimes!

There were occasions when both the partners experienced images. In such cases two kinds of situations were encountered. The first deals with expected results where their conversation resulted in both visualising predictably similar images, such as a sloping roof. Mental imagery seems to have supported their dialogue, strengthening their co-ordination. The second type refers to a situation, where the words prompted vivid images for the speaker as well as the listener. But what SF visualised was very different from what YR saw in his 'mind's eye'. As expected, this affected the teamwork, as the teammates appeared to be going in different directions. For example, the exhibition spaces visualised by YR and SF were very different in nature. It was only later that they converged into a common understanding.

It was equally surprising to note that on some occasions, even though the words used in the conversation seemed to have potential for prompting images, the team members' did not report any imagery experience. For example, words like 'Jehangir Art Gallery' [00:15:20] (a well-known local art gallery) and 'sculpture' [00:11:23] failed to invoke any imagery in either of the team members. The words and images share an interesting relationship. They trigger and expand the image canvas of the partner, giving him an opportunity to explore new areas, but the operation of the trigger remained unpredictable.

There are several instances which show that the partners visualised more than what they conveyed through words. For example, when YR talked about the staircase, lounge and tourists [00:07:45], he did not convey the rich experience of imagery to his partner. These details, had they been conveyed, would have enormously enhanced the entire experience of shared imagery. We will return to this topic when concluding the paper.

21.3.2 Imagery Experiences and the Development of Design Thinking

It was interesting to note the metamorphosis of the final image of the building project as it emerged through the 75-minute session. In the initial part, the images that the teammates experienced were seen as individual fragments, but slowly they acquired coherence and connectedness. It seems that, as the process reached the final stages, these visual image fragments seemed to fall in place like a jigsaw puzzle. For example, when both the subjects started with the discussion about a staircase, lounge and exhibition spaces, images that they initially reported occurred more as independent fragments. But as the design progressed, the image fragments were integrated into a coherent building space. It was also interesting to note the differences between the initial and later imagery experiences. The images in the initial part of the design process contained people performing activities, while the architectural spaces remained as background settings. As the session proceeded further, the people vanished from the imagery and it was dominated more by architectonic spaces.

Though the process was dominated by the words used in the conversation, the online access to each other's sketches seems to have played a unique role not only in team interaction but also in prompting (or sometimes suppressing) imagery. Do the sketches influence the mental imagery, teamwork, and the design process in any way? If yes, can it be better understood?

21.3.3 Online Access to Sketches

Attempts by the team members to understand each other's ideas, and their questions, were often based on access to each other's sketches and diagrams. From a very early stage, both the partners understood each other's ideas about the relationship of spaces from their simple bubble diagrams [00:10:00]. Interestingly, the strategies and idiosyncrasies of the sketching did not hinder the sharing of information. Online access to sketches seems to play an important role in understanding each other's ideas. For example, when asked to sketch the architectural ideas as perceived during the third pause, their independently drawn sketches of the building showed many common features.

To a large extent, taking cues from the speech strings and sketches available on the screen ultimately depended on the willingness of the partners. For instance, when SF switched from the idea of 'Trimurti' to sketch a cross-shaped plan [01:38:44] in the third segment of the session, YR's image of the building in his mind seems to have remained unaffected. Two different inferences can be drawn from this. On one hand, it can be taken as a positive approach, as YR remained rooted to the basic idea, while he let SF explore a new direction. On the other hand, it could be seen as a negative development, as YR lost an opportunity of exploring a new viewpoint, and perhaps missed out on widening the canvas of his mental imagery.

To sum up, sketching was used as an active tool for communication and interaction right from the beginning. This took place in spite of the fact that one of the partners worked basically in 2-D while the other adopted a 3-D mode of sketching. Online access proved to be very useful in the synchronisation of their work. However, in the current experimental set-up, it is difficult to directly establish the influence of sketching on the experiences of mental imagery.

21.4 Precedents, Imagery... the Unexplained

There appears to be a perceivable logic in the way precedents were recalled and used. Words often seem to prompt teammates to access precedents. For instance, in the second segment of the design session [00:04:30], the word ‘recreation’ evoked the image of a bar in their minds. This seemed to have its roots in their exposure to cities, similar cultures and income groups. Though irrelevant to the location and functional context, a certain fascination for these images made the partners develop them in great details. Similarity in imagery experience can be attributed to similar cultural and professional background.

The retrospective reports suggest that the precedents often appear in the form of real world images. These images seem to show ambience of light and shade, which suggest a particular time of the day. In the earlier reported images accompanying words like lounge, staircase and activities, YR experienced images as if they were a part of early evening. YR narrated a strange experience where the precedent was seen as a drawn plan. Perhaps YR recalled the precedent the way he had experienced it and it did not evolve into a 3-D image. These events must be seen in the context of architectural design problems which tend to have a strong spatial dimension and a site to build on.

There were instances, when the member who dwelled on a precedent had an elaborate image in his ‘mind's eye’, but he could not evoke any image in his partner’s mind, despite detailed description. For example, when SF said: ‘This reminds me of a museum done by Arther Errickson.’ [01:07:45], in spite of the speaker’s excitement, no image was prompted in YR's mind. It is not easy to explain this on the basis of familiarity. Let us take another example to understand the point. SF mentioned the strong symbolic character of ‘Trimurti’ (central historic statue in the caves) [01:06:20]. Although SF dwelt on this extensively, this surprisingly, did not influence YR, nor did it prompt him to alter his current images. Precedents had a positive effect only when they were shared through the description of the experiences, or when one of the members motivated the other to expand his imagery and enhance the progress of the team.

21.4.1 Unexplained Experiences

During the pauses, the subjects were asked to guess what the other team mate was experiencing in his ‘mind's eye’. Surprisingly, YR could make reasonable guesses about the imagery features of SF. For example, YR made a correct guess about the time of the day in SF’s imagery world, in spite of the fact that the protocols reveal no clues of these details.

There are other similar instances, where they were asked to independently sketch the current status of the design during the third pause. To our surprise, both showed a circular cut-out in the slab over the lounge. This is difficult to explain, except by attributing it to both having trained and practised in similar environments.

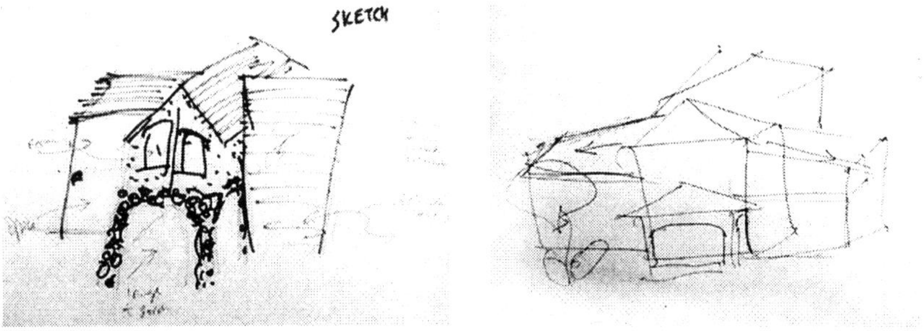

Figure 21.1. Left image shows SF's visualisation of the current idea. The right image shows YR's perception of SF's image at that point of time.

When visualising the structure, one of the architects typically experienced his own presence on the site and in the spaces that they had created. He often moved inside and outside these spaces. In fact, one of the teammates almost guessed the position and level from which his partner was visualising and experiencing the spaces in his imagery world. These experiences appear similar to a walk-through in a virtual environment. Interestingly, some of these spaces were experienced in very specific light conditions.

21.5 Conclusions

Our observations on imagery are based mainly on reports during pauses, and, wherever possible, they are cross-referred to the protocols of the session. Though the discussion occasionally drifted to other aspects of the teamwork, the focus was on the imagery experiences. We do not claim to have clear answers to all the questions on imagery in team effort that we raised at the beginning of the paper. But the results do offer some pointers.

The fact that it was an architectural problem that was being handled has influenced the nature of imagery experiences. But for a few exceptions, it is quite clear that imagery, and in turn the words that prompted it, were central to the development of the initial approach at an individual level, and to a limited extent, at a team level. Most of the imagery that accompanied the design precedents appears to have been motivated by words in the conversation, but the actual experiences of the imagery remained unpredictable. For instance, some of the words, in spite of their potential to invoke images, failed to prompt imagery even though the subjects were familiar with the precedents.

As the session progressed, the imagery experiences also changed. The initial discussions were accompanied by images in which people and activities were more prominent. As the spatial issues got resolved, the images were dominated by shared architectonic spaces and people within these spaces almost vanished. Perhaps this convergence could be attributed to the fact that they had online access to the development of each other's sketches.

It is interesting to note that experiences of concrete imagery of spaces and building forms were common even at a very early stage. Surprisingly, these experiences occurred long before the spatial issues were even touched on, almost as early as the discussion on the brief started. Most of this imagery was built on 'stereotypes' of architectural and interior spaces. It is difficult to predict if cross-cultural teams would share these stereotypes. Though the flow of conversation in the discussion prompted the experiences of stereotypes, much of the richness in the imagery remained private to the teammates.

Unfortunately, many of these private experiences remained unreported in the conversation, and the opportunity to jointly develop these ideas and images was lost, a problem that team design has to address. The team members often did not report all the rich imagery experiences in the conversation. This not only prevented sharing of ideas but also prevented the partner from taking clues and developing his own creative canvas. There are several reasons that could explain this. First, they may not be aware of their own imagery experiences as they often naturally accompany the conversation; second, perhaps they did not consider the experience important enough to share; and third, even when the attempt was made, the words fell short of capturing the richness of the images.

However, it would be too premature to come to the conclusion that the partners should convey their imagery experiences to each other. The images flow naturally with the thought process. It is not clear if the burden of describing imagery experiences would slow down the speaker's natural flow of ideas and even hamper the development of the thought.

21.6 References

Anderson R and Helstrup T (1993) Multiple Perspectives on Discovery and Creativity, in the Minds and on Paper. In: Roskos-Ewoldesen B, Intos-Paterson MJ, Anderson R (eds.) Imagery, Creativity and Discovery. North Holland, pp223-251

Athavankar U (1997) Mental Imagery as a Design Tool, Cybernetics and Systems. 28 (1):25-42. (Initially published in Trappl R (eds.) Proceedings of the Thirteenth European meeting on Cybernetics and Systems. Austrian Society of Cybernetics Studies and the University of Vienna, Austria, April 19-12, 1996, Vol. II, pp382-387)

Badke-Schaub P, Frankenberger E, Kurfursten A (1999) Analysis of design projects. Design studies 20:465-480

Cross N, Cross AC (1995) Observations of teamwork and social processes of design. Design studies 16:143- 170

Finke R (1990) Creative imagery, discoveries and inventions in visualisation. Lawrence Erlbaum, New Jersey

Goel V (1995) Sketches of Thought. MIT Press, Cambridge MA

Kaufmann G, Helstrup T (1993) Mental imagery: Fixed or multiple meanings? Nature and function of imagery in creative thinking. In Roskos-Ewoldesen B, Intos-Paterson MJ, Anderson R (eds.) Imagery, Creativity and Discovery. North Holland, pp223-251

Peng C (1994) Exploring communication in collaborative design: co-operative architectural modelling. Design studies 20:537-551

Roskos-Ewoldesen B, Intos-Peterson MJ, Anderson R (eds.) (1993) Imagery, Creativity and Discovery. North Holland

Scrivener SAR, Harris D, Clark SN, Rockoff, Smyth M (1993) Designing at a distance via real time designer-to-designer interaction. Design studies 14: 261-282

Singh A (1999) The potential of mental imaging in architectural design process. In: Proceedings of International Conference on Design And Technology Educational Research and Curriculum Development, IDATER 99. University of Loughborough, England, pp230-236

Sommerwald DH (1996) Communication role that support collaboration during the design process. Design studies 17: 277-301

Chapter 22

Interaction and Entrainment in Collaborative Design Meetings

F.J.M. Reid and S.E. Reed

22.1 Introduction

Many aspects of social life are temporally organised, and various rhythms and cycles are known to be mutually entrained in individual behaviour (e.g., Kelly, 1988). The concept of entrainment originated in the study of circadian and other biological rhythms, where it refers to the capturing of an endogenous rhythm by an external pacing cycle or event, so that the endogenous rhythm is modified towards the phase and periodicity of the entraining process. The present paper extends the concept of entrainment to group-level processes of cognition and social interaction in engineering design teams.

Recent studies of design collaboration have emphasised how the evolution of design ideas is governed by social interaction processes, particularly the patterns of interaction within collaborative design teams. For example, Brereton *et al.* (1996) observed that design discussions proceed through successive cycles of *focusing in* on issues and *transitioning between* topics. In the design group they analysed, individual engineers developed partial solutions to the design problem by pointing to and handling the design object (a bicycle), whilst other group members played supporting roles, listening, asking questions, offering encouragement, and eventually criticism. These rhythmic bursts of conversational activity—sometimes increasing into a flurry of turn taking, whilst at other times decreasing in pace with long pauses and a reduced tempo of speaking turns—have been a prominent feature of descriptive models of team discussion (e.g., Stasser and Vaughan, 1996).

But what determines the periodicity of these rhythms in design discussions? One possibility is that participation waxes and wanes as the team generates, develops, and evaluates a succession of partial solutions to the design problem. In her analysis of the think-aloud protocols of professional architects, Goldschmidt (1991) found a regular pattern of shifts between two modalities of design argument, *figural* and *conceptual*. Figural arguments are predominantly visual design ideas, intimately though not exclusively linked to the freehand sketching of partial solutions to the design problem. Conceptual arguments are primarily verbal, and deal with requirements, specifications, and constraints on the emerging design. Goldschmidt found approximately equal numbers of figural and conceptual arguments in her protocols. However, she also observed a rhythmic pattern of switching between argument modalities, so that clusters of arguments of one modality were almost always followed by one or a few arguments of the other modality. Goldschmidt's analysis of the same design team protocol studied by Brereton *et al.*, suggested that alternation between argument modalities was a natural feature of design team thinking, with individuals implicitly or explicitly adopting specialist roles in this process (Goldschmidt, 1995).

22.2 Design Team Discussions

The general hypothesis we advance in this paper is that speaking and turn taking cycles are entrained to these alternating modes of team design reasoning. Specifically, we posit a process of *cognitive entrainment*, in which rhythmic oscillations between periods of predominantly figural or predominantly conceptual reasoning capture and entrain cycles of speaking and turn taking in team design discussions. To test this hypothesis, we collected time-stamped video recordings from six teams of engineering design students discussing realistic design consultancy problems in the final year of their training. Each team consisted of about six designers drawn from different specialisms (mechanical and materials engineering, manufacturing, design technology, *etc.*). Half of the teams (Teams A, D, and J) designed a lightweight, low-cost portable river crossing system for use by aid workers, whilst the remaining teams (Teams E, H, and K) designed an electrically adjustable changing bed for use in schools and clinics. Each team was presented with customer requirements, market analyses, supplier and manufacturing information, and large-format paper pads and sketching materials.

We recorded team discussions from wide-angle and overhead cameras, and analysed these to identify the occurrence of figural and conceptual design arguments, and the distribution and rate of speaking and turn taking within each team. Each discussion was divided into consecutive 30-second time intervals, and argument categories adapted from those developed by Goldschmidt (1991) were used to classify design arguments expressed during these intervals. *Figural arguments* were defined as utterances and actions conveying single, coherent ideas relevant to potential design solutions, for which a visual action (sketching, pointing to a sketch, or figural gesturing) was necessary—or at least facilitatory—to the communication of this idea to other team members. Typically, one or more workspace actions (sketching, pointing, or gesturing) accompanied figural arguments, although these arguments could be expressed, arguably less effectively, by speech alone. *Conceptual arguments* were defined as utterance or action units directly related to the design problem that could be understood without recourse to a visual action, such as design requirements, product or materials specifications, and constraints on the design. To check reliability of coding, a randomly selected team protocol was independently coded by the two authors, and an intercoder agreement of 83% and a highly significant kappa agreement coefficient ($\kappa = .75$, $z = 18.87$) obtained.

Speaking and turn taking were measured in two ways. *Turn rate* summarised the tempo of design-focused discussion during each time interval, and was based on whether consecutive design arguments (of either type) were produced by the same or different speakers. High turn rates indicated periods of highly interactive design reasoning, characterised by brief contributions and a brisk exchange of arguments, often between several team members. Low turn rates indicated less interactive periods in which individual team members held the floor more or less continuously while others offered brief comments or questions. Finally, a measure of *participation rate* was used to summarise the distribution of design argumentation among team members. This measure compared the number of arguments contributed by each speaker within each time interval with the number they would be expected to produce assuming equal rates of participation among team members. High scores reflected more equal participation within the specified time periods, whilst low scores signified periods in which one or a few speakers dominated the design discussion.

22.3 Frequency Domain Analyses

To uncover the temporal structure of these design discussions, exploratory spectral frequency analyses—using the Tukey-Hamming spectral window and a smoothing bandwidth of five 30-second time intervals—were carried out on all four measures separately for each team discussion. Spectral analysis employs the Fourier transform to assess the cyclicity of time-based measures, and determines the periodicities of any cycles it detects. A function called the spectral density is used to describe these cycles. The spectral density measures how much variance is accounted for by each frequency cycle (from 0 to .5) underlying the sequence of observations. Spikes in the spectral density function at particular frequencies point to regular cycles against the background of random noise. Peaks in the spectral density function over a band of adjacent frequencies signify a less regular pattern of cycles varying over a range of periods.

The majority of team spectra showed broad peaks in the .05 to .1 frequency range indicating a broad band of irregular, low frequency rhythms underlying team discussions. The period of a cycle is the time it takes for the behaviour to repeat itself, from peak to valley and back again to peak, and is the reciprocal of the frequency of the cycle. Thus, the period of these oscillations in design reasoning and group interaction ranged from about 5 to 10 minutes. Figure 22.1 provides an illustration of the spectral density functions for a typical team (Team A) in the present study.

To examine the harmonisation of these cycles, we inspected the coherency spectrum for each pair of variables within the six design discussions. The coherency spectrum resembles a series of pairwise correlation coefficients, and indicates whether two cycles peak at similar frequencies. Although a number of teams showed coherency peaks in the shorter frequency bands, the majority of pairwise coherency spectra showed significant coefficients in the .05 to .1 frequency range.

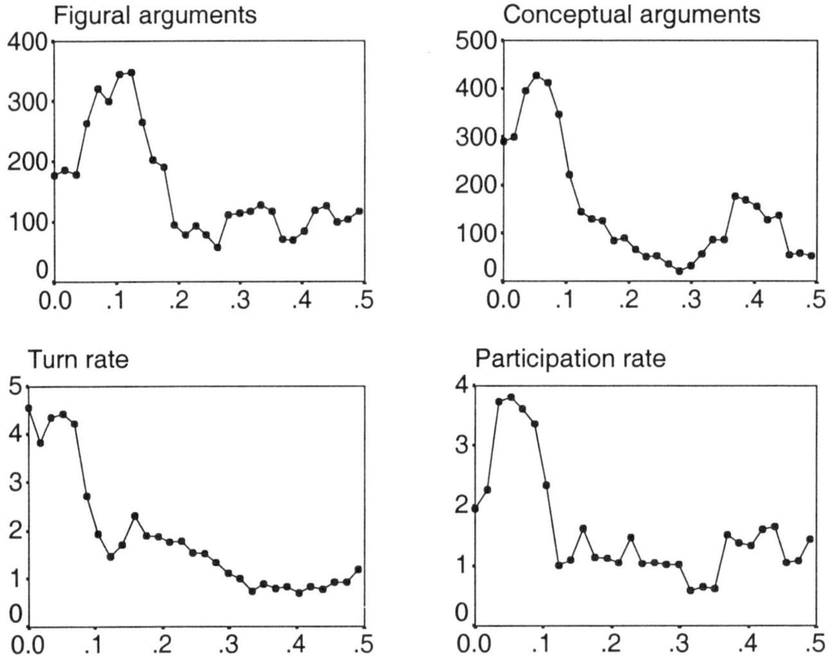

Figure 22.1. Spectral density functions (ordinate) for each frequency cycle (abscissa) of the four dependent measures for Team A

To complete our frequency domain analyses, we examined the phase spectrum of the cross-spectral functions for each pair of measures. The phase spectrum coefficient is a measure of the extent to which the frequency components of one cycle are synchronised with, lead, or lag behind the same frequency components of another cycle. If two cycles share a common periodicity, but are phase shifted so that one leads the other in time, then an entrainment process may potentially be present.

Our analysis revealed that figural argument cycles peaked ahead of turn rate cycles in all six design discussions, and also peaked ahead of conceptual argument cycles in five of the six design episodes. Conceptual arguments, turn taking, and participation cycles were all phased in synchrony with each other. This analysis suggests that figural reasoning plays a prominent *lead* role in the majority of design discussions, not only entraining the phasing of group participation and turn taking cycles, but also entraining cycles of conceptual reasoning.

22.4 Modelling the Pooled Time Series

Whilst our spectral analyses point to the rhythmic patterning of team discussion, they do not have sufficient power to test statistically the extent of entrainment between design argument and team interaction cycles. To accomplish this, a *pooled time series analysis* (Moore *et al.*, 1994) was carried out on the six team discussions combined end-to-end to form a single time series. The basic form of a pooled time series analysis is a linear regression model that treats each 30-second time period for each design episode as a separate case. The pooled time series in the present study provided a total of 353 consecutive time periods. Figural and conceptual argument sequences vary across these time periods and across consecutive episodes, and the purpose of the pooled time series model is to predict changes in group interaction measures associated with these sequences.

First, we needed to remove from the analysis gross differences between team discussions in levels of figural and conceptual reasoning, turn rate, and participation rate. To do this, a series of regression analyses were carried out using a set of dummy variables to represent the six team discussions. These differences accounted for small but statistically significant proportions of the variance on three of the four measures, with values of R^2 ranging from .03 (for figural arguments) to .09 (for conceptual arguments and turn rates). These dummy variables were therefore entered first in subsequent regression analyses to absorb overall differences between design teams in mean levels on these measures, and remove them from subsequent analysis.

Further regression analyses were then applied to partial out the cyclical components of team discussions identified in the spectral analyses reported above. This was accomplished by fitting a series of auto-regressive models in which each measure was regressed on itself at successive time period lags. Each regression model therefore tested for serial dependencies within the four measures, *i.e.* the extent to which scores for each 30-second time period could be predicted from the scores for the immediately preceding time period (lag 1), the time period two steps removed (lag 2), three steps removed (lag 3), and so on. At each stage, we inspected for statistically significant changes in R^2 associated with the addition of successive time lags to the model.

Adding a single time lag was sufficient to account for serial dependency in figural arguments ($R^2 = .22$), turn rate ($R^2 = .14$), and participation rate ($R^2 = .12$). However, two time lags were required to account fully for serial dependency in conceptual arguments ($R^2 = .20$). By adding these lagged variables to the dummy team variables, cyclical components of team discussions were removed for each dependent measure in turn, and subsequent analyses carried out on the standardised residuals.

Summarising the progress of the pooled time series analysis, we found that regression models consisting of five dummy team variables together with one, or at the most two, lagged variables were sufficient to account for differences between teams on each of the four measures in the present study, and for their predictable patterning over successive 30-second time periods. By fitting regression models comprising these variables to each dependent measure in turn, these trends and cyclical components were removed, leaving the standardised residuals for the final test of the cognitive entrainment hypothesis.

22.5 Structural Equations Analysis

We tested the cognitive entrainment hypothesis by applying structural equations modelling to these standardised residuals. This entails first constructing a plausible causal model of the relationships between our measures of figural and conceptual reasoning, turn taking, and participation. To construct this causal model, we first inspected cross-correlations between the four standardised residuals. As might be expected, turn and participation rates were highly correlated ($r = .74$), whilst figural and conceptual arguments were inversely related over time periods ($r = -.41$). Whilst low negative correlations were found between figural argumentation and team interaction measures, conceptual argumentation was moderately strongly associated both with turn rate ($r = .36$) and participation ($r = .38$). Two significant or near-significant first-order lagged correlations also remained after removal of the cyclical components underlying these measures. These suggested an *inhibitory effect* of figural argumentation on conceptual argumentation ($r = -.19$) and turn rate ($r = -.10$) in the ensuing 30-second time period.

These results confirmed the conclusions reached in the frequency domain analysis, and suggested a causal model in which *alternating cycles of design reasoning entrain patterns of team interaction characterised by differential turn and participation rates*. In this model, peak levels of figural argumentation act to inhibit conceptual reasoning and turn taking within the same and immediately following time periods (*i.e.* at lag 0 and lag 1). However, as figural argumentation declines, conceptual reasoning takes over, fostering a brisker turn rate and higher levels of group participation. The path diagram shown in Figure 22.2 represents this causal model, in which facilitatory paths are shown by solid lines, and inhibitory paths by broken lines.

Structural equations analysis was then carried out to test this causal model. In fact, this model proved an excellent fit to the data, with an acceptable normed fit index (.995) and a non-significant chi-squared goodness-of-fit ($\chi^2 = 2.32$, $df = 3$). The path coefficients for all but two of these paths—those predicting turn rates from conceptual argumentation at lag 0 and figural argumentation at lag 1—reached significance in the predicted direction (see Figure 22.2). A revised model that omitted these two paths obtained only a marginally diminished fit index (.984) and a non-significant change in chi-square, so these links could therefore be removed from the model with little explanatory loss. They are shown as faint lines in Figure 22.2.

To test the causal priority of figural over conceptual argumentation, and of both categories of design argument over team interaction measures, we then *reversed* the causal direction of the paths linking these measures in a revised model, and tested its fit to the data. This reversed model obtained an acceptable fit index (.955), but a highly significant and unacceptable chi-square goodness-of-fit ($\chi^2 = 17.23$, $df = 1$). Thus, a model in which causal influence flowed *from* team interaction measures *to* design argument measures proved overall to be a significantly poorer fit. This confirmed our supposition that periods of predominantly figural or conceptual reasoning determined patterns of team interaction in our design meetings, rather than the reverse.

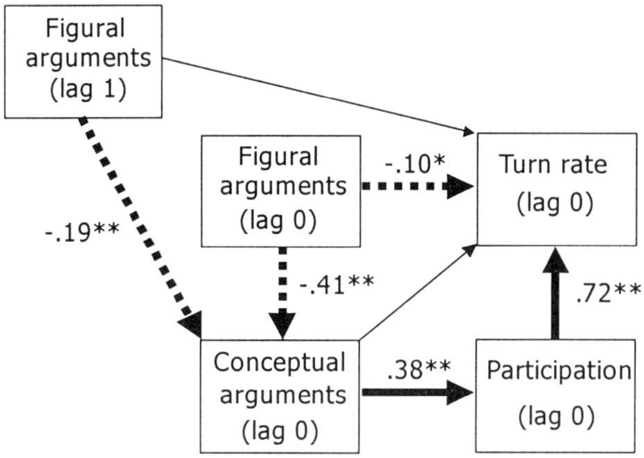

Figure 22.2. Path diagram of causal relationships between design reasoning and group interaction measures, showing significant facilitatory (solid lines) and inhibitory (dashed lines) path coefficients (* $p < .05$, ** $p < .01$, two-tailed). Non-significant paths are shown as fine lines.

22.6 Cycles of Engagement and Disengagement

Our results provide support for the main hypothesis of this paper—the cognitive entrainment hypothesis—that periodic shifts in mode of design reasoning entrain patterns of team interaction. Taking the findings of the frequency domain and time series analyses together, the evidence points to a rhythmic structure underlying these design discussions in which alternating but irregular cycles of design reasoning entrain patterns of team interaction characterised by differential turn and participation rates. Our analyses suggest that figural and conceptual reasoning both cycle unevenly over a 5 to 10 minute period, but are offset in their phasing so that they peak and trough on an alternating basis. Patterns of team interaction oscillate in rhythm with these cycles—peaks in conceptual reasoning are synchronised with peak levels of team participation and turn taking, whilst peaks in figural reasoning coincide with troughs in these measures.

The present study suggests that design discussions alternate periodically between highly interactive sequences of conceptual arguments in which a number of designers exchange brief speaking turns, and extended 'megaturns' of predominantly figural arguments by individual designers, typically accompanied by sketching, pointing, or gesturing. Figural reasoning appears to play a lead role, and elsewhere we have suggested why this might be the case in the design episodes analysed here (Reid *et al.*, 1999). Specifically, the thoughtful and introspective character of these speech sequences—in which engineers struggle to verbalise their design ideas but at the same time direct their gaze, not towards the listener, but to the sketches they are creating—may signal a special state of conversational disengagement. The result is that listeners hesitate to interrupt whilst the speaker is engaged in creative visualisation, even after long pauses in their speech, and may actually require prompting by the original speaker to resume conversation.

These cycles of engagement and disengagement have profound implications for the design of technologies for supporting collaboration between designers working at some distance from each other. Although the majority of the workspace actions observed in the present study could readily be supported by an audio link coupled to a virtual sketchbook (e.g., Tang and Minneman, 1990), the cycles of figural reasoning we observed clearly require the additional capability to monitor the direction of a colleague's gaze and their focus of attention. To do this, listeners need visual access to the shared drawing surface and to a view of the speaker that provides information on their attentional focus. Here we might expect that a *seamless workspace*—a workspace in which designers not only see and manipulate images on a shared drawing surface, but also simultaneously see each other and what each is doing (e.g., Ishii *et al.,* 1995)—would facilitate visual contact and conversational engagement in dispersed design meetings. However, it would also make engagement more difficult to avoid. Our present findings imply that any enhancement of the shared workspace which prevents designers from visibly disengaging from their colleagues is likely to perturb not only the process of visualising design ideas, but also the subtle and precise interactional cues associated with the production and exchange of speaking turns. Finally, our observations also have implications for the management of team meetings and the training of design engineers. It is now well known that novice engineers adopt distinctive problem solving strategies that differ systematically from those of experienced designers. Whilst expert designers take a breadth-first view of the design problem and solution space, searching for underlying principles to guide design ideas, novices tend to engage in continuous generate-and-test activity until an ostensibly acceptable design solution emerges (Smith and Leong, 1998). The design discussions of the pre-expert engineers in the present study were dominated by this oscillating pattern, indicating the presence of potent discussion dynamics pushing design teams towards this natural interaction rhythm. Design managers and educators will need to be aware of these dynamics, and take appropriate steps to emphasise the importance of managerial activities in organising and focusing design discussions.

22.7 Acknowledgements

The work described here was funded by the Department of Psychology, University of Plymouth. We would like to thank Eric Mingo and his colleagues from the School of Manufacturing, Materials and Mechanical Engineering, University of

Plymouth, for their generous co-operation in this study, and the final year students who enthusiastically took part.

22.8 References

Brereton MF, Cannon DM, Mabogunje A, Leifer LJ (1996) Collaboration in design teams: How social interaction shapes the product. In: Cross N, Christiaans H, Dorst K (eds.) Analyzing design activity. Wiley, Chichester, pp 319-341

Goldschmidt G (1991) The dialectics of sketching. Creativity Research Journal 4:123-143

Goldschmidt G (1995) The designer as a team of one. Design Studies 16:189-209

Ishii H, Kobayashi M, Grudin J (1995) Integration of interpersonal space and shared workspace: ClearBoard design and experiments. In: Greenberg S, Hayne S, Rada R (eds.) Groupware for real-time drawing: a designer's guide. McGraw-Hill, London, pp 96-125

Kelly JR (1988) Entrainment in individual and group behavior. In: McGrath JE (ed.) The social psychology of time: New perspectives. Sage, Newbury Park, Cal, pp 89-110

Moore KJ, Osgood DW, Larzelere RE, Chamberlain P (1994) Use of pooled time series in the study of naturally occurring clinical events and problem behavior in a foster care setting. Journal of Consulting and Clinical Psychology 62:718-728

Reid FJM, Reed S, Edworthy J (1999) Design visualization and collaborative interaction in undergraduate engineering teams. International Journal of Cognitive Ergonomics 3:235-259

Smith RP, Leong A (1998) An observational study of design team process: A comparison of student and professional engineers. Journal of Mechanical Design 120:636-642

Stasser G, Vaughan SI (1996) Models of participation during face-to-face unstructured discussion. In: Witte EH, Davis JH (eds.) Understanding group behavior: Consensual action by small groups, vol 1. Lawrence Erlbaum Associates, Mahwah NJ, pp 165-192

Tang J, Minneman S (1990) VideoDraw: A video interface for collaborative drawing. In: Proceedings of the ACM/SIGCHI Conference on Human Factors in Computing, pp. 313-320

Chapter 23

The Influence of Co-designers on the Generation and Evaluation of Solution Alternatives

L.J. Ball and T.C. Ormerod

23.1 Introduction

A principle that pervades the prescriptive design literature is that designers should systematically generate and evaluate a range of solution alternatives to a design problem or sub-problem. For example, Cross (1994, p. 24), in reviewing established engineering-design methods (e.g., as advanced by Pahl and Beitz, 1984) states that 'the emphases here are on performance specifications logically derived from the design problem, generating several alternative design concepts by building up the best sub-solutions and making the rational choice of the best of the alternative designs'. Pugh (1991) likewise makes strong claims for the utility of exploring solution alternatives when he directs designers as follows: 'Remember, you need as many ideas as you can possibly generate - single solutions are usually a disaster' (p. 69); and 'Consideration must be given to a wide range of alternatives without prior commitment to any particular alternative' (p. 221).

Of course, a prescriptive model of effective design practice is quite different to a descriptive account of what takes place in real-world design. Understanding the extent to which the generation and exploration of alternatives actually arises in design remains important for both theoretical and applied reasons. At the theoretical level, understanding solution-development processes - and *how* and *why* they deviate from the norms of prescribed design practice - is vital for a comprehensive account of the nature of design expertise. From an applied perspective, an analysis of solution-exploration in design is valuable for devising methods for design education and techniques for computer-based design support.

In light of the importance of understanding solution-exploration activities in design, the present paper has three key aims. First, it sets out to examine whether claims concerning the importance of multiple solution exploration in design are justifiable from a psychological perspective. Our arguments hinge on a distinction between routine and non-routine design problems (*cf.* Gero, 1990), with our key proposal being that exploring alternatives for the former problems is typically unnecessary for effective design activity whilst for the latter it is vital. Second, the paper aims to review several studies that have explicitly focused on examining the nature of solution-generation and solution-evaluation activities in design. The majority of this research concerns non-routine design situations and demonstrates that individual designers often encounter difficulties arising from failures to pursue alternative design concepts. One limitation of existing research, however, has been its predominant focus on de-contextualised, laboratory-based design activity by single designers in socially-impoverished environments. Whether findings about the limited exploration of alternatives that derive form laboratory-based research with individual practitioners also extend to multiagent design in full-blown, people-rich contexts is an important empirical issue that only appears to have been addressed by a few studies to date. The third aim of this paper is, therefore, to overview the results of an observational study of team design work that we have recently undertaken, which set out to uncover the influences of co-designers (including design managers) on the pursuit of design alternatives. The study centred on a rich time-slice of team design activity being undertaken in the course of the normal day-to-day activity of a major international design company.

23.2 The Utility of Exploring Design Alternatives

The idea that exploring multiple design alternatives has high utility derives from an economic consideration of the design process. In essence, since financial cost (in terms of time and effort) can be high if chosen design ideas subsequently turn out to be misconceived, it is advisable for the designer to defer fixing upon initial solution ideas until a range of (possibly better) options have been fully explored and evaluated. Moreover, it is especially important to defer any premature commitment to initial design ideas in the early, conceptual stages of design since a high-level design solution may subsume weeks or months of detailed design work. It is worth noting that the prescriptive literature rarely claims that designers should strive to generate *optimal* design solutions, since it is recognised that the space of possible solution concepts is very large (and therefore computationally intractable) and that important design requirements (e.g., aesthetic ones) may be subjective and non-quantifiable. Still, an emphasis is placed on deriving the *best* concept from a range of possibilities (e.g., see Pugh's, 1991, method of 'controlled convergence').

From a psychological perspective, Simon (1981) concurred with the prescriptive design literature and made a strong case for the importance of designers pursuing alternative solution ideas. Simon also acknowledged that the solution-search process cannot be an optimising one because of the inherent limitations of the human information-processing system (e.g., in terms of available working-memory capacity). Simon further argued that the impossibility of attaining optimal solutions results in designers applying a 'satisficing' principle by which they strategically search for *acceptable* solutions and stick with these after only moderate exploration of the full search space.

Although both the prescriptive and psychological literatures have attached general importance to the search for alternatives in design, Ball *et al.* (1998) have qualified these arguments by invoking a common distinction (*e.g,* Gero, 1990) between routine and non-routine design problems - where the former are familiar to the designer and have readily-retrievable solutions that are known to be effective, while the latter have novel elements and unknown design solutions. Ball *et al.* argue that since routine design involves the application of known solutions then the generation and evaluation of further alternatives would be wasteful of time and effort. In contrast, non-routine design involves exactly the kind of problems where the exploration of multiple solution alternatives should take place (albeit with choices being driven by the application of a rational satisficing principle).

23.3 Evidence for the Pursuit of Alternatives

Anecdotal reports abound concerning the structure of solution-generation activities in design, and often claims are made that the pursuit of multiple design alternatives is a common aspect of professional design practice. Such anecdotal evidence, whilst offering salient pointers to the *possible* nature of design cognition, clearly requires validation and generalisation through rigorous empirical research. In this respect, only a limited number of published empirical studies appear to have systematically addressed the issue of whether alternative solution options are commonly explored in non-routine design. One pioneering study by Kant (1985), investigated designers developing algorithms for computational geometry, and revealed that individuals rapidly developed a single 'kernel' idea which was subsequently refined. This evidence for a lack of exploration of alternative concepts in design was echoed in other design studies during the 1980s in relation to both software engineering (e.g., Jeffries *et al.*, 1981; Adelson and Soloway, 1986) and mechanical engineering (e.g., Ullman *et al.*, 1988). Ullman *et al.*'s work, for example, demonstrated that designers became fixated on initial solution ideas for *all* parts and sub-parts of the developing design artefact, and revealed little evidence for the consideration of alternative design solutions. In addition, Ullman *et al.* observed that when weaknesses in the original design concepts were later uncovered by designers, they subsequently solved these by 'patching' the design rather than by discarding it and developing a new concept.

These observations concerning solution development with non-routine design tasks were generalised to the domain of electronics engineering in a more recent series of studies conducted by Ball *et al.* (e.g., 1994, 1998). In their initial study, Ball *et al.* (1994) recorded concurrent diary entries from electronics-engineering students independently tackling complex and temporally-extended design projects as part of their final-year degree assessment. Data revealed that these designers rarely generated or modelled alternative solutions but focused instead upon initial (often sub-satisfactory) ideas that were improved iteratively until they reached a state of adequacy. In their subsequent study, Ball *et al.* (1998) obtained concurrent verbal protocols from six expert designers attempting a novel integrated-circuit design problem. The data again revealed a general failure to search for alternative solutions, a marked inclination to stick with early solution ideas even when these were showing deficiencies, and superficial modelling and assessment of competing alternatives when such options were actually considered.

The empirical evidence reviewed so far seems to support a view of solution-exploration behaviour in non-routine design that deviates from the prescriptive view of design as requiring the pursuit of multiple solution alternatives at all levels of design abstraction. It is noteworthy, too, that the studies outlined transcend a number of different design disciplines, problem types and levels of design expertise. Before making premature claims as to the ubiquity of limited solution-exploration in design, however, it is worth considering that - with the exception of Ball *et al.*'s (1994) study - the focus of the research outlined above has involved designers attempting only realistic 'simulations' of design problems under highly constrained laboratory conditions. Although studying design in this way affords a useful degree of control for the researcher, it simultaneously eliminates much of what is central to real-world design activity. The point is, design activity as practiced in commercial companies tends to be: (i) extended over considerable amounts of time (often months and sometimes even years); (ii) contextualised by other people (e.g., co-designers, team managers and clients); and (iii) directed toward design goals that really matter (*i.e.*, the costs of real-world design failure can be very high). Such factors, whether in isolation or interaction, may well have a profound impact upon the structure of the ongoing design process.

One study that made a serious attempt to explore the pursuit of alternatives in design without undermining the real-world situatedness of activity is reported by Olson *et al.* (1996; see also Olson *et al.*, 1992). Their research focused on real design teams working in early software-design meetings. They studied 10 design meetings from four projects in two organisations. All communications were analysed to uncover the structure of design activity, including the generation and evaluation of alternatives. Analyses revealed considerable commonality across meetings in terms of how people spent their time and in the sequential organisation of behaviours. Of particular interest were the observations that: (i) over 21.5% of design time was spent generating and clarifying alternatives (which made the solution exploration the largest category of design activity); and (ii) for 80% of design issues (*i.e.*, design questions or problems) two or more alternatives considered, with the average being 2.5 alternatives per issue. It appears, therefore, that Olson *et al.* have uncovered good evidence for enhanced solution-generation activities in group design compared with individual design in socially-impoverished contexts. One proviso in relation to Olson *et al.*'s findings is their observation that only two thirds of the alternatives generated were explicitly evaluated (*i.e.*, 1.6 alternatives per issue). However, this result still reflects a marked improvement on the meagre exploration of solution alternatives observed with single designers.

One possible interpretation of Olson *et al.*'s results is that the greater rate of solution-exploration observed compared with studies of individual designers may reflect a basic arithmetic pooling of the solution-exploration activity of the separate individual team members (e.g., two designers working alone would, between them, be likely to generate more unique solution ideas than a single designer working alone). Another possibility is that there are important social processes associated with the dynamics of team design that facilitate the generation and evaluation of alternatives. One relevant factor may be the role of design-management strategies in team design. Team managers may proactively motivate the pursuit of alternatives by exhorting group members to seek out design options, or they may rigorously critique putative solutions. Even without such managerial input, design team design meetings - like other professional meeting contexts – may have naturally emerged as important cultural fora for the open airing of opinions and for the critical appraisal of proposed solution ideas. The present study aimed to shed further light on the social dynamics of solution-exploration activity in team design meetings. Our analyses were particularly focused on quantifying solution-generation and evaluation activities across different members of the design team.

23.4 The Study

23.4.1 Research Methodology

Our researcher (LM) videotaped a single design meeting involving all members of a small, established design group working in a major international company in the field of industrial design. The data collection associated with this meeting formed a self-contained fragment within a larger ethnographic research programme being conducted by the authors and LM within three design groups in four different design companies over a one-year period. The particular design meeting that forms the focus of the present paper took place during the second week of LM's sojourn with the industrial design group (*i.e.*, at a point where she had established herself as a familiar individual within the designers' day-to-day work environment).

The meeting lasted 1 hour and 42 minutes and its focus was to review work that had been undertaken for an external client on the conceptual design for a plastic moulding to cover a pressed-steel, fire-proof casing that would house a gas-meter. The required moulding was a sophisticated and multifaceted artefact which included a display section, some form of access drawer or door to a power supply unit, and a removable section to accommodate the possible addition of a swipe-card, pre-payment module. Issues concerning aesthetics, image and usability were particularly dominant in the design discussion, reflecting the design group's desire to maintain a recognisable house style that they had developed over a number of years. The design task itself was ill-defined and complex, with only a few key requirements and constraints having been passed on from the client. Indeed, the requirement to have a plastic frontage for a fire-proof steel unit involved particular complexities (associated with fixing the two housings together) which were further compounded by the client's requirement that the fixing should be based around a 'diagonal-split' concept. At least half of the meeting was devoted to reviewing how a visually effective functional solution to this fixing issue might be achieved.

The meeting involved all six members of the group, that is, the Team Manager, the 'Project Champion' who had been assigned the lead role in designing the artefact, and four other designers. During the meeting LM sat with the rest of the group and engaged only minimally in the ongoing discussions. Her main activity involved making ethnographic field notes that included details of seating positions, apparent group roles, information resources utilised during discussions, and key events arising during the course of ongoing design activity (see Ball and Ormerod,

2000, for a broad-based discussion of findings associated with our ethnographic stance on design research in this, and other, design companies). The meeting took place in the design group's open-plan office (the manager usually occupied an adjacent personal office). The group's activities were simultaneously open, friendly and supportive at one level and focused, professional and critically incisive at another level. Design meetings were an established working practice that formed a crucial aspect of the daily activity of this group. It was clear from interviews with the designers and the manager that these meetings were viewed as an important critical-review and decision-making forum where good design ideas could be steered and improved and where weaker design ideas could be rejected.

23.4.2 Data Coding and Reliability Checks

The videotape of the design meeting was audio-transcribed into continuous written text that was segmented according to the originator of particular utterances. The text was then coded according to the Questions, Options and Criteria (QOC) formalism developed by MacLean *et al.* (1991) for capturing the essential elements of a design rationale. Within the QOC scheme, *Questions* describe key design issues or problems, *Options* describe possible solution alternatives to design issues, and *Criteria* describe dimensions on which options can be evaluated (e.g., cost, size, power consumption *etc.*). The Q, O and C elements of this formalism are isomorphic with the Issue, Alternative and Criterion elements of the coding scheme adopted by Olson *et al.* (1992, 1996) for coding design-meeting transcripts. Our use of QOC in preference to Olson *et al.*'s issue-based terminology was pragmatic in that it enabled us to maintain coding consistency with a larger body of data that had already been collected as part of our ongoing research programme (see Ball *et al.*, 2000, for discussion of the relative merits of QOC- and Issue-based schemes).

Applying the QOC scheme to the meeting transcript involved scrutinising the text for embedded instances of Questions, Options and Criteria. Options and Criteria were always explicitly stated during design activity. Questions, however, were often implicit within the ongoing discussion - although they were always readily inferable from surrounding contextual information. During the coding process all implicit Questions were converted into explicit Question statements. A coded extract from the meeting transcript, presented below, captures a few minutes of verbalisation produced by the Project Champion who was leading the conceptual design of the meter-housing. The coded extract depicts a constellation of QOC elements and includes a central Question that was explicit.

> *'The challenge is to do something that fits this - this visual house style .. but copes with the problems dictated by that – by that split* ***[Question-3]****. One thing I've done is this step-up here* ***[Option-3.1]****. This is done for partic- this is done to get the fixing - this boss here above the level of the flange, so that you come on here with a spanner to tighten the thing up, this isn't in the way* ***[Criterion-3.1]****. You could just take it straight across and have a taller boss* ***[Option-3.2]****. But that's expensive* ***[Criterion-3.2]*** *and also, they want to try and make this structure strong enough so they don't have to put in an extra plate across those two bosses* ***[Criterion-3.3]****.'*

A reliability check of transcript-coding accuracy was undertaken. This involved the first author coding the full transcript, with an interval of 14 months intervening between the first and second coding events. High reliability was achieved (*i.e.*, 91% of QOC elements were coded identically between first and second passes through the transcript). The 9% rate of reliability error did not reflect any mis-coding of QOC elements but rather arose from coding omissions between coding events (e.g., some of the Criteria that had been identified during the first coding were overlooked during the second coding).

23.4.3 Results and Discussion

Table 23.1 shows the number of initial mentions of distinct QOC elements across the design meeting, broken down according to whether the elements originated from the Project Champion, the Team Manager or the remaining four co-designers. Options and Criteria are described as being 'distinct' relative to their associated Question. Thus any mention of 'component costs' as a criterion in relation to two *different* questions would lead to two separate scores being entered into Table 23.1.

From this table it can be seen that the Project Champion dominates the production of initial QOC elements with almost 70% of all first mentions of Questions, Options or Criteria being produced by him. Intriguingly, too, the vast majority of the Options considered during the meeting (*i.e.,* 22 out of 25) were generated by the Project Champion. Still, though, there remains a significant role for the rest of the design team in outlining QOC elements, with by far the greatest amount of activity centring on the generation of new Criteria that had not been mentioned by the Project Champion. The relatively high frequency of Criteria generated by the Team Manager and remaining designers appears to emphasise their roles as critical appraisers of Options generated by the Project Champion (*i.e.*, Criteria are typically associated with evaluation activities – *cf.* Olson *et al.*, 1996). Interestingly, the Team Manager generated nearly as many novel Criteria as did the four remaining co-designers in combination, which attests to his more central evaluative influence within the ongoing discussion.

In the present study the mean generation rate of Options per Question was 2.78 (with the range being from one to six Options per Question). This is comparable to the mean rate of 2.5 alternatives per issue noted by Olson *et al.* (1996) for their analysis of 10 design meetings. Unlike Olson *et al.*'s (1996) study, however, where, on average, only 66% of alternatives were evaluated, the present study revealed that over 95% of Options were followed up with the generation of at least a single evaluative Criterion. The average rate of generation of Criteria per Question was 7.4 in the present analysis (range = one to 17 Criteria per Question), again supporting the high level of evaluative activity being undertaken by the team.

Table 23.1. The distribution of initial mentions of distinct Questions, Options and Criteria across members of the design team

Meeting Characteristics	Total for Project Champion	Total for Team Manager	Total for Remaining Designers	*Combined Total for Team*
Questions	8	0	1	*9*
Options	22	0	3	*25*
Criteria	40	13	14	*67*
QOC Total	*70 (69.3%)*	*13 (12.9%)*	*18 (15.8%)*	*101*

Table 23.2 summarises the distribution of repeat mentions of Options and Criteria across members of the design team (where any element could be repeated from one to N times). Unlike the data presented in Table 23.1, these results indicate a greater balance between the repetition of Options and Criteria by the Project Champion when compared with the rest of the design team. This result confirms the subjective impression noted by the ethnographer that the co-designers are contributing heavily to the collaborative development and rigorous evaluation of the Options that were forwarded by the Project Champion. One further difference between Tables 23.1. and 23.2 relates to an apparent contrast in the design roles of the Team Manager and the remaining co-designers – where the former tends to be more inclined to generate novel design Criteria than to re-iterate previously-mentioned Criteria, and the latter tend to be more inclined to re-iterate Criteria than to generate them. Indeed the Criteria that the Team Manager did repeat were primarily unique ones that he had mentioned in the first place. The fact that the generation of novel design Criteria dominated the behavioural repertoire of the Team Manager suggests that he may have been taking a broader and more penetrating evaluative stance on developing design ideas than other co-designers.

Table 23.2 The distribution of repeat mentions of Options and Criteria across members of the design team

Meeting Characteristics	Total for Project Champion	Total for Team Manager	Total for Remaining Designers	*Combined Total for all Designers*
Repetition of Options	12	0	3	*15*
Repetition of Criteria	33	8	32	*73*
Repeated OC Total	*45 (51.1%)*	*8 (9.1%)*	*35 (39.8%)*	*88*

23.5 Conclusions

Most existing research investigating solution exploration in design has focused on individuals working in contrived laboratory contexts where they are isolated from design colleagues, including design managers. In this paper we have reported an observational study of a company-based, industrial-design team engaged in a meeting to discuss progress on an ongoing conceptual-design project. The data indicated that in this collaborative context a variety of social and managerial influences appear to come into play that function to increase the breadth and depth of solution exploration over what is typically observed with individual designers.

One intriguing aspect of our results in this respect is that the generally high mean number of Options that were generated per design Question derives from the activity of the Project Champion himself, and does *not* reflect the direct input of novel solutions by the rest of the design team. One way in which the meeting context might facilitate the lead designer's quantity of solution-generation activity is motivational in nature. The point is, the Project Champion is likely to know very well from prior experience that a high value is placed on the generation and evaluation of alternatives within the design-review forum, and that he is likely to be challenged to justify ideas in relation to evaluative criteria that are thrown at him by the Team Manager and other co-designers. Faced with this scenario, the sensible stance for the Project Champion is clearly to pre-empt as much as possible any potential embarrassment that may ensue from the presentation to the team of a minimal range of (potentially weak) design alternatives that he is unable to support in relation to key design Criteria. As such, the Project Champion appears to enter the design meeting well armed with alternative Options for design issues, and a good knowledge of the relative strengths and weaknesses of these Options in relation to design Criteria. Indeed it is notable that the Project Champion had at least one CAD image (usually in perspective view) for each design Option that he presented to the team for discussion. In the actual meeting itself, further benefits to the design process subsequently derive from the generation of novel design Criteria by the rest of the team (especially the Team Manager) and the thorough evaluation of Options against Criteria. Through these various mechanisms, then, it appears that premature commitment to single design options is a phenomena that gets 'managed out' of team design (*cf.* Ball and Ormerod, 2000).

Whether our results are specific to the meeting that we observed, or whether they generalise across other team meetings and to other groups in other companies, remains an important empirical question that we are currently investigating. A further important step for research on the exploration of alternatives in team design would be to assess whether unmanaged collaborative design activity outside of design-review meetings also shows evidence for important strategic elements that are focused on increasing the breadth of options considered and the depth of their evaluation. Clearly, too, there are a variety of factors which may impact upon the solution-generation and solution-evaluation processes of design teams, including the cohesiveness of the group, managerial expectations concerning the role of design-review fora, and how well established the Project Champion is within the team. Still, our preliminary results, together with those of Olson *et al.* (e.g., 1996) suggest that the enhanced exploration of design alternatives may form an essential element – possibly even the *raison d'être* – of collaborative design meetings.

23.6 References

Adelson B, Soloway A (1986) A model of software design. International Journal of Intelligent Systems 1:195-213

Ball LJ, Evans JStBT, Dennis I (1994) Cognitive processes in engineering design: A longitudinal study. Ergonomics 37: 1753-1786

Ball LJ, Lambell NJ, Ormerod TC, Slavin S, Mariani JA (2000) Representing design rationale to support innovative design re-use: A minimalist approach. Automation in Construction, in press

Ball LJ, Maskill L, Ormerod TC (1998) Satisficing in engineering design: Causes, consequences and implications for design support. Automation in Construction 7: 213-227

Ball LJ, Ormerod TC (2000) Putting ethnography to work: The case for a cognitive ethnography of design. International Journal of Human-Computer Studies, in press

Cross N (1994) Engineering design methods: Strategies for product design, 2nd edn. Wiley, Chichester

Gero JS (1990) Design prototypes: A knowledge representation schema for design. AI Magazine 11: 26-36

Jeffries R, Turner AA, Polson PG, Atwood ME (1981) The processes involved in designing software. In: Anderson JR (ed.) Cognitive skills and their acquisition. Erlbaum, Hillsdale, NJ, pp 255-283

Kant E (1985) Understanding and automating the algorithm design process. In: Proceedings of the International Joint Conference on Artificial Intelligence. Morgan Kaufmann, Los Altos, CA, pp 211-219

MacLean A, Young RM, Bellotti VME, Moran TP (1991) Questions, Options, and Criteria: Elements of design space analysis. Human-Computer Interaction 6: 201-250

Olson GM. Olson JS, Carter MR, Storrøsten M (1992) Small group design meetings: An analysis of collaboration. Human-Computer Interaction 7: 347-374

Olson GM, Olson JS, Storrøsten M, Carter MR, Herbsleb J, Rueter H (1996) The structure of activity during design meetings. In: Moran TP, Carroll JM (eds.) Design rationale: Concepts, techniques, and use. Lawrence Erlbaum Associates, Mahwah, NJ pp 217-239

Ormerod TC, Mariani JA, Ball LJ, Lambell NJ (1999) Desperado: Three-in-one indexing for innovative design. In: Sasse MA, Johnson, C (eds.) Proceedings of the Seventh IFIP Conference on Human-Computer Interaction—INTERACT '99. IOS Press, London, pp 336-343

Pahl G, Beitz W (1984) Engineering design. The Design Council, London [Original German text, 1977, Springer-Verlag, Berlin].

Pugh S (1991) Total design: Integrated methods for successful product engineering. Addison-Wesley, London

Simon HA (1981) The sciences of the artificial, 2nd edn. MIT Press, Cambridge, MA

Ullman DG, Dietterich TG, Stauffer LA (1988) A model of the mechanical design process based on empirical data. Artificial Intelligence for Engineering Design, Analysis and Manufacturing 2: 33-52

Chapter 24

Small Group Design Activity and Requirements on Collaborative Technologies

S. Turner and N. Cross

24.1 Introduction: Focus and Approach

In the last decade or so, many collaborative technologies have been developed which aim to allow people in geographically distributed locations to work together as if sharing the same space. Such tools range from the simple asynchronous resources of email and file transfer, to fully-featured collaborative virtual environments allowing virtual groups of designers to interact with virtual designs in real time. Despite many reported practical experiments with such tools, it is far from clear which flavour of collaborative technology best suits collaborative design. This study takes the approach of studying one particular form of design collaboration – the small group design meeting – and briefly discusses how such interaction could be technologically supported in virtual space. In particular, we will focus on the use of shared design workspace.

Many formal studies of design activity have been based on protocol studies of individual designers in experimental situations. However, design practice in the real world is more usually a small-group activity carried out over an extended period of time, with the group members involved in a variety of design projects and other tasks. Although this kind of activity is more difficult to study in a formal way, it is necessary to conduct such studies in order to understand more fully this normal design activity and to verify the findings of the other kinds of studies. This paper reports on such a real-world study. A three-person team of designers (within a larger organisation) was studied for a two-week period, whilst they worked on a software engineering project. This work complements studies such as the design team extensively studied in the 'Delft Protocols Workshop' (Cross *et al.*, 1996) and the work of Tang (1991). The work reported here concerns meetings that were largely focussed on informal review and associated design development. The analysis follows Olson *et al.* (1992) in their approach to the observation of software design meetings, and uses their classification scheme for design process.

24.1.1 The Design Team in Context

The software designers for this case study were the three members (project manager and two senior software engineers) of the 'Mallard' project team at a small software house specialising in research and development. Mallard was a collaborative project that aimed to provide visibility for evolving software products, and an integrated product development process and tools. The team's project was a configurable graphical animation workbench for the visualisation of specification and design notations. At the time of the case study, a specification had been completed and the team were working on the high level design of the animation tool. Activities during the two weeks of the study consisted of periods of individual work, some of which related to other project matters, interspersed with team meetings. The team's desks were close together, in part of a large open-plan office. There was frequent informal communication between team members. However, sustained discussions tended to disturb others, so these were held in the meeting room and provided the data for this study. This small room was equipped with a large table, chairs and two portable whiteboards.

In engineering practice, design meetings are islands in a stream of other activities. Moreover, decisions are rarely reached in the narrow context of the task-in-hand, but must take account of organisational constraints and work undertaken on other parts of large projects. An evolving design frequently exists before the start of a meeting, and is usually documented, while numerous other documents embody design history or details of related parts of the project. Equally, information must be captured during the meeting for the designers themselves for others in the next stage of development. Mallard displays many of these characteristics. All this documentation inevitably means many different paper resources are referenced, and often annotated, both individually and collaboratively. The interweaving of group and individual tasks also influences the type of design work undertaken in meetings. The Mallard meetings all drew on individual work by team members. The protocols show that typically, one of the team would explain the ideas he had developed, while the others questioned, argued and sometimes suggested alternatives.

24.1.2 Data Collection Methods

Five Mallard design meetings were video-recorded, resulting in over 7.5 hours of tape which were transcribed as verbal protocol and form the basis of the ensuing analysis and discussion. Data could not be collected from all meetings: the team could not be expected to schedule all their meetings for when the researcher was available and indeed this may have distorted work patterns. The choice of meetings was therefore opportunistic, but there is no reason to think that the selection was atypical. The meetings studied were held in the usual room, with the usual range of furniture and equipment, supplemented by the video camera. The camera was positioned to take in the whiteboards and the team members as they sat around one end of the meeting room table. The only special arrangement for the benefit of the study was that no one sat at the end of the table nearer the camera.

24.2 Representation and Workspace

A range of physical media was used in each Mallard meeting:

- whiteboard text and graphics, both finished and under construction;
- text and graphics in documents brought into the meeting;
- sketches and notes in notebooks/pads, and on the documents themselves.

However, there are other, transient representations in the ephemeral media of speech and gesture and, of course, the cognitive representations belonging to each team member. The current design therefore, and transformations upon it, are at any one point spread among a number of distributed representations. Activity in design meetings is thus an instance of distributed cognition, defined by Hutchins (1995, p.49) as: '...computation observed in the activity of the larger system [that] can be described in the way that cognition has been traditionally described - that is, as computation realized through the creation, transformation and propagation of representational states.'

Here the 'larger system' is the design team, the 'representational states' the various textual notes, diagrams, sketches, and less concrete forms mentioned above. The consequence is that participants have to maintain awareness of, and navigate between, all these different parts of the cognitive task. Examination of the videotapes and the associated verbal protocols shows how the physical representations were distributed among the different workspaces, where a workspace is any medium used to develop or represent the design or elements thereof.

The whiteboard was used in two modes: (i) to represent design elements deriving from an earlier meeting or session of individual work - individuals might add material to the whiteboard before the others arrived, or transfer it from notes during the meeting, explaining as they drew; (ii) as an exploratory tool for sketching new ideas and modifying old ones.

Printed documents, both formal project documents and individual work-in-progress, sometimes with notes made in reviews outside meetings, were used for reference to specifications and to earlier versions of the design, and sharing of material was necessary from time to time, since not everyone always had material to hand. However, as well as grounding discussion, documents were also a resource for active design work, when individuals annotated them with modifications and suggestions during the meeting.

Personal notebooks/pads were used for comments and notes, often detailed and sometimes explicitly intended for use as a group resource for activities after the meeting. Members also used them as an *aide memoire* for describing ideas.

An analysis of the focus of attention during meetings was carried out to establish the relative proportions of time spent in each workspace resource, and the sequence of transitions. Each tape was viewed and for each 1-minute segment it was noted whether the group's attention appeared to focus on the whiteboard or on any paper documents. (The unit of 1 minute was pragmatically determined by the somewhat coarse-grained time-stamping feature of the recording equipment used.) The focus was inferred from behavioural indicators: the site where writing or drawing was being carried out; the direction of gaze; the referent of gestures or deixis such as 'over there' or explicit verbal reference. Clearly not all gaze, gestures and deixis could be disambiguated, but sufficient information could be extracted to support focus identification. If the focus shifted during the segment, both were noted. Similarly if one team member was consulting documents while the others were looking at the whiteboard, both were again noted. In all meetings, there were periods when attention was divided between resources or switched rapidly between them, and overall, the team spent approximately half the aggregated meeting time focusing on primary workspace resources. A single observer, the first author, carried out focus classification. Each tape was viewed twice and coding checked so as to minimise 'drift'. The tapes were initially coded over a period of several weeks, then checked in a single session. Inevitably, with increased familiarity with the material, some segments were re-categorised on the second pass through the material.

Unsurprisingly, those meetings where discussion is more exploratory tend to focus around the whiteboard, while if documents are under review they claim the major share of the team's attention. Those meetings where both whiteboard and documents are exploited are of particular interest. In the meeting where the greatest number of instances of dual focus occurs, they account for almost 47% of the total. In such cases, attention switches rapidly between the dynamic design representations on the whiteboard and (usually) more static representations in documents. Such switches are common in the protocols. They are often mediated by deictic gesture, but space does not permit further discussion of this aspect of group interaction.

Some indication of the validity of these observations for other domains is provided by a comparison with the pattern of workspace use in the video and protocol data for the Delft workshop (Cross *et al.*, 1996). This example of product design - a device to carry a backpack on a mountain bike - was undertaken by a group of three experienced designers. From a design brief, the designers had to devise and produce a costed concept design with supporting sketches, within a two-hour time limit. The group had several shared workspaces at their disposal:

- documents containing the design brief;
- large sheets of paper, placed on the table;
- a whiteboard, of the sort from which copies may be made;
- a sample backpack and mountain bicycle.

An analysis of focus patterns is provided by Radcliffe (1996), who observes how work moves easily between different workspaces over the two hours of the design task, and that workspaces are often used in parallel by group members. Early work focuses mainly on documents, then attention moves to the whiteboard and artefacts, reverting to documents and sketches on the shared paper as the design is finalised. Closer inspection also shows that group members sometimes use corners of the shared paper or the whiteboard for independent work in a similar fashion to the use of private notepads by the Mallard team. The video data also show that the artefacts provided are sometimes used in tandem with the whiteboard to explore ideas physically as they are developed in sketches or text. For both groups, then, we have an overall picture of people moving between workspaces in a fluid and flexible way, sometimes focusing entirely on one workspace, sometimes working in parallel using different resources, and sometimes using one medium to supplement the affordances of another.

24.2.1 Ownership of Workspaces

It might be expected that an individual's notes and documents are private, whereas public space is confined to explicitly shared areas - the whiteboard in this instance, but others would be flipcharts, large sheets of drawing paper and so on. And indeed ownership of the different workspaces follows this pattern for most of the time. However, on a number of occasions, the boundaries between public and private are more fluid. Private space is occasionally shared, predominantly by pointing at someone else's material, and interestingly, the sharing is generally not initiated by the owner of the workspace. This is only the subject of comment among group members in one case - in all other instances permission to enter private space appears to be implicit and ephemeral. The overall frequency of such boundary shifts is:

- pointing at material on another person's document or notes (25 instances);
- glancing at another person's document or notes (5 instances);
- writing or drawing on another person's document or notes (2 instances);
- pointing at own document to show another person (1 instance);
- picking up another person's document or notes (1 instance);
- writing on own notes and showing result to another person (1 instance);
- converting own document into a group resource (1 instance).

24.3 Group Design Process

Design process analysis has been employed by a number of authors, e.g., Guindon (1990), Olson *et al.* (1992). Most authors use categories derived from design theory and from work on design rationale such as Moran and Carroll (1996). Core design categories usually include: identifying issues or problems; proposing solutions to these problems; identifying criteria for deciding between the solutions; deciding upon a solution. In addition some authors add categories for ancillary activities of the following general types: summarising progress; 'walking through' solutions. These categories can apply to both individual design work and work undertaken as a group project. In order to include activities specifically related to work in groups, three extra categories are commonly used (and most authors also include 'digression' and a very necessary 'other' category): seeking and providing clarification; managing the meeting process; managing the wider design project.

The design process categories used in the present analysis derive from Olson *et al.* (1992), thus allowing comparisons to be made between the current results and those reported in that study. The activity categories are: Issue, Alternative, Criteria,

Clarification, Summary, Walkthrough, Goal, Meeting Management, Project Management, Digression and Other. An analysis of the time spent on the activities across the five meetings shows a broadly consistent profile, although a smaller proportion of time was spent on design alternatives in meeting 4 (when the main subject was the structure of the design document) and the time devoted to walkthroughs decreases over the series. The Pearson correlations between the content of individual meetings are high, ranging from 0.84 (between meeting 1 and meeting 4) to 0.99 (between meeting 2 and meeting 3). Much more time is spent on the exposition of design alternatives than on any other activity, even in meeting 5, when the design concepts are being finalised. However, it should not be inferred from this that no progress has been made by the end of the last meeting. A consideration of the overall purpose of each meeting and of the contents of the verbal protocols indicates that the different sessions developed different areas of the design and at different levels of detail.

A complementary explanation of meeting content is that the team appears to have adopted what Potts and Catledge (1996) have termed the 'Inquiry Cycle', whereby design ideas are first developed in activity outside meetings, then presented and discussed and finally refined after such discussion. This could account for the comparatively large amount of time spent on explaining alternatives at the expense of applying criteria, a ratio of just over 5 to 1. Examining the sequence of different activity categories appears to indicate a depth-first, solution focused approach similar to that identified by Guindon (1990) (analysing software engineers) and discussed by Singley and Carroll (1996) in their proposed taxonomy of design reasoning. However, it should be noted that this has not been subjected to statistical analysis. The correspondence between the pattern of activity displayed by the Mallard team and the teams of software designers studied by Olson *et al.* (1992) is only moderate. Across all categories the Pearson correlation between the Mallard and Olson *et al.* category means is 0.63. In contrast to Mallard, Olson *et al.* note regular sequences of issues raised, followed by suggested alternatives and the application of criteria. However, without original data from that study, it remains uncertain whether the cause of the discrepancy is an artefact of coding practice, a real difference in design style or some other factor.

24.4. Conclusion: A Scenario-based Review

In recent years, scenarios have become a well-used tool in the analysis and design of human-computer systems. They have many roles, from the documentation of existing practice, through validation of understanding with users, to requirements generation and the envisionment or evaluation of future options (Carroll, 1995). Accordingly, a scenario has been structured to encapsulate the findings of this small group study, projected into a situation where the designers are no longer co-located, and fleshed out with background detail (space does not allow further discussion of this here). The scenario is then used to consider how existing technologies might support distributed design activities and to note what organisational issues might arise in their introduction.

24.4.1 An Illustrative Scenario

Company X is a small software house with premises spread over several sites. This scenario concerns the work of a team of three engaged on the design of a new case tool. The project has been running for some weeks, employing a pattern of individual work interspersed with co-ordination and review meetings. These latter draw on original project documentation and paper design documents of varying degrees of formality generated by members of the team. Few explicit formal sign-off procedures are mandated, although at various specified stages QA documentation is produced which must bear the team leader's signature. Otherwise, design developments or changes are implicitly authorised by the team leader during meetings. Project teams are normally co-located, but in this case one person has recently been seconded to another site to work part-time on another project. The sites are networked, providing email, shared file servers and a company intranet. All software designers are necessarily highly familiar with the PC technology and communication tools that form their everyday working environment, and all have an individual machine. Here it is the absent team member who suggests that the application sharing and videoconferencing technology already owned by the company for the purposes of demonstration to clients could be put to good use in saving him tedious travel through the conurbation where the company is located.

24.4.2 Potential for Technological Support

Tools already in use by the team will help to support co-ordination when the team members are working separately, with increased use of email and phone in substitution for casual face to face contacts. For more substantial interaction, application sharing may be helpful. The sharing and integration of the results of distributed individual work will be facilitated by one of the applications providing structured, shared workspaces, such as Lotus Notes. None of this is likely to meet with resistance, since no significant change in working practice is involved, and members of the team are generally keen to experiment with new tools – it is a normal part of their job. It is opportunistic contact – the casual glance at a diagram over someone's shoulder which triggers a new design suggestion – which will be lost. At present, the only potential means of supporting this would be the type of 'office share' video link which has so far proved to be relatively unsuccessful in the support of technical tasks.

As for meetings themselves, the potential of current systems must be considered in the context of the rich and intricate patterns of workspace we have described above. At present only partial solutions exist and these tend to be neither elegant nor sufficient, particularly for moving between workspaces. Furthermore, for the near future some of these workspaces will be paper-based, which raises the perennial problem of integrating paper and electronic media. Workspaces are by default either shared or private, but occasionally private workspaces are shared. So far, technical solutions go only part way to supporting the easy and flexible management of public/private boundaries displayed by groups who habitually work together. Finally, in the absence of clearly articulated gesture, meetings will be more effortful as designers will be required to spell out explicitly change of reference both within and between workspaces. At present, deictic gesture, indeed gesture of any sort, is not a medium which is supported effectively by technology, even in the form of collaborative virtual environments, although recent applications such as that described by Reynard *et al.* (1998) are beginning to address the issue. Techniques which afford the close integration of directly shared workspace and body image, e.g., Ishii and Ohkubo (1990), Scrivener *et al.* (1994), appear to have worked well for design dyads, but have not supported larger numbers or multiple workspaces. Thus at the current state-of-the art, meetings entailing substantive design work between remote participants are unlikely to be fully satisfactory, while routine progress reviews may be adequately supported by asynchronous means.

24.5 References

Carroll JM, Ed. (1995) Scenario-Based Design: Envisioning Work and Technology in Systems Development. Wiley

Cross N, Christiaans H, Dorst K (eds.) (1996), Analysing Design Activity. Wiley.

Guindon R (1990) Designing the Design Process: Exploiting Opportunistic Thoughts. Human-Computer Interaction 5 (2-3): 305-344

Hutchins E (1995) Cognition in the Wild. MIT Press

Ishii H, Ohkubo M (1990) Design of TeamWorkStation: A Realtime Shared Workspace Fusing Desktops and Computer Screens. In Gibbs S and Verrijn-Stuart AA (eds.) Multi-User Interfaces and Applications. Elsevier, pp131-142

Moran TP, Carroll JM (eds.) (1996) Design rationale: Concepts, techniques and use. Lawrence Erlbaum Associates

Olson GM, Olson JS, Carter MR and Storrøsten M (1992) Small Group Design Meetings: An Analysis of Collaboration. Human-Computer Interaction 7: 347-374

Potts C, Catledge L (1996) Collaborative Conceptual Design: A Large Software Project Case Study. Computer Supported Cooperative Work 5 (4): 415-445

Radcliffe D (1996) Concurrency of Actions, Ideas and Knowledge Displays Within a Design Team. In Cross N *et al* (eds) Analysing Design Activity. Wiley, pp343-364

Reynard G, Benford S, Greenhalgh, C and Heath, C (1998) Awareness Driven Quality of Service in Collaborative Virtual Environments. In Karat C-M *et al* (eds.) Proceedings of CHI'98. ACM Press, pp464-471

Scrivener SAR, Clark SM, Keen N (1994) The LookingGlass Distributed Shared Workspace. Computer Supported Cooperative Work 2:137-157

Singley MK, Carroll JM (1996). Synthesis by Analysis: Five Modes of Reasoning that Guide Design. In Moran TP and Carroll JM (eds.) Design rationale: concepts, techniques and use. Lawrence Erlbaum Associates, pp241-265

Tang JC (1991) Findings from observational studies of collaborative work. International Journal of Man Machine Studies 34 (2):143-160

Chapter 25

Comparing the Process Characteristics of Graphic and Sentential Idea-generation Techniques for Design Groups

R. van der Lugt

25.1 Introduction

Almost all documented techniques for stimulating creativity are based on sentential modes of expression while for designers visual expression, and especially sketching, is a key activity in the process of originating new product ideas (Fish and Scrivener, 1990; Goldschmidt, 1991). Earlier research suggests that there are substantial differences between the sentential and graphic idea-generation processes (Van der Lugt, 1999).

Including sketching into creative problem-solving appears to bring some kind of evaluation mechanism into the divergent thinking process. To be able to make an idea sketch, basic judgements have to be made about shape and form. In addition, the sketching process takes more time than the process of writing down an idea in abbreviated form. This gives the designers time to involve themselves in a quick reflective cycle while sketching their ideas. Several of such cyclical descriptions of the idea sketching process have been suggested, such as 'imaging-presenting-testing' (Zeisel, 1981), and 'present-test-cycle' (McKim, 1980).

We propose that graphic idea-generation techniques are, to a greater effect than sentential techniques, based on a cyclical process of idea production and reflection. Sentential techniques are predominantly based on a process of rapid association and quantity of production. In effect this means that for graphic idea-generation techniques, the number of ideas per time unit will be smaller, but there will be more use of the knowledge gained, signified by stronger building on earlier ideas.

To investigate whether this proposition holds ground, we compared the characteristics of the idea-generation processes of two associative idea-generation techniques. One of these techniques was taken as representative of graphic idea-generation, the other as representative of sentential idea-generation. We conducted four experimental meetings with design groups. In each meeting ideas were generated with both the graphic and the sentential idea-generation techniques. The order in which these techniques were applied was varied. The method used for analysing the data is a modification of linkography (Goldschmidt, 1998), which looks at the linking between ideas as a means for understanding the idea-generation process.

25.2 Two Representatives of Graphic and Sentential Creative Problem-solving Techniques

Two creative problem-solving techniques are selected to represent their categories of modes of expression. We concentrate on associative idea-generation techniques, in contrast to confrontative and structural techniques (Buijs, 1980). The core mechanism of associative techniques is freely building on each other's ideas. Association is the basic mode of operation in creative problem-solving meetings.

In this experiment we used a slight adaptation of the brainstorming technique (Osborn, 1953), called 'brainstorming with post-its™' (Isaksen *et al.*, 1994). In contrast to regular brainstorming where the facilitator writes down the ideas on a flipchart, during brainstorming with post-its™ group members write their own ideas on large post-its™ and then hand them to the facilitator. For ease of reading, in this paper we use the term 'brainstorming' whenever we refer to the 'brainstorming with post-its™' technique.

The representative of graphic creative problem-solving techniques used is named braindrawing. This technique is a visual modification of the more widely known brainwriting technique (Geschk *et al.*, 1973). Designers individually sketch their ideas on sheets of paper. After a few minutes, they briefly explain their idea sketches, pass on the sheets, and then continue to sketch ideas, based on the ideas that are on the sheet already. Usually, about four to five rounds of generating ideas take place. In between rounds, the facilitator may emphasise the 'building on other ideas' guideline.

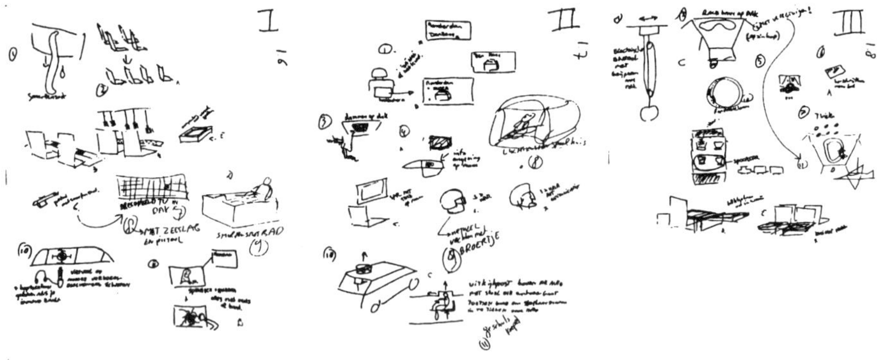

Figure 25.1. Three out of five flipcharts with idea sketches produced using braindrawing in the first experimental meeting

25.3 Method

The differences between graphic and sentential creative problem-solving techniques were explored in four experimental meetings. In each meeting both the braindrawing and the brainstorming techniques were applied. The meetings were moderated by two experienced professional creative problem-solving facilitators. Each meeting was videotaped. Ideas were recorded on flip charts that were posted on the walls. In each meeting, an observer took note of potential links between ideas and indications for these links. In each meeting a researcher was present to handle the video cameras, to control the timing, and to provide directions for the facilitator when needed.

25.3.1 Experiment Description

25.3.1.1 Task

An assignment was developed by keeping in mind both the requirements for tasks for design experiments (Dorst, 1996), and the requirements for problems to be suitable for creative problem-solving, such as an explicit need for many, varied or unusual ideas and, a clear understanding of the problem to be solved (Isaksen *et al.*, 1994). The following design assignment was developed with these requirements in mind: 'How to make travelling by car fun for children?' The designers were asked to generate ideas for products for a particular multi-functional family car with a flexible interior.

In order to provide enough of an understanding of the problem, large cards with key information were provided, covering, among other things: car characteristics; ergonomics; and playing behaviour of children. Before each meeting, the researchers briefly explained the design task and its context with the use of these cards. During the meetings the cards were placed on the table so that they were available for the designers to refer back to.

25.3.1.2 Participants

Each group consisted of five designers. These were fourth or fifth year product design students who were involved in a course in facilitating creative problem-solving meetings. The four groups were matched in terms of the designers' problem-solving styles (Kirton, 1989), gender, and sketching abilities.

25.3.1.3 Meeting Plan

Each meeting took two hours. The process plan for each meeting included a warming-up activity, a round of exploring the problem, two rounds of generating ideas, and time for group discussion about the idea-generation techniques used. This paper concentrates on the two idea-generation steps. In one of the steps brainstorming was applied, and in the other, braindrawing. The order of application of these techniques was varied. Each idea-generation step took fifteen minutes.

In our analysis, the two idea-generation steps in each meeting were treated as separate segments. This resulted in four braindrawing- and four brainstorming segments. In the design of the experiment, interaction between the two idea-generation steps was reduced in three ways. First, by having a slightly different problem for each segment (in the first round designers generated ideas for a three-year old age group, and in the second round for a nine-year old age group); secondly, by installing a break between the two idea-generation rounds; and finally, by removing the produced flipcharts with ideas after the first idea-generation round.

25.3.2 Method of Analysis

The flipcharts and videotapes were transcribed into a table to capture the key data. An example of such a table is provided below, covering some ideas generated during the final round of braindrawing in Meeting 1 (Subject explanations are translated from Dutch).

Table 25.1. Sample of the data table from Meeting 1

No	i.d.	Label	time	picture	Subject explanation of idea
44	II/C/10	Seat rises into see-through turret	60.07		C: I thought in the back, if you make a chair that with a button raises and a turret raises automatically above the roof ... that you can look over the whole car, that you can see the signs with city names better, which you can type on a keyboard .. that you have some kind of a cockpit.
45	II/D/11	Gun turret	60.27	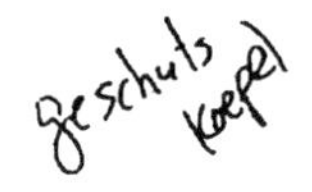 (gun turret)	D: Yes! ... that one *(points at Idea 44)* combined with that one *(points at Idea 35: virtually shooting other road users)*. *(inaudible)* *(laughter)* ...something like a gun turret.

46	III/D/10	Looping race track	59.57		D: I myself had a looping racetrack .. yes, simply such a thing which is easier to place into the car *(compares to Idea 8: drag-race track)*. You can put it in front of you and then making laps, racing against each other.
47	III/D/11	Racetrack for planes on roof	61.15	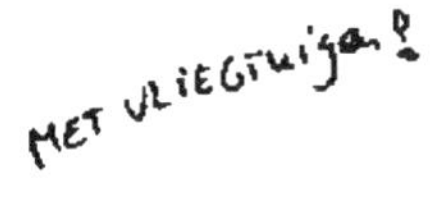 (with planes!)	*(written close to Idea 5: Racetrack upside down on ceiling)* D: .. and I liked that one very much as well *(points at Idea 5)* .. that idea for on the ceiling .. and you could do that with little airplanes, because they are in the air anyhow...
48	III/D/12	planes hang on strings so that they glide in bends	61.38		*(arrow from idea 47)* D: ... and you could align them nicely in two tracks next to each other and hang them from a string so that they really glide in the bends

The main reason for organising a group meeting is for the group members to interact in their problem-solving efforts. This makes the 'building on each other's ideas' guideline for divergent thinking (Osborn, 1953) especially relevant to this research project. Investigating the ways in which the designers build on each other's ideas provides direct process clues regarding the functioning of the techniques applied. Linkography (Goldschmidt, 1998) is a research approach which specifically addresses the ways in which designers build on each other's input. We have adapted this method for application in our research of creative problem-solving meetings. For a more elaborate description of the research method we refer to earlier papers (Van der Lugt, 1999; Van der Lugt, 2000).

For each idea, direct connections or 'links' with all earlier ideas are determined by gathering and evaluating evidence of connections. Evidence can be found within the content of the ideas, which we refer to as 'content indicators'. Evidence that is not part of the content of the ideas is referred to as a 'context indicator'. Context indicators can, among other things, consist of: gestures or remarks made by the designers when they explain their ideas; physical action when conceiving of the idea; or connecting symbols on the flipcharts. We believe that making explicit use of context indicators while constructing link systems enhances the reliability of linkography as a research method. By no means do we intend to suggest that context indicators can be found for all links. Early research suggests that context indicators are detected for about 75% of the links.

As an example, in Table 25.1, there is a link between Idea 45, 'gun turret', and Idea 44, 'seat rises into see-through turret'. The content indicator consists of the strong likeness of the ideas. By simply mounting toy guns onto Idea 44 the observation turret transforms into a gun turret. There are two strong context indicators: First, designer D shouts 'Yes!' in reaction to the see-through turret idea, and then directly starts explaining his idea. Second, designer D points at Idea 44 and mentions 'that one (points at Idea 44) combined with that one (points at Idea 35: virtually shooting other road users)'. Together this is more than enough evidence for accepting a link between Ideas 44 and 45.

For each of the eight segments, linking was performed twice by the same person with a time interval of at least one week. The discrepancies between these two link systems were then re-evaluated, which provided the final link system. For verification purposes, an independent judge was asked to fill out a link system for two of the segments. Inter-rater agreement was determined using Cohen's Kappa. Good levels of agreement were found (K= 0,73 and K=0,63) between the link systems of the two raters.

Once the link systems were finalised, link indices were calculated. The link indices provide insight into the general characteristics of the meeting process, regardless of the specific qualities of the separate meetings, such as the number of participants or the number of ideas generated. The link indices are:

The *link density (LD)* divides the total number of links in an area by the total number of ideas in that area. The link density is an indicator for the interconnectedness of the ideas, and thus for the way in which the 'seeking combinations' guideline is followed.

The *self link index (SLI)* consists of the number of links that the designers develop with their own ideas, divided by the total number of links within that segment. The self link index provides an indication for how much group members build on their own ideas. In an ideal group situation, the participants build strongly on each other's ideas, so the SLI will be low. If the SLI is high, the process consists to a higher extent of a combination of individual problem-solving processes. In a full group process with five participants, a SLI of about 0,2 is expected.

The *link type indices (LTI)* provide insight into the kind of connection that exists between two ideas. We use three categories of links: 1) Supplementary (small and auxiliary change between ideas); 2) Modification (structural changes while maintaining existing line of thought); and 3) Tangential (Indications for a link are present, while no direct function of the earlier idea is present in the current idea). The link type indices show the proportions of the types of links present. The link type indices show whether the idea-generation technique used leads towards a process of gradual improvement or towards a process with many wild leaps.

25.4 Results

The objective of this study is to investigate whether the proposition regarding the structural differences in process between graphic and sentential idea-generation techniques can be maintained. Therefore, the four braindrawing segments and the four brainstorming segments are compared. The results are presented in Table 25.2.

Table 25.2. Comparing the brainstorming and braindrawing results

	MEETING 1		MEETING 2		MEETING 3		MEETING 4	
	BS*	BD*	BS	BD	BS	BD	BS	BD
Number of ideas	72	51	85	51	60	48	44	42
Link density	0,92	1,14	0,85	1,27	0,73	1,10	0,77	1,48
Self link index	0,29	0,37	0,21	0,25	0,30	0,30	0,29	0,34
Link type indices:								
Supplementary (S)	0,14	0,10	0,07	0,32	0,20	0,19	0,09	0,35
Modification (M)	0,52	0,55	0,58	0,38	0,50	0,47	0,47	0,37
Tangential (T)	0,35	0,34	0,35	0,29	0,30	0,34	0,44	0,27

* BS = brainstorming and BD = braindrawing

Strong differences are found in the number of ideas generated, link density, and self link index. For all four meetings, the link density for braindrawing is higher than for brainstorming, while the number of ideas for braindrawing is lower. This confirms our conception that the braindrawing process is not so much based on producing a large quantity of ideas, but more on building ideas based on the information gained earlier in the process.

The Self Link Index is higher for braindrawing in all four meetings, but the differences are relatively small. This signifies that during braindrawing participants are somewhat more involved in their individual problem-solving processes. This may be related to the fact that to some extent, braindrawing invites the designers to involve themselves in a reflective conversation with their own drawings (Schön and Wiggins, 1992).

The Link Type Indices do not provide consistent differences between brainstorming and braindrawing over the four meetings. In Meetings 2 and 4 however, there is a notable difference in link types. For both meetings, the amount of S-type links is considerably higher while braindrawing, and both the amounts of M-type and T-type links are considerably lower. The implications of this finding are limited, as there is not a similar difference between the brainstorming and braindrawing segments in Meeting 1 and 3, but it may suggest that braindrawing is to a larger extent based on a process of gradual improvement.

25.5 Discussion

These findings support the preliminary hypothesis that a different type of idea-generation process occurs between sentential and visual idea-generation techniques. Future research will focus on achieving a deeper understanding of the qualities of these differences. For this, the general indicators presented here are not sufficient. We will have to start looking within the link systems for recurring patterns of links and link types. We will also look at the differences in qualities of the ideas generated and relate them to the characteristics of the linking processes that lead to these ideas.

25.6 References

Buijs JA (1980) Innovatie en interventie (innovation and intervention, second edition). Kluwer, Deventer Netherlands

Dorst K (1996) Describing design: a comparison of paradigms. Doctoral dissertation, Delft University of Technology

Fish J, Scrivener S (1990) Amplifying the mind's eye: Sketching and visual cognition. Leonardo 23 (1):117-126

Geschka H, Schaude GR, Schlicksupp H (1973) Modern techniques for solving problems. Chemical Engineering August:91-97

Goldschmidt G (1991) The dialects of sketching. Creativity Research Journal 4(2):123-143

Goldschmidt G, Weil M (1998) Contents and structure in design reasoning. Design Issues 14(3):85-100

Isaksen SG, Dorval KB, Treffinger DJ (1994) Creative approaches to problem-solving. Kendall & Hunt, Dubuque IA

Kirton M (ed.) (1989) Adaptors and innovators: styles of creativity and problem-solving. Routledge, New York

McKim RH (1980) Experiences in visual thinking (second edition). Wadsworth Publishing Company, Belmont CA

Osborn AF (1953) Applied imagination. Scribner's, New York

Schon DA, Wiggins G (1992) Kinds of seeing and their function in designing. Design studies 13:2:135-156

Van der Lugt R (1999) Developing a graphic tool for creative problem-solving in design groups. In: Goldschmidt G, Porter WL (eds.) Proceedings of the 4th international design thinking symposium, on design representation. Massachussetts Institute of Technology, Boston

Van der Lugt R (2000) Developing a graphic tool for creative problem-solving in design groups. Design Studies. In press

Zeisel J (1981) Inquiry by design. Cambridge University Press

Chapter 26

Making a Drama out of a Process: How Television Represents Designing

P. Lloyd

26.1 The Design Process on Television

'But listen to the music, he's evil!' Homer Simpson says to his wife Marge as they watch a television documentary about an alleged stalker (The Simpsons, episode 2F06). The quote reveals the often indeterminate relationship between 'truth' and televisual representation, the music being used to imply the nature of someone's character. With an increasing amount of designing on television, often reaching a considerable audience, it is important to critically analyse the ways in which designing is being represented. Are programmes about designing consistent with what design researchers have been finding in recent years? Is the design process being misrepresented in any way?

To look at the ways in which the medium of television represents the design process this paper focuses on a series of three programmes about the design process, 'Designs on Your... Bra/Loo/Car', first transmitted in June and July 1998 on Channel 4 in the UK. These programmes have several features that make them attractive for analysis. Firstly, each programme contains one particular design process and a good deal of primary – 'fly on the wall' – source material. Secondly, they contain existing clients who are trying to develop actual products and that means there is an argument for them being considered as 'real' design processes.

This paper provides a qualitative analysis of the above programmes and is structured in four main parts. The first part looks at how television literally *re*-presents the 'real' process. The second part looks at how the various participants in the process are characterised. The third part concentrates on the particular relationships between the managing director and the designer. The final part focuses on how the designers make their design arguments. The paper concludes that analysing designing on television can provide a properly critical account of designing, and an account based on publicly accessible data. More specifically the analysis reveals contradictions in how television represents designing, with designers explicitly represented as 'individual geniuses', even though the programme also illustrates the importance of other participants in the design process, and shows how reliant the designers are on those others.

26.2 The 'Designs on Your...' Programmes

In the first programme of the series, first broadcast 23rd June 1998, the product designers Richard Seymour and Dick Powell focused their attention on re-designing the bra. The organisation that they worked with was Nottingham based lingerie company Charnos, who had asked them to design, 'a bra that is far simpler in its concept, perhaps can be offered in fewer sizes, particularly suitable for bigger bust sizes... and [that] we can sell at a sensible price, and make a modest return' [*Bra*, sequence 10, 7:35 – 8:22]. The designers consult experts on both female anatomy at Nottingham University and Materials at Adidas, they look at the current market, prototype a series of ideas, and come up with their idea for a new concept in bras, 'the Bioform'. This is a three dimensional structure that replaces the under-wire of the traditional bra. The programme ends with all parties happy about the way things have turned out. The design has now been patented, and a product is due for release sometime in 2000. This would seem to be the most successful collaboration of the three.

In the second programme of the series, first broadcast 30th June 1998, the designers work with Shires, 'a traditional northern family firm' specialising in sanitary ware. They wish to produce a, 'modern classic [bathroom] suite with a hint of revival, a little bit of hybrid, not totally modern, but there has to be that little bit of classical style' [*Loo*, sequence 11, 7:05 – 9:18]. The designers again look at the current market, consult an ergonomist expert in the field, travel to Japan to view the 'state of the art', and suggest a number of improvements to the present toilet (the rest of the bathroom suite receives short shrift). The organisation, although finding the design suggestions interesting, feel them to require too much development work and there is then a series of meetings to preserve the 'purity' of the original design ideas. The programme ends at the product launch of the new 'Icon' bathroom suite. The organisation appear very happy with the results, although the designers are somewhat less enthusiastic, 'it hasn't moved forward in a way in which it could have done' [*Loo*, sequence 52, 48:12 – 48:48].

The final programme of the series, first broadcast 7th July 1998, looked at ways of improving the electric car. A Norwegian company – Pivco – provide a brief for the designers, 'to expand [the existing product] into a range of products in different niches, but basically having the core of the vehicle... the same, just dress it to different niches' [*Car*, sequence 14, 7:42 – 8:14]. From here the designers look at the infra-structural problems of producing electric cars, ask consumers what problems they have with existing cars, travel to Chrysler in Detroit to look at their recent developments in 'clean' fuel technology, before presenting a series of concepts to meet the original brief. The development team, who have only just completed the current product, are a little sceptical about the concepts and the structural changes that they would require. The programme ends indefinitely with Pivco finding the concepts interesting, 'I think there are definitely several ideas that Richard and Dick have brought to the table that I think we would like seriously to look into' [*Car*, sequence 62, 47:52 – 48:49]. No products or even prototypes are produced.

26.3 Method

The three programmes, each lasting 50 minutes (3000 seconds) and broken into three parts by two advertisement breaks, were transcribed verbatim. *Bra* consisted of 8691 words, *Loo* consisted of 7830 words, and *Car* consisted of 9030 words. The programmes were then sub-divided into 'sequences', each sequence consisting of a series of related camera shots, for example of a design meeting, an interview, a technical process, or some archive material. A sequence was judged to have come to an end when there was a definite move to a different subject and a corresponding change in the type of camera shot. *Bra* comprised 60 sequences, *Loo* comprised 54 sequences, and *Car* comprised 62 sequences.

With the transcripts divided into sequences, a coding scheme was applied to reveal some basic statistics and distinctions (Lloyd, 2000). Each sequence was coded in a number of ways. The first classification determined the type of camera shot presented. These were divided into *real world* shots, where some 'real' action to do with the design process was taking place; *talking head* shots, where a person talked to the camera; *illustrative* shots, that demonstrated an aspect of what was being talked about, or showed something tangential to the design process; *archive* shots, utilising, for example, old newsreel material or advertisements; and *mood* shots, where the idea seemed to be to convey a general mood with, for example, a lingering shot of dark night, or blue sky. These classifications were broken down further and analysed quantitatively. The following analysis is primarily qualitative.

In all subsequent quotation from the programmes the type of shot is indicated in brackets directly after the sequence number; 44 (rw) thus indicates that sequence 44 was a 'real world' sequence. This gives some indication of the type of source for the quote. It remains to point out that it is impossible to make any inferences about the *actual* design processes the programmes were based on. It is only possible to make inferences from the content of the programme. One would hope that there was a reasonably close relationship between the two, but this cannot be assumed. It is impossible, then, to say whether one process was better, or more efficient than any other; it is only possible to say that they are represented *as if* one was better, or more efficient than another.

26.4 Re-presenting the Process

The raw material that has been shot for the programmes (called 'the rushes') has to be edited. This means that decisions have to be made about how much material to show, what kind of material to show, and in what sequence to show it. These decisions determine what the programme is ostensibly about - however indeterminate *that* might be. In telling the three design stories, the programmes adhere to a roughly chronological sequence. *Bra* does this overtly with five captions throughout the programme, starting with September 1997 (3:33) and ending with June 1998 (45:11).

Against this, however, some scenes are clearly not edited in a chronological sequence. The most striking example of this occurs in *Bra* [17(rw), 14:36 – 18:05]. Having generated several ideas for the new product the two designers return to bra manufacturers Charnos to present their ideas and ask about new problems they've come across. In a 22 second sequence of images the camera tracks from one designer talking, to some drawings on a table, and then back up to the Charnos design chief talking. During the sequence the conversation is constant, suggesting that the sequence is a 'real' slice of action. This is not the case. A careful viewing reveals that when the first designer is talking he is doing so in front of a window. Outside it is clearly broad daylight. As the camera tracks back up again from the table, day has clearly turned to night. This happens within 22 seconds and along with a seamless flow of conversation. In fact throughout the 106 second sequence the sky outside the offices switches between light and dark. It would seem the programme editors are attempting to polarise the process, keeping the viewer focused on 'important' aspects of the design rather than jump from aspect to aspect.

The problem runs deeper. In deciding whether a scene is important enough to make the final cut the editor is informed by the endpoint of the design process - or at least the point at which the cameras ceased to film. Now this is a very different situation than the designers themselves are in, giving the sense that the design process is being understood 'backwards'. That is to say that comments early in the process become valuable or 'important' when they make direct reference to the (then unknown) endpoint. So, for example, in *Car* [38(th), 19:27 – 19:56], the Pivco design chief, Jonathan Etherington, reveals that the car structure:

> *'...has taken hundreds of hours on very expensive computers [to develop], it looks very simple but it is actually a very complicated piece of engineering to make sure that structure there absorbs all the energy in the crash test, and at the moment it's working successfully and we really hope we don't have to change that.'*

Of course that is exactly what the designers suggest they do do, and in doing so the comment above portends what is to come. A slightly different example of this occurs in *Bra*. In trying to think of possible materials for a new bra one of the designers remembers the Aerobie, 'it was like a frisbee... a polycarbonate structure' [*Bra*, 17(rw), 12:50 – 14:36]. This insight contributes to the development of a prototype. However, the sequence before this (*Bra*, 16(i), 12:20 – 12:50) pre-figures sequence 16. It shows two young boys, standing on either side of an advertising hoarding for 'Wonderbra', throwing an aerobie to each other. Here we have a scene that portends a future state of affairs in the design process, but it is a wholly invented scene. It is highly unlikely that the two boys 'just' happened to be caught on camera throwing an aerobie to each other in front of a bra advertisement. Instead the programme makers have decided that the 'aerobie' idea forms an important part of the design process and have chosen to accentuate it. Again the endpoint informs the content of earlier sequences, and an implicit theory about design is produced.

It is also informative to look at the mood that the programmes attempt to create. Specific mood sequences accounted, on average, for less than two per cent of the total programme time yet, as one documentary maker has pointed out, 'our primary response is to a mood' (Vaughan, 1999). These sequences, for example, contain images of night rushing by from a train [*Bra*, 26(m), 26:29 – 26:34], clear blue sky [*Bra*, 52(m), 44:39 – 44:57], or a panorama of a town at night, with a dog barking in the background [*Loo*, 38(m), 34:30 – 35:07]. These scenes work in several ways. Firstly they create a feeling of space in the programme, a pause in the action that allows the narrator a quick summary. They give the sense that there are quiet, reflective, moments in the design process, as well as intense discussion, a point made by Bryan Lawson in his book 'Design in Mind' (1994). More obviously though, they create associations. Night turning into day makes us think that the danger is past, while blue sky is associated with opportunity and exploration.

Some associations are created less subtly. In all three programmes music plays a significant part. An average of 13 sequences per programme had musical backing. Indeed the music was specially commissioned for the programmes. Again it is tempting to think, what has music got to do with designing? Music, however, is an effective vehicle through which to transmit mood. Adding music provides the viewer with a better indication of how things went. Fast music gives the sense of things happening quickly, slow music, that things are not turning out quite right. Pop music gives a sense of new, constantly changing things, while classical music, a sense of stability and even conservatism.

Coding the musical sequences in two classifications, *type* – classical, jazz, pop; and *tempo* – fast, medium, slow, revealed some interesting results. *Bra*, the most successful process for the designers, starts at a slow tempo and ends at a fast one, with pop being the most frequent choice of music. *Loo*, a disappointment for the designers, starts at a quick tempo and ends slowly, with classical being the most frequent choice of music. *Car*, a process that never gets past the concept stage, starts at medium tempo and slows slightly. These are obviously crude distinctions but do seem to catch the mood of a sequence or programme.

Particularly obvious uses of music occur in the sequences that show the two designers, mainly at the beginning of the programmes. The narration characterises them as 'men with a mission' [*Loo*, 1(i), 0:00 – 0:35] , and 'driven by passion' [*Bra*, 2(rw), 0:57 – 1:57]. To accentuate this idea the music that often accompanies them in the sequence works on the association of private detectives and secret agents, fighting for the good of mankind and making the world a better place. We get dramatic Mission Impossible-type music in *Loo* [1(i), 0:00 – 0:35 and 12(m), 9:18 – 9:25] and James Bond-type music in *Bra* [18(rw), 18:05 – 20:19]. There is a humorous element here, but again it appears that the decision to include music into the programmes was not one taken lightly. It performs a vital function for the viewer, but a function that tends towards 'making sense' of the design process for the viewer by over-simplifying character.

26.5 Characterisation

In two of the programmes, *Loo* and *Car*, there are two people who, at least in the programmes, share the characteristics of the 'bad guy'. These are the people working for the participating organisations responsible for the design content of the new product. The design chief in *Loo* speaks 1047 words (13% of the total words spoken in the programme), while the design chief in *Car* speaks 835 words (9%). It is what they say that reveals them as 'bad guys'. Early on both start off by giving a warning. In *Loo* the design chief says, 'head-turning is no use this is a commercially realistic brief...' [*Loo*, 11(rw), 7:05 – 9:18], while the design chief in *Car* says, 'we really hope we don't have to change [the car structure]' [*Car*, 38(th), 19:27 – 19:56]. This gives the impression that they are people who already have their minds made up about what they expect to see.

Progressively through both programmes they become the characters who question the concepts put forward by the designers. In *Loo* the design chief comments after seeing some of the designers' ideas, 'the self-cleaning toilet and self-cleaning basin, yes there's possibilities... [but] I don't think people are prepared to pay for it' [*Loo*, 18(rw), 16:21 – 18:29]. Later on he says of the concepts, 'I think there are some positive ideas here... but I think it's really a case of time' [*Loo*, 28(rw), 25:49 – 27:38], and at another meeting, 'we actually think that some of these are good ideas but technically, at this moment in time, I think they're fraught with danger' [*Loo*, 34(rw), 30:56 – 33:21]. In *Car* the design chief comments to his managing director, while looking at the designers' concept drawings, 'I think what [the designers] have done is they've actually exaggerated the dullness of the existing [car] and made this [new concept] look even better' [*Car*, 56(rw), 42:19 – 44:10], and later on in a meeting with the designers 'what I'd like to do is just to point out one or two of the details that we actually think are..., give us a concern with your design' [*Car*, 57(rw), 44:10 – 46:25].

The impression one gets is that these characters – 'bad guys' – are holding up the creative process. Not only that, but in doing so they are preventing the designers from achieving their professed aims of, 'making things desirable by making them better' [*Loo*, 4 (th), 2:05 – 2:36] and providing 'benefit to the consumer' [*Loo*, 28(rw), 25:49 – 27:38]. The implication is that these characters are holding up some sort of self-evident progress and stopping us all from having better lives (hence the 'bad guy' tag). The idea that the designers are portrayed as the 'good guys' is evidenced in a number of ways. Earlier it was mentioned that in several places 'men on a mission' type music accompanied some of the sequences featuring the designers, suggesting men struggling for the good of mankind. There is a mood sequence that shows the designers driving past the camera on large motorbikes [*Loo*, 12(m), 9:18 – 9:25] again implying both individuality and power. And in various places there is reference to the designers meeting a challenge, '[design] is about confronting an obstacle and finding a way through it, around it, over the top of it, and so that is a challenge' [*Loo*, 19(th), 18:29 – 19:17], and fighting battles, 'design needs to fight the battle that makes the product as good as it can be for the people for whom it was intended' [*Car*, 58(th), 46:25 – 47:23].

The designers also describe their process in these programmes as, 'starting with a relatively clean sheet of paper' which gives them, 'an opportunity to re-think' [*Car*, 27(th), 15:05 – 16:06] existing designs. This is necessary because, as one of the designers says in *Bra*, 'the overwhelming feeling I've got is 'what a stupid product' [*Bra*, 15(th), 11:53 – 12:20]. The idea here is that using reason to 're-think' existing products, providing consumers with 'clear benefits', means that the 'challenges' can be met, and therefore the 'battles' fought with the organisations currently producing 'irrational' products. The image is of the traditional enlightenment genius, cutting through swathes of irrationality with the sword of truth. The programme narration accentuates this view. The designers are described as 'top' and 'award-winning' [*Loo*, 1(a), 0:00 – 0:35], 'ground-breaking' [*Bra*, 2(i), 0:57 – 1:57], and 'most influential' [*Car*, 2(th), 0:27 – 1:26].

The representation of designers as 'geniuses', solely responsible for making the world a better place is receiving a growing amount of criticism. In the course of several papers Richard Coyne has written, from a hermeneutic point of view, about the mistakes of viewing designers in this way (1993, 1995, 1997). But the programmes go further than merely representing the designers in this way, since they actually show (perhaps unwittingly) how this view cannot be sustained in any coherent way.

26.6 The Managing Director and the Designer

The managing directors of the organisations participating with the designers all figure to a small degree in the programmes. In *Bra* the managing director says 496 words (5.5%), in *Loo* the managing director says only 90 words (1%), and in *Car* the managing director says 329 words (3.5%). The content of what they say tends to be fairly neutral, canvassing opinion and making practical suggestions rather than getting involved in discussions. In *Bra* the managing director asks those at a meeting 'are we defending an idea that we are less enthusiastic about progressively?' [*Bra*, 51(rw), 42:48 – 44:39]. In *Loo* the managing director inquires, again in a meeting, 'so what do our friends think?' [*Loo*, 41(rw), 35:47 – 39:30]. In *Car* the managing director suggests a future plan, 'if we can, in a way, try to move in this direction and combine it with the present frame, then we are very close to a product' [*Car*, 49(rw), 32:37 – 34:02]. In each meeting they remain calm, even while there are disagreements between others.

The managing directors, however, have responsibilities and this is what distinguishes them from the designers. They clearly have to think about the product in commercial terms - that is in terms of risk and cost in relation to benefit - and not just in terms of benefit (making the world a better place). In fact if, as the designers suggest at various points, their concepts will obviously benefit consumers, it is the managing directors who 'allow' it to happen. In other words it is the managing directors who ultimately 'make the world a better place', not the designers. The power they posses is illustrated in *Bra* when the managing director delivers his decision about the sequence of prototypes developed through the programme, 'I have discussed it with the chairman and the group chief executive, we're all on board with wanting to take this further...' [*Bra*, 54(rw), 45.11 – 48:07]. In the other two programmes, such a clear decision is not presented. In *Loo*, the original concepts are slowly altered, frustrating the designers, while in *Car* the managing director plays down the designers' significance, I think there are definitely several ideas that Richard and Dick [have] brought to the table that I think we would like to seriously look into' [*Car*, 60(th), 47:52 – 48:49]. In all the programmes, although the designers put a strong case for their designs forward, they are ultimately reliant on the decision of the managing director.

This is a point made convincingly by Adrian Forty in the last chapter of his book *Objects of Desire* (1986). He contends that it is mostly entrepreneurs who determine which consumer products are allowed into society. He writes that, 'although designers prepare designs, the responsibility for carrying them out rests with the entrepreneur' (p.241). It is *their* decision that results in a design becoming exposed to the public consciousness. Certainly there are more complex elements at work in the three programmes. In a way the designers are side-stepping this rickety route to a world of better design through entrepreneurs by directly appealing to the viewing audience. The programme remains, however, as a more general illustration of the relative powerlessness of designers.

26.7 Making the Design Argument

The designers are far from passive however. In looking at how the designers make their various arguments to design chiefs and managing directors (and the viewing audience) the series of programmes show how it is possible to present a relatively coherent-sounding message, yet contradict this message with some of the content.

As was stated before the designers' professed aim in designing was of: 'making things desirable by making them better' [*Loo*, 4(th), 2:05 – 2:36]. Invariably, in the programmes, this meant making the products easier to use. In *Loo* the designers are asked what they see as the difference between 'gimmicky and design' [*Loo*, 28(rw), 25:49 – 27:38]. 'That's easy', one of the designers replies, 'if it's a benefit to the consumer then it's not a gimmick'. However several seconds before, the other designer, explaining one of the concepts, had said, 'Very slim across this axis here, maybe quite broad across the back, we called this the Mackintosh chair when we were sketching it because of this very, very long slim form'. Now it is less obvious that a toilet looking like a Mackintosh chair will provide an obvious benefit to the consumer.

For these designers the aesthetic argument is second to the ergonomic argument. But not by much. In *Bra*, all that is needed is the ergonomic argument. The managing director, initially sceptical, reflects on the process half way through the programme, 'I think the scepticism has eased back because we think ... this idea actually fulfils a need and is an improvement in the fit of the bra' [*Bra*, 24(th), 25:58 – 26:29]. At the end of the programme the model comments, 'it's very comfortable' [*Bra*, 53(rw), 45:11 – 48:07]. In the other two programmes, however, the 'obvious benefits' aren't quite as obvious to the participating organisations. In *Car*, after suggesting various concepts to make the electric car easier to use, the designers suggest that it needs 'attitude'. One of the designers summarises halfway through the programme, 'what's becoming clear is that we're not going to get this beautiful vehicle with attitude with this slightly jaunty, perky shape...' [*Car*, 50(rw), 34:02 – 36:38]. The organisation are not convinced and the design goes no further. In *Loo*, after scepticism from the organisation about the benefits of their proposals, the designers try a more aesthetic approach. They compare their foam models with the organisation's foam models, themselves modified versions of the designer's original concepts:

> *'The other thing to watch for ... is the simplification of the form, its become more elegant but there's less on it, there's less curlicue, less twisty turny... yet it's proportionally purer, you notice that straight away, actually, if you look from one to the other, especially around these rear sections. It's immediately apparent.' [Loo, 41(rw), 35:47 – 39:30]*

If the 'benefit argument' - that the world will be a better place - fails to convince then the 'aesthetic argument' - at least the world will be a more beautiful place - rapidly takes its place. Although this seems to partially succeed in *Loo*, it is bolstered by a reference to the television programme itself. During a meeting in which the design concepts have politely been put to one side, one of the designers, making a last attempt at persuasion says, 'there is just one other factor, by the time this TV programme goes out, anybody watching the programme will know this idea' [*Loo*, 34(rw), 30:56 - 33:21]. Using the television programme to convince the organisation that they should adopt their concepts is certainly logical if the designers, as they consistently seem to, do believe that their benefit is self evident. Yet if this is true why can't the organisation see these benefits for themselves?

There is a further level of argument. The 'story' developed through *Loo*, pitching the good-guys (the designers) against the bad-guy (the chief design representative), centres mainly on the bad-guy telling the designers what people will, or won't buy. Little evidence is shown in the programmes for his view. At several points during the programme the design representative dismisses the design concepts because, 'I don't think people are prepared to pay for it' [*Loo*, 18(rw), 16:21 - 18:29]. It is a bald statement, and one that is difficult to argue against when it is based on an intuitive understanding of the market – as the programme seems to suggest – rather than a set of more objective figures. However, in the final sequence of the programme one of the designers uses this line of argument, 'I actually believe... that consumers actually do want a better loo, they would buy it, it would be more successful' [*Loo*, 54(th), 49:02 - 49:20].

The argument for the design concepts slips from benefit, to aesthetics, to television's role, finally to an intuitive understanding of the market. The designers, represented in the programmes as making the world a better place for us, begin to look a little more opportunistic. The programme makers may have intended this and, if so, is a subtle point well made. It highlights something about designing that is often not taught in design schools, at least not explicitly. And that is the importance of effective rhetoric.

26.8 Concluding Remarks

Analysis of designing on television does have several advantages over traditional design studies involving fieldwork in design and manufacturing organisations. Firstly, without any contractual obligations to the designers or organisations participating, the researcher can provide a properly critical account of designing. Secondly, the relationship between data and interpretation is made explicit – television being public data – and hence discussible. Finally, and perhaps most importantly, it is possible for television to represent designing in a certain way yet contradict it's own representation. Qualitative field data is always a 'representation' of sorts, but is rarely analysed *as* a representation. This paper has shown how the designers are represented as individuals employing reason to make the world a better place. In fact, if anyone is responsible for making the world a better place, the programme content points towards the managing directors. The paper has also shown how the designers produced several specious arguments in attempting to convince the organisations to develop their concepts, while professing to use only onc.

26.9 References

Coyne RD, Snodgrass A (1993) Co-operation and Individualism in Design. Environment and Planning B: Planning and Design 20:163-174

Coyne RD, Snodgrass A (1995) Problem Setting Within Prevalent Metaphors of Design. Design Issues 11:31-61

Coyne RD (1997) Creativity as Commonplace. Design Studies 18:135-141

Forty A (1986) Objects of Desire, Moffat, UK

Lawson B (1994) Design in Mind. Butterworth Architecture, Oxford, UK

Lloyd P (2000) Analysing the Design Process on Television. In: Scrivener S, Ball L, Woodcock (eds.) A Adjunct Proceedings of CoDesigning. Coventry University, Coventry

Vaughan, D (1999) For Documentary, University of California Press, Berkeley

Chapter 27

Matching Descriptions of Team Design

M. Mazijoglou, S.A.R. Scrivener and R.C. Valkenburg

27.1 Introduction

In recent years there has been a growth in empirical, observational studies of design-activity. Often, these studies involve the video recording of design-activity in a studio or studio-like environment comprising drawings, information sheets, *etc.*; in other words a media-rich environment. Such studies tend to fall into the category of qualitative research as they are concerned with an identified phenomenon in which the researcher is central to the sense that is made (Banister *et al.*, 1994). Given its reflexive nature, it is an important principle of qualitative research that other researchers are able to challenge a given researcher's findings from other interpretative frameworks. This can be assisted if interpretations are grounded in, and can be traced back to, the raw data.

However, design-activity data, of the kind described above, is voluminous, heterogeneous and not easily approached in its raw state. Hence, there is an argument for organising this data in ways that minimise interpretation while maximising accessibility. Mazijoglou and Scrivener (1998) present and illustrate such a method for augmenting design-activity data that incorporates structured transcriptions of verbal and non-verbal data which are related to drawing production and workspace resources. These linked transcriptions, together with the raw data, comprise the Rich Picture, a data-ordering scheme that indexes data without loss of information. They argue that Rich Picture provides a useful resource for the inductive interpretation of design-activity data. First, it assists interpretation by supporting the navigation of data. Second, when an inductive interpretation is linked to the Rich Picture, subsequent researchers are in a stronger position to question such an interpretation as it provides routes back to the raw data on which it is based.

Purcell *et al.* (1996) observe that there is a move away from the traditional approach to analysing protocol data, in which it is segmented by inferring categories, to one of 'imposing an externally derived structure onto the protocols...based on existing analyses and models of the design process'. Nevertheless, they observe that the literature 'reveals commonality and overlap between the categories used and tantalising correspondences between the results that have been obtained', suggesting that correspondences between results produced by induction and deduction strengthens the validity of findings. However, whether an interpretation of design-activity is arrived at by bottom-up or top-down analysis, the requirement that each interpretation is accessible to reinterpretation still applies.

We have suggested above that the Rich Picture supports both interpretation and reinterpretation. However, because it imposes a structure on the data it also has potential limitations. First, the structure might have the effect of biasing interpretations towards ones that fit it. Secondly, navigation of the data through the Rich Picture might unnecessarily constrain theory driven interpretation, as might the requirement to connect the interpretation to it in order to facilitate future reinterpretation of results. These uncertainties can only be explored by examining the extent to which a Rich Picture representation of design-activity data aligns with independently derived higher-level interpretations of the same data. In this paper, we present and discuss the results of a study designed to test the match between the data-driven, bottom-up Rich Picture description and a theory-driven, top-down description of the same design-activity data.

27.2 Discontinuities in Discourse Dynamic

The Rich Picture structures individual design acts into larger units (e.g., drawing strokes into drawings and turns in the conversation into chunks). The elements of the Rich Picture were not determined *a priori*, but emerged from observation of design-activity data. Since Mazijoglou and Scrivener's aim was to produce a data-ordering scheme to which different meanings could be attached, the primary rule governing the derivation of the Rich Picture's elements was that interpretation should be minimised. The Rich Picture does not capture the meaning of the design-activity but provides a data structure which, together with the raw data, is intended to aid the interpretation process.

Space here does not allow for an elaboration of the Rich Picture and instead the reader is referred to Scrivener and Mazijoglou (1997) for a fuller account of the representation. Fortunately, this should not impede understanding as in this paper we focus on the verbal component of the Rich Picture, describing Mazijoglou and Scrivener's (1998) method for segmenting the discourse into 'chains'.

The verbal transcription process begins with the video recording of a design session. Then, the design discourse is transcribed to a text format in which it is broken down into elementary units called turns, *i.e.*, individual contributions to the discourse. Figure 27.1 shows part of a conversation between three designers labelled 'I', 'J', and 'K'. It is obvious from this segment that a turn is a continuous utterance by one speaker followed or interrupted by the utterance of another speaker. Each turn is labelled with the speaker identifier and in parenthesis, 't' (to signify 'turn') and a unique number indicating the order of the turn in the discourse. Thus, Figure 27.1 commences with the 429th turn spoken by designer, J. This fragment is taken from a two-hour design session that produced 2114 turns.

```
J   (t 429)   it looks like it's got some sorta pivot here so that it
              collapses down a bit yeah but er that kinda presupposes
              a solution in my opinion so I think if we get something
              that's big and bulky then we need to evaluate it and
I   (t 430)   how we could fold it up
J   (t 431)   yeah make it smaller
K   (t 432)   mm mm
J   (t 433)   OK so now uh go back to attaching is that where we are
              joining concepts for joining (laugh)
K   (t 434)   concepts joining concepts OK so
```

Figure 27.1. A fragment showing turns in the discourse

To segment the turn-structured discourse into larger but more manageable chunks than the discourse as a whole, the transcript is read while simultaneously listening to the video recording which captures the dynamic of the dialogue and the stress and emphasis given to utterances. The discourse flows, but from time to time interruptions in the dynamic of this flow can be observed. The next step, having identified these interruptions, is to review the discourse before and after them to establish whether what the designers were doing before the interruption is different to that which follows it. To confirm the apparent shift in focus of attention the text at the point of interruption is examined for evidence of an attempt by one of the designers to intentionally shift the focus of the team's attention, e.g., Figure 27.2.

```
J   (t 429)   it looks like it's got some sorta pivot here so that it
              collapses down a bit yeah but er that kinda presupposes
              a solution in my opinion so I think if we get something
              that's big and bulky then we need to evaluate it and
I   (t 430)   how we could fold it up
J   (t 431)   yeah make it smaller
K   (t 432)   mm mm                                  ONSET OF CHAIN 9
-----------------------------------------------------------------------
J   (t 433)   OK so now uh go back to attaching is that where we are
              joining concepts for joining (laugh)
K   (t 434)   concepts joining concepts OK so
-
-
-
I   (t 691)   yeah and rotate it down and then there this one clam
              back there                             ONSET OF CHAIN 10
-----------------------------------------------------------------------
J   (t 692)   can I interject here?
I   (t 693)   yeah
J   (t 694)   on our weight spec if you just wanna sorta look at these
              they have weights yeah they're between four hundred...
```

Figure 27.2. A fragment of transcript illustrating intentional discontinuities

In the discourse preceding that shown above, the designers were discussing the features of their design. Designer J interrupts the discussion to consider the implications of a joining concept and then explicitly attempts to shift the team's focus of attention to the production of alternative joining concepts. The attempt is successful as the team turns to discuss joining concepts. Later, J interrupts and again shifts the team's focus of attention in a new direction. These intentional discontinuities segment the linear discourse into discrete sequential chunks, called 'chains'. The identification of the features characterising intentional discontinuities is achieved by bottom-up analysis of the data and without reference to a predetermined view of the phenomena. While interpretation is unavoidable (*i.e.*, judgement must be made about whether focus-of-attention shifts at an interruption and whether this was intentional) it is kept to a minimum and the researcher is not required to undertake an in-depth analysis of the task related content of the discourse.

27.3 Describing Design as Reflective Practice

Valkenburg has developed a method for describing team design in terms of predefined activities derived from Schön's theory of design as reflective practice (Schön, 1983). Within these activities (naming, framing, moving and reflecting) the designers make an inextricable connection between what they are doing (the action) and the design content on which the doing is focused (the knowing). Schön refers to this as 'knowing-in-action'. The Reflective Practice description method (*cf.*, Valkenburg and Dorst, 1998, for a detailed elaboration) divides team design protocol data into smaller team activities. A team activity is a part of the discourse in which the team discusses one subject (e.g., a solution or materials) or acts with the same goal (e.g., generating ideas, creating a survey of ideas). These team activities are then coded according to the reflective-practice activities and labelled by the content and aim of the activity. In this way an overview is provided of the succession of the design project - the 'flow' of activities within the team.

In reflective practice four activities are distinguished; naming, framing, moving and reflecting. Designers' work by naming the relevant factors in the situation, framing a problem in a certain way, making moves toward a solution and reflecting on those moves and the frames used. When the team is explicitly pointing to parts of the design task as being important the activity is coded as *'naming'*. During the naming activity the team is looking for relevant objects in the design task. When the team frames a problem or solution to explore further on, then the context is coded as a *'frame'*. The frame is a context for the subsequent activity; it is something to hold on to and to focus on while designing. Consequently, it can only be recognised in a sequence of naming, moving and reflection activities. Valkenburg and Dorst (1999) also claim that the development of frames within the team design project reflects an important 'mental structure' of the team. Frames reflect to the way designers approach the problem at a specific moment, and offer a framework for further activity. Experimental activities like generating ideas, making an inventory, sorting information, combining ideas, or comparing concepts are coded as *'moving'*. Through moves, the team explores the suitability of the current frame. A move is always characterised by a verb, identifying the activity, complemented with the content of the activity. An explicit reflection on earlier activity in order to decide what to do next is coded as *'reflecting'*. 'Reflecting' exhibits critical reflection of the team on their earlier actions and frames.

In order to recognise the four different activities the user of the method must look at what the team is doing and the goals they have in mind. The original video of the team design project is repeatedly viewed and segmented into activities. A new activity is an observable switch in what the team is doing or what they are focussing on. Each team activity is then coded as a naming, frame, move or reflection, and described in the words used by the team. The result is a description of the 'flow' of activities that can be represented visually in a chart, Figure 27.3.

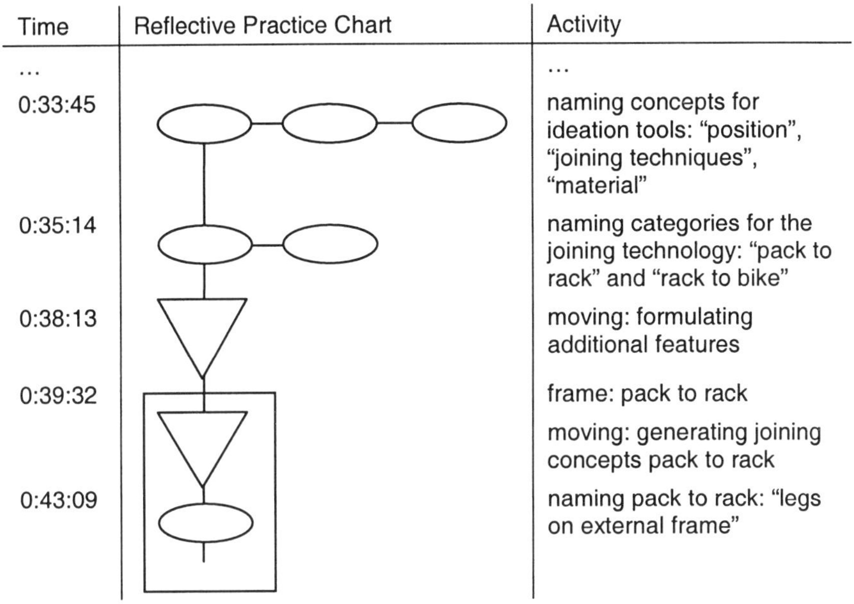

Figure 27.3. A fragment of a Reflective Practice coding of design-activity

27.4 Matching Design Descriptions and Results

Two ways of describing or segmenting team design discourse have been introduced, the first a data-driven, bottom-up approach contributing to the Rich Picture of design-activity, the second a theory-driven, top-down approach that organises the design discourse into meaningful chunks that fit theoretically defined categories. In Section 27.1, we observed that Mazijoglou and Scrivener (1998) argue that the Rich Picture of design-activity data is a useful resource for interpretation, whether inductive or theory driven, as long as this preliminary structuring of the data does not interfere with interpretation. Furthermore, we noted that these uncertainties could only be explored by examining the extent to which a Rich Picture representation of design-activity data aligns with independently-derived higher-level interpretations of the same data.

To test for such alignment, we applied both methods to the group design session data gathered for the Delft Protocol Workshop (Cross *et al.*, 1996). Here a three-person team of designers worked for two hours on the design of a fastening device to allow a backpack to be carried on a mountain bike. We then compared the outcomes of the two methods by examining the correspondence between the onset of changes in focus of attention, *i.e.*, chains in the Rich Picture representation, to the onset of activities in the Reflective Practice description, *i.e.*, framing, moving, naming and reflecting. The representations would have good correspondence if a high proportion of Rich Picture events matched Reflective Practice events.

Table 27.1. Matches between the Rich Picture 'chains' and Reflective Practice representation 'activities'

	Activities			
	Naming (6)	Framing (5)	Moving (51)	Reflecting (13)
Chains (25)	2	-	16	5

The Rich Picture produced 25 events as compared to 75 determined by Reflective Practice analysis. Intuitively, this was expected as the latter involves a higher level of interpretation than the former. Later we will explore the validity of this assumption, but here we restrict our discussion to the extent to which Rich Picture events match Reflective Analysis events. Table 27.1 shows that there is a high correspondence between the two representations. As we can see, 23 of 25 chains, *i.e.*, 92%, match with a naming, moving or reflecting event. Two chains fail to show any correspondence with the start of a naming, framing moving, or reflecting activity. In the first case the discontinuity occurred because the experimenter made a comment that caused the designers to shift the focus of attention to matters implied in the experimenter's utterance. In terms of the Reflective Practice, the team, prior to the 'interruption', was discussing the problem and continued to do so after the interruption. However, since the experimenter's utterance was designed to shift the teams focus of attention to a different aspect of the problem and since the designers did in fact proceed to consider this aspect the event registers as the start of a new chain. In the second case, after reviewing the design session it was concluded the chain was a highly questionable chain (*i.e.*, both the evidence for an explicit attempt to change and an actual change in focus of attention was weak).

Furthermore, Table 27.1 shows that most of the matching chains (70%) match with moves and none of the chains match with a framing activity. Again, this was intuitively expected, as determining frames in team designing requires an interpretation by the coder of the way in which the problem situation is progressed by the team. Frames can therefore only be detected through an analysis of the next activities of the team and an interpretation of the team's cognition. The Rich Picture approach, as stated in Section 27.2, detects changes in activities but does not capture the meaning of the design-activity.

27.5 Discussion

We have found that discontinuities in Rich Picture closely match activity shifts in Reflective Practice. However, they are clearly not the same thing as in at least one instance a valid discontinuity does not produce an activity shift, and 52 activity shifts are not related to discontinuities as defined in Section 27.2. As noted above, intuitively, the higher degree of interpretation associated with a Reflective Practice analysis might be expected to yield more events than the Rich Picture. However, no logical or empirical grounds for this being the case have been offered, and it is equally reasonable to assume that changes in activity register will register as discontinuities in discourse. To understand this the nature of design discourse discontinuities needs to be examined in greater detail.

In Section 27.2 we explained how discourse discontinuities are identified. However, this procedure is the result of an inductive analysis that sought to make sense of the design-activity data. First, discontinuities in the design-activity were perceived when listening to and looking at the design-activity. Subsequent analysis revealed that these perceptible interruptions were associated with utterances about what the designers were doing, why they were doing it, and how they were doing it (Scrivener and Mazijoglou, 1997). Furthermore, such utterances included a proposal of what to do next. Most had a common form comprising two parts: an element that stops current activity and an element that suggests a direction for future activity. Sometimes, the former component involves reflection, e.g., ' we're going to go with the rack [reflecting] lets' go with talk about braze-ons these braze-ons [redirection]'. Alternatively, the word 'OK', which implies unvoiced reflection, is used to interrupt the current activity, e.g., 'OK [stop], let's look at some of the rear mounts [redirection]'.

It was concluded that when the team acknowledged these utterances perceivable discontinuities were generated, perhaps because activity is interrupted and some time is needed for the designers to reorient themselves. Implicit in this, is the sense that the perpetrator intends to reorient the team as a whole since the direction for future activity is verbalised. (The exception to this is when an external factor, the experimenter, causes external interruptions to the dynamic flow of the discourse resulting in an 'unintentional' shift in focus of attention.) This 'sense making' process produced the method described in Section 27.2. Many activity shifts in the Reflective Practice analysis of the data do not generate discontinuities because they are not bounded by the linguistic constructions conveying the intentionality described above. On the other hand, any activity can cause a discontinuity if it contains a redirecting phrase.

However, three issues require further explanation. First, given the above discussion, why is it that not all 'reflecting' activities generate discontinuities? The answer to this is quite simply that they do not always include the explicit verbalisation of a future course of action and consequently do not shift focus of attention. Also, even when they do include such a verbalisation, the team does not always take up the new direction. In essence the proposal is ignored, at least for the time being. From a Reflective Practice perspective, reflections are explicit 'learning moments' within the team. So although subsequent activity may look the same as that preceding reflection, learning changes the situation and consequently a new move is always coded as commencing after a reflection. In contrast, in Rich Picture analysis such events are essentially failed attempts to shift focus of attention.

Second, we have argued from a Rich Picture perspective that every utterance that sums up past activity (even 'OK') is a form of explicit reflection, and that every utterance implying a new course of action reflects unvoiced reflection. Why is it then that not all chains are preceded by a Reflective Practice 'reflecting'? ('Reflecting' precedes only five of the eighteen chain-correlated, naming and moving events). The answer here is that in Reflective Practice analysis, 'reflecting' is defined as an explicit reflection on earlier activity that exhibits critical reflection of the team on this activity (*cf.*, Section 27.3). Hence, the presence of explicit reflection is essential to the identification of a reflecting activity. Clearly, while explicit, the word 'OK' does not convey critical reflection. This also explains why a focus of attention directing utterance *per se* is not registered as 'reflecting'.

Finally, why is it that Reflective Practice analysis generates more events than Rich Picture analysis? As well a detecting deliberate, voiced and intentional shifts, Reflective Practice analysis can discriminate shifts in focus of attention that are only revealed in the changing pattern of discourse. Thus, although there may not be utterances telling us explicitly that the team is changing attention, content analysis reveals chunks of similar utterance, the boundaries between them signalling change in focus of attention. Such cognitive shifts are not explicitly registered in verbalisation or discourse dynamic, which explains why Reflective Practice analysis is likely to yield a larger number of segments than Rich Picture analysis.

27.6 Conclusion

Our aim in undertaking the work described here was to test whether it is possible to match a bottom-up, data-driven description of group design-activity onto a top-down, theory-driven description. The matching process, initially undertaken separately by the authors, was helpful in verifying and validating the chains identified by Rich Picture analysis. The in-depth discussions on different ways of describing the same data provide useful and separate insights into the data and the assumptions underpinning each method. The Reflective Practice approach captures cognitive shifts within the design session, and the progression from one move to another, or to reflection, *etc.*, represents this cognitive shift. The Rich Picture approach also captures cognitive shifts but here the shift is very deliberate in its formulation and very noticeable within the overall process. The degree of match between cognitive shifts is of particular interest since the segmentation has been arrived at using separate methods, one top-down and theory-driven and the other bottom-up and data-driven.

The process of relating the two approaches has reinforced both approaches and to a certain extent the validity of each approach. Valkenburg and Dorst's theory-driven method validates the meaningfulness of the perceptible discontinuities that ground Mazijoglou and Scrivener's method, the latter approach providing secondary validation of the objectivity of the former. Furthermore, the matching of the two methods has helped to set the limits of interpretation when deriving the Rich Picture. For example, Mazijoglou and Scrivener (1998) attempted to go beyond chains to sub-chains by analysing the content of the discourse and grouping on content similarities between utterances. Not only was this a different form of analysis, *i.e.*, grouping on similarity rather than separation by discontinuity, it also involved a higher level of interpretation, *i.e.*, of the task-related meaning of utterances. Since the meaning of utterances depends on viewpoint, they found it very difficult to arrive at a coherent segmentation without first establishing such a viewpoint, *i.e.*, an *a priori* theory. The work described in this paper helped Scrivener and Mazijoglou to acknowledge this problem by revealing instances in the segmentation where judgement about the content of the task had been used to segment the discourse (and thus to discard sub-chain analysis). Additionally, cross comparison can help to maintain the coherence of interpretation when producing a given description, as both matches and mismatches must be justified.

Overall the Rich Picture description of the data studied aligns well with events produced by Reflective Practice analysis. While each method delivers a limited and focused view of design-activity the triangulation of methods increases confidence in the coherence of each. Furthermore, the results support the idea that a bottom-up data-driven description of design-activity could function as a skeleton onto which to attach theory driven descriptions without distortion. Thus, the Rich Picture approach offers a useful and robust bottom-up initial structuring of design-activity data prior to top-down interpretation of it. We would expect that other kinds of description organised in terms of cognitive shifts should map onto the Rich Picture. Having found one top-down approach that maps onto the Rich Picture, it would informative to determine whether the same is the case for other interpretative frameworks.

27.7 References

Banister P, Burman E, Parker I, Taylor M, Tindall, C (1994) Qualitative methods in psychology. Open University Press, Buckingham

Cross N, Christiaans H, Dorst K (1996) Introduction: the Delft protocol workshop. In: Cross N, Christiaans H, Dorst K (eds.) Analysing design-activity. John Wiley and Sons Ltd., Chichester, UK, pp 1-16

Purcell T, Gero J, Edwards H, McNeill T (1996) The data in design protocols: The issue of data coding, data analysis in the development of models of the design process. In: Cross N, Christiaans H, Dorst K (eds.) Analysing design-activity. John Wiley and Sons Ltd., Chichester, UK, pp 225-252

Mazijoglou M, Scrivener SAR (1998) The rich picture of design-activity. Automation in Construction 7,2/3:157-175

Schön DA (1983) The reflective practitioner. Basic books, New York

Scrivener SAR, Mazijoglou M (1997) Regulation in team design and its implication for collaborative CAD. In: Proceedings of IFIP formal aspects of collaborative CAD. Sydney, Australia, pp 187-214

Valkenburg RC, Dorst K (1999) Frame cognition. In: Proceedings of the 4th international design thinking research symposium on design representation. Massachusetts Institute of Technology, pp II-231

Valkenburg R, Dorst K (1998) The reflective practice of design teams. Design Studies 19: 249-271

Chapter 28

Analysing the Activity of Multidisciplinary Teams in the Early Stages of Conceptual Design: Method and Measures

A.J. Milne

28.1 Introduction

Economical 'intelligent' technologies emerging from the fields of electrical engineering, materials science, and computer science are making product design an increasingly multidisciplinary activity. Collaboration among discipline specialists is particularly important in the early phases of the design process, when alternative technology strategies are considered and initial solution approaches are planned. In the design context for this research project, subjects were asked to consider not only the design of new technologies, but the incorporation of these technologies into physical spaces to support new forms of human interactions – clearly an ambiguous, and complex, design challenge.

The work described here sought to develop a repeatable process for characterising the information-handling and social-interaction patterns of multidisciplinary design teams during very early phases of conceptual design, extending earlier work that considered individual designers (Baya, 1996). This paper will describe the analytical approach, including results from its application. The intent here is not to advocate this as the best means of analysing conceptual design events, but illustrate the kind of insight one can gain from rigorous studies of its kind.

28.2 Creative Design vs. Prescribed Methods

In developing new products whose fundamental technologies cross disciplinary boundaries, creative activity during the problem framing and conceptual design phases is an important component of the overall design effort. The simple 'brainstorming' technique, originally introduced by Osborn (1953), is an approach many organisations still use to foster creativity. The principles of brainstorming – including encouraging a large quantity of ideas, building on the ideas of others, and deferring judgement – establish a permissive environment for generating new ideas. They do not describe the specific mechanisms by which these concepts develop, nor do they suggest particular techniques for actively stimulating thinking in new and/or unusual directions. As design teams become more multidisciplinary in composition, it becomes important to understand the nature of the early phases of design and develop successful strategies for directing productive design sessions towards strategic design goals.

At Stanford University, the Center for Design Research (CDR) has been working with the Stanford Learning Laboratory (Learning Lab) to explore design processes related to one particular representative, interdisciplinary, design problem. The context for this investigation has been the design of globally-distributed learning spaces, a domain that presents design questions cutting across the fields of architecture, educational pedagogy, human-computer interface, digital media, and telecommunications. Researchers and designers at Stanford are interested in exploring how the combination of new electronic technologies and physical design features can support varied forms of teacher-learner and learner-learner interaction. The challenge from a design process perspective is one of how to involve specialists from each of the critical design fields and encourage them, without impinging on their individual creative practices, to work together to explore learning-space design from each of their vantage points.

The CDR/Learning Lab team organised an event that was to encourage appropriate design discussions. The team structured the event in accordance with Osborn's brainstorming principles while at the same time trying to guide the discussion towards topics of interest to the Learning Lab. When the outcomes did not address the original design objectives, the organisers recognised the need to examine more closely the nature of design team activity before sponsoring additional design events. Along with this, they also sought to develop performance measures to evaluate follow-on events as they iterated towards a successful design approach. A focus of this project was to devise a structured method for analysing activity patterns and measuring team performance in the conceptual phase of the design event.

28.3 Design Event Structure and Outcomes

In July 1998, the Stanford Learning Lab sponsored a 2½ day charrette to address the challenge of designing new forms of learning space. An organising group, consisting of the author and colleagues at the Learning Lab, recognised that the design of such spaces is increasingly a matter of more than merely the architectural design of the physical aspects of space. Designers need to concurrently consider the variety of technologies that might be embedded in the physical space as well as any that might impact the nature of the activities that might go on within the space. The design problem, therefore, was recognised as being highly interdisciplinary, perhaps requiring a new approach to the design process. The organising group selected a charrette as an appropriate initial design approach. The term 'charrette' was taken from the architectural community, but is perhaps less-well known in other design contexts. The author's working definition is given here: **Charrette (n.):** An event in which designers and/or others are asked to respond to a design brief in a limited amount of time; often involves public presentation of the responses to solicit rapid feedback.

Planning for the charrette was influenced by Schön (1983), who referred to the early state of a design problem as a 'mess' that designers must organise, and Dorst (1997), discussed the notion of the 'throwness' that designers encounter in being subject to the situation of their design activity. Both authors asserted that designers develop their own approach to defining the problem after encountering a design opportunity. Their observations seemed to suggest that the initial statement of the Learning Lab's design brief could be manipulated to focus the charrette participants' energies towards particular aspects of the design that were particularly interesting.

The charrette's organisers attempted to influence the participants' perception of the design situation through several vision documents and historical pieces on the University that they distributed to participants prior to the charrette. The organisers hoped these documents would indirectly encourage discussion about issues that included: facilitating transitions between public and private activity, designing spaces that support hybrid physical and 'virtual' communities, creating a 'permanently unfinished' building that supported frequent experimentation, and encouraging interdisciplinary engagement among globally-distributed student and faculty groups. The vision documents discussed these and other issues, elaborating on the kinds of questions associated with each, establishing the values espoused by the Learning Lab, and describing a variety of activity types that might be supported in the spaces. It also provided a list of possible architectural design elements whose character might be incorporated into proposed solutions. These elements included 'learning piazza', design loft, grand stair, and circus tent.

Pahl and Beitz (1996) and Cross (1989) discussed the fact that design seldom happens in a completely abstract sense, that in fact designers make use of antecedents and 'test solutions' as they explore the relevant design issues. The Stanford organizers issued invitations so as to ensure that the participant group included industry researchers who would, they presumed, bring into the discussion various forms of technology that might act as 'test solutions' in the work sessions. The participants were divided into four groups and asked to respond to any two of five ambiguously stated design challenges by developing usage scenarios for the learning spaces. Specifically, they were told, 'Each group should select two (2) of the following abstractions and develop design concepts that depict varied ways to implement the characteristics that are described.' The two abstractions selected by the groups studied in the project reported here (coincidentally, the two groups selected the same two abstractions) were:

1. Spaces that concurrently provide for different levels of formal and informal interaction in public and private venues. Such spaces should allow for fluid transitions between different modes of interaction, e.g., moving from a spontaneous discussion into a more formal project environment.
2. Spaces that are in some manner 'transparent' or otherwise 'viewable.' Here the idea would be to provide some form of linkage to ongoing, dynamic learning activities to exhibit the varied activities of the Learning Lab. The challenge

may lie in blending actual 'living' environments with the concept of an exhibition space without inhibiting the activities themselves.

It was hoped that the scenarios generated in the charrette would provide some insight about the most critical design requirements, as perceived by the participants. The charrette produced a number of creative design concepts and associated scenarios. These outcomes dealt mainly with 'macroscopic' design features, e.g., site, massing, and character of a building. There was little significant examination of the human-scale design features and the integration of technology and physical elements, critical design aspects that the charrette planners had hoped to explore. The departure from the intended outcomes motivated a more detailed study of the charrette to understand the techniques teams used to develop their design ideas.

28.4 Research Approach

The author chose protocol analysis as the research strategy for this investigation. During the charrette, the group worksessions held on Day 2 of the charrette had been videotaped. After an initial review of the tapes for all four of the groups, the author selected two groups for study, designated Blue and Red, and transcribed the videotapes for the morning worksessions from these groups. These transcripts were segmented according to a set of guidelines, separating the protocol into discrete elements of activity. The author then classified these segments according to a coding scheme in preparation for later quantitative analysis of the teams' activity patterns. The coding scheme was based in part on the work of Baya (1996), who studied the information-handling behaviours of individual designers along several dimensions. For the purposes of this research effort, the author began by using Baya's three 'Information Handling' activity classifications: Generate, Access, and Analyse. Based on the charrette data set, the author added three new 'Team-Handling' activity classifications: Elaborate, Verify, and Navigate. These activity classifications are defined in Table 28.1. A more detailed discussion of this classification scheme can be found in Milne and Leifer (2000).

Table 28.1. Classification scheme for coding charrette protocols

Generate	adds new information to information space from an unidentified source
Access	references information from an identifiable source
Analyse	takes action which changes an attribute of the information segment
Elaborate	provides additional information related to another segment, adding detail
Verify	links information handling & social interaction, adjusts to group feedback
Navigate	takes action connected to the social handling activity within the group

The coding scheme distinguished between information-related activities and process-related activities. Information-related activities consisted of the Information-Handling and 'Elaborate' classifications, the latter being included since it was an activity closely related to developing information despite being inspired by the team context. Process-related activities were taken as the Verify and Navigate classifications, as these dealt more closely with interaction between team members. The separation of activity types facilitated the extraction of those segments associated with information activities from the rest of the protocol.

To examine more closely how concepts were built, the information-related segments that introduced information into the worksession (Generate, Access, and Elaborate; excluding Analyse segments) were extracted from the rest of the protocol. These segments were then used to construct a visual representation of the connections between design suggestions, a representation that highlighted 'critical' links in which a significant level of idea 'building' is apparent as new design concepts are proposed and discussed. The visual representation used was a modified version of the Linkograph, a tool described and applied by Goldschmidt (1992). In this scheme, the information introduction segments were listed in the order they appeared in the protocol. Each segment was compared against all previous segments to determine whether it built upon an earlier segment in the protocol. A segment 'built' upon another if it: addressed the same thought (as evidenced by repeating words/phrases); continued to develop a common theme or topic; made an explicit reference to an earlier discussion point (e.g., 'as was stated earlier...'). When a relationship was identified between two segments, a corresponding box was filled in on the Linkograph. The resulting construct gave a visual representation of the concept-linking patterns that could be used as a tool to support further quantitative and qualitative analysis, as described later.

28.5 Results from the Structured Analysis

In the project reported here, two of the four workgroups were chosen for study, and the morning portion of their worksessions (constituting approximately 66 minutes of videotape for each group) were transcribed to form the data set for the protocol analysis. The two groups selected for the study, Blue and Red, appeared to have very different approaches to the design task from a process standpoint. The Red group spent the first 45 minutes of the protocol taking turns (in sequence) to present their background and personal thoughts on the design challenges, while the Blue group began generating ideas in a non-structured order after only 90 seconds. The Red group exhibited a total of 43 speaker transitions in the first 66 minutes of the protocol, while the Blue group had 282 for the same period. Finally, the Blue group made use of flipcharts, large sheets of paper, and personal notepads to record – or 'capture' – and later reuse elements of the discussion and personal thoughts as the session progressed, while the Red group relied solely on verbal exchange.

A quantitative analysis and comparison of the Blue and Red groups' information/team handling activities indicated that, despite apparent differences in their approach to the worksessions, they exhibited similar activity behaviours in terms of the coding scheme. Figure 28.1 shows a comparison of the percentage of segments that fell into each category. The figure compares the percentage of each group's first 600 segments, coded according to one of the activity classifications. The fixed number of segments represents 66 minutes of the Red group protocol and 45 minutes of the Blue group protocol. The figure shows similarities in activity between the groups, despite apparent differences in their approach.

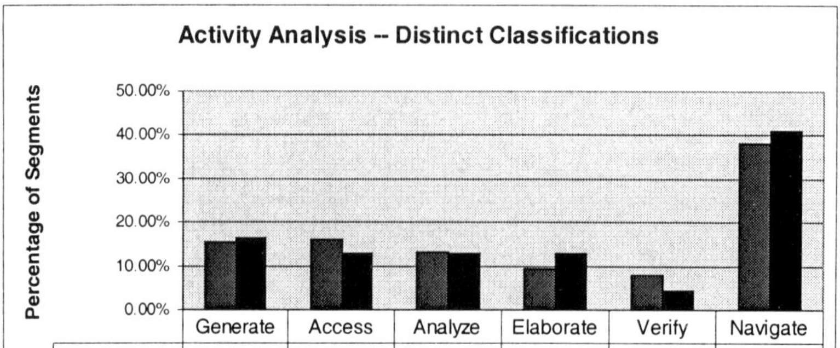

	Generate	Access	Analyze	Elaborate	Verify	Navigate
Red Group	15.51%	16.01%	13.04%	9.41%	7.92%	38.12%
Blue Group	16.24%	12.90%	12.74%	12.90%	4.30%	40.92%

Figure 28.1. Analysis of Red and Blue group activity by major classification

A detailed discussion of these results is given in Milne and Leifer (2000). For the present discussion, we note that this analysis shows the relative emphasis on information-related activities and team-interaction activities for the two groups. This comparison demonstrates that the structured-activity analysis can yield a result that was not readily apparent from an observational review of the videotape data, emphasising the value of the structured-analysis approach. Since the classification categories are focused on activities rather than outcomes, the coding scheme is not tied to a particular design domain, making it suitable for use across a range of disciplinary contexts as well as multidisciplinary design efforts.

Figure 28.2 shows a representative portion of the Linkograph representation for the Blue group data; the actual Linkographs, each analysing 250 individual segments, are too large to reproduce in this paper. Once complete, a visual inspection of a Linkograph reveals several trends. Wherever a segment is associated with several connection points (filled boxes) in a vertical line, the Linkograph indicates the idea in the segment built upon those expressed in several earlier segments. A horizontal line of connection points, on the other hand, indicates that a particular segment influenced several others later in the protocol. The vertical connections represent what Goldschmidt (1992) termed 'backlinks,' while the horizontal connections represent 'forelinks.' The quantity, size, and dispersion of backlinks and forelinks suggest a measure of performance for creative design activity. Goldschmidt established a threshold criterion that identified a 'critical link' as a backlink or forelink that contained three (3) or more segment connections. Analysing the data from the charrette, this threshold continued to be a useful demarcation for studying the data.

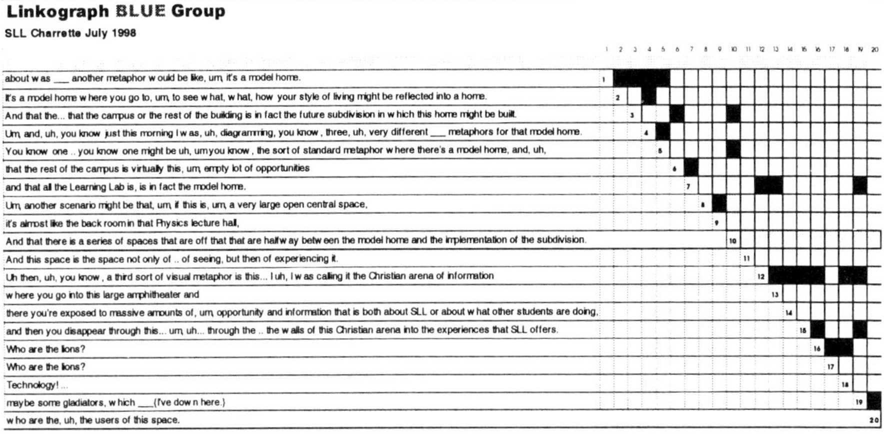

Figure 28.2. Partial section of Linkograph for Blue group

Quantitatively speaking, the Blue group displayed a total of 45 critical backlinks, while the Red group generated 16 backlinks. The total number of forelinks for the Blue group was 57, while for the Red group it was 27. Both measures show that the Blue group did more 'concept building' in the sessions. Visual inspection of the completed Linkographs also showed that the connection points for the Red group tended to be close to the diagonal, while those for the Blue group tended to be more widely dispersed. This indicated that the Red group participants were building primarily on thoughts that were most recently discussed, while the Blue group was more apt to build on ideas expressed in segments throughout the protocol.

While the Linkograph does not provide insight as to causal relationships for the different measures for the two groups, the technique could be applied to future controlled experiments to determine whether the differences in interaction patterns or the use of information-capture techniques influenced these quantitative outcomes. The critical-link analysis provides a performance measure for this form of team design activity. It identifies design concepts that are deemed 'successful' in the sense that one or more individuals value them enough to devote time trying to develop the concepts, or at least to promote them at various points in the protocol.

28.6 Implications for Studying Early Design Phases

The intent of this paper was to advocate the structured analysis of the early phases of design in order to examine more closely the critical methods by which designers negotiate the creative process in a multidisciplinary context. With this objective, the approach could not be domain-specific, nor could it interfere with the natural creative modes of the subjects under study. The results of the research project described here demonstrate the application of one such structured approach, highlighting some of the performance measures and research results that arose from the analysis.

While the activity-classification scheme applied in this project allowed for a structured analysis and comparison of group activity, it would require further development before it could be a useful tool in its own right. In particular, more robust classification guidelines are needed to guarantee reliability when using multiple individuals to code transcripts. The classification scheme does, however, serve as a filter for extracting segments for use in the linkograph analysis. The linkograph, in turn, provides for both a visual representation of the data and quantitative analysis of concept-linking performance, yielding both an intuitive and measurable understanding of concept development during a group design activity. While guidelines for identifying concept links are more closely associated with content for this tool, the general principles used in this work appear to be suitable for wider application without significant modifications.

The analysis tools applied in this project suggest performance measures that could be used to characterise team design activity in future studies. The amount of emphasis placed on various activities in the coding scheme could measure the extent to which each is supported or encouraged in the face of planned experimental interventions. The quantity and dispersion of critical links could act as a performance metric that evaluates experimental interventions meant to improve the degree of concept-building behaviour in design sessions.

28.7 Future Work

At Stanford, the CDR/Learning Lab team continues to plan process interventions to support both co-located and distributed design sessions associated with the design of learning spaces. Building on the results from the first charrette, CDR and its partners hope to organise research projects that will:

- investigate whether design-capture techniques could increase concept linking by recording ideas as they are introduced and making them available in some way throughout the duration of a design activity;
- study the relative importance of information- and team-handling activities for geographically distributed design teams with electronic connections;
- evaluate the effectiveness of new technologies and interfaces in supporting informal design activities.

The author plans to apply a modified form of the analysis approach presented here in the study and evaluation of future multidisciplinary design activities. This strategy will, it is hoped, provide a consistent means of evaluating outcomes across a series of design events while building knowledge concerning the nature of creative design practices.

28.8 Acknowledgements

The author gratefully acknowledges the funding and facility support of the Stanford Learning Lab and the Center for Design Research in completing this study. This research was conducted under the supervision of Prof. Larry Leifer, Professor of Mechanical Engineering in the Design Division at Stanford University.

28.9 References

Baya V (1996) Information Handling Behaviour of Designers During Conceptual Design: Three Experiments. Ph.D. Unpublished dissertation, Stanford University

Cross N (1989) Engineering Design Methods. John Wiley & Sons, Chichester

Dorst K (1997) Describing Design: A Comparison of Paradigms. Ph.D. Dissertation, Delft University of Technology

Goldschmidt G (1992) Criteria for Design Evaluation: A Process-Oriented Paradigm. In: Kalay, Y.E. (ed.) Principles of Computer-Aided Design: Evaluating and Predicting Design Performance. John Wiley & Sons, Inc., New York, pp 67-79

Milne A, Leifer L (2000) Information Handling and Social Interaction of Multi-Disciplinary Design Teams in Conceptual Design: A Classification Scheme Developed from Observed Activity Patterns. In: Proceedings of the ASME Design Theory & Methodology Conference (To be published September 2000)

Osborn A (1953) Applied Imagination: principles and procedures of creative thinking. Scribner, New York

Pahl G, Beitz W (1996) Engineering Design: A Systematic Approach, 2.ed. Springer, New York

Shön D (1983) The Reflective Practitioner: How Professionals Think in Action. Basic Books, New York

Chapter 29

Tracking the Design Trajectory of an Innovative Electronics Engineering Project

F.J.M. Reid, P.F. Culverhouse, A.P. Jagodzinski, C. Burningham and R. Parsons

29.1 Introduction

The research described in this paper was carried out at GEC-Plessey Semiconductors (GPS) in Plymouth, England, and describes a study of industry practice and problems associated with the management of engineering design teams within the company. GPS designs and fabricates complex, silicon, integrated circuits—for example, embedded microprocessors for cellnet telephones—and competes successfully in this volatile global market. GPS is a 'fast cycle time' company (Meyer, 1993), applying total quality management principles in the use of self-managing work teams, continuous process improvement, and participative management (Jablonski, 1991). Its workforce is therefore accustomed to working in an environment of rapid change.

Nevertheless, senior staff within the company recognised that the management of design activities is often problematic. Since under-management can lead to low design productivity but over-management to a loss of creativity, the dilemma facing the company pivoted on the optimal level of control that should be placed on design engineers. This dilemma is especially acute in the early stages of design, which are concerned with turning what can be quite vague and conflicting aims into hard engineering targets. Owing to the increasing complexity and size of artefacts this has through necessity become a team process, in turn requiring a high degree of flexibility and adaptability in the interpretation and exchange of ideas. An essential feature of this team process is that flexibility and adaptability are achieved by what is primarily a social process of interaction and negotiation between designers, in which conflicting aims and constraints are juggled to produce an acceptable definition of the design problem and its potential solutions (e.g., Brereton *et al.*, 1996).

The focus of the present paper is therefore on the interplay between technical and social challenges confronting design teams as they struggle to translate preliminary customer requirements into detailed functional design specifications. Here we describe a study—part of a larger investigation of design team management within the company (Reid *et al.*, 2000)—which tracked the course of an electronics engineering project throughout these early stages of design. Our goal was to follow a real design team through the entire problem definition phase, and to identify how the team responded to the various engineering and organisational demands made upon it during the course of design development.

29.2 The Interview Diary Method

We were able to study a design team within GPS that was developing a new RISC processor, an innovative programme with high levels of uncertainty, both in the technologies to be exploited and the software tools to be used in the design process. As it turned out, the duration of this programme was 33 weeks. In keeping with the naturalistic, non-intrusive approach adopted within the larger project, we set out to develop methods for studying the project team that attempted to uncover the understandings and meanings used by engineers and managers to structure and coordinate their work activities. In the absence of clear theoretical models of design innovation, we needed to remain open-minded about the functioning of design teams within the company, but gather information about the developmental course of the project as it unfolded over this extended period of time.

Tracking project development over this 33-week period required a framework that would allow us to explore some categories of information consistently on a week-by-week basis, whilst remaining open to new information over the life cycle of the project. We accomplished this by having one of the research team (CB) complete a weekly *interview-diary* with two key engineers in the RISC project team (engineers G and H). We developed the interview diary method to enable week-by-week comparisons of engineers' perceptions of the project, using regular brief interviews to gather this information. These interviews took place at the end of each week's work, lasted between 30 and 60 minutes, and followed a consistent format in which the engineers were invited to reflect on progress during the week, as well as identify and elaborate on significant developments in the project. Each engineer was given every opportunity to focus on matters they perceived as significant.

The interviews also incorporated a series of prompts designed to explore team response over the project life cycle to key indicators of project progress identified in interviews with GPS project leaders in an earlier stage of the study (Reid *et al.*, 2000). These key indicators are shown in Table 29.1.

Audio tape recordings totalling about 50 hours of interview material were collected in this phase of the project. This material was not transcribed in its entirety. Instead, a synopsis of the material, consisting of approximately 500 brief verbal summaries of the two engineers' responses to each of the seven prompts for each week of the project was prepared. These synopses provided the raw material for two parallel but independent analyses of the RISC processor project.

Table 29.1. Key project indicators used in the GPS interview diaries

1. Decisions	Whether decisions made by the project manager, the individual engineers, other company staff, or outside contractors servicing the project, were beneficial or adverse
2. Breakthroughs	Technical breakthroughs achieved (or constraints encountered) at any level of the project
3.Communications	The quality and extent of company and team communications, including the extent of team participation in project decision making
4. Knowledge	The acquisition (or deficit) of individual knowledge or skills
5. Information	The degree to which each engineer had ready access to or was awaiting relevant information or expertise, internal or external to the company
6. Progress	Perceived progress on the project, typically as judged against management Gantt charts and PERT project plans
7. Morale	The current outlook of each engineer, whether optimistic or poor, and current climate within the design team

29.3 Problem Tracks and the Engineering Challenge

The first analysis was carried out by an expert electronics engineer (PC), who set out to identify the engineering obstacles encountered by the RISC processor design team. This required careful and repeated scrutiny of the diary interview synopses, alongside company Gantt charts and PERT project plans, for evidence of technical and organisational problems emerging during the course of the project, and the time course and path taken by the design team in their attempts to resolve them. This analysis identified six *problem tracks* (see Table 29.2) emerging at different points during the project life cycle.

This analysis revealed that the RISC processor programme required the design team to grasp new concepts and technologies and to adapt to shifting management strategies and customer requirements. The design was seriously delayed by shortcomings in computer support tools, and although the core team was located at one site, project management was split across geographically remote sites for much of the programme, presenting obstacles to communication and decision making. Furthermore, at the request of an influential customer, the project time scale was foreshortened by about one third just a few weeks into the programme. Despite this, the project was completed only one week late, albeit to a reduced design specification.

Table 29.2. Problem tracks emerging during the RISC processor project

A change midstream in the information model of the system to be developed.
A management problem arising from the customer changing the project timescales.
Problems arising from inadequate software tools, which crashed frequently and lost data.
A requirement to use software development tools which no one in the company had experience using.
The need to task an engineer with no prior experience to develop the graphical user interface.
A series of hardware and company infrastructure failures.

This was clearly a difficult but not atypical project for GPS, and presented the design team with a volatile mixture of problems concerning customer specification, relevant technology, support tools, timescales, team communication, and decision making.

A number of observations emerged from this first analysis of the interview diaries. Firstly, the six problem tracks identified above peaked and troughed at different points during the life cycle of the RISC processor programme, and were marked by a series of *project milestones*, which emerged as psychologically salient and much talked about events in the engineers' interview diaries.

Secondly, problems were seldom resolved quickly and painlessly. Instead, many dragged on for several weeks, as engineers explored their ramifications, and attempted to contain and confront the challenge they presented. Thirdly, different facets of the problem tracks often followed separate time paths, as the impact of the problems rippled outwards within the team, and upwards through the management hierarchy. Finally, it was possible to characterise the total *engineering challenge* facing the RISC processor designers as the combined intensity of these six problem tracks during each week of the programme.

This analysis suggested three phases to the project, the first occupying about 10 weeks, during which the engineering challenge confronting the team steadily climbed as engineers assimilated the detailed technical nature of the product specification. During a middle phase of about 15 weeks, major problems were identified, assessed and solutions developed. As tensions rose, conflicts became apparent, and more problems were uncovered than solutions developed. It was at the midpoint of this second phase that the engineering challenge reached its highest point. During the third 'round-out' phase, solutions were finally implemented, the challenge declined, and the project was successfully completed.

29.4 Design Trajectories and the Shared Project Space

Our second analysis aimed to uncover underlying trends in the design team's perceptions of project milestones and the engineering challenge associated with the RISC processor project. To accomplish this, the diary interviewer (CB) independently rated the two engineers' responses to each of the seven key indicators, using a numerical scale ranging from 0 (to indicate absent, deficient, or adverse decisions, breakthroughs, communications, knowledge, information, progress, and morale) to 5 (indicating the maximum positive presence of these indicators). These ratings were pooled to form a matrix of observations in which each engineer's weekly scores on the seven indicators formed a separate case, providing a series of weekly snapshots of the RISC processor project from each engineer's vantage point.

This matrix of observations was then subjected to a common factor analysis (with $N = 66$), which yielded three factors with eigenvalues exceeding unity, together accounting for about 66% of the variance. A varimax rotation revealed that the first factor (33% of the variance) loaded on the decisions, breakthroughs, and information indicators, whilst the second factor (17% of the variance) loaded on the morale and communication indicators. The third factor (15% of the variance) loaded exclusively on engineers' perceptions of the progress of the project.

Taken together, these three factors define the *shared project space* of engineers G and H, and represent their combined psychological map of the RISC processor project. Figures 29.1 and 29.2 depict their weekly snapshot views in this shared space, plotted for consecutive weeks against the first two factors for ease of presentation. Each axis is labelled to show the key indicators that loaded significantly upon them. The paths these snapshots follow over the projects 33-week life cycle trace the separate *design trajectories* followed by each engineer, as their perceptions of the RISC processor project change over time.

Figures 29.1 and 29.2 indicate that both engineers traverse the factor space during the project life cycle, and with some backtracking and minor deviations, both follow a similar sequence. In the early weeks of the project initial progress is accompanied by declining morale (lower right hand quadrant), followed by a rise in morale as decisions and technical breakthroughs continue to be made (upper right hand quadrant). There then follows a striking deflection for both engineers across the project space into the lower left hand quadrant, as obstacles to progress mount and morale dips during the middle period (weeks 12-18). As these obstacles are surmounted, each engineer's trajectory reverses, and moves into the upper right hand quadrant at project completion.

By superimposing the engineering problem tracks on the shared problem space, we were able to explore whether these changes in design trajectory were contingent upon increases and decreases in the engineering challenge associated with developing the RISC processor. To illustrate these contingencies, the project milestones recorded in the diary interviews leading up to and following the 180 degree deflections in weeks 18 and 19 (engineer G) and week 17 (engineer H) have been superimposed on Figures 29.1 and 29.2.

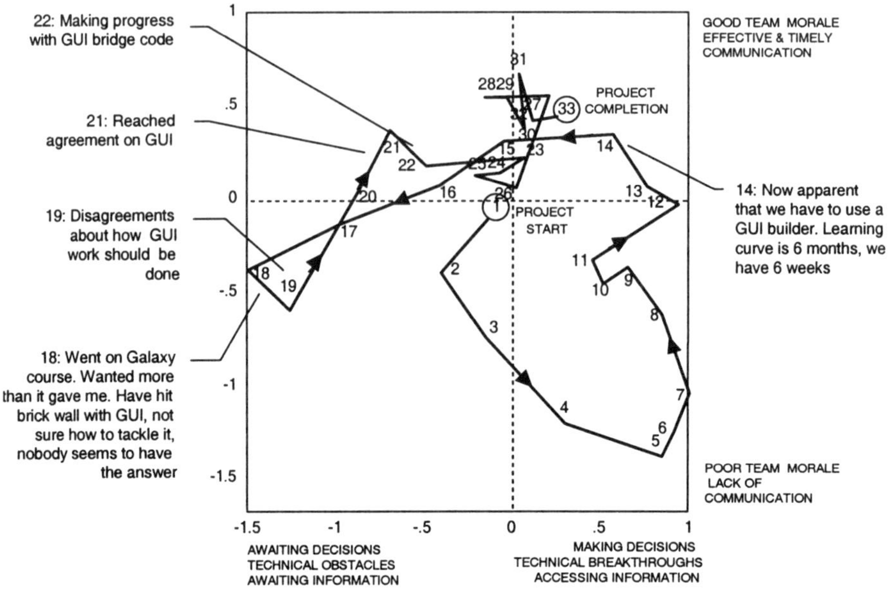

Figure 29.1. Project trajectory for engineer G, showing events associated with the development of the graphics user interface (GUI) (From Reid *et al.*, 2000)

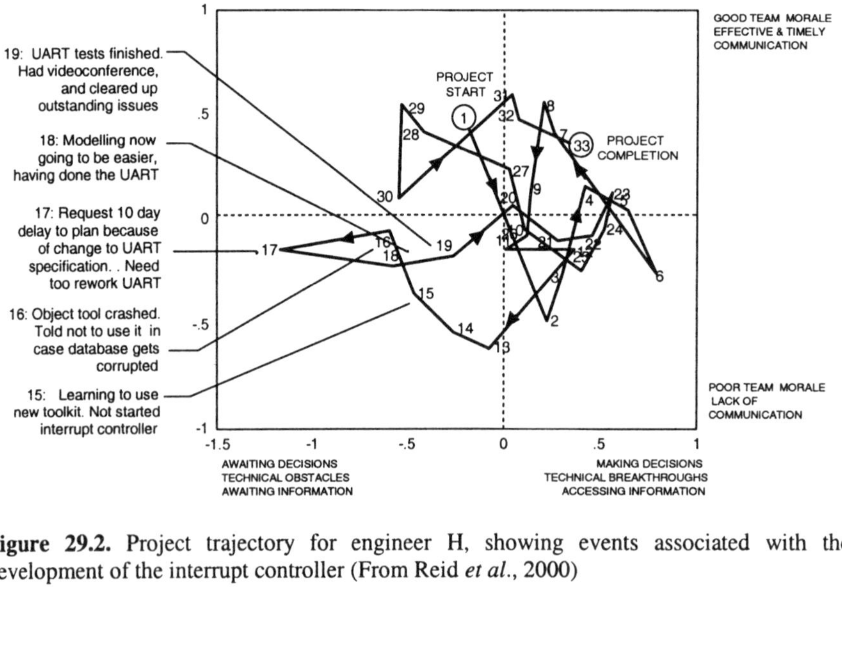

Figure 29.2. Project trajectory for engineer H, showing events associated with the development of the interrupt controller (From Reid *et al.*, 2000)

The challenge for G at this time was closely linked to the development of the graphical user interface. This required G to use specialist software with which he was entirely unfamiliar, necessitating specialist training and tough decision making. Although occurring at roughly the same time, the turning point for H was contingent on a quite different problem track associated with problems arising from the software tools in use. Here, the shortcomings of the software toolkit supplied by the company prevented progress on critical components of the processor, and it was not until specifications for these components were revised that forward progress became possible.

29.5 Managing Design Contingencies

A number of remarks can now be made about these analyses. By showing how the engineering challenge confronting a design team unfolds over time, we have highlighted the dynamic character and varying implications of design team management over the course of a single project. We have also shown that it is possible to construct a shared psychological map of the project, and to use this map to track the progress and morale of each engineer through the project life cycle. Furthermore, variations in engineers' progress and morale and can be traced directly back to the engineering problems they encounter, and the milestones they pass in resolving them.

We believe that these analyses point not only to the possibility of linking design team responses to events impinging on the team, but also to the need to develop a coherent decision framework to relate these contingencies to appropriate tactical choices in the management of design teams. In fact, a more detailed analysis reported elsewhere (Reid *et al.*, 2000) of the rich and complex data collected during this project revealed that in some aspects of project management, a lack of such a decision framework resulted in mismatched or inappropriate choices of tactics to meet contingent events. For example, we observed instances of radical changes in the task environment, including changes in the specification of requirements, and a reduction in the available time, and in each case there was no direct, effective response in terms of changed team tactics. On occasions there were lags of several weeks before it became clear that different team structures or management styles were required to cope with such changes.

One reason why these problems have arisen is that existing methods for managing design have difficulty keeping pace with the increasing complexity resulting from rapid technological changes. For the company studied here, the problems we have identified can only get more difficult as technological complexity and market pressures increase. This conclusion is borne out by other work. For example, Pinto and Mantel (1990) observed that project failure was influenced by a lack of control over such contingent events. Similarly, support for a decision framework is provided by Shtub (1996) who suggests that both the critical path method and the programme evaluation and review technique (PERT) lack the flexibility needed to handle changing situations as a project progresses. Unfortunately little progress has been made to date in establishing such a framework, although an ability to adapt to contingent events is seen as crucial by Shenar (1998) in his discussion of project management styles.

The present study suggests that engineering challenges are most successfully addressed by the adoption of a flexible, dynamic approach to design team co-ordination, in which moment-to-moment demands can be met by appropriate design management tactics. An important part of this approach would be the rapid identification by engineers and project managers of day-to-day changes in the circumstances of their project, and secondly, correct identification of the actions which are required. In other words, design engineers need reliable indicators of the changing circumstances the team is encountering, which can be linked by a coherent decision framework to particular changes of tactics in the way in which they organise the team. Elsewhere, we describe the development of a prototype software tool to operationalise this management decision framework (Parsons *et al.*, 1997). For the present, we conclude by noting that future research will need to identify in more detail the nature of such indicators and responses.

29.6 Acknowledgements

The work described here was funded by the Engineering and Physical Sciences Research Council (EPSRC) under grant reference number GRH 43526. We would like to thank GEC-Plessey Semiconductors of Roborough, Plymouth for their generous co-operation in the conduct of this research.

29.7 References

Brereton MF, Cannon DM, Mabogunje A, Leifer, LJ (1996). Collaboration in design teams: How social interaction shapes the product. In: Cross N, Christiaans H, Dorst K (eds.) Analyzing design activity. Wiley, Chichester, pp 319-341

Jablonski JR (1991) Implementing Total Quality Management: An Overview. Pfeiffer, San Diego

Meyer C (1993) Fast Cycle Time. Free Press, New York

Parsons R, Jagodzinski AP, Reid FJM, Culverhouse PF, Burningham C, Evans J (1997) A multidisciplinary approach to identifying and establishing support for the management of the team process of design. In: Riitahuhta A (ed.) World Class Design by World Class Methods: ICED 97 (Proceedings of the 11th International Conference on Engineering Design vol 2). Tampere University of Technology, Tampere, pp 393-398

Pinto JK, Mantel SJ (1990) The causes of project failure. IEEE Transactions on Engineering Management 37:269-276

Reid FJM, Culverhouse PF, Jagodzinski AP, Parsons R, Burningham C (2000) The management of electronics engineering design teams: Linking tactics to changing conditions. Design Studies 21:75-97

Shenhar A (1998) From theory to practice: toward a typology of project management styles. IEEE Transactions on Engineering Management 45: 33-48

Shtub A (1996) Project segmentation: A tool for project management. International Journal of Project Management 15: 15-1

Chapter 30

One Step Forward and Three Back: A Study of the Patterns of Interdisciplinary Conceptual Design

J.L. Steele, S.G. Macmillan, S.A. Austin, P. Kirby and R.J. Spence

30.1 Introduction

Design activity, particularly at the early stages of a project, is recognised as being dynamic, highly iterative and non-linear. However, under the rigours and pressures of the contemporary project environment, designers are being urged to undertake early design activity in a far more programmable, and thus manageable, fashion. Within this environment iterative, or cyclic, design progression is often criticised, with the concept of 'going round in circles' being one that is generally discouraged (Hickling, 1982). However, design is a learning activity and, owing to the complexity of contemporary building projects, it is often only by moving ahead to improve knowledge of the problem, before taking a step back to re-address a problem with improved understanding, that the design process can progress (Lawson, 1980). This is possibly the most commonly recognised type of iterative design progression among design researchers and practitioners. However, there are many other types of iterative design progression that are common to early-stage design activity. For the last two years the MDP (Mapping the Design Process during the conceptual phase of building projects) research project has endeavoured to improve understanding of conceptual design activity. It has been undertaken at the Department of Architecture, University of Cambridge, in close collaboration with Loughborough University and a number of construction industry firms. The research team has gathered empirical evidence which suggests that, although every design project is unique, there are commonalties within the iterative structure of periods of design progression across projects. This paper describes briefly the genesis of a framework for conceptual design before describing the results from two of the seven design teams studied to track design progression. The maps of design activity are then discussed and analysed to unravel and reveal the iterative trends within conceptual design progression.

A literature survey, encompassing upward of 200 texts starting with the early design methods literature that lead to the formation of the Design Research Society in the mid-1960s, provided a wide and far-reaching outline of the pre-existing knowledge of design. This, along with reviews of process models both within and beyond construction (Macmillan *et al.*, 1999a), interviews with designers about case histories, and observations of workshops where interdisciplinary teams of designers were observed during the concept phase of a design project, enabled a preliminary framework for concept design to be devised (Macmillan *et al.*, 2000). This comprised: i) a standard framework describing five design phases that are generic from one project to the next; and ii) at the lowest level, a structured set of 12 generic design activities in which project-specific tasks, knowledge, and data could be stored. The framework was developed to be flexible and adaptable, to accommodate different types of project, client, and design environment, while still offering a structure to which project-specific sub-models can be connected.

This framework was utilised subsequently as a means of tracking the conceptual design progression of seven individual interdisciplinary teams. Six of the teams were monitored over the course of two 'Designing together' workshops (Austin *et al.*, 1999 and 2000; Steele *et al.*, 1999). The first involved designers from a single multi-disciplinary organisation collaborating in the research; the second involved designers from each of the MDP project's industrial collaborators.

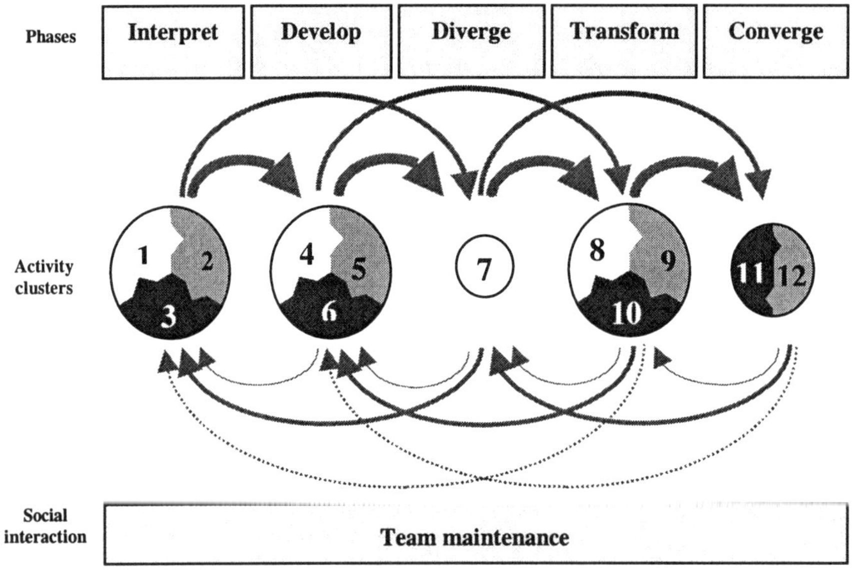

Figure 30.1. Reinterpreted conceptual design framework model

The design exercise within the two-day workshop involved the design of a window façade system for the re-cladding of 1960's office buildings.

Upon analysing the patterns of design progression of the workshop teams it became apparent that iterations across the activities and phases of the design processes that were recorded during the workshops fitted within a higher level of iteration representing the entire conceptual design phase. In light of this finding the preliminary design framework model was developed into a more realistic representation of the conceptual design phase (Figure 30.1).

The seventh and final team was tracked over the course of a conceptual design workshop on a live project in industry. This workshop, which aimed at developing a concept proposal for an airport terminal, also allowed a preliminary web-based design support system to be trialled; details of this component of the research are provided elsewhere (Steele *et al.*, 2000). Two of the seven patterns of design progression that were produced are illustrated and discussed in the remainder of this paper.

30.2 Patterns of Design Progression

Traditionally, models of the design process comprise sets of stages, phases or activities that are followed linearly during design activity; an example of which is the RIBA Plan of Work for Design Team Operation (1969), which, although developed some 30 years ago, remains the most widely referenced model of building design. Typically, this type of model appears to imply that the components of design are of broadly equal duration and importance.

However, when the conceptual design framework is used to track the design progression of interdisciplinary teams in practice, a more complex set of relationships between activities emerges. The two maps of design progression, which are discussed in the following section, provide insights into the nature of interdisciplinary conceptual design activity. Additionally, a number of trends have become apparent which suggest that, although the patterns of design progression are unique to particular teams within particular working environments, elements of design activity, and the reasoning behind them, are ubiquitous.

30.2.1 Experimental Workshop Team

Although a linear sequence of phases was pre-defined by this experimental workshop team it became apparent that the design actually progressed linearly but in a number of iterative bursts. Two iterations were performed to establish requirements while developing a design strategy, after which a period of concept generation and transformation took place. Two further iterations were undertaken to arrive at the final proposal - one to generate and choose the primary concept and another, to conceive and crystallise sub-elements of the proposal. The team members collaborated successfully throughout the exercise with little, if any, confrontation between members.

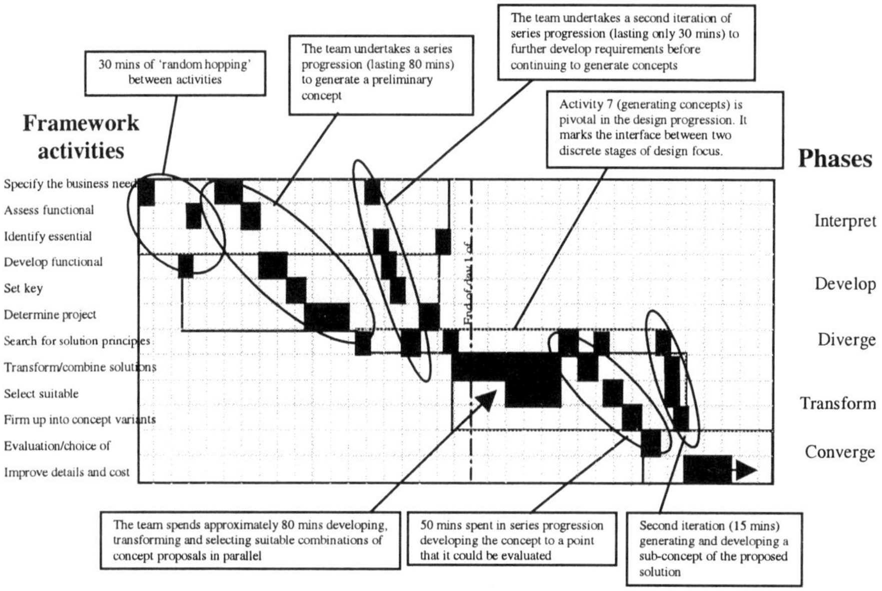

Figure 30.2. Pattern of progression of the experimental workshop team

30.2.2 Live-workshop team

Initially the team addressed a number of activities in parallel that focused on achieving the phase objective of interpreting the client's need. This was followed by a brief exploratory step to generate concepts before stepping back to resume a fairly linear progression. The pattern of progression portrays an iterative spike mid-way through the design activity. This was a needless iteration that occurred as a result of an oversight on the part of an individual. Thus, the design time spent performing this iteration could well have been avoided. The design team comprised individuals that had worked together previously on projects of this type for the same client. Consequently, there was a good team dynamic from the outset and no real confrontation between members.

30.3 Trends Within the Patterns of Progression

The two patterns of design progression that have been described and illustrated are, like the other maps produced, unique in that the same holistic pattern is never repeated. However, in analysing sections of design progression it is apparent that there are certain generic sub-patterns of design activity.

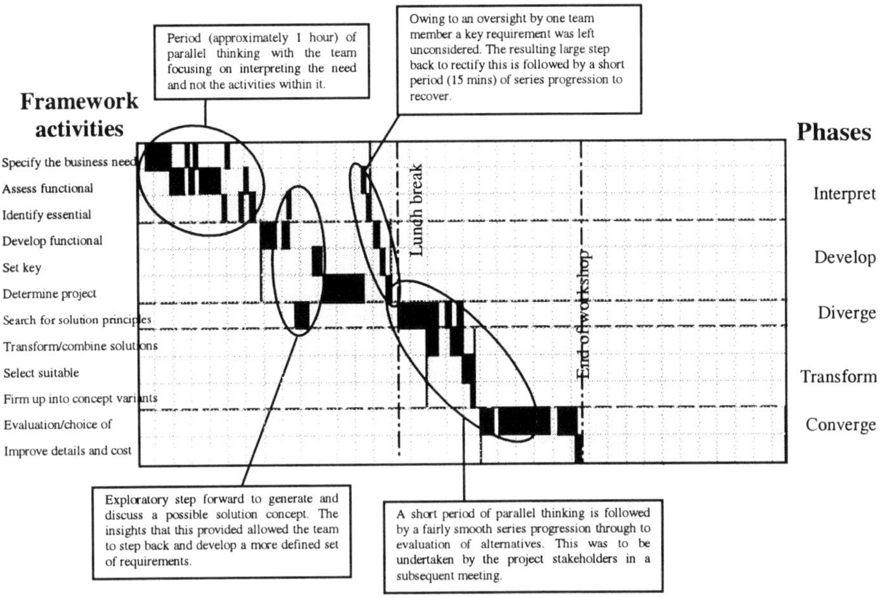

Figure 30.3. Pattern of progression of live-workshop team

30.3.1 Series and Parallel Progression

In analysing Figures 30.2 and 30.3 it appears as if steps between activities mean exactly that - that each of the activities are considered in isolation of the others for that period of design time. It may be more beneficial for the design team to think of activities as objectives that it must strive to reach (ways of thinking to reach an objective), rather than as steps that can be made in sequence if certain tasks are performed at each step. Of course, if this idea is applied, exactly what each member cogitates becomes irrelevant as long as the focus of their individual thinking aims at achieving the activity objective – their thought processes will be very different but their objective in thinking is synchronised. Thus, depending on the manner in which the steps were taken (two types are identified in Figure 30.4) it is suggested that the team are focusing on a different level of objective, *i.e.*, they are working toward the phase, rather than the activity, objective – in effect, different periods of design time are spent focusing on different levels of the framework.

Figure 30.4 illustrates two very different types of iteration. Instead of regarding all steps between activities in the same manner it may be valuable to differentiate between these periods in terms of: (i) parallel thinking and (ii) series thinking periods. Assuming this is the case then Figure 30.4a could be better represented as Figure 30.4b. The patterns of design progression can be misinterpreted if it is assumed that taking a step forward automatically means that the last activity has been completed and is no longer being considered. Although this does occur during series progression it is not always the case. However, it is important to recognise that the dominant form of design progression was observed to be steps forward sequentially (series progression), followed by a leap back over several activities before entering series progression again.

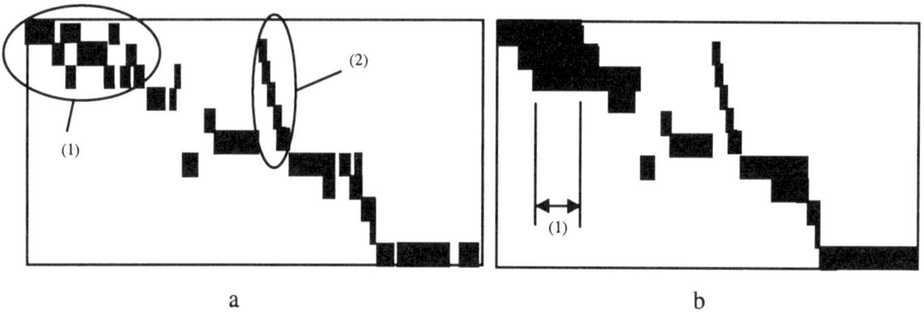

Figure 30.4. Differentiating between parallel and series focus

30.3.2 Speculative Progression: Stumbling Vs Exploration

Typically, if there is a large jump forward over a number of activities (*e.g.*, 2 – 7) it is followed by a similarly large step back (7 – 2, 7 – 3). This type of sporadic leaping can been described as speculative progression and is driven by speculative thinking (Figure 30.5). This type of progression, which is common in the patterns of design progression, suggests that either: i) the team members have progressed hastily and, upon realising that they did not do enough background investigation to achieve the objective, are forced to step back to the original activity; or ii) the step was taken, for example, to attempt a solution to improve problem definition before stepping back to address the earlier activity with improved knowledge. If the latter is the case, then it is assumed that the problem must be ill-defined and, as a result, there is some underlying rationale behind the sporadic stepping between activities. If the former is the case the latter will occur, to some extent, by default. However, this represents a stumbling progression in design terms and it could be described as neither effective nor efficient design behaviour – it is purely a symptom of having designers who do not understand fully the high-level design phases and activities.

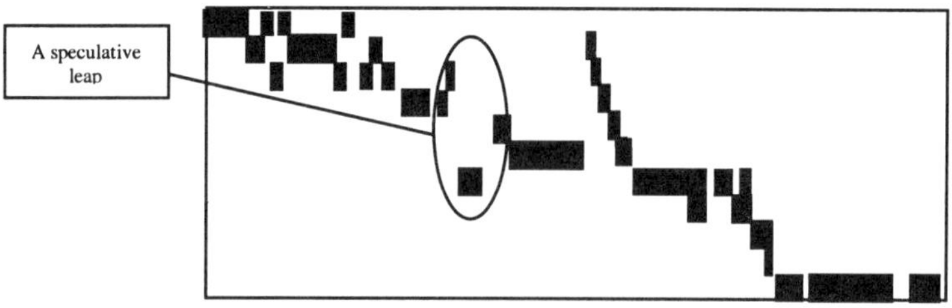

Figure 30.5. Speculative design progression

30.3.3 Iterative Steps

When progression has been made either using series or parallel thinking, a large backward step over a number of activities is typically followed by some further series or parallel thinking/progression. It was common to see this pattern of iteration repeat until an appropriate design was generated (see Figure 30.2).

When this iterative design progression occurred it was apparent that, in all cases, the gradients of the iteration were different to the gradients of the first occurrence of design activity. Although it might be expected that the iteration would always be steeper than the first occurrence of design progression (as some learning should have occurred) this was not the case. Generally, when a team progressed rapidly through a series of activities, any iteration tended to take longer to perform than the first occurrence, *i.e.*, a reduced gradient was apparent. This type of progression, be it series or parallel activity focused, tended to lead to a lack of certainty in proposals and increased the possibility of oversights. Conversely, when a team progressed slowly through a number of activities, any iteration through those activities tended to be undertaken more quickly, *i.e.*, an increased gradient was apparent. In outlining the differences between the types of iteration it must be noted that no data has been gathered to suggest that one is more beneficial then the other. Furthermore, excluding the case of parallel lines, it is impossible for the gradients to be anything but flatter or steeper than the first progression. However, this latter type of iteration could be deemed more beneficial for team members as it provides short periods of reflection on design activity thus improving their understanding of solution and problem as they progress (see Schön (1983) for details of this concept of 'reflection-in-action').

30.4 Effects of Problem Definition on Patterns of Design Progression

Problems can be categorised in terms of their level of definition or degree of complexity. Rittel and Webber (1973) have described this as the 'wickedness' of the problem. Ill-defined, or wicked, problems require, typically, exploratory (speculative) design progression and large amounts of iteration in order to be solved. Conversely, well-defined problems, typically, require far less iteration (although it is still a very necessary component of the design activity) and speculation in order for a suitable proposal to be developed. The difference between the types of iteration that are performed to solve these types of problem rests in the size of the 'leaps' between activities. The previous section has identified differing types of design progression that characterise periods of conceptual design activity. At times, step progression from one activity to the next does not necessarily mean that the team will stop performing the latter activity, it may mean the team merely bring another activity into consideration in parallel with any number of the prior activities.

30.4.1 Iteration Bandwidth

If the activities being performed in parallel over a period of time are recognised as such (Figure 30.4) a band can be drawn across the conceptual design phase which describes the activities that are under consideration over any given period of time. This is described as the 'iteration bandwidth'.

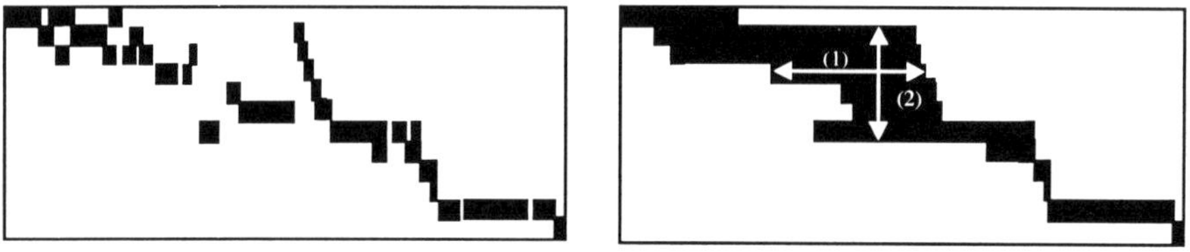

Figure 30.6. The concept of the evolving bandwidth – based on the design activity of the live workshop team

Depending on the manner in which a team progresses and, more importantly, iterates, the width of this band may not remain constant but instead narrow and widen as the design activity progresses. If the analogy of a concertina is used, expanding and contracting over a period of time, it may be simpler to envisage the evolving band enveloping more or fewer activities as the team progresses. If this is related to the previous discussions on reasons for iteration it is apparent that the narrower the bandwidth becomes the fewer the number of activities that are enveloped, and thus addressed or considered, in parallel (vertical (2) dimension in Figure 30.6) over a given period of time. Additionally, the horizontal dimension (1) at any point on the pattern of progression relates to the period of time over which a single activity or phase is considered (sporadically of constantly).

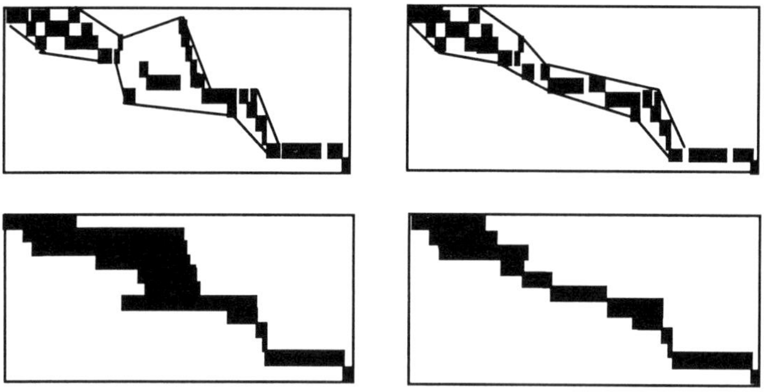

Figure 30.7. A reducing bandwidth (notional) resulting from improving understanding of the problem type – moving towards being well defined

Given a working environment where a team has worked together previously on a similar project, and will do so again, it is fair to assume that knowledge of the problem will evolve, thus reducing its complexity. The members of the live-design team verified this notion during a post-design discussion by stating that; 'This is an airport and we are all airport people and we kinda know where we are going...this is why the map [pattern of progression] is so smooth'. Thus, it is not unreasonable to suggest that teams that have worked together previously on a certain type of project for a certain type of client, could fine-tune their design progression around characteristic iterations. This, in theory, could allow the bandwidth to be tightened over the course of a number of projects (Figure 30.7), thus removing any wasteful iteration from future design activity.

30.4.2 Applying the Iteration Bandwidth Concept in Design Management

Our study to date has focused on the observation of design teams and the description of their processes. However, it may be possible to apply the findings more prescriptively to help designers become more reflective about their design progression and, potentially, to work more effectively. It is to be expected that teams addressing ill-defined problems would need to progress far more speculatively with large amounts of iteration (wide horizontal bandwidth dimension), addressing large numbers of activities in parallel (wide vertical bandwidth dimension), while those addressing fairly well-defined problems would expect to exhibit a narrower vertical bandwidth dimension, though not necessarily a narrower horizontal dimension. It may be possible to look at a design team, their experience on a particular project type, in a particular working environment, and set expected (characteristic) iteration bandwidth dimensions. If a design manager were to monitor the design team based on the characteristic iteration bandwidth it would be possible to identify when the team step outside of it. Reasoning for the uncharacteristic step could be sought and then fed forward into the next project process of that type.

This concept is based around attempts to manage conceptual design by encouraging iteration within prescribed boundaries based on descriptions of previous design activity. In effect, bandwidth could be a mechanism that allows the iterative loops to be revised and reviewed over a number of projects, with the loops designating the appropriate expected bandwidth based around the model of conceptual design (shown previously in Figure 30.1).

30.5 Concluding Remarks

The Mapping the Design Process project has provided an opportunity to monitor the design activities followed by interdisciplinary teams of designers during the concept phase of building projects. We have devised a simple graphical means of recording and displaying the pattern of progression through the activities that the teams followed. We have also used this to study and analyse these patterns in terms of the gradients and bandwidths of iterative working.

It seems highly appropriate that graphical methods are being used to study design, since design is often a visually-based process. And indeed, the designers who have participated in the project do seem interested in these patterns, which they can readily assimilate. These ideas about iteration, bandwidth and gradient might be passed back to designers to encourage them to reflect on their own processes, and help design teams manage their own teamwork processes more effectively.

The notions of phases and activities of conceptual design have been embodied in a prototype web-based interactive system that can be run over the internet. This was tested in workshop seven. This support tool for conceptual design focuses on the gates between the various activities, and provides a database for recording design decisions taken during each of the phases. It also contains Team Thinking Tools, should designers need help to broaden the solution space by generating more concepts, set priorities, or choose between competing alternatives. Its development and testing has been reported elsewhere (Steele *et al.*, 2000) and one of the industrial collaborators is proposing to develop it further as part of their in-house management of design.

Finally, only seven design teams have been monitored and, in every case, in workshops. Six teams were working on artificially defined problems in a training workshop. In the seventh case, the team worked on a live project, but again during a short workshop. We believe that the patterns we have identified are of considerable interest, but we have insufficient data to draw robust and generally applicable conclusions.

30.6 Acknowledgements

This work has been undertaken as part of a project entitled ‘Mapping the Design Process during the conceptual phase of building projects’. The research is funded under research grant GR/L39292 by the EPSRC and by industry (AMEC Design, BAA Plc., Hotchkiss Ductwork Ltd, Hutter, Jennings and Titchmarsh, Matthew Hall, and Pascall & Watson), with additional support from the Department of Civil and Building Engineering at Loughborough University.

30.7 References

Austin S, Steele J, Macmillan S, Kirby P, Spence R (1999) Using training workshops to map interdisciplinary team working. In: Proceedings of Chartered Institute of Building Services Engineers National Conference, 1999. Harrogate, UK, pp 38-49

Austin S, Steele J, Macmillan S, Kirby P, Spence R (2000) Mapping the conceptual design activity of interdisciplinary teams. Accepted for publication in Design Studies

Hickling A (1982) Beyond a linear iterative proces’. In: Evans *et al.* (eds.) Changing Design. John Wiley, Chichester UK

Lawson B (1980) How Designers Think. The Architectural Press Ltd., London UK

Macmillan S, Steele J, Austin S, Kirby P, Spence R (1999) Mapping the early stages of the design process - A comparison between engineering and construction. In: Proceedings of the 12th International Conference on Engineering Design 1999. Munich, Germany, pp 1175-1178

Macmillan S, Steele J, Austin S, Kirby P, Spence R (2000) Development of a conceptual design framework. Accepted for publication in Design Studies

RIBA (1969) Plan of work for Design Team Operation - Royal Institute of British Architects. Original Edition

Rittel H, Webber MM (1973) Dilemmas in a general theory of planning. Policy Sciences 4:155-169

Schön DA (1983)The reflective practitioner: how professionals think in action. Basic Book New York

Steele J, Austin S, Macmillan S, Kirby P, Spence R (1999) Interdisciplinary Interaction during Concept Design. In: Proceedings of the 15th Annual Conference of the Association of Researchers in construction management 1999. Liverpool, UK, pp 297-305

Steele J, Austin S, Macmillan S, Kirby P, Spence R (2000) An internet-based system to support interdisciplinary and inter-organisational collaborative conceptual design. In: Proceedings of the Chartered Institute of Building Services Engineers National Conference 2000. Dublin

Part III Studies of Collaborative Design

Summary

The studies of collaborative design reported in this Part are connected by several dominant themes that emerge despite the fact that the studies have cut across a variety of design domains (e.g., architecture, product engineering, electronics engineering and software engineering), have involved a range of methodologies for acquiring and examining data (e.g., observational and experimental analysis, examination of retrospective reports and interview diaries, and discourse analysis), and have examined an array of collaborative-design situations (e.g., face-to-face design-review meetings, video-linked design work, and team-design sessions).

One connecting theme concerns the importance of understanding the nature and determinants of patterns and sequences of activity in collaborative design. Reid and Reed examine the cyclical characteristics of communication in collaborative design and provide support for their 'cognitive entrainment' hypothesis which specifies that patterns of team interaction oscillate in rhythm with periodic shifts in modes of design reasoning. They show that peaks in conceptual reasoning are synchronised with peaks in team participation and turn taking, whilst peaks in figural reasoning coincide with troughs in these measures. Van der Lugt's data also provide evidence for cyclical aspects of design. Graphic idea generation is shown to involve a slow cyclical process of reflective consolidation of earlier ideas, whereas sentential idea generation involves rapid and associative concept development with high quantities of conceptual output that build only minimally on earlier ideas. Milne, too, is interested in the idea-generation processes of collaborating designers, and the way that ideas evolve over the time-course of the design session. His comparison between two design groups demonstrates how their approaches differed in terms of the type and quantity of concept building evident in their design work. In one group, designers were building primarily on recently discussed ideas, whereas in the other group, designers were more inclined to develop ideas introduced throughout the session. What determines the different activity patterns observed by Milne remains an important avenue for future research. Finally, Steele, Macmillan, Austin, Kirby and Spence outline how their records of the conceptual design activity of interdisciplinary teams can be analysed to unravel the common iterative trends that arise within conceptual design progression. A number of cyclical trends are identified including ones associated with 'speculative design progression', where designers take a large jump forward over a number of activities and then take a similarly large step back. Such speculative progression is ubiquitous across design groups and may be associated with either recoveries from unintentional and over-hasty design development or the intentional, exploratory generation of solution ideas that are used to inform problem understanding.

A second common topic concerns the important role of visual thinking in collaborative design. The papers by Reid and Reed and by van der Lugt demonstrate the central involvement of visual cognition in idea-generation. Gill, Deshmukh and Athavankar also focus on visual aspects of design. Their evidence demonstrates how a designer's imagery and their attendant design work may be influenced in important ways by the verbalisations and sketches of co-designers through processes of association and cross-cueing. Their results additionally indicate important changes in imagery experiences over the design session. There are also links to issues of visual thinking in design in the study by Turner and Cross, which demonstrates how design teams make sophisticated and flexible use of a multiplicity of representational forms within shared workspace resources.

A third generic element of these papers concerns methodological aspects of design research, with many studies (e.g., those by Milne, Ball and Ormerod, Turner and Cross, and van der Lugt) involving quantification of the relative proportions of different activities and the time spent engaged in them. Quantitative abstractions from raw data are clearly essential for developing theoretical accounts of collaborative design. Indeed, it is noteworthy that many contributions use near-identical coding schemes and analysis techniques that derive from the recent design literature - which attests to the emergence of well-proven analysis methods in this field. Mazijoglou, Scrivener and Valkenburg are also interested in methodological aspects of collaborative design research. They assess the validity of their 'Rich Picture' scheme which organises collaborative design discourse into a series of discrete chunks (called 'chains') which reflect intentional discontinuities in verbal data. They show how their predominantly atheoretical Rich Picture description of a data-set has key aspects of isomorphism to an independently-derived, theory-based description of the same data-set based around the concept of 'Reflective Practice'. Their results support the idea that a bottom-up description of design activity can function as a skeleton onto which to attach theory-driven interpretations.

A fourth theme connecting these papers concerns the potential advantages to the design process that are afforded by collaborative activity as opposed to individual activity. Gill, Deshmukh and Athavankar reveal how collaboration broadens the 'imagery canvas' within the session in a potentially fertile manner. Ball and Ormerod present evidence that design-review meetings may serve to enhance the generation and critical evaluation of alternative design options compared to what is typically observed in individual design contexts.

Finally, the critical role of managers and clients in shaping collaborative design emerges as a key finding in several of the studies. Reid, Culverhouse, Jagodzinski, Burningham and Parsons analysed their interview diaries, taken week-by-week over a 33-week period, to examine the interplay between technical and social challenges confronting members of a design team as they grappled with translating preliminary customer requirements into a functional design specification. The analyses reveal how various technical and social challenges (e.g., relating to changing customer timescales and hardware and company infrastructure failures) shape the trajectory of team members in terms of their perceptions of progress and morale. Reid and colleagues argue that their findings underline a need to develop a coherent decision framework that can relate design challenges and team responses to appropriate tactical choices on the part of team managers. These proposals hint at potential weaknesses in the effectiveness of team management processes in commercial design contexts which arise from the increasing complexities inherent in contemporary design practice. Ball and Ormerod's study emphasises certain positive dimensions of managerial input in design-review meetings. Their analyses indicate that the team manager may have a specialised role to play in driving the thorough and systematic evaluation of design options through the generation of novel design criteria. Lloyd's examination of the way in which television represents the design process reveals an interesting opposition between the characterisation of designers as 'individual geniuses' and the design chiefs in the client companies who are portrayed as 'bad guys' who already have their minds made up about the designs they expect to see. The observation, however, that throughout the programmes these chiefs become the characters who continually question the validity of the designers' concepts, suggests that they may have a crucial evaluative role in grounding the product design in commercial and technical realities comparable to the role of the team manager in Ball and Ormerod's study.

Studies of collaborative design are still in their infancy, and the findings reported here certainly require replication with larger samples of participants. In addition, there is rich scope for further explorations of the generalisability of detailed findings across different tasks, teams and contexts. Finally, there is considerable room for future studies to assess the applied implications of the findings reported in these contributions. Indeed a majority of the reported studies reflect on how best to support and enhance processes of collaboration - especially between team members working at a distance and communicating via technological media. Broad issues relating to mediated communication in collaborative design form the focus of the Part IV, whilst specific issues surrounding *computer*-mediation in collaborative design are covered in Part V.

Part IV

Mediated Communication

Part IV Mediated Communication

Introduction

Argyle (1991) identifies three modes of cooperation: on tasks; for the development of relationships; and in coordinating shared activity. He argues that, 'communication, both verbal and non-verbal, is one of the most important joint activities, needed for both tasks and relationship', and goes on to point out that communication itself cannot occur without cooperation. Communication is both a cooperative shared activity and one that is central to most forms of cooperation (e.g., on the completion of a shared task for external reward).

Developments in telecommunications, computer networking and collaboration-technology have stimulated research into the nature and role of communication in collaborative activity. Generally speaking, the research question has been framed as follows: given the potential of this technology to enable mediated communication over time and distance how can it be deployed to best effect? Three equally valid approaches to this question can be observed. The first takes the view that new communication technology (e.g., videoconferencing) should improve on old communication technology (e.g., telephone and fax), and the second, that this technology should be at least as good as face-to-face communication, while the third argues that new communication technology can do better than face-to-face communication. In the first approach, old telecommunications technology is the reference point against which new telecommunication methods are measured; in the latter two approaches it is face-to-face communication. We shall discover evidence of all of these approaches in the papers in this Part.

Research in mediated design communication was stimulated by reports of an observational study of listing, drawing and gesturing in face-to-face design (Tang, 1989). Tang, and his colleagues at Xerox PARC, went on to develop telecommunication tools that supported much of the behaviour observed in face-to-face communication. While this approach set the tone for much subsequent research, it is not without its dissenting voices, most notably those of Hollen and Stornetta (1992) who argue that modelling telecommunications on face-to-face communication blinds us to the potential of new technology.

The papers in this Part of the proceedings are all concerned directly or indirectly with mediated design communication. The first, by Gabriel and Maher, describes a comparative study of three communication environments: face-to-face, computer mediated using videoconferencing, and talk-by-typing. In each environment the designers were able to produce and share drawings. Gabriel and Maher appear to be interested in the potential of talk-by-typing as a medium for capturing design discourse for later retrieval and use. Intuitively, talk-by-typing seems an impoverished medium of communication compared to the other two environments. However, they argue that the data being communicated in design may be of more importance than the method by which it is conveyed, and hence the impact of talk-by-typing on design communication might be much less than one might imagine. Indeed, it may have the benefit of forcing communication to be more direct, precise and concise. Implicit in Gabriel and Maher's argument is the idea that computer-mediated communication can improve on face-to-face communication, or that the spin-off benefits (e.g., semantic archiving) will more than compensate for communication impoverishment.

Most of the remaining papers in this Part focus on non-verbal design communication. It is generally assumed that design would be impossible, or unacceptably difficult, without the support of visual modelling and communication tools. Observation supports this assumption: designers routinely create and interact with drawings, models and workspace artefacts, *inter alia*. The following papers pose or evoke a number of questions about non-verbal communication and interaction in collaborative design. For example, is it necessary for collaborative design? If it is, does it merely support or augment verbal communication, and are new, enhanced forms of non-verbal communication and interaction possible?

Reed and Reid argue that speech, sketching, and bodily gestures mutually interact to give shape to emerging ideas in face-to-face meetings. However, they observe that, as yet, little is known about how designers combine these actions with speech to develop shared ideas in collaborative design. They describe an experiment designed to investigate whether workspace activity (*i.e.*, sketching, figural gesturing and pointing) provides such a communicative function over and above that of speech alone. In their experiment, two groups of co-located design dyads undertook two tasks. Dyads in one group were prevented from seeing each other's drawings by means of an opaque screen located between them. Reed and Reid anticipated that the performance of this group would be seriously affected by the removal of shared visual access, compared to the group with visual access, and also that they would have to work harder at verbal *conversational grounding* – the two-way exchange of positive evidence of mutual understanding.

Like Reed and Reid, Scrivener, Chen and Woodcock are interested in the role played by visible action in supporting verbal communication in collaborative design. However, they focus on methods for uncovering design communication needs. They postulate that collaborating designers will choose what they perceive to be the most effective and efficient means of communication at their disposal in order to complete a task. In a multiple medium-mechanism (*i.e.*, video-gesture, whiteboard-gesture, *etc.*) environment, they argue, shifts between medium-mechanism (*i.e.*, between video-gesture and video-drawing) should be explicable in terms of gains in communication efficiency and/or effectiveness. Analysis of these transfers, they continue, will yield insights into visual communication needs that can be employed in designing and redesigning communication environments. In support of their argument, they report an analysis of the data emerging from a study where design dyads collaborated using a computer-mediated communication environment to complete a design project over a four-week period.

Reed and Reid, and Scrivener *et al.*'s work is typical of earlier work on the role of visible actions in collaborative design insofar as it focuses on drawing and gesturing (although the latter's work does encompass workspace artefact-related actions). In contrast, Logan and Radcliffe's paper describes a design setting where drawing is virtually non-existent. In a Rehabilitation Engineering Centre (REC), rehabilitation engineers, physiotherapists and occupational therapists work together and with the client to design 'assistive' devices. Design occurs during team-client sessions as participants discuss, experiment, mock-up and try out ideas. Recognising the potential value of videoconferencing as a means of conducting sessions between remote centres, Logan and Radcliffe aim to develop techniques that will allow the REC team to capture sufficient and accurate data to manufacture 'assistive' devices in isolation from the client. Theirs is a paper of two parts: the first, an analysis of video recorded REC design sessions, focuses on the relation between categories of talk (e.g., design proposal) and visible workspace activities (e.g., constructing), and the coupling of artefacts and actions; the second presents the findings from ten rehabilitation engineering videoconferencing sessions.

While Gabriel and Maher's approach is concerned with going beyond, and Reed and Reid, and Logan and Radcliffe's with matching face-to-face design communication, Scrivener *et al.*'s approach take existing telecommunication tools as the reference against which to measure improvement. Likewise, Woodcock, Lee and Scrivener's paper can be seen as reflecting this latter approach. Design at a distance is already a reality using high-tech and not so high-tech modes of mediated communication. First, Woodcock *et al.* summarise the results of an analysis of the documentary records of 21 international design projects conducted between 1993 and 1997, and funded by the China External Trade Development Council. Their approach focuses on uncovering the problems arising in the projects due to, or exacerbated by, distance and the international nature of the activity. They also describe the findings from a follow-up interview study of a sample of participants involved in the projects.

Wagner's paper brings us back squarely to visual communication. She describes a combined ethnographic- and interview-method study of the forms and uses of what she terms 'persuasive artefacts' in mediating collaboration in architectural design. Notwithstanding its debt to ethnographic method, Wagner's research is guided by a teleological imperative: she is concerned with change that opens up new possibilities. For while valuing the useful emphemerality of designers' visualisations, she notes that this emphemerality has its disadvantages. She proceeds to outline a computer system designed to preserve the ephemerality of these 'persuasive artefacts' while introducing additional functionality that helps to account for the complexity of a developing project.

Finally, Maher and Simoff develop the themes of non-verbal communication and 'beyond face-to-face collaboration' in architectural design. First they describe the concept of 'design within the design' in which the designers inhabit a representation of the design. In a sense, the designers inhabit their own imaginary world, seeing it grow as the project progresses, and interacting with it and each other throughout. For team-based design, this involves developing the design model as a collaborative environment. They describe the implementation and testing of such an environment, before presenting an analysis of design-team communication occurring in it during a 'design within the design' session.

Argyle M (1991) Cooperation: the basis of sociability. Routledge, London

Hollan J, Stornetta S (1992) Beyond being there. In: Proceedings of CHI'92. ACM Press, Monterey CA, pp 119-125

Tang JC (1989) Listing, drawing and gesturing in design: a study of the use of shared workspaces by design teams. PhD Thesis, Department of Mechanical Engineering, Stanford University

Chapter 31

An Analysis of Design Communication with and Without Computer Mediation

G. Gabriel and M.L. Maher

31.1 Introduction

With recent developments in CAD and communication technologies, the way we visualise and communicate design representations is changing. A matter of great interest to architects, practitioners and researchers alike, is how computer technology might affect the way they think and work. The concern is not about the notion of 'support' alone, but about ensuring that computers do not disrupt the design process and collaborative activity already going on (Bannon and Schmidt, 1991). Designing new collaborative tools will then have to be guided by a better understanding of how collaborative work is accomplished and by understanding what resources the collaborators use and what hindrances they encounter in their work (Finholt *et al.*, 1990).

Designing, as a more abstract notion, is different from having a business meeting using video conferencing. In design it is more important to 'see' what is being discussed rather than 'watch' the other person(s) involved in the discussion. In other words the data being conveyed might be of more importance than the method with which it is communicated (see Kvan, 1994). Similarly, we believe that by using text instead of audio as a medium for verbal communication, verbal representations can then be recorded alongside graphical representations for later retrieval and use. In this paper we present the results of a study on collaborative design in three different environments: face-to-face (FTF), computer-mediated using video conferencing (CMCD-a), and computer-mediated using 'talk by typing' (CMCD-b). The underlying aim is to establish a clearer notion of the collaborative needs of architects using computer-mediation. In turn this has the potential for assisting developers when designing new collaborative tools and for assisting designers when selecting an environment for a collaborative session.

31.2 Computer-mediated Collaborative Design

In order to compare computer-mediated collaborative design (CMCD) and face-to-face (FTF) collaborative sessions between architects, we first need to look at how architects collaborate in FTF environments, the media they employ and the communication channels they utilize in order to convey design representations to their partner(s). When working FTF, architects have been observed to hold certain preferences for the way they organise their design and creative environments, and what 'traditional' tools they choose to use, whether designing alone or collaborating with colleagues (Carter, 1993). Some architects might prefer to work with thick pencils scribbling 2D sketches on butter paper (Gross, 1994; Kvan, 1994). Others might sketch as well as start working with 3D volumetry. Sometimes they hastily proceed to build 3D massing models, made of polystyrene or cardboard (Visser, 1993). This enables them to acquire an enriched 'experience' of the space they are working with and makes it easier to communicate their 'idea' to other parties involved in the design. However, the continuous development of computer and telecommunication technologies, has seen architects increasingly using these mediums for communication as well as work. Hence architecture as a profession is becoming dependent on computers not only in ways of documenting designs, but also in the form of representing and communicating design ideas between various parties, from colleagues to clients and to the general public. Research into communication channels used in CMCD environments has shown that there is little agreement on whether audio and video channels are essential in such ventures as well as what constituted the appropriate channels (Maziloglou *et al.*; 1996, Olson *et al.*; 1997; Vera *et al.*, 1998). A popular view held by some researchers is that adding audio, video and graphics is somehow expected to make the medium more 'real' (Sudweeks and Rafaeli, 1995). According to Greenberg *et al.* (1992), some researchers maintain that *tele-presence* is an the alternative to FTF collaboration, where distributed participants in a collaborative venture are given the feeling that they are present in the same meeting room. Whether or not seeing one's partner has an effect on performance seems to be highly dependent on the type of task performed (Olson *et al.*, 1997). On the other hand, Vera et al (1998) observed a slight decrease in low-level design as opposed to high-level design in text-based compared to audio and video computer-mediated experiments.

31.3 The Study

In our study we investigate collaborative communication in a design session with two architects. We use a method similar to a protocol analysis in which we collect and transcribe the utterances of the designers during the design session and analyse the *design communication protocol* using a coding scheme. Protocol studies on design activity date back to that of Eastman (1970), where he studied architects in the late 1960s. The protocol analysis method continues to be an accepted way to study design, although most of the studies look at single designers (Akin, 1986; Goldschmidt, 1991). Recently, the conventional, single-subject, method of protocol analysis has been used to analyse team design activity, (Cross *et al.*, 1996; Vera *et al.*, 1998). We conducted twenty-six one-hour experiments using fifty-two 5th and 6th year architecture students. The participants were paired and each pair participated in only one experiment from any of the three categories using the same brief. We conducted eight experiments in each of the FTF and CMCD-a categories and ten in the CMCD-b category. One brief was designed for all three categories in order to reduce the variables. The eight FTF sessions were carried out in a room containing a central table with participants sitting on either side. Each pair was given four A1 tracing sheets with a pair of black and blue felt pens in order to account for the sketches. They were each given a copy of the brief as well as extra A4 copies of the site plan and section, which can be used to trace over. A Sony™ Hi-8 CamCorder connected to VHS in the same room, was placed at an angle to capture both verbal and graphical interaction between the participants. Two rooms separated by a third larger one were used for the eighteen CMCD sessions. Each room was equipped with a Silicon Graphics O2™ (SG) Unix workstation. The two SGs were connected, along with the observer's terminal (in the central room), by a high speed Local Area Network (LAN). The CMCD-a sessions used computer-mediated audio and video with a shared electronic whiteboard (they used the Inperson™). The CMCD-b sessions used a chat-like environment to talk to each other by typing messages, and a shared electronic whiteboard (also using Inperson™) without the video channel.

31.3.1 The Design Brief

In each experiment the two designers were given the same brief, as shown in Figures 31.1 and 31.2. All the experiments were 1 hour long. All experiments were both audio and video taped to facilitate the transcribing and later stages of analysis.

```
A City based painter recently acquired a site on top of a cliff in
an inner-west suburb of the city. He stumbled across the location
by taking the wrong turn one day and ending up in a cul-de-sac, on
top of a boulder with breath-taking views (Figure 2). To the owner,
a dwelling represents more than a shelter or a place to live in. He
prefers to think of it as a space comprising certain functions,
some of which are living, working and entertaining. Far from being
a novel idea, the house as a shelter that combines the working and
living environments dates back a few centuries. Numerous
contemporary architects have relished such unique opportunities to
investigate and develop their own architectural theories. The brief
set out by the owners along with their teenage son (19), and
daughter (17), was a simple list. Functions such as an entertaining
area, a decent sized naturally lit workshop and roof terrace
overlooking the cliff were among the items included. The owners
require that the design be unique while reflecting and enhancing
the natural attributes of the site.
```

Figure 31.1. The text of the design brief

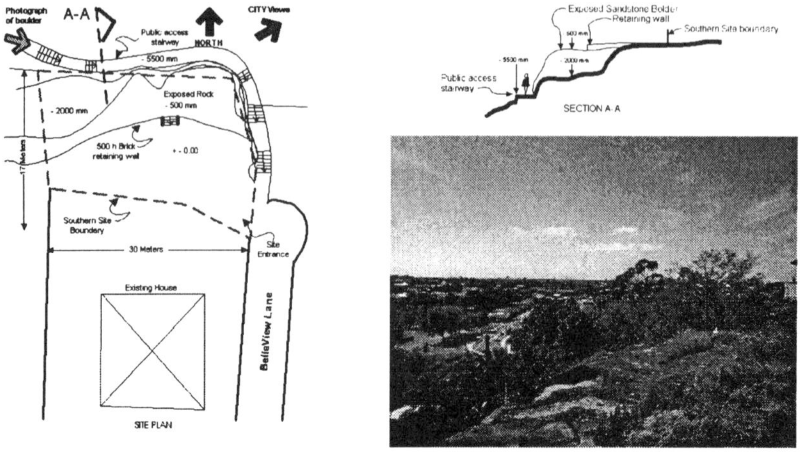

Figure 31.2. The brief site plan, section (NTS) and photograph (taken from site)

31.3.2 The Coding Scheme

We considered four different coding schemes from separate research projects. The first, (see Sudweeks and Albritton, 1996) categorises communication types as follows: *Informal control of communication, formal control of communication, socio-emotional communication, conceptual communication, task communication.* The second coding scheme investigates the amount of time spent in computer mediated collaborative sessions on 'introducing *new ideas and clarifying those ideas'* (see Olson *et al.*, 1997). The third coding scheme on the other hand classifies interaction between FTF and Video-conferencing technologies by investigating *'Interruptions, overlaps, hand-overs and dominance'* (see O'Connail and Whittaker, 1997). Part of the fourth coding scheme investigated *'low level design'* versus *'high level design'* in computer-mediated design sessions with full and limited communication channels (for further details see Vera *et al.*, 1998).

Our coding scheme is formed of four major classifications and in turn some of these are further broken down into sub-categories, illustrated in Figure 31.3. These classifications are:

1. *'Communication control',* a theoretically- and externally-derived structure which would help identify possible differences, advantages and disadvantages between the three design communication mediums (FTF, CMCD-a and CMCD-b. Communication control includes statements made by the designers to hold the floor, to interrupt, to acknowledge presence, and to hand over communication to the other person.
2. *'Communication technology',* a data-derived structure, looks at discussions held between participants related to the use of the tools and the collaborative environment.
3. *'Social communication',* a data- and externally-derived structure, looks at the amount of time spent on social talk in the three collaborative media. Examples

of social talk are: making a personal statement about the person running the experiment, commenting on the way a person looks in the video window, talking about what the other person is wearing.

4. *'Design communication'*, a data-, theoretically- and externally-derived structure distinguishing between 'design ideas', 'design scope' and 'design task'. This category helps investigate whether there were any variations in the way the design itself was discussed between the 3 different media. (For a more detailed account see Gabriel and Maher, 1999).

These categories are not intended to be exhaustive, but to indicate, through analysis, the relative amounts of communication in each category when comparing FTF and computer-mediated collaborative design. We are particularly interested in whether computer mediation affects the ability to discuss design issues (that is, whether we would see less design communication), and whether there are significant differences in the way communication control occurs in the different collaborative environments.

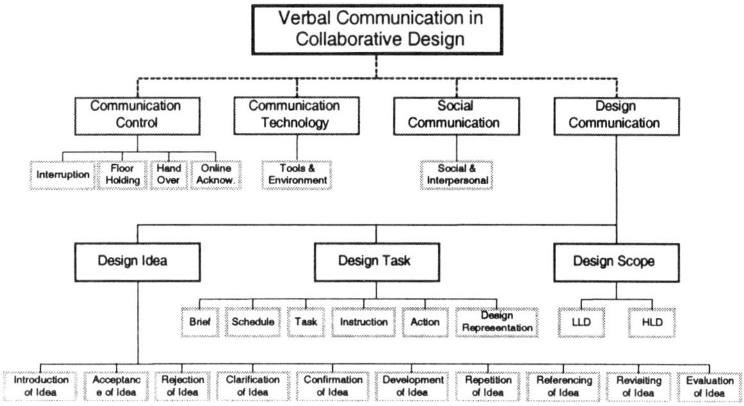

Figure 31.3. A hierarchical tree of the coding scheme

31.4 Results and Discussion

All 26 experiments were transcribed so that each utterance by a single designer was considered as a single 'text unit'. Two independent coders coded each utterance. One or more types of communication were assigned to each text unit. An arbitration process followed, resulting in single unified coded transcripts.

Figure 31.4 shows the average distribution of the percentage[1] of the coded text units, across the 4 primary coding categories in all 3 categories of experiments, revealing some important variations. For example, in the CMCD-b sessions (no

---

[1] We show the results as an average of the percentage of text units in each category in order to normalise the values. Some sessions had a large number of text units and others had a relatively small number, even though each session was one hour long.

audio or video) there were fewer text units dealing with communication control and more text units concerned with design communication than in FTF or CMCD-a. However, to further measure these differences and assess the impact of the different communication channels, we briefly characterise our observations and analysis of the coded sessions into 2 principal classifications. These are differences in *communication* and differences in *verbal design representation*. To begin with, we compare differences in communication across the 3 categories of experiments through the first 3 primary categories of the coding scheme: *communication control*, *communication technology* and *social communication*.

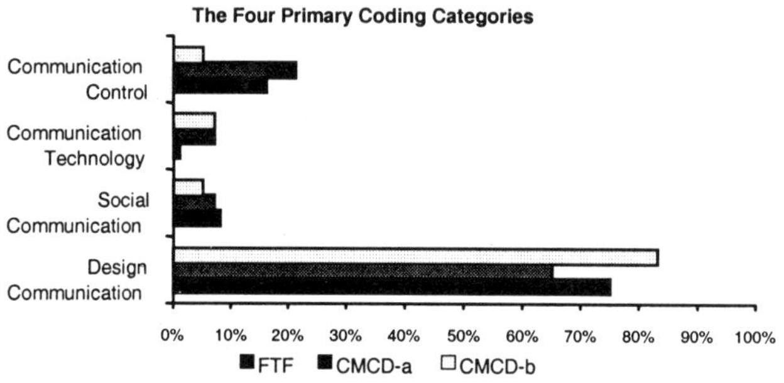

Figure 31.4. Mean percentage of the 4 primary coding categories across the 3 categories of experiment

Communication Control varied markedly between the 3 categories of experiments. We were expecting differences mainly between FTF and CMCD sessions. We expected that the participants in CMCD sessions would have higher levels of interruptions in an attempt to overcome the novel collaborative environment and the remoteness created by the communication medium. However, in both FTF and CMCD-a the verbal (audio) communication was 'spontaneous' and at times participants were observed competing for the 'conversational floor', which resulted in higher levels of interruptions thereby obstructing the flow of communication. However contrary to our expectations there were fewer interruptions in the CMCD-a category compared to FTF since the mean percentage of utterances coded under the interruption sub-category was 41% FTF to 31% CMCD-a. Communication was less spontaneous in CMCD-b with participants getting straight to the point using shorter and clearer utterances, and essentially no interruptions or floor holding.

Discussions about *communication technology* did not vary much between the CMCD categories, where the designers experienced similar difficulties, mainly with the tools of the shared whiteboard. However *social communication* presented us with an unexpected result since despite the removal of the audio-video channels, CMCD-b users had similar levels of social interaction compared to the other 2 categories (8% FTF and 7% CMCD-a, versus 5% CMCD-b).

Differences in *Verbal design representation* are directly linked to the *design communication* principal coding category. Figure 31.4 indicates that CMCD-b had the highest percentage of text units coded as *design communication*. Figure 31.5

shows the differences in how *design ideas* were discussed in the different categories of collaboration. There are nearly twice the amount of text units concerned with *introduction of idea* in CMCD-b compared to the other 2 categories (10% FTF and 9% CMCD-a versus 28% CMCD-b, out of total text units coded under design idea).

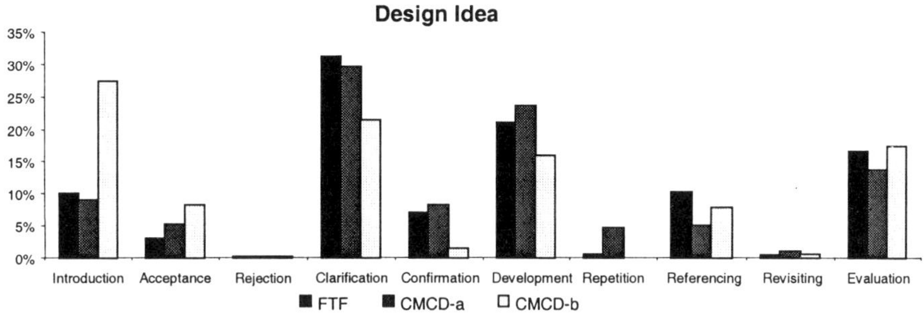

Figure 31.5. Mean percentage of design idea across the 3 categories of experiments

With no interruptions participants in CMCD-b appeared to communicate their ideas as if they were designing by themselves. That is they kept introducing new ideas, whether they had a response or not, as if they were 'thinking aloud'. Therefore rather than stopping at the first idea coming to mind, participants continued to introduce new ideas, recording them in writing. However, the high percentage of new ideas being introduced means that there was a lower percentage of text units in confirmation of ideas, rejection of ideas, and revisiting ideas.

In summary, our expectations were that the major differences in collaborative communication would occur between the FTF sessions and the CMCD sessions. Our results show that in many ways, there is a strong similarity between FTF and CMCD-a and major differences between CMCD-b and the other two types of collaboration. This implies that a significant factor in the way designers communicate is whether the designers are communicating by speaking or by typing, not whether the designers are physically present in the same room or using computer mediation. The results do not imply that one way of communicating is better than another is, only that the differences are significant.

31.5 Acknowledgements

The authors wish to thank all the students and people who participated and helped in the experiments including Silicon Graphics Australia and Terra Plane Press, Fay Sudweeks for here valuable discussions on communication processes, Simeon Simoff for his help with the statistical results, and finally Laura Ceccherelli for her involvement in the coding and arbitration processes, as well as her valuable critique on this work.

31.6 References

Akin Ö (1986) Psychology of architectural design. Pion Limited, Brondesbury Park London

Bannon, L.J. and Schmidt, K. (1991) CSCW: four characters in search of a context. In:: Bowers JM and Bendford SD (eds.) Studies in computer supported cooperative work. North Holland, Amsterdam

Carter K (1993) Computer support for designers: back to the drawing board. International symposium, creativity and cognition. Loughborough University (Pre-prints), Loughborough

Cross N, Christiaans H and Dorst K (eds.) (1996) Analysing design activity. John Wiley & Sons Ltd., Chichester New York

Eastman CM (1970) On the analysis of intuitive design processes. In: Moore GT (eds.) Emerging methods in environmental design and planning. MIT Press, Cambridge MA, pp 21-37

Finholt T, Sproull L and Kiesler S (1990) Communication and performance in ad hoc task groups. In: Galegher J, Kraut RE and Egido C (eds.) Intellectual teamwork: social and technological foundations of cooperative work. Lawrence Erlbaum Associates, Hillsdale New Jersey, pp 291-325

Gabriel GC and Maher ML (1999) Coding and modeling communication in architectural collaborative design. In: Ataman O and Bermudez J (eds.) Media and design process: ACADIA '99. Association for Computer Aided Design in Architecture, Salt Lake City

Goldschmidt G (1991) Dialectics of sketching. Creativity Research Journal 4:123-143

Greenberg S, Roseman M, Webster, D and Bohnet, R (1992) Issues and experiences designing and implementing two group drawing tools. In: 25th Annual Hawaii International Conference on the Systems Sciences. Hawaii, Vol 4, IEEE, pp 139-150

Gross MD (1994) The fat pencil, the cocktail napkin, and the slide library. In: Proceedings of ACADIA 94 National Conference. St Louis, MO

Kvan T (1994) Reflections on computer-mediated architectural design. In: Transactions on Professional Communication. IEEE, 37

Maziloglou M, Scrivener SAR and Clark SM (1996) Representing design workspace activity. In: Cross N, Christiaans H and Dorst K (eds.) Analysing design activity. John Wiley & Sons Ltd., Chichester New York, pp 389-416

O'Connail B and Whittaker S (1997) Characterising, predicting, and measuring video-mediated communication: a conversational approach. In: Finn KE, Sellen AJ and Wilbur S (eds.) Video-mediated communication. Lawrence Erlbaum, Mahwah NJ, pp 107-131

Olson JS, Olson GM and Meader D (1997) Group work with and without video. In: Finn K E, Sellen AJ and Wilbur S (eds.) Video-mediated communication. Lawrence Erlbaum, Mahwah NJ, pp 157-173

Sudweeks F and Albritton M (1996) Working together apart: communication and collaboration in a networked group. In: Keen CD, Urquhart C and Lamp J (eds.) Proceedings of the 7th Australasian Conference of Information Systems (ACIS96) Vol 2. Department of Computer Science University of Tasmania, Hobart Tasmania, pp 701-712

Sudweeks F and Rafaeli S (1995) How do you get a hundred strangers to agree: computer mediated communication and collaboration. In: Harrison TM and Stephen TD (eds.) Computer networking and scholarly communication in the twenty-first-century University, SUNY Press, New York

Vera A, Kvan T, West R and Lai, S (1998) Expertise and collaborative design in computer-human interaction. (CHI'98): Conference on Human Factors in Computing Systems 'Making the Impossible Possible'. ACM SIGCHI, Los Angeles CA, pp 502-510

Visser W (1993) Collective design: a cognitive analysis cooperation in practice. In: Roozenburg, N (ed.) Proceedings of International Conference on Engineering Design (ICED93). Heurista Zurich, The Hague

Chapter 32

Sharing Design Ideas: Conversational Grounding in Collaborative Design

S.E. Reed and F.J.M. Reid

32.1 Introduction

An important objective of research on collaborative design has been to understand how designers use freehand sketches and other modes of design representation to develop and communicate their ideas during design meetings. Some years ago, Tang (1991) carried out an important observational study on the role of visible sketching and other workspace actions (such as gesturing, pointing, and listing) for storing design information, expressing ideas and mediating interaction between designers working alongside each other. This study stimulated considerable interest in the problem of designing virtual workspace technologies for enabling designers working at different locations to see, point to, draw, or write down design ideas whilst simultaneously holding a conversation with their colleagues (e.g., Heath *et al.*, 1997; Ishii *et al.*, 1997). However, whilst the trend of recent work has been to explore the relationship between freehand sketching and the cognitive aspects of design activity (e.g., Purcell and Gero, 1998), little is yet known about how designers combine visible workspace actions with speech in the collaborative process of building on, and developing shared design ideas. One strategy for exploring this is to examine the impact on designers of removing visual access to the sketches produced by their design partners. In this paper, we describe a study that investigates how *conversational grounding*—the two-way exchange of positive evidence of mutual understanding (Clark, 1996)—is achieved when designers are able to sketch their own ideas, but are unable to see their partner's sketches.

It is our view that speech, sketching, and bodily gestures mutually interact to give shape to emerging ideas in face-to-face design meetings, each activity contributing to and modifying the meaning conveyed by other activities. Support for this view is provided in a recent study by Kelly *et al.* (1999), who show how speech and gesture combine to make communication more understandable than by either speech or gesture alone. More generally, Clark's (1996) distinction between *methods of signalling* helps us understand the range of communicative activities available to designers, and their artful fusion of linguistic and non-linguistic activities to form composite signals during design interaction. These methods fall into three broad categories: *describing-as* (mainly verbal communication, but also formal drawings, diagrams, and other materials containing symbolic content), *demonstrating* (depictive freehand sketching, iconic or figural gesturing); and *indicating* (referential pointing to sketches or other materials in the workspace). Depending on the nature of the design problem, designers can be expected to deploy these methods differentially to establish the mutual belief that they have generated sufficient common ground for design ideas to be shared and evaluated. In the present paper, our focus will be on design problems for which demonstrative and indicative aspects of workspace activity predominate.

More direct evidence of the interplay between sketching and speech comes from a study of realistic engineering design meetings by Reid *et al.* (1998). This study found evidence for an implicit conversational rule discouraging interruptions whilst individual designers elaborated their design ideas using freehand sketches. No such barrier to interruption appeared to operate during conceptual phases of design discussion, in which customer specifications and constraints on the emerging design were the focus of discussion. This suggested that engaging in visual thinking—and the sketching activity that typically accompanies it—may signal a special state of conversational disengagement, in which designers direct their attention away from their colleagues and towards the sketches they are creating. However, an alternative interpretation of this finding is that designers actively sketch to facilitate conversational grounding, and that colleagues refrain from interrupting whilst they observe the developing sketch in order to gain a better understanding of the ideas being expressed. This would indicate a complex relationship, not only between visible workspace activity and conversational grounding, but also between the visual and communicative nature of the design task itself.

32.2 Collaborative Design Interactions

To explore these questions, we carried out an experiment in which participants worked on two different design problems either in full view of each other (and each other's sketches), or within each others' hearing but without shared visual access. By removing shared visual access, we expected not only to make conversational grounding more difficult to achieve, but also to observe the mix of signalling methods that participants deploy in order to adapt to the lack of opportunity to observe each other's sketching behaviour (Clark and Brennan, 1991).

We therefore set out to investigate a series of questions. Firstly, if shared workspace activity provides a communicative function over and above that of speech alone, removing the opportunity to observe each other's sketching activity should seriously undercut participants' ability to achieve shared solutions to the design problem. Secondly, the precise impact of removing shared visual access will depend on the nature of the design problem. In the present study, design problems differed in the extent to which they required predominantly *indicative* or predominantly *demonstrative* methods for establishing common ground, as determined by extensive pilot studies carried out by the first author. The indicative problem involved manipulating items of furniture on a room plan drawn on squared paper, with the aim of accommodating an additional piece of furniture (a dresser) without removing any existing pieces from the room. The demonstrative problem was an engineering task in which participants were required to design a lightweight, low-cost portable river crossing system for use by aid workers.

Pairs of students from the University of Plymouth worked on these two problems, presented in counterbalanced order. Six pairs worked face-to-face (the *FTF* condition), sitting on either side of a desk where they could work independently or share the drawing materials supplied with each problem. The remaining six pairs worked at the same desk, but shared visual access was denied (the *VAD* condition) by the erection of a wooden screen between the two participants. This screen divided the work surface into two equal-sized work areas, but allowed participants to continue to converse normally. Time-stamped video recordings of the design interactions were collected from a ceiling-mounted enclosed-dome camera positioned over the work surface. Participants' speech and workspace activities were then categorised from these video recordings using the codes depicted in Table 32.1.

32.3 Conversational Grounding and Workspace Activities

To examine the effects on these two design problems of removing shared visibility, the frequencies of workspace activities and speech codes were first expressed as percentages (to remove differences in the amount of time allocated to the two problems), and subjected to a series of mixed-design analyses of variance, each with visibility condition as a between-subjects factor, and speech and workspace activity codes as within-subjects factors. Contrary to initial expectations, there were no differences between visibility conditions on measures of performance (for the dresser problem) or solution similarity within design pairs (on the river crossing problem). The immediate conclusion that one might draw from this is that sketching and other workspace activities are not essential for effective communication in design problems of this kind. However, the increase in the *display costs* associated with the loss of visibility in this condition (Clark and Brennan, 1991) would be expected to encourage participants to deploy alternative techniques for grounding, e.g., by formulating utterances that express ideas more clearly without recourse to sketches, gesturing, or pointing.

A further finding is that when analysed alongside all speech codes, both communicators and addressees appeared to engage in equal amounts of workspace activity *regardless of visibility condition*, indicating that shared workspace activity was not perceived by participants as having a uniquely communicative function. We are inclined to conclude from this that communicators in the present study engaged in workspace activity—whether visually shared or not—primarily *to assist their own design thinking*, whether this entailed thinking up new ideas, expressing these ideas, or exploring how well their partner understood them.

Table 32.1. Workspace activity and speech categories used to code video recordings of the design interactions in the present study

The following *activity* categories refer to workspace activities performed by participants whether or not these are visible to their partner.		
1.	Sketching	Makes drawing marks on the page
2.	Figural gesturing	Gestures in the air clearly meant to depict shape and/or motion
3.	Figural pointing	Traces and/or points around a sketch without making marks on the page
The following *speech* categories refer to requests or offers of evidence of understanding achieved through explicit utterances or speech intonation.		
4.	Communicator seeks evidence of addressee's understanding	e.g., 'Do you understand what I mean?'
5.	Communicator seeking evidence of own understanding	e.g., 'Is this what you mean?'
6.	Communicator provides evidence of addressee's understanding	e.g., 'You understand (do not understand) what I mean'
7.	Communicator provides evidence of own understanding	e.g., 'I understand (do not understand) what you mean'
8.	Non-grounding contributions	Speech contributions other than those specified above

For addressees, there was a tendency (though not significant) to carry out more workspace activity when shared visibility was denied, suggesting that they needed to pin down the ideas presented by their partners in the form of a sketch or other external representation. In the FTF condition, however, addressees tended to carry out less workspace activity (though again not significantly) suggesting that the opportunity to observe the communicator's activity was sufficient for them to gain sufficient understanding of their partner's ideas.

32.4 Sketching and Figural Pointing

Our second expectation was that grounding effort would increase when shared visibility was denied. However, this was not generally supported. Nevertheless, a significant task-by-condition interaction ($F(1,10) = 14.43$, $p = < .01$) revealed a tendency for participants to seek more evidence of their partner's understanding in the VAD condition, but only for the predominantly indicative dresser problem. This confirms earlier findings (Clark and Wilkes-Gibbs, 1986) that little effort is required to establish that objects and locations have been mutually grounded on indicative design problems, providing participants have visual access to a shared workspace. For the predominantly demonstrative river crossing problem, however, considerable grounding effort was required regardless of visibility condition. We believe these differences arise from the differing signalling requirements of the dresser problem, with its emphasis on the orientation and location of objects, and the river crossing problem, with its emphasis on constructional materials, mechanical arrangements, and conceptual ideas. The contrast between the indicative nature of the dresser problem and the demonstrative character of the river crossing problem is readily grasped by comparing the example solutions shown in Figure 32.1.

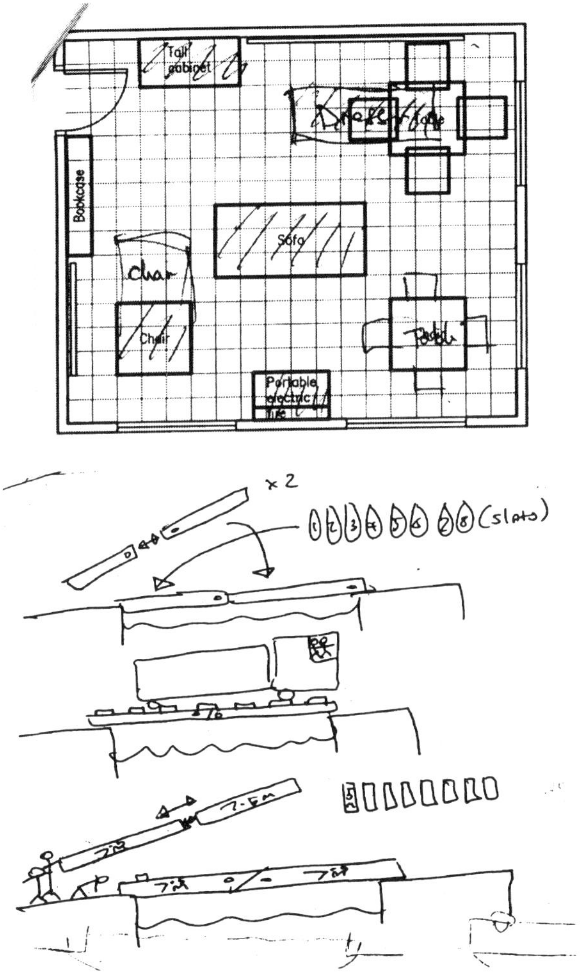

Figure 32.1. Examples of attempted solutions to the predominantly indicative dresser problem (upper panel) and the predominantly demonstrative river crossing problem (lower panel)

Closer examination of the relationship between conversational grounding and workspace activities revealed evidence that supported our earlier study (Reid *et al.*, 1998). In particular, a significant interaction between design role and visibility condition was found on the river crossing problem ($F_{1,10} = 11.08$, $p = < .01$). As depicted in Figure 32.2, communicators in the FTF condition were more likely than addressees to engage in sketching activity whilst seeking evidence of their partner's understanding, implying that *communicators expect addressees to benefit from observing their workspace activity* and use this activity to facilitate grounding efforts. However, whilst communicators in the VAD condition were *less* likely to sketch, their addressees were *more* likely to sketch, presumably in an effort to capture the communicator's design ideas on paper. This result underlines the positive communicative function of sketching for the communicator when seeking evidence of understanding on this type of design problem.

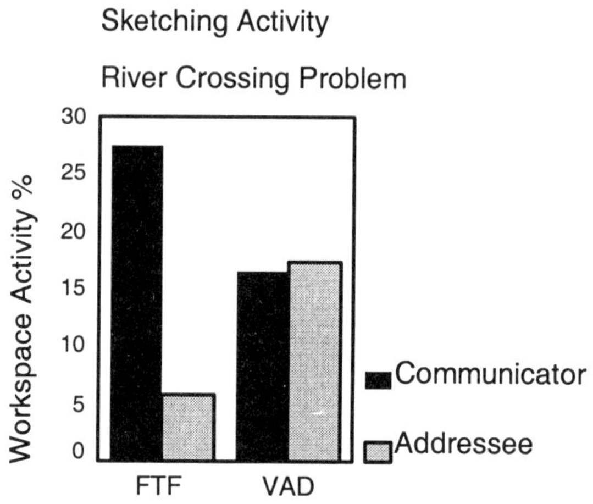

Figure 32.2. Sketching as a proportion of total workspace activity on the river crossing problem during exchanges in which the communicator sought evidence of their partner's understanding

It is worth noting that although there was more sketching overall on the dresser task, the predominant workspace activity used by the communicator to seek evidence of the addressee's understanding was figural pointing ($F_{2,20} = 35.88$, $p = < .001$), with very little sketching or gesturing. We believe this indicates that most sketching takes place when trying to solve the problem, and that idea sharing occurs after this formative sketching has been completed. On the river crossing problem however, sketching is actually more appropriate for idea sharing, as it conveys how things will look and work, a function which is less relevant to the orientation and positioning of items of furniture in a room plan.

As figural pointing was the predominant workspace activity associated with the indicative nature of the dresser problem, this measure was used as a comparison with sketching activity on the river crossing problem. As shown in Figure 32.3, the pattern of activity is similar except when shared visibility was denied, with communicators engaging in significantly more figural pointing here than in sketching on the river crossing problem, $t_{1,11} = 3.24, p = < .01$. This interpretation fits well with the previous finding that a greater grounding effort is required for this predominantly indicative problem when shared visibility is denied, and that this is accompanied by figural pointing to assist the communicator in achieving satisfactory grounding of objects and their locations.

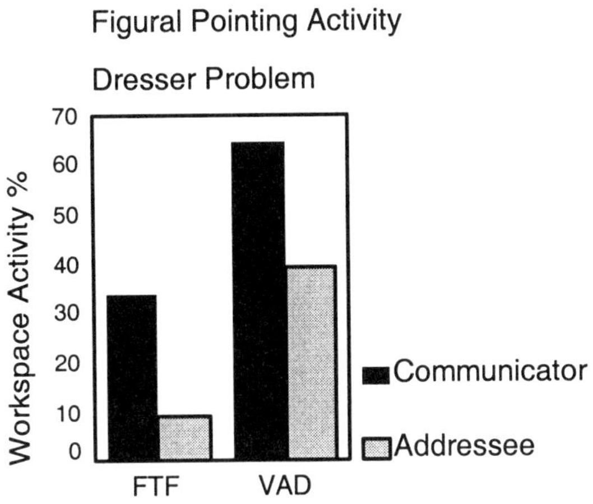

Figure 32.3. Figural pointing as a proportion of total workspace activity on the dresser problem during exchanges in which the communicator sought evidence of their partner's understanding

Two final results merit consideration. First, we observed significantly more misunderstandings on the dresser problem when visual access was denied ($F_{1,10} = 5.04$, $p = < .05$), which explains the high levels of 'seeking evidence of partner's understanding' found for this problem and condition. Finally, figural gesturing and writing only took place in the river crossing problem. This further underlines the key differences in common grounding between the two types of problem studied here—the dresser problem, with its requirement for the precise orientation and location of objects, and the river crossing problem, with its emphasis on the affordances of materials and mechanical arrangements.

32.5 Conclusions and Implications for Collaborative Design

It is clear from these results that the interaction between conversational grounding and workspace activity does indeed vary over types of design problem, collaborative role, and visibility condition. Interestingly, our findings suggest that the purely communicative functions of workspace activities may be *less* important than previously thought. Design performance, the extent to which solutions were similar for each pair of participants, and the prevalence of workspace activity did not generally differ as a function of visibility condition. For one thing, designers seemed able to adapt their communications to express their ideas more clearly when they were aware that their partners were unable to directly observe what they were pointing to or sketching. By helping them externalise and develop their ideas, these kinds of workspace activities may turn out to be more important to the *individual designer* than to the process of design communication.

Workspace activities nevertheless do play an important communicative role in supporting conversational grounding, but only when designers believe that their activity is observable by their partners. During these phases, addressees were more likely to refrain from activity and concentrate on observing the communicator's activity in order to ensure understanding. However, when shared visibility was removed, addressees continued to engage in parallel workspace activity. Under these conditions, we believe addressees were *echoing their partner's design ideas* in order to capture and retain the ideas they were expressing. This echoing occurred on both types of problem when visibility was removed. There was, however, one major difference between the two problems—whilst communicators *reduced* their problem-related workspace activity on the river crossing problem, they *increased* it on the dresser problem. We believe in the latter case that communicators found it helpful to point at and around objects and locations even though they knew that the addressee was unable to see these actions, precisely because these indicative actions supported their own efforts in achieving mutual understanding.

Because of its design, the present study was limited to an examination of the workspace activities of the two participants whilst either was speaking. As a result, we were unable directly to test the operation of the implicit 'no interruption' rule reported by Reid *et al.* (1998). Of course, it may be the case that both explanations for the workspace asymmetries we have observed hold true—the addressee refrains from interrupting whilst their partner is sketching because their partner's disengagement signals the relevance of an implicit rule, *and also* because they wish to observe their partner's sketches and gestures to obtain a clearer understanding of what they have in mind. For the present we can surmise that there will be occasions when observing another's workspace activity in relation to sketches and diagrams is important, and that if this is not possible, then the opportunity to converse naturally becomes indispensable.

These changing patterns of conversational grounding and workspace activity have important implications for the development of virtual workspace systems for supporting collaborative design. Firstly, system developers can clearly capitalise on the differential visibility needs of the design problems being undertaken — for visual design problems requiring high levels of demonstrative representation support will need to be provided for depictive freehand sketching, whereas more indicative, diagrammatic problems can readily be addressed using systems that offer only simple gesturing and pointing support. Secondly, even when concurrent visual access to a shared workspace is unavailable, designers will continue to need workspace facilities in order to echo the design ideas of their colleagues. Future research needs to examine more closely how sketching and figural pointing assist the individual designer to think, express ideas, and negotiate shared design understanding.

32.6 Acknowledgements

The study reported here forms part of a PhD programme of work supported by a University of Plymouth studentship awarded to the first author. Both authors wish to thank the students who took part in this study.

32.7 References

Clark HH (1996) Using language. Cambridge University Press, Cambridge

Clark HH, Brennan SA (1991) Grounding in communication. In: Resnick LB, Levine JM, Teasley SD (eds.) Perspectives on socially shared cognition, APA Books, Washington DC, pp 127-149

Clark HH, Wilkes-Gibbs D (1986) Referring as a collaborative process. Cognition 22:1-39

Heath C, Luff P, Sellen, A (1997) Reconfiguring media space: supporting collaborative work. In: Finn KE, Sellen AJ, Wilbur SB (eds.) Video-mediated communication. Erlbaum, Mahwah NJ, pp 323-347

Ishii H, Kobayashi M, Arita K, Yagi T (1997) Iterative design of seamless collaboration media. In: Finn KE, Sellen AJ, Wilbur SB (eds). Video-mediated communication. Erlbaum, Mahwah NJ, pp 435-455

Kelly SD, Barr DJ, Church RB, Lynch K (1999) Offering a hand to pragmatic understanding: The role of speech and gesture in comprehension and memory. Journal of Memory and Language 40:577-592

Purcell AT, Gero JS (1998) Drawings and the design process. Design Studies 19:389-430

Reid FJM, Reed S, Edworthy J (1998) Design visualization and collaborative interaction in engineering teams. International Journal of Cognitive Ergonomics 3:235-259

Tang JC (1991) Findings from observational studies of collaborative work. International Journal of Man-Machine Studies 34:143-160

Chapter 33

Using Media-mechanism Shifts to Uncover Design Communication Needs

S.A.R. Scrivener, C.D. Chen and A. Woodcock

33.1 Introduction

In early work, Tang (1989) and Bly (1988) determined the requirements for remote design support systems by studying collaborative design in face-to-face settings, focussing on communication with drawings and gesture. Later studies addressed other design communicative materials, such as objects (Harrison and Minneman, 1996), physical and CAD models, and the changing studio environment itself (Minneman and Harrison, 1997). These researchers regarded the communicative media and behaviour of face-to-face working as necessary to effective design communication. Yet, studies comparing face-to-face and multimedia collaboration environments failed to show significant effects due to impoverished communication or, at best, contradictory results (*cf.*, Whittaker, 1997), suggesting that the media and behaviour of face-to-face working are not necessary conditions for successful computer mediated collaborative distal design. Indeed, Kvan *et al.* (1997) postulate that social and situational factors have little effect on shaping the measurable outcomes of collaborative design. Designers adapt to or adapt a communication environment to overcome social and situational impoverishment. Earlier, when summarising the learning from a number of distal collaborative design studies, Scrivener and Clark (1994) concluded that motivated designers exploit the resources at their disposal to complete a task, even when this involves radical changes in work and communication methods. Hollan and Stornetta (1992) also argue against face-to-face, or 'being there', as a model for designing telecommunications. They draw a distinction between needs, media and mechanisms, where communication needs are independent of the medium used. Mechanisms are ways to meet communication needs that are enabled by a medium. While needs are media independent, mechanisms are closely connected to specific media. New mechanisms, they argue, are required for new media.

Nevertheless, the behaviour observed in face-to-face design must have communicative value. Otherwise, we would have to accept that designers waste much effort on purposeless activity. Scrivener and Clark (1994) postulate that designers will choose what they perceive to be the most efficient and/or effective means of communication at their disposal in order to complete a task. In other words, they adapt to achieve the best result they can in a given environment. A face-to-face telecommunication model may blind us to the possibilities offered by new technology, but ignoring it may result in inefficient and ineffective communication systems. On the other hand, a method is needed that will uncover design communication needs without forcing us to replicate the pitfalls of face-to-face communication. Here, we argue for such an approach based on a model of designer behaviour in multimedia communication environments which postulates that designers select from the communication resources available to them those most efficient and/or effective for the task in hand.

33.2 Maximising Efficiency and Effectiveness

In Chen *et al.* (1999) we describe an investigation of the visual communication needs of design teams when engaged in the conceptual design stage of an industrial design task. The task was based on the Delft Protocol (Cross *et al.*, 1996). Briefly, three design dyads were asked to design a fastening device to enable a backpack to be carried on a mountain bike. Each dyad conducted 4 one-hour weekly sessions over a computer mediated communication system (with individual work between sessions). The system enabled communication via a whiteboard and video conferencing system supporting five media-mechanism combinations, *i.e.*, whiteboard-drawing and -gesture, video-drawing, -gesture and -object manipulation. All discourse and interaction was video recorded. Artefact-related utterances were then identified and every word was coded as a part-of-speech, *i.e.*, adjective, adverb, noun, preposition, pronoun, and verb. Discourse about visual objects was further categorised in terms of whether or not it was accompanied by any form of visualisation. Every part of speech co-occurring with visualisation in this visually supported discourse was labelled. We found that adverbs, nouns and pronouns were coupled with visualisation more often than adjectives, prepositions and verbs.

The data were analysed to determine whether this pattern of visual coupling was an artefact of the task, the medium, *i.e.*, whiteboard or video, or the mechanism, *i.e.*, drawing, gesture or object (Chen *et al.*, 1999). In fact, the pattern of visual coupling does not vary significantly with task or medium. In contrast, the pattern of visual coupling does vary with mechanism; e.g., the proportion of visual coupling with verbs is greater for gesture than drawing. However, if determined by mechanism, *per se*, the pattern of visual coupling in whiteboard drawing should be similar to that in video. It turned out that this was not the case, the pattern of visual coupling was very different between whiteboard and video drawing. Hence, we concluded that the visual coupling pattern reflected visualisation need; *i.e.*, some parts of speech benefit more than others do from visualisation. Given this interpretation, the variability of visually coupled parts of speech over medium-mechanism combination is the consequence of designers selecting the most appropriate combination at a given moment to satisfy communication need. Hence, these data support the postulate that designers use available communication resources to satisfy visualisation needs by selecting, moment by moment, those most efficient and/or effective for the task in hand (Scrivener and Clark, 1994).

33.3 Analysing Medium-mechanism Shifts

Following this line of argument, shifts between medium-mechanism combination should be explicable in terms of relative efficiency and/or effectiveness. Furthermore, explication of why shifts are made from one medium-mechanism combination to another will offer insights into the merits of the latter which may be used to enhance the performance of the former through redesign. In the remainder of Section 3 we explain: how the occurrence of a shift to, and the selected medium-mechanism are identified; how the purpose served by the shift is assigned and the benefit in efficiency and effectiveness gained by the shift is assessed, as compared to competing medium-mechanism combinations.

33.3.1 The Purposes Served by Visual Coupling

The purpose of an utterance is different to the purpose served by visual coupling. The former is the reason for the communication while the latter concerns the way that specific visual information supports the communication. Analysis of verbal transcripts can identify the purpose of an utterance, whereas examination of the video recording of communication channel usage is essential in determining the purpose of visual coupling. Tang (1993) observed how designers used gesture to facilitate communication. They concluded that gesture has three functions: to store information, to express an idea, and to mediate interaction. Harrison and Minneman (1996) studied the role of objects in design discourse, analysing relationships among designers, designer's gestures, and the object referenced in a design environment. They observed that gesture may be used to clarify or specify something, concluding that objects provide a rich source of communicative information in group design. Based on the above, five purposes of visual coupling in design discourse were identified and defined with reference to dictionary sources, as follows:

1. *Clarify*: dictionary sources define this term as meaning to 'make more comprehensible' and 'to clarify something means to make it easier to understand, usually by explaining it in more detail'. Using this definition, visualisation coupled with a term was identified as clarification if the designer used a medium-mechanism combination to communicate an attribute of a thing, such as its shape, or a relation between two things.
2. *Specify*: is defined as to 'identify clearly and definitely' and to 'give information about what is required or should happen in a certain situation'. In

visual coupling, 'to specify' means to identify clearly and definitely what is being discussed thus isolating it from the other drawings or objects present. What distinguishes it from clarification is that no new information is introduced.

3. *Emphasise*: means to 'give emphasis to', emphasis being defined as the 'special importance, value, or prominence given to something'. It is also given the meaning 'to indicate that it is particularly important or true, or to draw special attention to it'. Again what distinguishes emphasis from clarification is that no new information is introduced. Nor is emphasis simply specification as it draws special attention to what is being discussed.
4. *Annotate*: is defined as to 'add notes to (a text or diagram) giving explanation or comment' and to 'annotate written work or a diagram means to add notes to it'. Generally, annotation adds nothing to what is being said and merely echoes verbalisation, serving primarily to store information for later reference.
5. *Identify*: means 'to associate someone or something closely with' and 'if you identify someone or something, you name them or say who or what they are'. In visual coupling, the distinction between indentification and clarification or specification is subtle. Like specification it does not involve the introduction of new information, but unlike specficiation it relates one visualisation to another.

33.3.2 Analysing Medium-mechanism Combination Shifts

In the following excerpt from one of the design sessions, two shifts of medium-mechanism combination occur during the design dyad's discussion of how to the strap of rucksack could be tightened by means of a device attached to a pole. (The terms coupled with visual information are underlined below.)

316	R	so you want something on this post there[WB-D], so we can move up in that direction[WB-D] or that direction[WB-D]
317	D	yeah
318	R	that'll take up[VD-G] the slack in the strap[VD-G]
319	D	no because, I mean, ok, you've got the post there[WB-D], good idea and you've got your strap[WB-D] here and you're pack's at[WB-D] the front[WB-D] and...

Here, term 'direction' at the end of Turn 316 is coupled with whiteboard-drawing (WB-D). The next term 'take up' in Turn 318, is coupled with video-gesture (VD-G). A further shift occurs in Turn 319. In Turn 318, the term 'strap' is coupled with video-gesture while the next visually coupled term 'post there' in Turn 319 is coupled with whiteboard-drawing. Thus, the designers shift from whiteboard-drawing to video-gesture and then back to whiteboard-drawing.

33.3.2.1 Identifying the Purpose of a Shift

In the first shift, Figure 33.1, the designer shown in the top left video window moves his upturned palm upward, simultaneously bringing his fingertips together before dropping his closed hand slightly. Given the context of the turn, the gesture was interpreted as meaning to take something and transform it, and to represent an activity related to the terms 'take up'. On inspection, the meaning of this turn could be understood by reading Turn 318 and adjacent turns without needing to refer to the gesture. Thus, this visual information was not necessary for the remote partner to understand what 'take up' meant in this context. Indeed, although the function of the pole was to 'take up', the gesture was not an illustration of the action of the pole or its effects. Hence, we concluded that the designer used the gesture to emphasise the meaning conveyed in his utterance, and the purpose of the switch from whiteboard-drawing to video-gesture was registered as being 'to emphasise'.

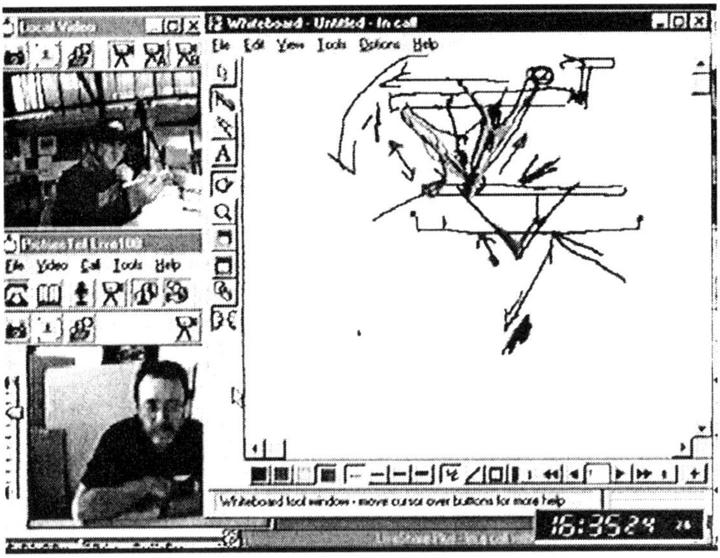

Figure 33.1. Designer in top left window shifts from drawing to gesturing

33.3.2.2 Assessing Medium-mechanism Shifts Benefits

So far, we have illustrated how medium-mechanism shifts are located and assigned a purpose. The proposition set out above states that a shift occurs because the destination medium-mechanism is judged as being more efficient and/or effective for the need in hand than the source combination or any of the other available resources. This being the case, it should be possible to explain why the one medium-mechanism is more efficient and/or effective than the others in our example. Moving from one medium-mechanism combination incurs work costs. Where the shift involves a change in medium the costs will be even greater. Yet the designer chose video-gesture over whiteboard-gesture. However, the latter would have required the designer to move his digital pen to and 'click' the pointer button on the whiteboard toolbar to select the pointer. He would then have had to pick the pointer up and drag it to achieve the desired effect. It is not difficult to see that this is likely to be less efficient than a direct hand gesture. We have argued that the purpose of the visual coupling was to emphasise the idea of 'taking up'. In this instance the constraints in whiteboard-gesture might lead to the communication being misunderstood. First, unless the whiteboard-gesturing was completed in a blank area it might be interpreted as 'to specify' a part of the drawing to which 'take up' applied. Second, whether in white space or not, it might take the meaning 'to clarify' if the gesture was understood as describing the moving of a thing or its motion. We would argue that these latter misinterpretations are more likely in whiteboard-gesture as it is constrained to one plane. Hence, an argument can be made on grounds of effectiveness as well as efficiency. Alternatively, the designer could have used whiteboard drawing to convey his intention, thereby incurring no medium-mechanism transfer costs. Here, it is difficult to make a case for transfer-efficiency gains; instead we must look for gains in effectiveness. First, whiteboard-drawing suffers from the same potential losses in effectiveness as those described above for whiteboard-gesture but with the added disadvantage that a permanent record of the communication is retained. This may have both immediate and long-term disadvantages. Drawing is clearly more concrete than transient forms of visual communication. A drawing has shape, size, line thickness etc., and these properties could be taken as meaningful, thus making it difficult to merely 'emphasise'. For example, video-gesture could be used to emphasise the idea that something should 'be thicker'. However, drawing, whether on whiteboard or paper, could be interpreted as meaning 'how thick' or 'where thicker'. Alternatively, the designer could have annotated the terms 'take up'. However, this would not have functioned to emphasise the action of taking-up. Emphasis by drawing, whether graphic or text, also has the effect of filling the drawing surface with redundant information. Thus, in this instance, a case can be made for gains in effectiveness of video-gesture over whiteboard-drawing. A shift from whiteboard-drawing to video-drawing would have incurred the same disadvantages in effectiveness as continuing with whiteboard-drawing, with additional costs in efficiency. First, there is the cost of medium transfer. Secondly, it would have been necessary to turn the camera to view the paper or to bring the paper to the camera. Finally, video-object suffers from even greater unwanted specificity than whiteboard-gesture or whiteboard- or video-drawing, and the greater effort required to locate and bring the appropriate object into view. This example illustrates that a case can be made on both grounds of efficiency and/or effectiveness for the selection of a particular medium-mechanism combination over other available resources to achieve a particular purpose at a given moment in a task. It also shows that the analysis of these transfers gives insights into the strengths and weaknesses of medium-mechanism combinations with respect to visual communication needs or purposes. The analysis described above was applied to all medium-mechanism shifts in all design sessions. In each case, the shift could be categorised by purpose and explained by gains in efficiency and/or effectiveness.

33.4 Visual Coupling in Medium-mechanism

Above, we have argued that users select from among available communication resources supporting visual coupling, that which is most efficient and effective for the task in hand. We have shown how medium-mechanism can be explained in terms of gains in efficiency and effectiveness. In fact, the analysis was not restricted to medium-mechanism shifts; the purpose of every visually coupled term was determined (see Table 33.1). We now consider whether these data support our argument.

Table 33.1. Visual coupling frequency against purpose in medium-mechanism

	Video (VD)			Whiteboard (WB)		
	VD-D	VD-G	VD-O	WB-D	WB-G	Total
Emphasise	2	346	1	39	1	389
Annotate	0	0	0	83	0	83
Clarify	237	405	107	1834	12	2595
Identify	15	12	8	12	0	47
Specify	157	32	72	853	1	1115
Total	411	795	188	2821	14	4229

First, it is notable that the overall distribution (*cf.*, the rightmost column) of terms to purpose is not even; terms are visually supported for some purposes more than others are. This pattern could reflect need, task dynamics, or medium-mechanism combination. That is, the need for visual clarification might be greater than the need for emphasis in communication, independent of task. Alternatively, for example, emphasis might be low relative to clarification because it is not necessary in the early stages of the task. Or perhaps visual coupling for a particular purpose is only possible via a particular medium-mechanism combination, e.g., annotation might be intrinsically bound to whiteboard-drawing (WB-D). Against the second interpretation is the fact that the purpose frequencies, as a proportion of total visual coupling within each session, are more or less constant over sessions. As we can see from Table 33.1, most medium-mechanisms are used to support most purposes. Furthermore, although whiteboard-drawing dominates clarification and specification, as the task progresses the use of video medium-mechanism combinations increases until by the final session it is greater than that of the whiteboard. In contrast, these observations are not in conflict with the view that variation in frequency of visual coupling to purpose reflects communication need.

Viewed from a perspective of the efficiency and effectiveness in the use of available resources to realise communications needs, we can observe the following from inspection of Table 33.1. First, whiteboard-drawing would appear to be the most effective and efficient resource, followed by video-gesture, video-drawing, video-object and finally, whiteboard-gesture. Clearly, the latter medium-mechanism combination is virtually useless. Second, across medium-mechanism combinations the distribution of visual coupling to purpose is not uniform, with some combinations showing strong dominance for particular purposes, e.g., video-gesture for emphasis and whiteboard-drawing for annotation. While we have argued that the task does not determine the pattern of visual coupling to purpose, it would appear to determine the use of medium-mechanism combinations. If the relative value of different medium-mechanism combinations did not vary with the task then an efficiency and effectiveness interpretation would predict that the most efficient and effective medium-mechanism combination for a given purpose would win out over all others. However, as noted above, the use of video medium-mechanism combinations for specific purposes increases over time relative to whiteboard combinations. Given our view of medium-mechanism usage, this would imply that the task affects what is judged to be the most efficient and effective resource at any given moment. Two likely determinates of task related changes spring to mind, *i.e.*, need and environment. For example, it might be that as the task progresses greater detail in drawing is required, hence the shift from whiteboard-drawing to video-drawing. Alternatively, for example, the existence of drawings on paper produced between sessions could account for the shift. As noted at the end of Section 33.3.2.2, determinates of medium-mechanism selection can be inferred from the analysis of medium-mechanism shifts.

33.5 Medium-mechanism Shifts

Table 33.2 shows the frequency of shifts to medium-mechanism combination against purpose, e.g., it shows that users shifted forty nine times to video-drawing from one of the other medium-mechanisms for the purpose of clarification.

Table 33.2. Frequency of shifts to medium-mechanism combination against purpose

	Shifts to Video			Shifts to Whiteboard		
	VD-D	VD-G	VD-O	WB-D	WB-G	Total
Emphasise	1	120	1	3	1	126
Annotate	0	0	0	3	0	3
Clarify	49	85	53	81	7	275
Identify	8	6	6	0	0	20
Specify	29	13	36	34	1	113
Total	87	224	96	121	9	537

We have argued above that the evidence supports the notion that users exploit the medium-mechanism resources at their disposal to meet communication needs as efficiently and effectively as possible. We have also observed that, generally, medium-mechanisms can be successfully employed to support each purpose, the exceptions being emphasis and annotation where respectively video-gesture and whiteboard-drawing dominate. Given our interpretation of the data, we should expect the shifts to a particular medium-mechanism to reflect its overall usage (*cf.*, Table 33.1). For example, it would be surprising if there was a higher frequency of shifts to specification than clarification for video-drawing, as this combination does not appear to be especially favoured with regard to the former compared to video-object, whiteboard-drawing, and overall usage. Conversely, video-gesture does appear to be particularly strong for emphasis compared to other combinations. It would be surprising in this case if the frequency of shifts to clarification was significantly greater than emphasis. Inspection of Table 33.2 reveals that neither of these is the case and that, overall, shifts reflect the relative usage of medium-mechanisms for purposes.

33.6 Design Using Medium-mechanism Shifts

We are not simply interested in understanding for understanding's sake. One of the primary reasons for carrying out the work reported here was the expectation that it might provide insight into how to design more effective and efficient communication environments. The identification of visual coupling purpose and explanation of why, in each case, the selected medium-mechanism was chosen for this purpose (as *per* Section 33.3.2) can contribute to design or redesign. For example, the analysis reveals the relative strengths and weaknesses of medium-mechanisms. The design goal should be to remove these differences, *i.e.*, to remove relative weaknesses and even to build on strengths. Let's say, for example, that we found that the shift from whiteboard-drawing to video-drawing was because of the use of paper drawings produced between sessions. While highly efficient in this instance, video-drawing might be less effective than whiteboard-drawing because of the camera resolution compared to display resolution. Whiteboard-drawing could be improved by making it easier for designers to capture and transfer paper drawings, and video-drawing could be improved by using a higher resolution camera. Each medium-mechanism would be enhanced so that when process factors necessitate shifts from medium-mechanism combination, high levels of efficiency and effectiveness could be maintained within an enhanced environment. Such an analysis of all visual couplings (*i.e.*, 4229 in this study) is very time consuming and is unlikely to be of practical value in design. However, a method based on the detection and analysis of shifts between medium-mechanism offers a more viable alternative (*i.e.*, in this study shifts between medium-mechanism represent 13% of visual coupling).

33.7 Conclusions

A corollary of Hollan and Stornetta's (1992) argument is that communication can only be studied in the context of particular media and mechanisms. According to our model, designers' behaviour in face-to-face working reflects the relative efficiency and effectiveness of the available communication resources in meeting communication needs. As such there is no standard to which other communication environments should be compared. Different behaviour in one compared to another do not necessarily mean that one is less efficient of effective than the other, and in any case such comparisons reveal little about communication need. Neither do we learn much about communication need by comparing differences in design goals or design outcomes, even if we are able to devise measures sufficiently sensitive to overcome designers' adaptiveness. Instead, we have argued that, for a given multimedia communication environment, communication needs may be uncovered by assessing why a selected medium-mechanism is more efficient and/or effective than other available medium-mechanisms. Furthermore, such assessment may stimulate ideas about how to redesign the environment to better exploit its potential to satisfy communication needs. In short, we would argue that the method provides a way of uncovering the communication needs satisfied in face-to-face working without limiting the potential for exploiting new telecommunications technology.

At present, we have not analysed the reasons for each medium-mechanism shift. Demonstrating the feasibility of this analysis is a key step in affirming the value of a formative evaluation and design method based on medium-mechanism shifts. Another important task is to establish how much knowledge is sacrificed by discounting analysis of all visual coupling. Finally, the method relies on the validity of the postulate that designers select from the communication resources available to them those most effective and efficient for the task in hand. We have tried to illustrate this qualitatively, but experimental verification is desirable and possible, and should be conducted as a matter or priority.

33.8 References

Bly SA (1988) A use of drawing surfaces in different collaborative settings. In: CSCW 88. Portland USA, pp 250-256

Chen CD, Scrivener SAR, Woodcock A (1999) Visualisation in collaborative computer mediated dialogue. In: Bullinger H, Ziegler J (eds.) Human computer interaction, Volume 2. Lawrence Erlbaum, Mahwah New Jersey, pp 226-231

Cross N, Christiaans H, Dorst K (1996) Introduction: The Delft protocols workshop. In: Cross N, Christiaans H, Dorst K (eds.) Analysing design activity. Wiley, England, pp 1-16

Harrison S, Minneman S (1996) A bike in hand: a study of 3-D objects in design. In: Cross N, Christiaans H, Dorst K (eds.) Analysing design activity. Wiley, England, pp 417-36

Minneman SL, Harrison SR (1997) Space, timestreams and architecture. In: Maher LM, Gero JS, Sudweeks F (eds.) Formal aspects of collaborative CAD. University of Sydney, Australia, pp 15-25

Hollan J, Stornetta S (1992) Beyond being there. In: Proceedings of CHI'92. ACM Press, Monterey CA, pp 119-125

Kvan T, West R, Vera O (1997) Tools for a virtual design community. In: Maher LM, Gero JS, Sudweeks F (eds.) Formal aspects of collaborative CAD. University of Sydney, Australia, pp 109-123

Scrivener SAR, Clark SM (1994) Experiences in computer-mediated communication. In: Proceedings of the ISAT School/IFIP TC 8/WC 8.5 Workshop. Biblioteka Informatyki Szkol Wyzszych, Poland, pp 122-154

Tang JC (1989) Listing, drawing and gesturing in design: a study of the use of shared workspaces by design teams. PhD Thesis, Department of Mechanical Engineering, Stanford University

Tang JC (1993) Findings from observational studies of collaborative work. In: Baecker RM (ed) Readings in group and computer-support cooperative work. Morgan Kaufmann Publishers Inc, USA, pp 251-270

Whittaker S, O'Conaill B (1997) The role of vision in face-to-face and mediated communication. In: Finn KE, Sellen AJ, Wilbur SB (eds.) Video mediated communication. Lawrence Erlbaum, Mahwah New Jersey, pp 23-50

Chapter 34

Videoconferencing to Support Designing at a Distance

G.D. Logan and D.F. Radcliffe

34.1 Introduction

Rehabilitation engineering is the application of engineering knowledge and skills to address physical difficulties experienced by people with disabilities. A Rehabilitation Engineering Centre (REC) employs rehabilitation (professional) engineers, technicians, physiotherapists and occupational therapists working in a team to match assistive technologies with an individual's requirement, in order to increase their independence and reduce handicap. A rehabilitation engineering team usually practices face-to-face with the client and the client's care providers to: (i) acquire information about the client's physical situation, functional ability, and performance of current assistive devices such as a wheelchair; (ii) experiment to understand how functional performance, postural control, *etc.*, is enhanced or exacerbated by various interventions; (iii) devise unique solutions to problems based on outcome of intervention and trialing various items of equipment. Much of the information upon which decisions are based is acquired by observation.

Videoconferencing offers a medium to facilitate rehabilitation engineering service provision to people with disabilities who live in remote locations (Shapcott *et al.*, 1998), or who are unable to travel because of their disabling condition. This paper reports an empirical study of a cross-discipline rehabilitation engineering team that was undertaken to gain an understanding of practice in this unique design environment. It also considers implications of the study's results when the REC team interact with a client using videoconferencing technology. This cross-discipline team services clients with severe physical disability whose issues include mobility, comfort and control of sitting posture, and pressure sores. The team designs and manufactures custom seating and related devices to ameliorate specific problems. In face-to-face practice design and manufacture occur as concurrent activities. Engineering drawings are not used. The technician may make simple sketches with basic dimensions. Designing occurs during the team-client sessions as participants discuss, experiment, mock-up and try ideas. Participants (REC team, client and care providers) talking, frequently employing hand actions in synchrony with their talk, and including artifacts[1] in their actions is common during the designing. One aim of research into practice is to develop techniques for the REC team to capture sufficient data with accuracy during the videoconference sessions with the client group, to manufacture seating in isolation from the client, with confidence of a good fit.

34.2 Characteristics of Face-to-face Interaction

A longitudinal study of a cross-discipline rehabilitation engineering group has provided useful insights into practice and the communication of technical information that has a bearing on how to apply new techniques like videoconferencing to rehabilitation engineering practice.

34.2.1 Data Gathering and Analysis

Data was gathered from studies of a cross discipline, rehabilitation engineering team carrying out its everyday work. Video recordings, including wide-angle shots and close-up shots, were made of 24 work sessions over a 3 year period. Methodology aimed to preserve the natural work setting as practicably as possible in order not to interfere with the practice of the seating clinic and to improve the reliability and validity of the data (Peräkylä, 1997). It was observed in each of the 24 videotapes that participants' communications concerning the work at hand frequently contained talking in association with actions. Sometimes participants made actions without talking. Seven of the videotapes with extensive segments of talking and actions were selected for detailed analysis. The total length of the segments in the seven videotapes was 2 hours 45 minutes. Data was extracted from the seven videotapes chronologically by transcribing the talk[2] of participants, describing the actions[3], noting the artifacts used, and timing the length of talking and acting.

While many artifacts were used, and many actions were applied to them, within a wide range of talk-types, common and recurrent actions were easily recognisable. Categories to code participants' actions used while communicating seating design

---

[1] An artifact is any physical object that a designer intentionally involves in communicating information to other participants of the design process, in this case other members of the rehabilitation engineering team.

[2] Talk transcription was restricted to oral output conveying information about the client, equipment, *etc*. Talking of a social nature, humour, *etc*., was not transcribed and was not included in analysis.

[3] Action is defined as purposeful hand and body movements performed with or without artifacts.

information were developed based on evidence, visible or audible, derived through observation of the videotapes, as proposed by Tartar (1989). Harrison and Minneman (1996) studied the use of objects by a team of three designers during a videotaped design activity and devised a five-category system for use of objects in design. This system was expanded to six categories and modified to cover activities observed in the rehabilitation engineering videotapes that Harrison and Minneman did not consider. The six action categories and their indicative types of action shown in Table 34.1 offered an objective classification scheme that could be applied by observing the participants' actions accompanying their talk, or occurring in silence. The types of action listed in the right hand column of Table 34.1 were developed by generalising into simple, unambiguous movements and actions all the actions observed across the videotapes.

Table 34.1. Categories developed to classify actions participants use

Action Category	Types of action observed
Constructing	Mock-up Hand shape Animation
Locating/Indicating	Touch Tap Trace around Scribe marks
Measuring	Hand-to-hand span Hand-to-object distance Thumb-finger distance Using conventional measuring instrument
Demonstrating Function (of equipment or the client's body)	Demonstrating a feature or situation Demonstrating a desired effect Applying force, push, pull, rotation, to show effect Pantomime (mimic a body orientation, function, action)
Examining	Handling an object to show its properties
Gesturing	Look at, glance towards an object, person Pointing to an object but not touching it Waving an arm at an object or person

A system to categorise talking was developed from evaluating videotape transcripts and reviewing the videotapes to find obvious groupings. Table 34.2 shows the final schema developed to code participant's talk. Reliability of coding was tested using three evaluators, two of whom where not associated with rehabilitation engineering. Refining of raw data was tested by re-transcribing and re-categorising videotapes. It was found that 10% of events required some re-assignment of data coding. The talk and action data extracted from the videotapes was stored in a relational database.

Table 34.2. Categories developed to classify talk participants use

Talk categories	*Defining features*
Design proposal	Contains data and concepts to specify a physical requirement to be constructed.
Dimension / location	An actual dimension or estimation of size, position, orientation of something that has design relevance and may be included in the completed construction.
Explanation	A response to a question or statement by another participant that provides additional information or should enhance understanding.
Information client	Information and data about the client's body that is important to the design
Information physical	Information about the physical environment of the client including equipment that is important to the design.
Comment	An unsolicited utterance that may provide additional information or be an opinion, or observation.
Question	Indication that something is unknown and should be known.

34.3.1 Results

Table 34.3 lists the use of talk, or talk combined with action according to the Talk-type in the 882 events[4] analysed. This comprised 319 events in which only talk was used, 41 events in which action occurred without talk, and 522 events where actions were combining with talk. Nearly 60% of communicating was performed using talk combined with action (here-after labelled Talk & Action). In events in which the talk was developing or relating a design idea the use of Talk & Action rose to 86%.

Table 34.4 shows the extent of action of each type that occurred for each talk-type. The top number in a cell is the total of the action-type for the corresponding talk-type in that row. The number in brackets in a cell is the action of the column's action-type for that row. This number termed the 'action value' is the normalized amount of action per talk-type event. The 'action value' provides comparison of action across talk-types with different totals of events.

---

[4] An event is defined as an episode of talking by a participant, with or without associated action, or an episode of action performed in silence. An event continued for the time a participant was speaking and/or acting.

Table 34.3. Number of events of each talk-type in the data

Talk-type	Number of Events	Percentage
Design proposal	161	19.2%
Dimension / location	53	6.3%
Explanation	97	11.5%
Information-physical	92	10.9%
Information-client	42	5.0%
Comment	222	26.4%
Question	174	20.7%
Total	841	100.0%

Table 34.4. Action-types used with talk in talk and action events

Talk-type (with number of events of, and number of actions in that Talk-type)	Action-type					
	Constr-ucting	Locate/ indicate	Meas-uring	Demon-strating function	Examin -ing	Gestur-ing
Design proposal events 137, action 293	86 (0.635)	104 (0.759)	20 (0.146)	25 (0.183)	22 (0.16)	36 (0.248)
Dimension/location events 45, actions 74	5 (0.111)	25 (0.556)	28 (0.622)	2 (0.044)	10 (0.222)	4 (0.088)
Explanation events 73, actions 132	7 (0.096)	46 (0.631)	13 (0.178)	21 (0.288)	21 (0.288)	24 (0.329)
Information-client events 27, actions 41	2 (0.074)	8 (0.296)	0	15 (0.556)	7 (0.259)	9 (0.333)
Information-physical events 56, actions 92	3 (0.054)	20 (0.357)	8 (0.143)	21 (0.375)	17 (0.304)	23 (0.411)
Comment events 94, actions 139	12 (0.128)	30 (0.319)	8 (0.085)	25 (0.266)	31 (0.33)	33 (0.351)
Question events 87, actions 113	6 (0.069)	40 (0.456)	5 (0.058)	9 (0.103)	29 (0.333)	24 (0.276)
Total action	121	273	82	118	137	153

The normalised cell value was compared with the average of the cells in the table, (*i.e.*, 0.274) to determine the existence of action patterns. Table 34.4 shows the characteristics of the data in bearing designing communication implications that include:

1. 'Locating/indicating' actions are important to all the talk-types; they have a wide spectrum of application and very high usage in the talk-types that could be

considered to offer explicit information.

2. 'Measuring' has a narrow spectrum of use. The high use of 'Measuring' action in talk-type 'Dimension/location' was expected but its lack of use in other talk-types is perplexing. It would reasonably be expected that 'Measuring' would feature strongly in 'Design proposal' because measuring provides specific detail about size and location. The zero use of 'Measuring' in 'Information-client' is a reasonable expectation because the physiotherapy and occupational therapy assessment when physical features of the client were examined was excluded from the videotape acts and vignettes analysed.
3. 'Design proposal' and 'Dimension/location' talk have high association with action-type groupings 'Constructing' and 'Locating/indicating', and 'Measuring' and 'Locating/indicating', respectively. In both talk-types the pair of action-types with very high use contrast with the below average involvement of the remaining action-types. The high 'action value' indicates these action-types appear to have specific association with these talk-types. For example, 'Constructing' actions are clearly associated with 'Design proposals', 'Measuring' actions are clearly associated with 'Dimension/ location' and 'Locating/indicating' actions are important for both Talk-types. The other action-types do not appear to be as important in communicating 'Design proposal' and 'Dimension/location' information. The dichotomy is not repeated in the other talk-types.
4. The high 'action value' for 'Locating/indicating' associated with 'Explanation' talk, and high 'action value' for 'Demonstrating function' associated with 'Information-client' talk, suggests information communicated in these talk-types is reliant on certain types of actions. These talk-types show near-average 'action value', for most of the other action-types, hence they are different to the case above for 'Design proposal' and 'Dimension/location' talk.
5. Gesturing has less specificity than the other action-types and enjoys almost universal application across talk-types. Its near-average 'action value' in six of seven talk-types suggests gesturing plays an important role, though at this stage of analysis the role is elusive. It may be being used as a social modifier to attract and maintain attention.

34.3.1 Artifacts

Artifacts were implicated in many actions. Approximately 83% of actions in the Talk & Action events involved an artifact. The artifacts employed by participants to assist their actions were usually the objects around which discussion was currently focused. Artifacts served to focus the attention of the clinic participants, stimulated question - answer dialogue, and drew out participant's experiential knowledge and considered opinion. Observation of many designing episodes, where it was evident there was close association of artifacts at hand with an idea or sequence of ideas proposed by team members, provided evidence of the roles of the artifacts in aiding the design process. These were: 1) to initiate discussion and discovery; 2) to stimulate generation of new ideas; 3) to verify information and understanding; 4) to develop and propose ideas.

Team members using artifacts, individually and in groups of two or more, gathered around the wheelchair discussing and trying out ideas, helped build a common (shared) reference and mutual understanding about design possibilities, design and manufacturing decisions (specifications) and achievable outcomes. The high usage of artifacts in the rehabilitation engineering work is an indication that artifacts add value, probably because they assist participants to increase the complexity of data communication through creating visual reference points to underpin oral descriptions. Face-to-face interaction of participants is strengthened by the ability to physically share an artifact. Team members can work together on a problem by asking questions, developing answers, all the while using the artifact(s) to test ideas and extend thinking to new solutions. The same benefit of using artifacts cannot be had in a video-mediated interaction because some participants, isolated physically, can see the artifacts in use but are unable to physically share them. Some alternate means to replace physical interaction may have to be developed to assist this aspect of designing by videoconference.

34.3.2 Significance

The research results show that rehabilitation engineering practice in a face-to-face environment has three activity elements: (1) the gathering of general information that has relevance to designing and the capture of specific design data; (2) the development of ideas based on information presented in question and answer induced discussion, physical examination and measurement of the client; and (3) the designing of solutions and trialing of these, on the spot, using artifacts and mock-up constructions. Practice is not necessarily orderly, though it is guided by a prepared document that aims to ensure specific questions have been asked and specific information acquired. A high degree of serendipity creeps into practice and there is considerable evidence that information and comments that focus on an artifact (part of a wheelchair, seat hardware, even the client) draw individuals and the team to new avenues of enquiry and new ideas (Logan, 1999). The elements may also overlap; for example an idea generated becomes a design solution that a participant rapidly mocks-up using a nearby artifact and proceeds to test for performance.

Another significant feature of the face-to-face interaction is the availability of all the action to all the participants simultaneously (assuming they are watching). Talk accompanied by physical actions performed by the speaker is a feature of face-to-face communication. The level of diction used in the rehabilitation engineering sessions is not high and there is great reliance on the action component to make the talk lucid. For example, 'It will need to be attached there.' lacks identification of what 'it' is and where 'there' is unless the accompanying actions are observable.

34.4 Observations of Communication Performance During Videoconferencing

Ten rehabilitation engineering videoconferences have been held over 2 years between REC and health facilities up to 1,000 miles away. Videoconferencing has been held over ISDN links at 128 kilobytes per second transmission rate. Equipment used has varied between large screen television systems and personal computer-based systems such as 'Picturetel PC100'. The sessions typically involve a client, care provider and physiotherapist at the remote site and three team members at the REC site. A video recording is made for after-session review and analysis. Some remote therapists have performed multiple videoconference sessions and have developed a rapport with REC team members and a confidence in participating in the sessions.

In videoconferences that have been held to date the issues that have emerged include:

1. Lack of understanding of the intent of the interaction and misunderstanding of outcome between remote therapist and the REC team. New participants were unprepared for this type of videoconference interaction.
2. Lack of direction of the interaction causing time wasting while explaining process. Normal turn-taking cues that guide face-to-face interaction do not work well in a videoconference.
3. Poor alignment of the video camera with the assembled remote group so that all actions and detail were not visible to the REC team. Remote videoconference systems lacked effective pan/tilt/zoom controls or the remote group did not know how to use them.
4. The remote group do not have wheelchair and seating hardware 'artifacts' to utilise that may assist them trying their ideas and testing REC ideas under direction.
5. The remote therapists are inexperienced in measuring the client and wheelchair and taking shapes of the client's body contours.
6. Diction issues similar to those occurring in face-to face talk place reliance on visual cues to convey the full meaning to an observer. The visual detail is less clear than in face-to-face contact.

34.5 Discussion

Communicating and designing in the rehabilitation engineering environment is quite different to most product development scenarios, particularly the absence of drawing and the heavy reliance on physical artifacts to indicate, locate, orientate and demonstrate what is being talked about. This contrasts with observations by Hummels *et al.* (1997) that most designers currently visualise their ideas with sketches on paper because this is a quick, intuitive and expressive way to generate, present and evaluate ideas. Face-to-face interaction between the REC team and client group has relied on design communication by acting out ideas using artifacts, constructing mock-ups of ideas and converting these sometimes impromptu mock-ups into aesthetic, manufactured items of equipment.

Using Videoconference communication to emulate the processes of the rehabilitation engineering session presents some challenges. The reliance of design data communication by the REC team through actions with artifacts, such as pointing to, orientating with other artifacts and assembling artifacts to mock up a construction is difficult to do in videoconferencing because: (a) the technically-minded participants are isolated from the client; (b) suitable artifacts are not usually available to the remote client group; (c) camera location with respect to the therapists and client may miss important visual detail.

Whittaker and O'Conaill (1997) suggest video might improve communication, because: (i) video supports behaviours hence it supplies non-verbal information; (ii) video provides visible information about the environment and the availability of other people; and (iii) video provides video as data about objects and events in a shared visual environment. The REC experience is that non-verbal behaviour information and data about objects and events is not easily evident. Videoconferencing does not provide a co-present environment for information exchange; this has to be engineered or contrived. Sellen and Harper (1997) suggested videoconferencing provides a common frame of reference that could be established by showing work-related documents and objects that focused discussion rather than transmitted detailed information. We believe the common reference can be implemented by:

1. Developing an explanatory document outlining what is to happen and why, the limits of the process (what it is hoped will be achieved), the role of the remote group, and instructions outlining the rules of conduct for videoconferencing.
2. Assembling a toolkit of equipment complete with user instructions that can be shipped to the remote videoconference site for use by the therapist and client group. The toolkit contains measuring equipment, an instructional video on measuring technique, and seating artifacts for the remote group to use either serendipitously or under instruction from the REC team.
3. Providing a digital video camera for the remote site to, (a) take images of the client in the home, school, and community environments that will influence design ideas and, (b) link into the remote site videoconference system to provide a high quality picture rendering detail visible at the REC site.
4. Run a test videoconference with the remote site therapist (at least the first time) without the client to establish rapport and basic understanding of how the client session will run and to check that basic issues like taking measurements and shapes of the client have been understood.

34.6 Conclusion

It is clear that setting up for a videoconferenced rehabilitation engineering session is different to a face-to face session. Contact must be made well in advance of the videoconference to develop understanding between the REC team and client group about the purpose of the session, and what might be achievable, and to prepare the remote group to participate. Disparate expectations of outcome will lead to conflict, disappointment and the possibility of videoconferencing being rejected as a tool before its potential is realised and exploited. The host (REC contact person) has an important role to ensure achievables can in fact be met by the REC team and its available resources. Reliance must be placed on the remote group to be competent at technical tasks like measuring. Education of remote therapists in the techniques to be applied need to be done in advance to avoid time consuming explanation during the videoconference and to minimise error.

34.7 References

Harrison S, Minnemen SL (1996) A bike in hand: a study of 3D objects in design. In: Cross N, Dorst K, Christiansan HH (eds.) Analysing Design Activity. John Wiley & Sons

Hummels C, Smets G, Overbeeke K (1997) An Intuitive Two-Handed Gestural Interface for Computer Supported Product Design. In: Wachsmuth I, Frohlich M (eds.) Gesture and Sign Language in Human-Computer Interaction. In: Proceedings International Gesture Workshop. Springer-Verlag, London

Logan GD (1999) An investigation of talk, actions and use of artefacts by a cross-discipline rehabilitation engineering team, PhD Thesis, The University of Queensland

Peräkylä A (1997) Reliability and Validity in Research Based on Tapes and Transcripts. In: Silverman D (ed.) Qualitative Research. Sage Publications, London. pp 201-220

Sellen AJ, Harper R (1997) Video in Support of Organizational Talk. In: Finn KE, Sellen AJ, Wilbur SB (eds.) Video-mediated Communication. Lawrence Erlbaum Associates, Mahwah, New Jersey. pp 225-243

Tatar DG, Foster G, Bobrow DG (1991) Design for conversation: lessons from Cognoter. International Journal of Man-Machine Studies 34:185-209

Shapcott N, Schmeler N, Pelleschi T, Malagodi M, Garand S (1998) Telerehab-telemedicine in Rural Areas: Assistive technology and Service Delivery. In:Proceedings 14th International Seating Symposium, Vancouver BC, Feb 1998 pp 175-178

Whittaker S, O'Conaill B (1997) The Role of Vision in Face-to-Face and Mediated Communication. In: Finn KE, Sellen AJ, Wilbur SB (eds.) Video-mediated Communication. Lawrence Erlbaum Associates, Mahwah, New Jersey, pp 23-49

Chapter 35

Communication and Management Issues in International Collaborative Design

A. Woodcock, L.-C. Lee and S.A.R. Scrivener

35.1 Introduction

In our earlier work (Lee *et al.*, 1999), we differentiated between three groups of problems that might arise in international collaborative design projects: problems which would normally arise within the context of group working, (e.g., the management of within-group conflict); problems which might be exacerbated through working at a distance (e.g., inability to clear up misunderstandings in project briefs); and problems which arise specifically because of the international quality of the endeavour. The latter relate to issues such as cultural misunderstandings (e.g., Baron and Walters, 1994; Desazars de Montgailhard, 1982), contractual interpretations (Dodgson, 1993), financial issues (e.g., Levcik and Stankovsky, 1979), and political and economic turbulence (e.g., Pandia, 1994).

In this paper we seek firstly to contextualise our research in terms of current design practice; secondly, to validate the results of the previous work using interviews with project participants; thirdly, to consider the issues raised in terms of the use of asynchronous and synchronous communication; and fourthly to look at the wider implications of these issues in relation to the needs of designers working in an environment which is becoming increasingly pressurised and techno-centred.

35.2 Context

To enhance economic growth, the governments of developing countries have started to play a proactive role in helping indigenous manufacturers achieve national and international competitive advantage. In Taiwan, this has been achieved by the creation, in 1979, of the China External Trade Development Council (CETRA), which in 1990 implemented the Euro-Design Coordination Project. This was to guide manufacturers in the design of their products in accordance with different market requirements. The Taipei Design Centres, TDCs, were established to raise the level of design in Taiwan and enhance the quality, image and competitiveness of Taiwan-made products on world markets. The TDCs were located in primary international markets to act as agencies to support and mediate between the designers and clients, and monitor the effectiveness of the international co-operative design projects.

The relationship between the end users, designers, TDCs and manufacturers is shown in Figure 35.1. In terms of the projects outlined here, CETRA (and especially its Design Promotion Centre) acted as mediator between the manufacturers and the TDCs (*i.e.*, manufacturers request help from CETRA who then forward their requests to the most appropriate TDC). The assumption behind the development of the centres is that designers located in the market heartland are better able to conduct market surveys and produce concepts meeting the requirements of the target population. This enables Taiwanese companies to acquire an in-depth understanding of the target market and their requirements, thereby gaining competitive advantage.

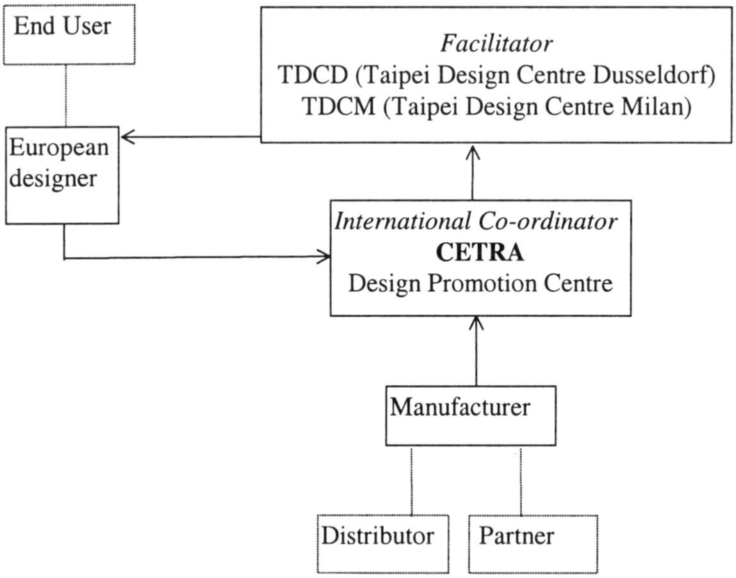

Figure 35.1. Initial position of members of the project team

The success of such ventures depends on the willingness of the partners to co-operate, the nature of the co-operative process, the effectiveness of communication methods and project management, and the validity of the belief in a close relationship between commissioned designers and the target population. Typically, these projects have two or three face-to-face meetings and then rely on distance communication methods for the remaining transactions.

By systematically analysing the written documents of 21 international design projects funded between 1993 and 1997, we were able to classify their problems as relating to time, information flow and communication, brief specification, project management, participation and language. We were also able to establish the relative importance of each of these problems in terms of their prevalence and severity across those stages of the design process for which we had documentary evidence.

Obviously there are major shortcomings with this methodology (see below). Ideally projects should use participant observation (e.g., Burns and Vicente, 1999), or be structured to allow the systematic collection, collation and analysis of all communications (both written and verbal) within the lifetime of the actual design life cycle (this is in fact the current stage of this research project). However, with confidentiality agreements and the sensitivity surrounding new product development and launch, developing ways of using project material retrospectively may, in some cases, be the only way of gaining an understanding of the manner in which projects are conducted and the effects of this on their ultimate success.

35.3 Summary of the Previous Study

In the initial study, the twenty-one projects selected were considered typical of the design projects sponsored by the TDC. They included the design of computer peripherals, communication and domestic equipment. The mean number of designers in each project was 6, with an average project duration of 5.9 months.

The majority of the available documentation spanned stages from project briefing through to construction development, *i.e.*, all the concept design stages were represented in all the projects. The documentation available for analysis included written project documents (mostly minutes or information circulated during face to face meetings) and faxes. Before discussing the results of the investigation it is worth noting the limitations of this type of data for retrospective analysis:

- documents were not produced for the purpose of uncovering problems in the design process, but at CETRA's request;
- some companies retained, or produced more documents than others;
- we do not have a complete set of documentation for any of the projects and do not know what has been omitted;
- from our own experience, day-to-day co-ordination problems in international projects are often resolved through email and telephone conversations. This material was not available for inspection.

In total, 143 problems were identified. The percentage distribution of these problems across categories and the main issues arising in each category are outlined in Table 35.1 below.

Table 35.1. Breakdown of problems emerging from project documentation

Category	Examples of issues arising	Frequency
Time management	External constraints	31%
Information flow and communication	Lack of feedback, corporate workflow problems	30%
Brief specification	Changes to specification	11%
Project management	Verification, clarification, changes to product brief, patent, delays	10%
Participation	Key members not present at meetings, changes of personnel	9%
Language	Semantics, translations	9%

No relationship was found between the number of problems documented and the length of the project, the number of actors, the origin of the designers (e.g., Taiwan, Germany, *etc.*), or project types. It was expected that such relationships would have been found - for example, due to different cognitive styles, linguistic problems existing between Asian and European countries, and the differing nature of the projects. That these differences were not found might have been due to the nature of the documentary evidence we were inspecting.

Each problem was also classified in terms of the impact it had on the overall conduct of the design project. This coding relied on the principal investigator's knowledge of the design projects and the documentation (in Chinese and English). Just over 70% of the breakdowns were classified as having a medium or high impact on the project. As far as we could determine approximately 50% of these were definitely resolved during the lifetime of the project - for the remainder, the nature of the source material made it difficult to judge whether the issues had been resolved, or to speculate as to the manner of their resolution. In our analysis we also tried to interpret the extent to which these breakdowns were a result of, or had been exacerbated by international working. We determined that just under 70% of the total number of problems were due to the inter-nationality of the project (28% were not, 2% indeterminate). This appears to be a high number of incidents. This may impede the design process because the time spent on resolution of problems cannot automatically be equated with making progress on the design itself.

35.4 Method

The follow up study comprised face-to-face interviews with five representatives of the original projects in Taiwan. Its aims were to validate the results of the previous investigation, further identify the nature of the problems that were occurring and determine the use of extant communication technologies in international projects.

The interviews included both structured and unstructured questions. The structured part was used to determine whether the difficulties we had identified from the documentation were in fact considered to be problems by participants in the actual projects and the relationship of these to different stages in the design process, and to different modes of communication. The interviews concluded with a series of unstructured, open-ended questions designed to provide wider information about the context in which design problems arise.

35.5 Results

35.5.1 Synchronous and Asynchronous Communication

The interviews showed that the fax was still the most preferred and frequently used method of communication, allowing both text and graphical information to be transmitted. Faxes also enabled more effective synchronous communication to take place over the phone. It is of interest to note that email was not used by two of the five respondents. Email attachments could have been used in a similar manner to faxes, if graphical documents were produced initially on computers. These would also have been in colour and higher resolution than fax transmissions. Video conferencing was only used by one of the respondents, teleconferencing and videophones by none of them. However, all respondents expressed a great interest in the use of these newer technologies. Face-to-face meetings were seen as important but frustrating if key people were unavailable. As most communication had taken place via faxes and minutes of meetings, it appears that the material on which we based the earlier analysis was appropriate.

35.5.2 Project Management

After having focused on general communication methods the interview concentrated on the types of problems which had been experienced in the earlier projects and whether these were typical of the problems experienced in similar international design ventures. The following is a summary of the issues, in order of importance (on a scale 1 to 5 where 5 is the most important):

- *Time management (rated 3.07).* Major issues related to projects falling behind schedule, sometimes due to external constraints, which required the rescheduling of activities. It is now quite typical for designers to work concurrently on a number of projects, with different personnel and at different stages of completion. This can lead to knock on effects. Notably, time consumed by face-to-face meetings was not seen as a large problem.
- *Brief specification (rated 2.87).* This was problematic, with issues relating to changes of product specification, inattention to user and client requirements and insufficient marketing information. Many projects suffered from a mismatch between what the designer did and what the clients and manufacturers expected of them (e.g., market research).

- *Participation (rated 2.69).* The main items of concern here included co-operation and communication with engineers, establishing a project team, and key persons not being present at project meetings (which hampered agreements).
- *Project management (rated 2.58).* Issues here related to a lack of understanding of different corporate strategies and an underestimation of project complexity.
- *Information flow and communication (rated 2.43).* These issues were not rated as problematic as the time issues. Communication *per se* was not seen as a problem. Understanding corporate workflow and coping with changing design ideas were both exacerbated through distance working.
- *Language (rated 1.08).* This was not seen as a major issue.

35.5.3 Methodological Matters

The interview study affirmed the overall categories that we had uncovered in the retrospective analysis of the project documentation. This confirms the utility of the project documentation as a starting point for our investigations. However, the relative weightings assigned to the problems were different in the two studies. This was possibly due to the nature of the material that we were examining (e.g., many information flow and communication problems might be resolved through phone conversations that are not documented). Alternatively, the experience/working environment of the respondents themselves might lead respondents to put a greater weighting on problems that occurred less often in the documentation, such as time management and changes to the product brief.

35.6 Discussion

It has been assumed that a 'well designed' teleconferencing system might replace face-to-face meetings because of savings of time, expense and inconvenience, *etc.*, (e.g., Kelleher and Cross, 1985). It is perhaps time to face up to the fact that this may simply not be true. Our work indicates that teams require and expect at least one face to face meeting and would in fact like more of them, not less. Face-to-face meetings are problematic; for example, in terms of their scheduling, the non-attendance of key personnel (albeit sometimes unavoidable), and the lack of preparedness of meeting participants. Some of these issues are unavoidable, some beyond the remit of professional practice. However, face-to-face meetings continue to have other affordances not offered by other means of communication. They allow people to 'escape' from office situations, receive perks, enhance the status of their project, become immersed in a different culture, feel part of a team (and understand other members' problems) and become entirely focused on the task in hand (mobile phones notwithstanding!). It is hoped that the last three of these could be amenable to the development of design support systems and methods.

Inadequate planning caused many of the problems that occur in the projects. Lean-teams mean that if one project slips or unforeseen problems occur, these will effect the scheduling of other projects. With distributed teams, members may not be aware of what is happening at other sites and the problems which are occurring on a day-to-day basis.

The informal communication that takes place during meetings and workplace interaction provides opportunities for people to actually discuss their workloads. Concurrent working on many projects now seems to be the norm, but piecemeal division of time requires careful planning so that work across different projects may be co-ordinated. Schedulers *per se* are not the answer.

A second category of problems related to not understanding corporate culture and strategy. Entering into agreements requires a series of key questions/ development of understanding which need to be addressed at the start. This requires a guided synchronous communication session, supported by information provided in advance by each partner. Guided by the findings from this work, and our knowledge of problems of distance working, it is our intention to develop a series of management aids to reduce problems especially associated with distance working and which are not fully met by current methods.

Figure 35.2 shows the distance between project members at the start of any particular project. There are many different ways in which this gap can be reduced, for example through using effective and efficient technology enabling members to work on the task in an efficient manner. Face-to-face meetings are still one of the most effective ways of moving people closer together so they develop a shared 'model' or understanding, as illustrated in Figure 35.2b.

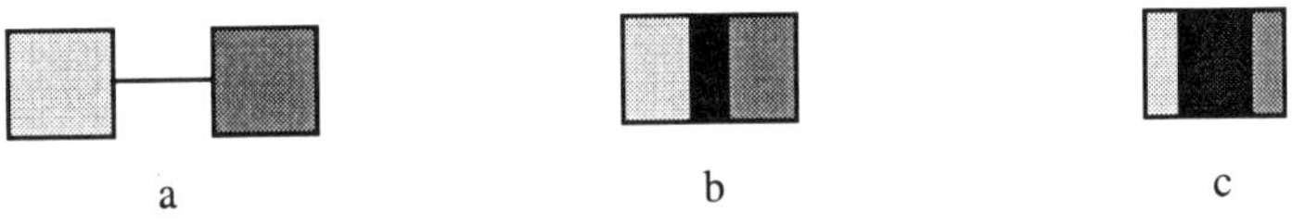

Figure 35.2. Enhancing effectiveness of international working: a. without communication; b. with effective communication channels; c. added 'methods and training' for co-operative work

However, of itself this is not sufficient. A shared understanding requires an immersion in some part of another's experience, and the present opportunities for this in international design are just too small to enable this to happen effectively.

The research reported here has concentrated on the retrospective analysis of design projects. The study presently being undertaken is one in which we are able to gather data throughout the design process - through the recording of video conferencing sessions, diary studies and log of other forms of communication. The rapid analysis of this material allows us to identify breakdowns in communication and interview the designers about these occurrences in the following session. By doing this we will show the ways in which the present channels of communication can be used more effectively to move designers through to the levels of shared understanding, idealised in Figure 35.2c.

35.7 Conclusions

Previously, we claimed that, '...as design researchers, we do not wish merely to study design behaviour, but to develop solutions to the problems that we uncover.' (Lee *et al.*, 1999). In terms of Figure 35.1, international co-operation will only be successful, if the distances between groups of individuals are reduced. This and other work conducted in the Visual and Information Design (VIDe) research centre (e.g., Chen *et al.*, 1999) emphasises the development of more effective and efficient communication systems as a means of reducing the 'distance' between the various stakeholders. However, in our earlier paper we also concluded that technology would not provide all the solutions. This study reconfirmed that problems may arise throughout the design life cycle which are not amenable to technological intervention.

Although technology enables rapid, inexpensive communication between individuals, it has not provided people with the incentives to respond to the communiqués they receive. Joint drawings can be developed and annotated but there is no guarantee that they are understood or will be followed. Likewise systems have been developed to ensure that interdependencies and contingencies are known prior to the commencement of the project, but design teams are often not able to meet the plans which have been made.

Bearing these points in mind we concluded our last paper with a call for the need to develop 'methods of nurturing and fostering co-operative working, which will enable the technology to be used to its maximum effect at all stages of the design life cycle' (*op. cit.*). The critical issue for co-designing projects would seem to be a managerial one which goes far beyond the piecemeal introduction of the latest hardware/software, but lies in the design of people-centred, socio-technical systems (see also Woodcock, 1999).

However, we would also advocate the introduction of training for designers wishing to practice co-operative design in an international arena. This should focus on time management and communication/computer skills for international design projects, cultural and work ethics, universal design issues and international law. This would enable designers and design managers to gain the necessary technical, social and administrative skills required to successfully complete design projects on time and to the satisfaction of clients and end users.

35.8 References

Baron A, Walters M (1994) The culture factor: Corporate and international perspectives. London, Institute of Personnel and Development

Burns CM, Vicente KJ (2000) A participant-observer study of ergonomics in engineering design: how constraints drive design process. Applied Ergonomics 31:73-82

Chen CD, Scrivener SAR, Woodcock A (1999) Visualisation in collaborative computer mediated dialogue. In: Bullinger HJ, Ziegler J (eds.) Human-computer interaction: communication, co-operation and application design. Vol 2. Lawrence Erlbaum Associates, New Jersey, London, pp 226-230

De Montgailhard JD (1982) Opportunities for improvement in industrial productivity. In: Fusfeld HI, Haklisch CS (eds.) Industrial productivity and international technical co-operation. Pergamon Press, New York, pp 11-16

Dodgson M (1993) Technological collaboration in industry. Routledge, London

Kelleher K, Cross TB (1985) Teleconferencing linking people together electronically. Prentice Hall, Englewood Cliffs, NJ

Lee L-C, Woodcock A, Scrivener SAR (1999) Breakdowns in the management of international design projects.6th International Product Development Management Conference. Cambridge, vol 2, pp 689 – 700

Levcik F, Stankovsky J (1979) Industrial co-operation between East and West. White Plains, N.Y

Pandia RM (1994) International projects: opportunities and threats. In: Cleland DI, Gareis R (eds.) Global project management handbook. McGraw-Hill, New York, pp 18-1 - 18-23

Woodcock A (1999) Human factors perspectives on globalisation and technological change. Information Society '99. Vilnius, Lithuania, pp166-177

Woodcock A, Scrivener SAR (1998) Groupware and CSCW. In Molina A, Kusiaka A, Sanchez J (eds.) Handbook of lifecycle engineering: concepts, tools and techniques. Kluwer Academic Publishers, The Netherlands, pp 573 – 596

Chapter 36

Persuasive Artefacts in Architectural Design and Planning

I. Wagner

36.1 Introduction

Architectural design and planning is a collective activity, embedded in a community of practice. At the heart of the work, therefore, is the need to share materials and to mobilize support. Sketches, scale models, images, plans, *etc*., play a crucial role in this process. They are created and used for expressing, developing, detailing, communicating, and presenting an evolving design concept (Henderson, 1995; Perry and Sanderson, 1998). One of the most important advantages of such visual and graphical material lies in its ability to create 'persuasion' and to invite others into a dialogue (Latour, 1986). We have explored architects' uses of inspirational objects (images, sound, video, animation, 3D objects) for forming ideas, expressing qualities, and for convincing others of the viability and power of an idea. Such inspirational material is open to diverse levels of interpretation and understanding, stimulating contradictory and surprising associations (Büscher *et al*., 1999). The availability of digital media increases the possibilities for architects to create and share visual materials. Techniques such as rendering, computer animation, and modeling provide a new visual language. At the same time, issues of how to combine digital material with tangible artefacts and how to support fluid transitions between both media come to the fore. This paper examines the role of 'persuasive artefacts'- both digital and tangible - for cooperation in architectural design.

36.2 Visual Material in Use

Ongoing long-term fieldwork in an architectural office is used to analyse the role of 'persuasive artefacts' in cooperative work. Fieldwork material has been collected over a time-span of almost two years through ethnographic observation of work combined with in-depth interviews. It covers a variety of building and urban planning projects in their different stages. We observed how the architects produced, modified, and arranged visual and graphical material in their physical and digital workspaces and collected examples of it, including the stories of use surrounding them. This paper focuses on three types of visual representations of design work, which demonstrate how artefacts mediate cooperative work.

36.2.1 Narrative Sketch

We frequently observed how the architects, in intense conversations with each other or in a meeting, produce a series of sketches, while talking through a design problem, identifying constraints, settling or affirming a principle, jumping between different variants of a solution. These 'narrative sketches' consist of two closely interwoven types of material – the sketch, either on transparent paper or directly on a plan, and the 'story'.

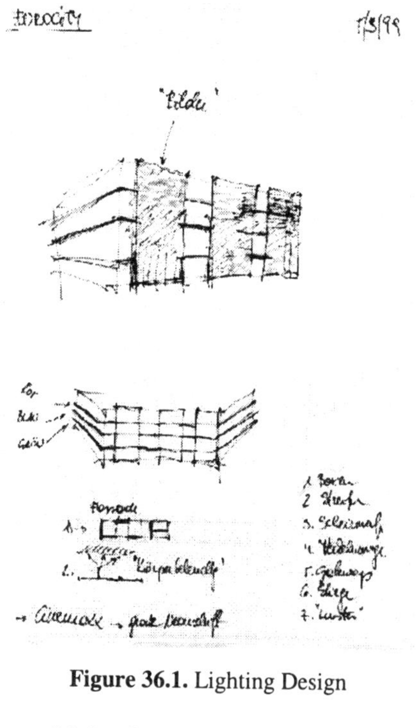

Figure 36.1. Lighting Design

This example (Figure 36.1) of a narrative sketch was created in a meeting of the architects with a lighting designer. Using differently coloured pencils for different types of lighting, the architect developed his notion of light as a changeable surface for *Eurocity* (a large cinema complex), describing elements such as light strips (on ceilings), light beams (on the façade), projections (from inside, outside), flooding the volumes with light. This combination of sketching and talking played an important role in mediating the cooperative development of the lighting concept. It fulfilled the following functions:

- the lighting designer got to know the building, 'walking through', pointing to particular places and elements;
- sketching and explaining ideas are tightly interwoven – the architect was thinking aloud, listening to questions and suggestions, and simultaneously visualizing the design concept;
- a common understanding of the concept, including some practical-technical problems was created – the coloured photocopies of these narrative sketches were then used as part of presentations to different audiences;
- a notation was developed, with different colours standing for different types of lighting – this notation was then used in all documents around lighting.

Colour copies of this set of sketches were exhibited on the walls of the office. The lighting designer used them for a presentation of his design to the client. Still, with the 'voices' of the participants missing, this representation is incomplete. Again and again we observed people in the office looking at such a narrative sketch, trying to remember what exactly the idea was, how exactly a joint should be constructed. It might be worthwhile to capture the act of sketching and explaining. This could be done in the form of a small video with the sketch being done with a pressure sensitive pen.

36.2.2 Conceptual Sheet

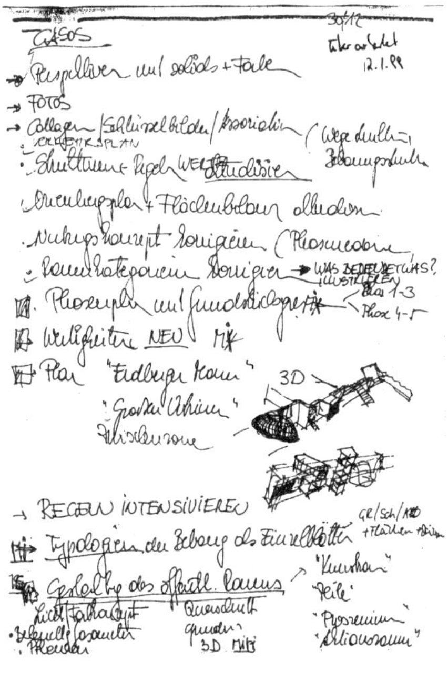

Figure 36.2. A complex visualization

As shown in the previous example, the ways in which representations of work are created are informal, and often intermingled with talk. This particular example of a 'conceptual sheet' (Figure 36.2) has been taken from an urban planning study. It was produced as part of a first planning session. It contains several elements:

- a first work plan – how to proceed;
- visual material that should be collected or created (pictures, collages, association images, shadow plans, *etc.*) - how to represent the design of the urban area;
- metaphors – how to talk about the urban area;
- a specification of methods – to define spatial qualities, to 'intensify rules', *etc.*;
- explanatory sketches.

A 'conceptual sheet' is not just a list but a complex visualization of a design idea and of open questions. It may include small sketches, arrows, references to material to look for or people to contact. People placed copies of the sheet on their desks. It was present in meetings, annotated, and used throughout the course of the project.

Another feature of architects' work practice stands out in this example: the mixing of text and images. There are many other examples of this mixing, such as highlighting a particular point on a plan or list by putting a red circle around it, adding detail (correct measures, specifications of material), or marking a part of a drawing or sketch with a post-it with some instructions for changes. The specific form such a document takes is an intricate part of architects' work practice.

36.2.3 Assemblies

While some of these visual artefacts stand for themselves and assume the role of 'key artefacts' for the design process, some of them melt into assemblies of materials, which tell a 'story', such as the story of the design concept or of a particular choice of material and product. This assembly of material (Figure 36.3) has been produced in support of the concept formation process for a large building project (*Pleasure Dome*). Several documents were brought together with a metaphorical description to illustrate 'themes' or design principles (Tellioglu *et al.*, 1998), including:

- a 3D visualization of the 'façade as coloured patchwork';
- association images (a painting by Richard Serra, the coloured glass in the cathedral of Chartres) that express some of the visual qualities the architect had in mind;
- a sketch in which the architect describes his idea of the movie theatres as 'stones that dip into water - above the surface of rough, rocky quality, below precious stones that glitter in water – silver, gold, ruby, emerald';
- the 'themes' that were formulated in a meeting – 'Playing with volumes', 'The 'Big wall', impregnated with colour and light'.

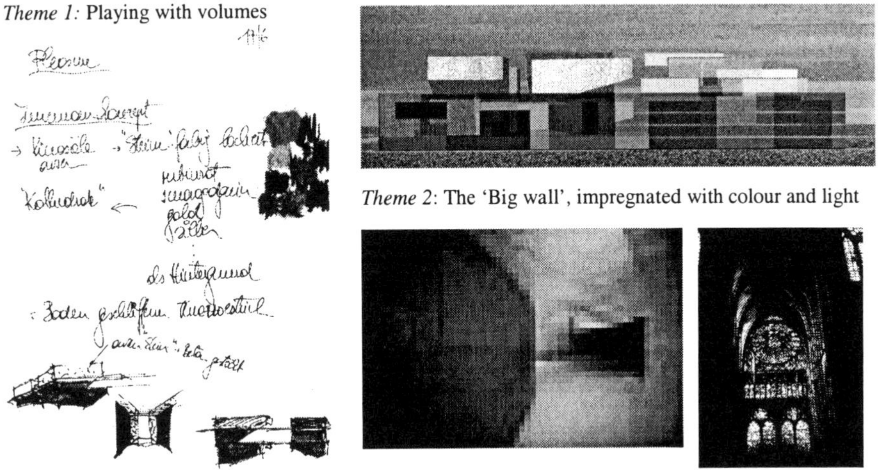

Figure 36.3. Assembly of themes, sketch, inspirational images, and 3D visualization

Placed in A4 format close to each workplace, this assembly, together with several large-scale 3D visualizations of the project, which had been produced for an exhibition, acts as reminders of the design idea. It is used to point out things to external visitors, and it provides the design team with a language, both textual and visual, for talking about the project.

This is an artful assembly. Mostly, assemblies of materials are created as an intricate part of ongoing work. An example is the search for a suitable product for the façade of the cinema complex *Eurocity*. A strong image guided this search process - the notion of 'wrapping the building into a membrane which uses light for producing an almost imperceptible metamorphosis, from hermetically shimmering in the morning to communicating the building's contents - projected cinematic images, people's movements - in the evening'. Several 3D visualizations were constructed. They helped to shape and express the idea of covering the façade with glass of an irregular textile appearance onto which light will be projected. Samples of textiles were scanned and additional properties of the material specified, as non-reflecting, easy to clean, available in 3x3m format, with particular insulating properties, affordable, *etc*. As such material cannot just be bought from the shelf, a variety of methods for producing it were investigated. Some of this research was taken over by the façade consultant who contacted several producers of industrial glass.

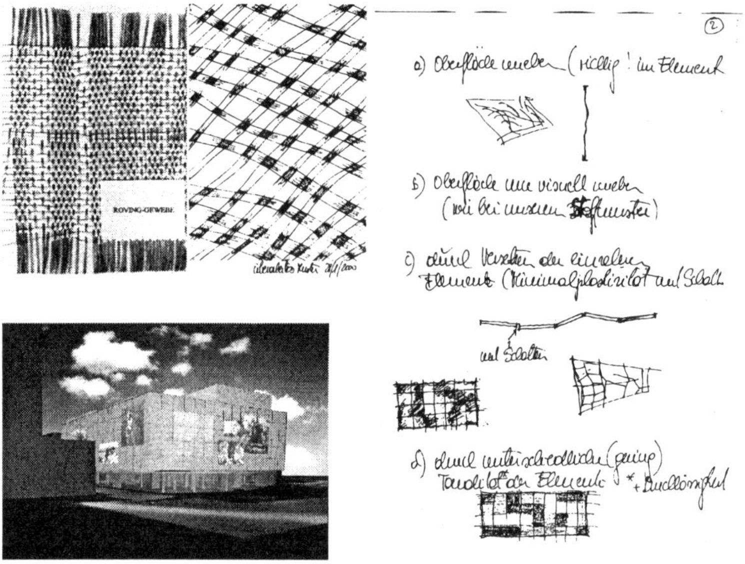

Figure 36.4. Assembly of samples, conceptual sheet, 3D visualization

For the architect working on the façade the skin is represented by an assembly of materials: inspirational resources of various kinds – images of glass with a textile-like surface, architectural projects, samples of textiles; information about research, collected in professional meetings or seminars, by browsing through journals; product information from catalogues and the Internet; information about materials and techniques; samples, assembled *ad hoc*, produced by a glass manufacturer, *etc.*

Most of these documents have been stored in folders on a shelf behind the architect's workplace, some are hidden in a pile of papers on his desk, and some (the images) are fixed on the wall behind him. When working on the problem, the architect might assemble part of this material (some of which is shown in Figure 36.4) on his desk or spread it out on the meeting table, in ways that depend on the task at hand.

Another frequently observed assembly involves the use of transparent tracing paper, where a blank transparent sheet of tracing paper is placed over a plan, and 'anchored' by entering some positional markers. The tracing paper is then used, for example, to experiment with design ideas. Such an assembly creates a precise relationship between a design sketch and (one or several) other documents

Assemblies represent materials 'in context'. One of their central features is their narrativity. The story of the search for glass starts with a set of images (some of them inspirational) first samples of textile structures, and a 'conceptual sheet'. More and more material is added as the search for a solution proceeds. It is invoked when the solution in mind has to be explained to significant others and detail is added.

Much individual *and* cooperative work is done through creating assemblies of materials. From our fieldwork we have many snapshots of people's desks when they configure materials for a particular task, and reconfigure for the next task. Some assemblies provide a good overview, including pointers to things that should be kept present. Some have a strong narrative character, helping people to remember the context and history of a particular detail.

36.2.4 Viewing Persuasive Artefacts 'in context'

A central characteristic of assemblies of materials in the physical world is their ephemerality. They tend to be arranged for the moment – a particular task, an instant in a conversation – to be reshuffled as people interrupt or take away a particular document for photocopying, or make a request, shift task, or simply come to the end of the workday. This is a positive feature in the sense that documents and objects can be assembled and arranged relatively easily, for different purposes. Yet the ephemerality of arrangements can also be a nuisance. When a job is done, many documents and objects are returned to their locations, others are thrown away, yet others lost or buried under the continuous flurry of documents and objects on the move.

A digitally-enhanced environment, with all its limitations and deficiencies compared to a physical one, may help overcome some of these constraints. The architects may want to place one and the same object in several arrangements at a time. They may want to be able to store the relationships connecting a set of materials, to search for a particular assembly representing a solution, which has been developed in an earlier project. They may want to view several assemblies at the same time, e.g., have all the material representing the search for glass (Figure 36.4) present, while at the same time looking at a series of detail drawings and connected product information. The physical environment of cluttered desks, stacks of folders, boxes, *etc.*, is not ideally equipped for this.

As part of ESPRIT Project DESARTE, we are developing a 3D environment for assembling and viewing materials 'in context' in the form of workspaces. Figure 36.5 shows a screenshot from the current *Manufaktur* working prototype. It shows an open workspace containing (references to) a set of objects. Double clicking on any of the document objects – CAD file, bitmap, Word, Excel, *etc.* - will launch it in its respective application with that document, and changes to it will be updated within 3D *Manufaktur* in near real time. The objects can be sized, moved, rotated, *etc*; light effects may be applied; documents can be made (semi-) transparent; organized into groups; and much more. Several workspaces can reside at once and be opened, closed, referred to, or looked at from another workspace.

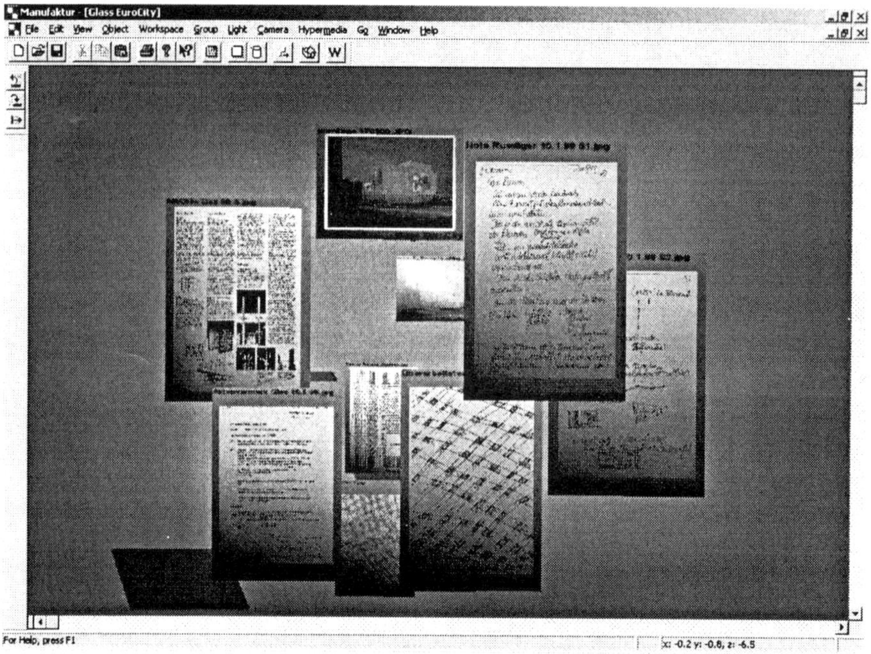

Figure 36.5. Configuring materials in a 3D *Manufaktur* workspace

The architect working on the *Eurocity* project has arranged an assembly of materials relevant for his current work in a *Manufaktur* workspace. In the centre he places 3D images of the design; below are scanned samples of glass textures, together with a drawing by a graphic designer and a letter from a manufacturer; to the right are two 'conceptual sheets' describing different techniques for producing glass; to the left a page from an architectural journal. Workspaces such as this one can be saved and shared with others. The material can be re-arranged in many different ways.

The main advantage of the *Manufaktur*, which has been described in more detail elsewhere (Büscher *et al.*, 2000; Mogensen and Grønbaek, 2000) is that it provides an overview of assemblies of (visual) materials and the parameters they represent, thereby helping to account for the complexity of a developing project. Seeing 'the whole thing' and being able 'to zoom into a detail' (of a particular document) and out again, supports the oscillation between precision and fuzziness, which is crucial for design work.

36.3 Ways of Persuasion

The prototypal artefacts described above mediate cooperative design in various ways. Indeed, it could be argued that manipulating these materials is not just a background to the work, but an integral part of it (Büscher *et al.*, 1999). The examples point the ways in which visual and graphical material helps:

- create a common understanding of a design idea or task;
- talk about a design in a rich, metaphorical way, supported by images to be pointed at and referred to;
- imagine qualities of space and appearance, which could not easily be communicated in words;
- act as reminders of design principles, approaches, methods, open questions, *etc.*;
- reserve the memory of a design solution and the arguing behind it.

Some characteristics of these artefacts stand out. One is the wide range of visual materials used in architectural design, from CAD drawings to sketches and inspirational images. Some of the material is precise and detailed but much of it is conceptual, metaphorical and under formation. While some of these artefacts are produced in digital format or converted to it, others are and remain tangible. Architects' valuation of tangible material has partly to do with the qualities of the process – they are created in intense interactions with others, incorporating specific qualities of scale and size, colour, touch, *etc.* Also, preserving the informal and mixed character of visualization of a 'conceptual sheet' may capture the complex reasoning behind a design problem better than a neatly structured task list.

Another characteristic of these artefacts is the intermingling of visual material with text. Many of these artefacts are created as part of talk. Others contain text as well as visual elements. Relating text (in our case metaphors, short notes, *etc.*) and image (sketch, 3D image, a scan of an object, *etc.*) requires multiple translations (Tomes *et al.*, 1998). We saw from the example of the narrative sketch that when there is no text (the voice having faded away), understanding the image may be difficult. These types of visual-verbal relationships are crucial for many design disciplines. As Mitchell argues: ' ... all arts are "composite" art (both text and image); all media are mixed media, combining different codes, discursive connections, channels, sensory and cognitive modes ' (1994, p. 95). So is writing in its graphic form is inseparable from the visual. Conversely, stories are created around images. The growing, metamorphosing idea of a design is represented in the material that is collected and produced, including the talk around it.

Most of the material we analyzed was not 'just' or even primarily created for presentational purposes. It is produced as an intricate part of the process of forming an idea, approach, and method. It helps in taking a step forward in the common understanding of a design. It supports the development, expression, and sharing of ideas with others. If visibly displayed in the architectural office, it is frequently looked at and referred to as part of the ongoing work. Likewise, the scale model used for controlling the spatial aspects of the evolving 2D CAD plans. The 3D images help stretch people's imagination of what the place may look like, before the details of material, colour, *etc.*, have been fixed. The sketch with 'precious stones that glitter under water' reminds people of the design idea. The inspirational objects (e.g., Richard Serra, Chartres) express some of the specific qualities the architects have in mind.

This interweaving of work and representation is visible in architects' use of their office as an exhibition space. The walls are decorated with visualizations from previous and current work. This way of making work visible reminds people of ideas to be pursued or further developed, of tasks to accomplish, of standards, *etc.* It is also is an important vehicle for peripheral participation in a project, allowing other members of the office or visitors to enter its context. In our fieldwork we observed that many of the progress markers of a project - a particular focus of current work, an important decision, the start of construction work, a crisis, or the beginning of a new project - were made visible on the wall.

Latour (1986) has argued that visualizations simultaneously support constructing the artefact, demonstrating its performance and its understanding by others. A particularly relevant aspect of 'persuasive artefacts' is their openness and interpretive flexibility (Lainer and Wagner, 1998). This is why metaphors and 'abstract' rather than 'realistic' visualizations - sketches, association images, 3D images - play a large role in design work. To create persuasion in a conversation an architect may arrange a variety of artefacts at hand to create a distinct impression. Also, assemblies of materials may be quite subtle, such as arranging artefacts within a spatial vicinity, layering or juxtaposing them, without giving a precise definition of their relationship.

The crucial role of 'persuasive artefacts' in design work is that they support open and rich ways of communicating, allowing architects to present their work in a form that invites cooperation and is open to the possibility of change.

36.4 Acknowledgements

This paper refers to several projects of the Architekturbüro Rüdiger Lainer: *Eurocity* (a cinema complex in Salzburg), *Pleasure Dome* (an entertainment centre in the Gasometer area in Vienna), and the *Gasometer Study*, an urban planning project. I am very grateful for the contributions of Monika Büscher, Rüdiger Lainer, Dan Shapiro and the other partners in EU ESPRIT LTR Project 31870 'DESARTE - The Design of Artefacts and Spaces', funded by the European Commission (http://desarte.tuwien.ac.at).

36.5 References

Büscher M, Mogensen P, Shapiro D, Wagner I (1999) The *Manufaktur*. Supporting work practice in (landscape) architecture. In: Bodker S, Kyng M, Schmidt K (eds) Proceedings ECSCW'99. Kluwer Academic Publishers, Dordrecht Boston London, pp 21-40

Büscher M, Christensen M, Grønbaek K, Krogh P, Mogensen P, Shapiro D, Ørbæk P (2000). Collaborative augmented reality environments: integrating VR, working materials, and distributed work spaces. In: Proceedings of Collaborative Virtual Environments 2000. San Francisco

Büscher M, Kompast M, Lainer R, Wagner I (1999) The architect's *Wunderkammer*: aesthetic pleasure and engagement in electronic spaces. Digital Creativity 10:1-17

Henderson K (1995) The visual culture of engineers. In: Star SL (ed.) The cultures of computing. Blackwell, Oxford, pp 197-218

Lainer R, Wagner I (1998) Reichhaltige Erzählungen: Themenbezogenes Arbeiten und die Kommunikation von Qualitäten. Architektur & BauForum 197 (November/December): 138-141

Latour B (1986) Visualization and cognition: thinking with eyes and hands. Knowledge and Society: Studies in the Sociology of Culture Past and Present: 1-40

Mitchell WJT (1994) Picture theory: Essays on verbal and visual representation. The University of Chicago Press, Chicago

Mogensen P, Grønbaek K (2000). Hypermedia in the virtual project room - toward open 3D spatial hypermedia. In: Proceedings of Hypertext 2000. San Antonio, Texas.

Perry M, Sanderson D (1998) Coordinating joint design work: the role of communication and artefacts. Design Studies 19:273-288

Tellioglu H, Wagner I, Lainer R (1998). Open design methodologies. Exploring architectural practice for systems design. In: Proceedings of PDC'98. Seattle, pp. 19-28

Tomes A, Oates C, Armstrong P (1998) Talking design: negotiating the verbal-visual translation. Design Studies 19:127-142

Chapter 37

Collaboratively Designing Within the Design

M.L. Maher and S.J. Simoff

37.1 Introduction

Design studios are places where designers work alone or collaboratively on design projects. Most design studios are now incorporating some collection of computer-based tools for handling electronic documents and communication. Following the traditional office paradigm large amounts of project data files (such as drawings, documents, spreadsheets, databases, manuals, forms, communications, schedules and discussions) move around the studio from one computer workplace to another, where they are processed on the individual designer's 'desktop'. The use of file-server technology is usually reduced to the most rudimentary operations of moving files from one shared disk to another. Sometimes the same information is unnecessarily duplicated, sometimes important files remain either locked on the personal computer or lost somewhere on a barely navigable list of shared directories on a file server.

We have developed and experimented with designers using computer-mediated collaborative environments (Maher, Simoff, and Cicognani, 2000; Maher *et al.*, 1997), which we refer to as *virtual design studios*. Virtual design studios can range from simply using email for project communication to a collaborative virtual world. Different metaphors can provide the conceptual basis for the development and use of a virtual design studio. For example, the virtual design studio can build on the *desktop* metaphor, popular in computer operating systems. The notion of a *place* is another major metaphor used for setting up virtual design studios. In this paper we focus on a case study of the use of the place metaphor in a design project.

Virtual places, which include virtual worlds and virtual reality applications, can be defined as *computer-mediated dynamic world models that create a sense of place*. The Internet has been accommodating more than a dozen different technologies supporting multi-user text-based, and two- and three-dimensional graphical virtual world*s*. When adopting the place metaphor, preparing a virtual design studio is much like designing a physical studio. The studio is set up to facilitate and support collaborative design activities.

One aspect of the virtual place metaphor is the establishment of the identity of the people in the place. In a physical studio, a person's appearance, personality, and knowledge become known through their interaction with others in the studio. This also occurs in a virtual place through the representation of each individual as an avatar or object ('character'[1]) that has various properties. An 'avatar[2]' (Damer, 1998) is a 3D model of the person and shows where they are, where they are looking, and what gestures they want to communicate. Object representations of a person include characteristics such as a verbal description and messages about their movements in the place, while links to web pages and publications help establish their identity and personality. The visual presence of the avatars brings a new dimension to communication in virtual places.

There are several approaches to developing a virtual design studio based on the place metaphor. The two approaches we have considered are the design office and 'desiging in the design'. The design office approach involves the development of a design studio as an office space with a collection of private and public spaces for the individual,s and tools for the management of the documents and models. Designing in the design involves the development of the model of the design as the collaborative environment.

We demonstrate this idea by describing a virtual design office and 'design within the design' scenario in the Active Worlds[3]. Active Worlds provides a central server on which the model of the world is stored. Active Worlds provides an Internet-based browser that allows users to navigate through the built environments of various virtual worlds. Accessed either as a visitor, or as a registered citizen, Active Worlds invites everyone to contribute to its construction by adding new buildings, special features – for example, teleporting tools – and objects. It is then possible to change the shape, colour, texture, location, and dimension of a specific object by using a building interface. The inhabitants of the virtual world can also define new areas, which can be compared to new cities.

37.2 The 3D Virtual Design Studio

Developing a virtual design studio as an office in a 3D-modelling environment involves a consideration of what people do and need in a design office and then translating that into a virtual environment. The desktop metaphor has created a prototype of what people need in a virtual working environment; however it does not create a sense of place or identity of the other people in the place. In Figure 37.1 we show the virtual design office developed as part of our project, and described in (Hong, 1999). The office is located within the 3D-modelling environment and includes a meeting room in the centre, with a walking area around

---

[1] 'Character' as a term is used in text-based virtual worlds.

[2] Avatar is an ancient Sanskrit term meaning 'a god's embodiment on the Earth' (Damer, 1998).

[3] http://www.activeworlds.com

the meeting room for viewing the development of the 3D models for various projects. The workspace environment includes also asynchronous and synchronous communication areas, and Web information area.

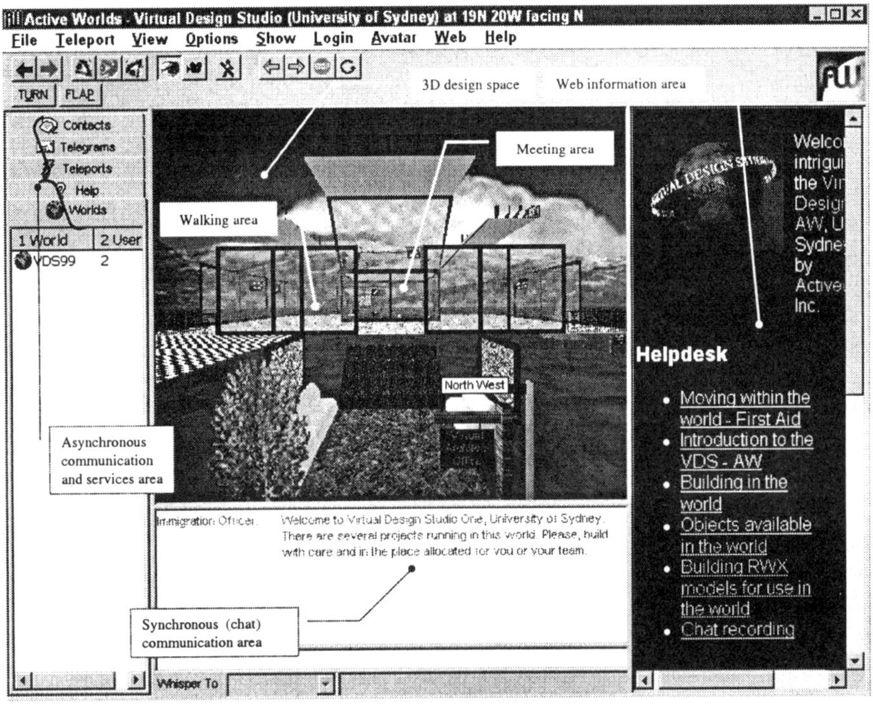

Figure 37.1. A virtual design office

The virtual design studio can also include a model of the product or building being designed. Although this is a relatively new approach, Woo *et al.* (1999) show how such an immersive 3D environment can be used to evaluate design alternatives. The major feature of this kind of virtual design studio is the development of the design within the collaborative, multi-user environment. Designers can work alone or collaboratively building the model and discussing the design as they view the model. There is only one representation of the model so there isn't a problem with simultaneous changes to different versions. There is a *continuum of the process* – a person does not shift environments when designing alone or collaboratively, and there is a *continuum of the workspace* during the design session - all working information about the design is accessed and shared through the same environment. Traditional 3D-modelling environments allow the designer to create a fly-through or animation for visualisation, presentation and discussion, but only a static 3D model can be modified.

Figure 37.2 shows a collaborative project in which the 3D model became the meeting place for the design team. The top left part of the figure is the meeting place with the 3D model of the building design, and the people are represented as avatars. Below the 3D meeting place is the talking interface that is like a chat room environment. The right part of the figure shows web pages that are linked to specific objects in the 3D model.

The project involved the design of a building for the Global Learning Centre at Stanford University. The design team consisted of graduate students in the Virtual Architecture course in the Faculty of Architecture at the University of Sydney, with a group of people at Stanford as the clients. The students were given a design brief outlining the intended use of the building, emphasising the requirement for flexible use of space. The students were also given a description of the existing building that would be modified for new use as a Global Learning Centre. Materials for the project were kept and accessed only in electronic form.

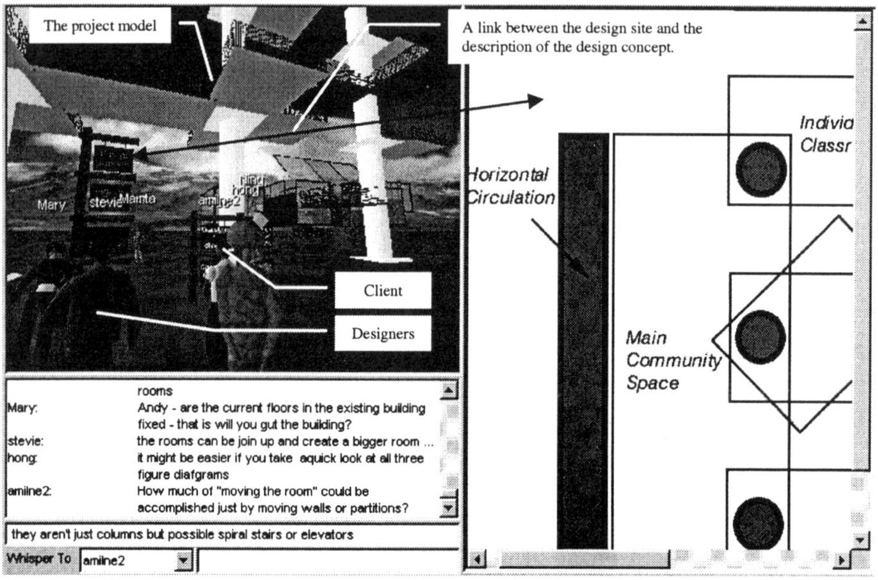

Figure 37.2. The Stanford project design site

There are a number of features that characterise the 'design within the design' in a virtual design studio. The first group are connected with the spatial location of the design elements. Designers organised design objects on the site, introducing simultaneously the spatial relations between the objects and between the designers and the objects. They have a common reference and orientation. The second group is connected with populating the space. Having an avatar representation of each designer 'within the design', collaborating designers are aware of the presence and, to some extent, about the activities of their partners. Collaborators can see and interpret each other's actions, the distances between themselves and the design objects, and finally every modification of the design. These features of the designing within the design may influence the design communication that occurs during design.

37.3 Communication During Team Meetings

Communication in the 3D-modelling environment depended on visual contact with others in the meeting through the avatars and discussion in a chat-like window where people 'talk by typing'. We analysed the discussion in order to characterise the concept of 'designing within the design'. Our analysis used the recorded discussion as a transcript and employed a combination of the coding schemes proposed by Simoff and Maher (2000) and the coding scheme introduced by Gabriel and Maher (1999b). Gabriel and Maher (1999a, 1999b) developed a coding scheme for analysing one hour design sessions in an experimental setting of two designers. Our analysed discussion is slightly different because the design team included more than two people and the team worked together before and after the one-hour discussion. The coding scheme of Gabriel and Maher (1999b) looks at various types of communication in design collaboration: communication control; communication technology; social communication; and design communication. Of particular interest is considering how the collaborative environment influences collaborative communication. In this case study, we extended the coding scheme with two categories: *Communication mode*, with 'Global' and 'Individual' as subcategories, to capture dynamics within a team; and *Navigation*, to capture the interaction with the environment. Another modification of the original coding scheme is the addition of 'Synchronisation' as a subcategory of the *Communication control* category, which depicts moments of synchronisation of the focus of all designers of the team (e.g., 'Can everyone see the concept drawing?').

The session analysed in this case study has 176 utterances. The diagram in Figure 37.3a shows that the designing within the design session is characterised by a high proportion of design communication with respect to the other communication categories. The dominant category in the design communication, as illustrated by the diagram in Figure 37.3b, is the communication of design ideas, combined with high-level (conceptual) design decisions. Gabriel and Maher (1999b) observed similar results in their CMCD-b session where designers used text-based communication. To some extent this means that the 3D presence within the design does not decrease the intensity and concentration of text-based communication, identified by Gabriel and Maher (1999b). The higher percentage of task management communication may be due to the teamwork and the length of the design project in comparison with the one-hour duration of the CMCD experiments in the Gabriel and Maher study.

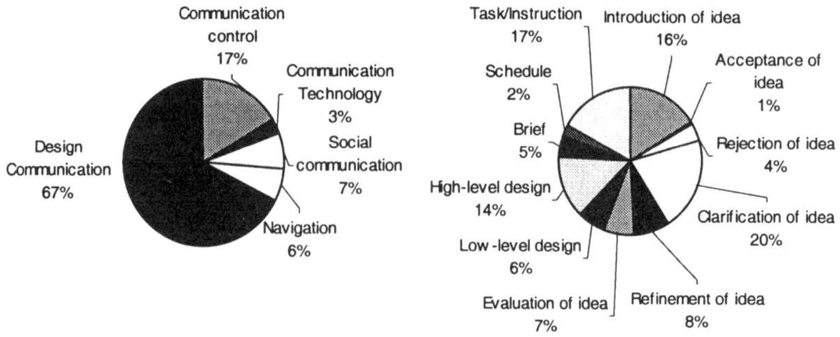

a b

Figure 37.3. Categories and amounts of communication in team meeting: a. across categories; b. design communication

The dynamics of design communications is shown in Figures 37.4 and 37.5. These graphs represent parallel timelines (each time point corresponds to an utterance) for each category of design communication. The graph in Figure 37.4 shows that designing within the design is characterised with fairly intensive introduction and clarification of ideas during almost the whole session. The fairly low final acceptance and rejection of ideas can be explained by the quick visualisation and illustration of the concepts in the 3D-modelling environment. Figures 37.5 shows that the design communication at the end of the session was focussed on task management.

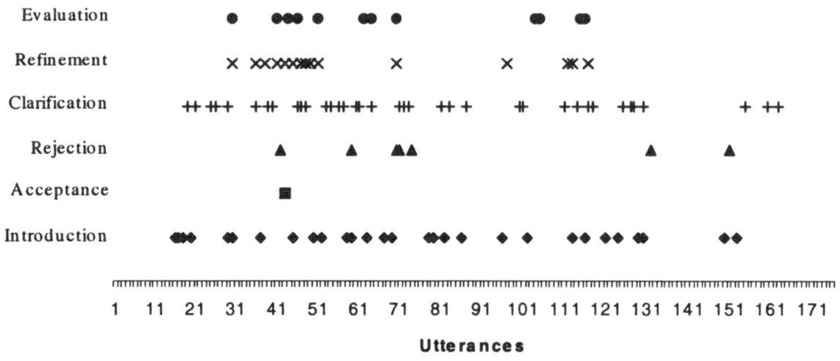

Figure 37.4. Communication of design ideas during the session

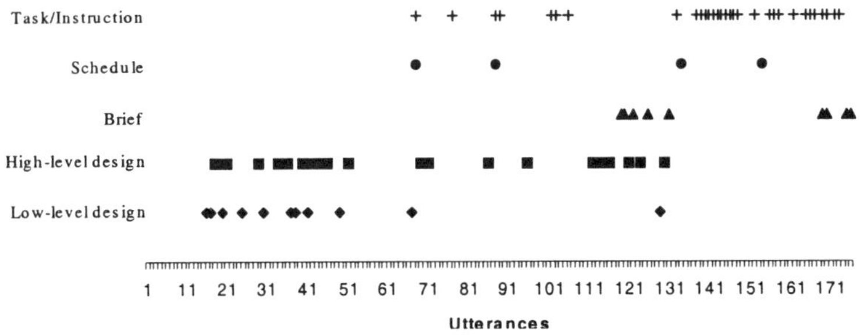

Figure 37.5. Communication relevant to the design scope and tasks

In addition to coding the types of communication, we considered the distribution of communication content by performing a text analysis of the transcript. The major focus of the design team was on the concept of a light construct with enhanced circulation. The list of most frequently used concepts, shown in Figure 37.6a, demonstrates that despite the available visualisation of geometric forms, designers need to explain and refine the semantics of these forms. For example, the horizontal circulation caused a major discussion (indicated by the relatively high frequency of related terms), while the idea of the vertical transportation came across fairly easily from the model (indicated by the relatively low frequency of the relevant keywords).

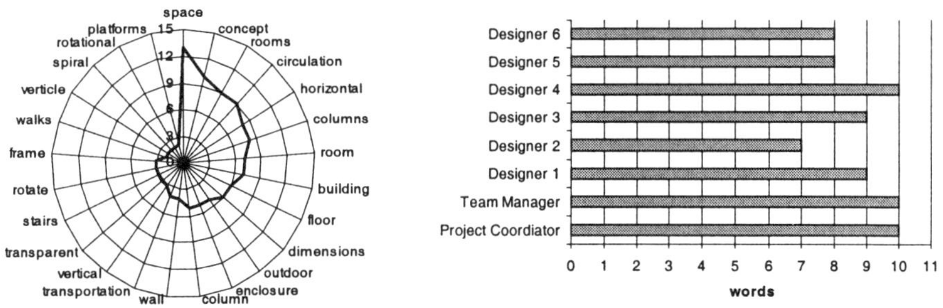

a b

Figure 37.6. Text statistics of the design session utterances: a. key concepts; b. average length of utterance

We expected that being within the design would allow the designers to explain design ideas with fewer words, based on simple references to the objects. The average[4] length of utterances is less than a dozen words. More than a third of the words in an utterance goes to the class of stop words (words that are part of the grammatical form and do not carry semantic meaning). To some extent this

[4] We used median as a central tendency estimator to avoid the biases of occasional very short and/or long utterances.

supports our initial hypothesis; however, further investigation is required for more rigorous conclusions.

We also considered the threads of conversation during the team meeting. Despite relatively short phrases, the communication within the design was fairly focussed. As shown in Figure 37.7 the first half of the session is characterised by low level of local threads between individual designers. The substantial increase in the individual threads in the middle is an indicator of potential asynchronisation of the design session, correctly detected by the project coordinator. The work of the team also had to be synchronised, when the attention of the designers was divided between the model of the design concept in the virtual world and an external illustration of design concepts (utterances 115-120).

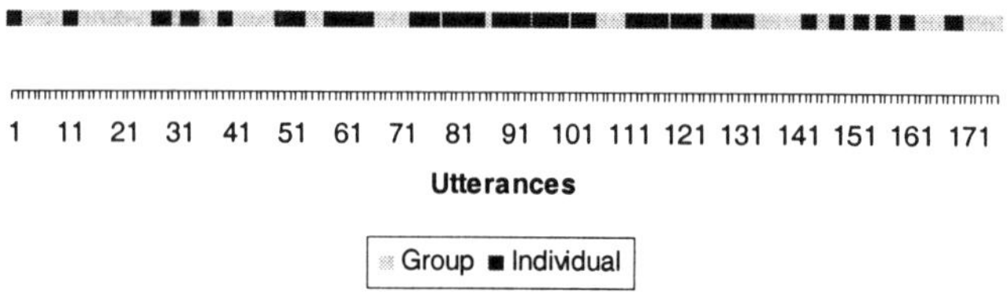

Figure 37.7. Changes in communication mode during the design session

37.4 Designing Within the Design

The use of a 3D model as the basis for a design project is not new. Many CAD systems are now emphasising a shared-model approach to design projects as well as sharing documents relevant to the project. The concept of designing within the design goes further than a shared 3D model by making the 3D model the meeting place for the design team. In addition to the model of the design, the avatars representing the design team are in the 3D environment. In this sense, designing within the design is a substantially different concept from the traditional design approach.

When designing in 3D virtual worlds, the initial emphasis seems to go towards the form, due to the existence of objects with 'prefabricated' forms. The design process is more like building than drawing. The collaborative potential in this type of environment is not well understood. Our observations of this type of environment indicate that the ability for the client to move around the design while discussing the project with the designers made it easier to understand the design. The traditional alternative is to discuss the design while looking at drawings or prepared 3D projections or animations.

In this paper we analysed the discussion that took place in the team meetings. Our initial results show that the discussion in this environment is heavily influenced by the type of talking supported by Active Worlds. That is, the participants talk by typing rather than by speaking. Our results show that there is a large percentage of the communication devoted to the introduction of new ideas, possibly due to the lack of interruptions in a talk-by-typing environment. We also noticed that the discussion was primarily focussed on high level decisions rather than the details of the design model.

An important conclusion is that the development of 3D collaborative modelling environments does not restrict the ability of the participants to have successful discussions about the design. The existence of the participants in a 3D model of the design enhances the design communication environment and allows changes to be made and visualised immediately. The added value of having the participants represented as avatars moving around the model is the ability to see where another person is looking while discussing the design.

37.6 References

Cicognani A, Maher ML (1998) Two approaches to designing virtual worlds. In: Proceedings of Design Computing on the Net 98. International Journal of Design Computing, http://www.arch.usyd.edu.au/kcdc/journal Vol 1

Damer B. (1998) Avatars. Peachpit Press, an imprint of Addison Wesley Longman.

Greenberg S and Roseman M (1998) Using a room metaphor to ease transitions in groupware. Research Report 98/611/02. Department of Computer Science, University of Calgary, Calgary, Alberta, Canada

Gabriel G, Maher ML (1999a). Does computer mediation affect design representation? Design Thinking '99. Cambridge, MA

Gabriel G, Maher ML (1999b) Coding and modeling communication in architectural collaborative design. In: Ataman O, Bermudez J (eds.) *ACADIA '99*. ACADIA, pp152-166

Hong K (1999) Architectural Design in Virtual Worlds. Honours Thesis. Faculty of Architecture, University of Sydney

Lin T-H, Protzen J-P (1997) Desktop design: a toolkit approach to collaborative design. In: Maher ML Gero, J, Sudweeks F (eds.) Formal aspects of collaborative design. University of Sydney, pp 237-251

Maher ML, Simoff S, Cicognani A (2000) Understanding virtual design studios. Springer-Verlag, London

Maher ML, Cicognani A, Simoff S (1997) An experimental study of computer mediated collaborative design. International Journal of Design Computing (http://www.arch.usyd.edu.au/kcdc/journal), Vol 1

Simoff SJ, Maher ML (2000) Analysing participation in collaborative design environments. Design Studies (to appear)

Woo S, Lee E, Sasada T (1999) Shared virtual space for evaluation of design alternatives. In: G Jingwen and W Zhaoji (eds) *CAADRIA'99*. Shanghai Scientific and Technological Literature Publishing House, Shanghai, pp 89-95

Part IV Mediated Communication

Summary

In the introduction, we identified a number of themes underlying this Part: the role of non-verbal communication in supporting verbal communication; the collaborative functions of artefacts; the potential of novel modes of visual interaction; the impact of design process factors; and approaches to the study of mediated design communication. In the following we will examine each of these issues in the light of the findings reported in this Part.

First, there is the question of whether and how visual workspace activity, *i.e.*, non-verbal communication, contributes to verbal communication. Drawings and other artefacts are produced and used so routinely during design that it would seem self-evidently the case that visible workspace activity is necessary to, is supportive of, and adds value to verbal communication. Therefore, Reed and Reid's results may surprise many people, for they find that collaborating designers are not only able to successfully complete tasks when denied visual access to each other's workspace, but that they also do so without significant reduction in performance as compared to designers working with visual access. So, for the tasks studied at least, it would appear that verbal communication is sufficient for effective collaborative design. Nevertheless, Reed and Reid also show that drawing and pointing have a value for both communicator and addressee in supporting conversational grounding in design communication. Likewise, Scrivener *et al.* describe a number of ways in which meaning is conveyed through the combination of non-verbal and verbal communication. At the very least, design communication is probably less effortful when non-verbal and verbal communication are used in conjunction. But does non-verbal communication 'add value' to verbal communication: are there some meanings that could not be conveyed at all using verbal communication alone? The answer to this question is probably, but we will have to look to future research to answer this question conclusively.

So far we have considered the communicational functions of visual artefacts in collaborative design. Additionally, visual artefacts serve cooperative functions. In other words, they support shared activity in specific ways, and in ways that are distinct from, and additional to, the ways in which they support individual activity.

Reed and Reid, for example, show how drawings contribute to shared-understanding by supporting conversational grounding. Logan and Radcliffe argue that three-dimensional artefacts contribute to collaborative working by initiating discussion and by verifying information and understanding. They also argue that artefacts may provide a common frame of reference; a physical locus that the designers can 'talk around'.

Wagner identifies artefact-functions that go beyond the dynamics of particular group design sessions. It has often been observed that drawings, for example, combine words and image: that they meld together aspects of what was shown and what was said in their production. Wagner emphasises a 'discourse' that is only partly evident in the artefacts of designing. She stresses how these artefacts are residuals of the design discourse, and how, as residuals, they embody the 'story' of a design project: the ideas presented, the arguments won and lost, the points of agreement, the inspirations, the current state of thinking, *etc*. In this sense, when displayed around the studio, they act as points of focus for inducting others into the evolving 'story' and as reminders and re-enforcers of shared understanding and commitment. These uses of design process-artefacts serve to emphasise some of the ways in which things often produced at a particular moment to serve a particular purpose are artfully used and re-used by designers to progress the design process.

Most of the above relates to those elements in the cited papers that report studies of 'conventional' workspace activity. These papers all share a common motivation for undertaking that work: they are all interested in how the knowledge uncovered might be employed in developing computer mediated communication and collaboration tools. (Indeed, this is a goal shared by all of the paper in this Part.) Some authors look forward to technology that supports modes of collaboration that go beyond those currently available. Pursuing the visual theme, Wagner describes a tool for constructing and interacting in new ways with 'assemblies'. Maher and Simoff argue that their implementation of the concept of 'design within the design' enables the simultaneous introduction of spatial relations between objects and between designers and objects. Furthermore, the representations of each designer 'within the designer' as an 'avatar' enables collaborating designers to be aware of each other and of each other's activity. While neither paper claims to confirm these 'benefits', they remind us that new technology may enable new and enhanced forms of visual communication and interaction.

The issues discussed up to this point are associated with moments of intense interaction, and the communication and collaboration occurring during these events. However, typically, design projects are completed over a period of time during which other factors may be of equal or greater importance than communicating, producing and interacting with workspace artefacts. For example, Logan and Radcliffe argue that preparation and education are necessary to achieving a common frame of reference in computer-mediated rehabilitation engineering. Likewise, Woodcock *et al.* show that many of the problems in international collaborative design projects are not due to communication *per se* (e.g., time management, specification of the brief, project management and the involvement and commitment of team members). They too stress the need to look beyond communication and workspace interaction. While they recognise that some of these problems may be ameliorated by the development of design support systems and methods, they also argue that the issues are ones whose successful resolution is dependent on managerial and educational intervention. In short, these papers remind us that communication and workspace interaction of themselves may not be the main barriers to efficient and effective mediated collaborative design.

As noted above, none of the papers in this Part is concerned exclusively with understanding for understanding's sake. Each has an implicit or explicit teleological motivation: the development of computer mediated collaborative design tools that either improve on existing telecommunication tools or that match, or go beyond, non-mediated modes of working, e.g., co-located and face-to-face working. Although the goal is often simple enough to state, finding the best means of achieving it is usually a less straightforward matter. Initially, researchers compared behaviour: for example comparing drawing and gesturing in face-to-face working to, say, a videoconferencing and shared drawing surface mediated design environment. Similar patterns of behaviour in each environment might be enough to conclude that mediated communication does not impoverish interaction. However, what does it mean if patterns of behaviour are different? Reed and Reid's work would suggest that it does not mean that mediated communication necessarily impairs performance, e.g., designers in their without-visual-access condition appear to adapt to achieve a similar level of performance in their with-visual-access condition. Behaviour and performance do not appear intimately coupled.

More recently, researchers have focused on analysing what it is that is communicated. Gabriel and Maher, Reed and Reid, Scrivener *et al.*, Logan and Radcliffe and Maher and Simoff's papers all include schemes for coding data in terms of communication content. However, while such coding schemes preserve more meaning than activity-based coding schemes, interpretation is still not straightforward. Gabriel and Maher's paper illustrates this point. In their study, they find the main differences between the face-to-face and talk-by-typing conditions to be in communication-control (*i.e.*, a lower percentage in CMCD-b compared to the other conditions) and design-communication (*i.e.*, a higher percentage in CMCD-b). If we assume that interruptions are negative and design-focussed utterances are positive interventions, this could be taken as suggesting that the CMCD-b environment enhances design communication as compared to FTF. On the other hand, if most interruptions are positive, then design might be impoverished in CMCD-b as compared to FTF. Gabriel and Maher present evidence that can be seen as supporting either one or the other interpretation. While performance measures might have assisted in determining which environment was the most effective in this instance, Reed and Reid's work reveals one of the problems with them: the difficulty of finding measures that are sufficiently discriminating over the kinds of task and task duration typical of such comparative studies. Furthermore, performance measures do not in themselves help that much in determining why one condition is better than another. This requires a theory of effective and efficient collaborative design (e.g., one that would allow interruptions to be understood as impairing design performance): in other words, what is needed is an understanding of communication and cooperation needs. One final pitfall when comparing differences between communication environments is worth mentioning. In Reed and Reid's study the behaviour of participants between conditions is quite different but their performance is similar. If we assume that performance is contingent on the satisfaction of communication and collaboration needs, then this would suggest that different linguistic (or other communication) mechanisms may be employed to satisfy them. Reed and Reid recognise this when they suggest that participants might employ alternative techniques for grounding. Thus, using Gabriel and Maher's data for the purposes of illustration, the needs satisfied by communication control might be transferred to design communication. Hence, changes could be due either to inability to satisfy needs or to the satisfaction of needs using other techniques.

So far we have focused on approaches aimed at matching or improving face-to-face design communication In contrast, those interested in enhancing existing mediated communication need not worry about face-to-face communication. Instead, they can focus on the problems that occur under mediated communication and on ways of removing these problems and thereby enhancing performance. Woodcock *et al.*'s work is an example of this approach. As mediated design communication increases in frequency and amount we should expect such approaches to become more prevalent. However, here too, there are a number of issues to bear in mind. The first is to find ways of getting below surface problems. For example, many of the deficiencies identified by Scrivener *et al.* do not appear as problems as such, but as on-the-spot user adaptation. The second is to find ways of assessing the relative importance of the identified deficiencies. The third is to avoid becoming so fixated on limitations as to be become blind to possibilities.

Finally, returning to an earlier part of the discussion, perhaps the most damning criticism of telecommunications design that relies on face-to-face communication as a model is that it both ensnares us in the past and blinds us to the future. Surely new technology will open up new and enhanced ways of communicating. We have seen in a number of the papers in this Part how designers readily adapt to achieve successful outcomes in novel communication and collaboration environments, some of which offer functionality that would not otherwise be possible (e.g., 'design within the design'). While we would not advocate forcing designers to accept deficiencies in the design of such environments, to insist that designers should not have to adapt would be quite wrong. For it is this adaptability that enables the new to be absorbed. By definition old ways of working will be inadequate for novel situations. The ability to change and adapt is fundamental to the successful handling and exploitation of novel situations. If the potential offered in a new communication environment is worth the effort designers will adapt, even if this means that some work becomes more effortful as a consequence. The design goal should be to make such systems as effective and efficient as possible at all times. This is why we will need to develop methods, such as that proposed by Scrivener *et al.*, that address communication in mediated environments, and that enable communication needs to be uncovered through the use of these environments and in a form that contributes to improvement through redesign.

Part V

Computer-mediated Collaborative Design

Part V Computer-mediated Collaborative Design

Introduction

Collaboration is recognised to be an essential aspect of contemporary, professional design activity, but its success depends crucially on the effective sharing of knowledge between team members, clients and potential product users. Such knowledge may be verbally-based (e.g., relating to design ideas and associated design requirements and constraints) or visually-oriented (e.g., idea sketches and CAD images). In addition, the communication of such knowledge between team members may be re-enforced by a variety of non-verbal cues and gestural acts. Part IV explored important issues relating to the nature and support of communication in collaborative design teams, with a particular focus on accommodating designers' communication needs when they are collaborating at a distance through technological media such as video-conferencing systems. This Part of the proceedings continues the theme of technology-mediated collaborative design but with a more specific focus on *computer*-mediated activity in networked design groups. Issues that are considered concern not only the optimum way to support effective communication between distributed designers, but also the important socio-cultural implications of distributed - and at times global - collaboration (such as the need to accommodate differing values, preferences and practices) that have been highlighted in previous Parts of this volume.

The first contribution, by Seitamaa-Hakkarainen, Lahti, Muukkonen and Hakkarainen, examines how the collaborative design activity of university students may be supported by a networked learning environment. The research is motivated by an assumption that computer-based, collaborative technology may significantly facilitate collaborative designing. The technical infrastructure for Seitamaa-Hakkarainen *et al.*'s study involved an in-house, networked environment (termed FLE) that incorporated tools for discussing ongoing design work and for sharing conceptual and visual ideas. Data were collected from design groups made up of first-year students who used this collaborative design environment over a 15-week period on a task requiring the development of clothing for premature babies. Data were examined using qualitative content analyses to determine the key design-process characteristics of the teams.

The contribution by Kolarevic and Ng maintains a focus on student designers but moves away from computer-mediated activities within local networks, to the collaborative activities of designers working over the web in geographically distributed teams. The virtual design studio (VDS) described in this paper is at the centre of an ongoing investigation by the authors of remote design collaboration that places a particular emphasis on the issues surrounding collective design authorship. In the reported study, students in Hong Kong and Singapore worked on an architectural design project over a single semester using traditional and digital media, a central database, the web and a video-conferencing system. Data derived in this study included survey-based feedback and progress-monitoring of the status of the VDS database. Such measures were aimed at enabling assessments to be made about the way in which design ideas had evolved and about the collaborative processes arising in computer-mediated design.

Like the previous contribution, the next two papers are also interested in issues of authorship in collaborative design and the potential of web-based tools to support dispersed design teams. Rodgers, Caldwell and Huxor's paper outlines a novel theoretical approach to the development of such tools based around a publishing metaphor ('AI as writing'), where tools are viewed as 'knowledge media' for the expression and communication of information. Rodgers *et al.*'s AI-as-writing approach has been implemented as a web-based support system for conceptual design (WebCADET) which facilitates knowledge capture and design transparency. Designers are able to write, amend and share knowledge (in the form of 'rule-texts') relating, for example, to ergonomic or cultural decisions about consumer products. Rule-texts that are edited by different 'authors' (e.g., to make design solutions appropriate for different markets) are stored as different versions in the system database.

The contribution by Zaychik, Sevy and Regli presents an overview of a virtual collaborative design studio (termed DCDS) which aims to support the computer-aided design of mechatronic artefacts by decentralised, collaborative design teams. The DCDS environment encompasses tools that allow design engineers to pursue the collaborative development of design solutions and to archive the design process within a design-knowledge repository. This latter aspect of Zaychik *et al.*'s DCDS system links with current research on *design rationale*, which concerns the documenting of the underlying reasoning behind the design of an artefact to aid better understanding of design issues as well as design maintenance, redesign and design reuse. The design rationale approach embedded in Zaychik *et al.*'s system is communication-centred (e.g., relating to email exchange) and is based around the automatic indexing and storing of design information. The authors illustrate the use of DCDS by describing three design scenarios in which the system can assist in the collaborative design process. These scenarios address three distinct aspects of collaborative design, that is, communication, access to the design-information repository, and adaptive design reuse.

Kälviäinen and Pontecorvo's contribution also emphasises the potential for digital technology to enable the creation of useful repositories of design information. Their paper, however, broadens the focus of this Part to issues surrounding computer-mediated collaboration between designers and the actual consumers of design products. Kälviäinen and Pontecorvo's research hinges on the idea that a valuable way to integrate consumers' intuitive preferences about the style and image of products within the design process is to assess their responses to a sizeable array of product-image variants created by a generative software system. The product-image generation system that these researchers present allows the designer to define a set of parameters (e.g., relating to aesthetic, structural and cost factors) that provide basic constraints on instances of the artefact. Consumers can then be presented with product variants that they are able to express preferences about. The pattern of a consumer's exploration of the space of design variants within the system enables the designer to draw inferences about how best to customise the product concept for different market niches.

Chapter 38

Collaborative Designing in a Networked Learning Environment

P. Seitamaa-Hakkarainen, H. Lahti, H. Muukkonen and K. Hakkarainen

38.1 Introduction

The purpose of the present exploratory study was to analyse how university students' collaborative design process may be supported by a networked learning environment. Designing is generally considered to be a form of complex, problem solving (Goel and Pirolli, 1992; Goel, 1995). Designing is an iterative process by nature; *i.e.*, design solutions emerge gradually as a process of structuring and restructuring, composing and decomposing the problem, defining and redefining constraints of designing, and generating and testing design solutions (Akin, 1986; Goel, 1995). Design constraints form the design context by defining, for example, the intended users and their special needs for the artefact, and the function of the artefact. Further, designing may be conceptualised as a process of searching through two problem spaces, *i.e.*, composition space (visual designing) and construction space (technical designing) (Seitamaa-Hakkarainen, 1997; Seitamaa-Hakkarainen and Hakkarainen, in press a). The visual and technical elements involved in designing must be considered and related to each other, within the constraints of the project in order to create a functional and aesthetic solution. Cognitive research on expertise indicates that novices tend to generate problem solutions without engaging in extensive problem structuring whereas experts focus on structuring and restructuring the problem space before proposing solutions (Glaser and Chi, 1988).

Design processes have been intensively studied by cognitive researchers during the past three decades (e.g., Akin, 1986; Akin and Lin, 1995; Goel and Pirolli, 1992; Goel, 1995; Goldschmidt, 1995; Seitamaa-Hakkarainen, 1997; Seitamaa-Hakkarainen and Hakkarainen, in press a; Seitamaa-Hakkarainen and Hakkarainen, in press b). Only a few previous studies specifically addressed collaborative aspects of designing. Recently, however, more and more studies strongly emphasise importance of analysing collaborative design processes (Cross and Cross, 1995; Perry and Sanderson, 1998; Valkenberg and Dorst, 1998). The relative importance of research on collaborative designing is partially facilitated by cognitive research that provides the users with tools for collaboratively building and sharing knowledge and solving problems together (Scardamalia and Bereiter, 1994). The emergence of computer-supported environments for collaborative learning and work encourages the exploration of possibilities for teamwork in designing.

Collaboration is an essential aspect of professionals' practical activity in the field of modern design (see, for example, Ferguson, 1992). An essential aspect of the professional design process is to share knowledge, such as design ideas, visual sketches and interpretations, among members of a design team. Moreover, designers frequently search for and use information from other experts or other related disciplines to help them to construct new knowledge related to the design topic. Social collaboration plays a particularly important role during the conceptual phase of designing, *i.e.*, while generating and articulating design through searching new information that would help to determine design constraints and produce a satisfactory design.

The present study is based on an assumption that the new collaborative technology may significantly facilitate collaborative designing. A fundamental aspect of the design of computer-supported collaborative learning environments is to provide users tools for posting knowledge into a shared working space, and provide users tools for progressive discourse interaction between the users (Scardamalia and Bereiter, 1994). The technical infrastructure of the present study was provided by a networked learning environment, called the Future Learning Environment developed by the Media Laboratory, University of Art and Design, Helsinki, and the Department of Psychology, University of Helsinki (Leinonen *et al.*, 1999). The FLE-Tools is designed to facilitate experts working with knowledge in collaborative design. FLE-Tools engages students in generating new design ideas and working collaboratively to develop their ideas. The system provides tools for discussing ongoing design processes as well as for sharing conceptual and visual design ideas.

FLE-Tools provides tools for collaborative designing, so that multiple actors can work and communicate asynchronously through visual representations in the virtual design environment. The users are able to access it from any internet-linked computer and post ideas and thoughts to the FLE-Tools database directly or using their standard office applications and productivity tools. The ‘knowledge’ posted may be in various formats, such as text, graphics or video. The environment consists of several modules that facilitate collaborative knowledge-building and aid the design process. It provides ‘thinking tools’ for organising and structuring these processes.

The main FLE-Tools modules include the Virtual WebTop, the Knowledge-Building Module, the Jam Session Module and the Library. The Virtual WebTop refers to a personal adjustable display window, which is automatically opened as the user logs onto the system. The Virtual WebTop contains graphical links, represented by folders, that provide information about the project and links to the Virtual WebTops of the other participants of the group. The Virtual WebTop is a place for the designer to store his or her own documents created by standard office applications in various formats.

The Knowledge-Building Module is a shared space for discussing topics, design ideas, and tentative solutions generated by the designer. All KB messages within a project are posted to the shared space, visible as lists of messages. Each KB discussion is accessible only to the users enlisted as participants of that specific project or course. In the conceptual phase of designing, KB supports generation, externalisation, and explication of design ideas, and the determination of the design constraints. A designer or student preparing a note for posting is asked, to facilitate the goal of knowledge-building, to label his or her note by choosing a Design-Thinking-Tools label (*i.e.*, design problem, working idea, comment, deepening of knowledge, summary, metacomment, and help) corresponding essential aspects of design process (Seitamaa-Hakkarainen *et al.*, in press c.). These scaffolds are designed to encourage students to engage in expert-like designing. By categorising the design notes posted to KB-module, the user is guided to specify his or her design idea, generate and articulate multiple working ideas and evaluate those critically, search for new information, comment on the fellow users' design process and share the whole process with the other members of the design team in question.

FLE- Tools contains the also Jam Session Module that is designed to facilitate collaborative designing by providing a space for collaborative construction of digital artefacts. The Jam Session Module supports asynchronous communication with drawing and models. The module enables a graphic representation of an object and, thereby, assists in making thinking visible. Designers may import their own digital artefacts (e.g., pictures, videos, text and software) to the session and let other users further elaborate their ideas and comment on the work in progress. Moreover, the other designers or students can continue and develop the imported design sketch, on the condition that all users use the same graphics software. In its simplest use, all visual documents are stored in Jam Session and it provides a means for recording and storing all the changes to the documents (*i.e.*, portfolio).

Finally, the Library contains course or project materials chosen by the tutor of the project. The Library is intended to provide tools for helping teachers and students to create digitised study materials, to collect interesting link addresses for folders, and also gain access to other libraries linked to the Internet.

The present study focuses on analysing first-year university students' collaborative design process as it occurred in a complex and authentic design task *i.e.*, designing clothing for premature babies. It explores how students worked in a learning environment designed for sharing and jointly advancing their design knowledge, determining and applying design constraints. The problem addressed in the study was to analyse whether and how working in the network environment supports participation in a collaborative design process.

38.2 Method

38.2.1 Research Procedure and Methods of Data Analysis

The data was collected from a 15-week design project entitled 'Clothing-design Project for Prematurely Born Babies' where students use the FLE-Tools - environment throughout the project. In the project the participant comprised 34 first-year, textile students teachers at the University of Helsinki, who had never before participated in a collaborative design project or used computer-supported learning environments.

The students were divided in three FLE groups, each consisting of approximately ten students. Further, the project was organised so that the students worked in three or four student design teams within each FLE course. The students were asked to design and produce functionally and aesthetically delightful clothes for premature babies. The entire design project consisted of four basic stages: 1) knowledge building and visual designing; 2) prototype making; 3) testing of the prototype (in a hospital); and 4) production of the clothing. In the first four weeks, the students worked independently, building knowledge of neonates, especially those prematurely born, by elaborating their design ideas, sharing new knowledge among their team members, and specifying their design problems. Each member of the design team designed one individual prototype that was tested and evaluated in a hospital environment. The end of the project had produced 200 pieces of clothes. The entire Knowledge-Building database consisted of over 350 knowledge-building messages, involving approximately 1200 design statements. The present paper focuses on analysing a part of the database that was produced by three design teams (nine students) during the stages of knowledge building and visual design (12 weeks).

The knowledge posted by the students to FLE's Knowledge-Building database was analysed using qualitative Content Analysis to examine the nature of the team-design process. The knowledge-building messages posted to the database were segmented into propositions representing one main idea (Chi, 1997). Two independent researchers were used to assess inter-coder reliability of the segmentation. The inter-coder reliability of segmentation was satisfactory (.88). Each segment representing the main meaning was coded and analysed using the computer program (ATLAS/ti) for qualitative content analysis. The students' productions posted to the FLE-Tools database were classified according to several variables that represented essential aspects of collaborative designing. The classification schemata consisted of the following categories: 1) design thinking; 2) design process phases; and 3) process co-ordinating. The design content variables consisted of 1) constraints; 2) composition; and 3) construction design.

The design-thinking category consists of: specifying design context; working idea; comment; new information; and metacomment. By specifying design context the design problem was determined (e.g., Does anyone know how big (size) prematurely born babies usually are?). Design progresses by articulating and elaborating different working ideas through which the final design emerges (e.g., I have been sketching a body-type of clothes for 35-cm babies). During the design process students also searched and provided new information concerning designs context or design elements (e.g., I found some nice pictures from www.premierstore.com. Go and look). Through metacomments these students assessed whether the design process was progressing in the desired direction, whether the appropriate methods were being used and shared in the accomplishments of members of the design team. Figure 38.1 presents classification schemata used in this study.

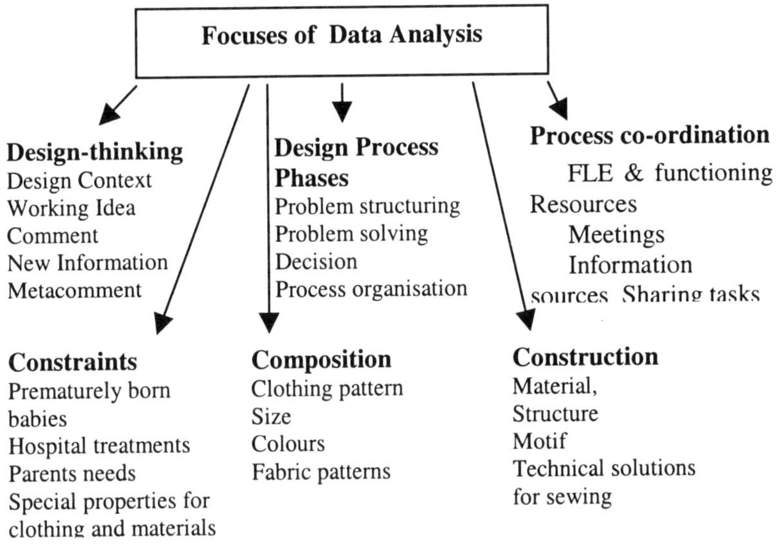

Figure 38.1. The classification schema of the design process

We used the design-process phase category to assess the students' participation in the different phases of design problem solving that are generally present in the design process. The category consists of problem structuring, problem solving, decision phases, and process organisation variables. We were particularly interested in whether the students would engage in intensive problem structuring and careful consideration of design constraints rather than producing immediate solutions, such as novices frequently do. Further, we were interested in whether students co-ordinated their collaborative design process and if so how they did it. Process co-ordinate variables included aspects about FLE and its functioning, resources (e.g., time and money available in the project), arrangements of meetings, information sources and sharing the tasks.

In order to examine how the students processed constraints, and what was the role of the visual and technical design elements in their designing we created three design content categories: design constraints; composition; and construction design. These variables refer to the actual design content: 1) design constraints (*i.e.,* babies, hospital treatments, parents needs, special properties for clothing and materials); 2) composition design (*i.e.,* clothing pattern, size, colours, fabric patterns), 3) construction design (*i.e.,* material, structure, motif, technical solutions for sewing seams and bottoms, for example, and sewing order). On average, the inter-coder reliability coefficient of classifying the data across these six variables was .84.

38.3 Results

An examination of FLE-Tool's database indicated that overall the students engaged in rather intensive work in the learning environment. We conducted detailed analyses for three different design teams in one course. These groups appeared to represent a significant degree of participation. The entire database of these three design teams consisted of 302 design statements. Table 38.1 shows the distribution of design thinking in three groups' design process.

Table 38.1. Design Thinking in three teams' design process

Variables	Team A		Team B		Team C		Total	
	f	%	f	%	f	%	f	%
Design Context	9	10	4	5	29	22	42	14
Working Idea	27	31	17	21	37	27	81	27
Comment	12	14	19	24	19	14	50	17
New Information	22	26	31	39	37	27	90	30
Metacomment	16	19	9	11	14	10	39	12
Total	86	100	80	100	136	100	302	100

An examination of the teams' productions indicated, that their notes frequently consisted of statement that represented different categories of design thinking. As much as 30% (f=90) of the statements produced by the three design teams represented new information; 27% (f=81) of statements represented working ideas; and 14% (f= 42) of the statements defined the design context. Moreover, the metacomments concerning the process were as little as 12% (f=39), and comments, 17% (f=51) of the statements produced by the three teams. In design-context statements students asked for information about premature babies, such as what kind of treatment the babies received, and in what kind of positions the babies sleeps, for example:

```
Do the premature babies turn around while sleeping? Do
they lie down on their back or on their stomach?
```

The students provided and shared new information about premature babies which they found from different information sources, such as, from the internet web-pages, books and research papers, parents of premature babies, their relatives or experts, for example. The working ideas were directly applied to the designing of clothing or other design elements. The comments were added to the other students' working ideas or evaluations of information acquired. In metacomments students evaluated how the design process was progressing, and how each of the students had dealt with the shared tasks, for example, in the following way:

```
We should send more messages. But now there is so much
other works going on at the same time (embroidery
course) but very soon we should concentrate more on
this, ok?
```

Figure 38.2 demonstrates how the different forms of the design thinking progressed during the 12-week period of designing. The students acquired new information as well as structured design context in thefirst six weeks However, the proportion of new information statements decreased whereas the working ideas increased substantially toward end of the period.

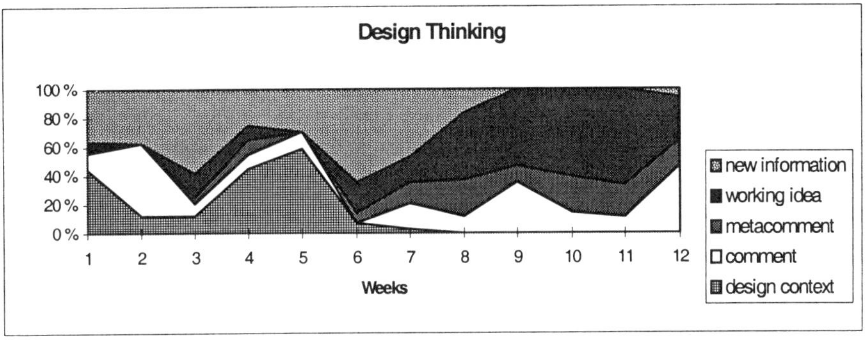

Figure 38.2. The design thinking during the12-weeks period

Further, in order to obtain comprehensive information about how the design was progressing the students' design statements were analysed in relation to different phases of their design process. The analysed database revealed that the three design teams most of their time structuring the design context, since 46% (f=141) of the statements posted to the database represented problem-structuring activity; 23% (f=68) represented problem solving; and only 10% (f=30) represented decision-making. The design teams focused on organising their design process in 21% (f=62) of the statements. Table 38.2 shows the distribution of the design process statements in the three teams' designing.

Table 38.2. Design process phases in three teams' designing

Variables	Team A		Team B		Team C		Total	
	f	%	f	%	f	%	f	%
Structuring	36	42	38	47	67	49	141	46
Solving	18	21	19	24	31	23	68	23
Decision	12	14	6	8	12	9	30	10
Process organisation	20	23	17	21	26	19	63	21
Total	86	100	80	100	136	100	302	100

A detailed examination of the students' design activity indicated that during problem structuring the students mainly dealt with new information. In problem solving, the statements represented students' own working ideas (*i.e.*, possible design solutions to different design elements). The statements presented in the decision-making phase represented mainly working ideas (*i.e.*, decision about the design solutions). The process organisation variable clearly distinguished from other variables, because in those statements students did not handle any design ideas or design constraints, but rather organised their working methods. For example, they share tasks in the following way:

```
Should we adopt a collective way of working or should
everyone do what they want and we'll see happens when we
meet?
```

Further, we were interested in whether students co-ordinated their collaborative design process and how they did it. The co-ordination of the process was as much as 35% (f=120), indicating that the students made a lot of effort to organise their actions and design methods of designing. In their co-ordinating statements students considered aspects of the FLE-Tools environment (difficulties in scanning pictures in Jam –module), resources of the project (money and timing), plans for searching new information from different sources, and also arranged face–to–face meetings.

From Figure 38.3 we can infer that the problem structuring activity was very intensive for eight weeks, and problem solving started from very beginning but intensified as problem structuring decreased. Decision-making as well as the process organisation increased toward end of the period. In the problem-structuring phase the students clarified the design context and design constraints, defined requirements for clothing and analysed the needs of the neonate's parents for clothing. Figure 38.3 represents how the design process progressed during the 12 weeks.

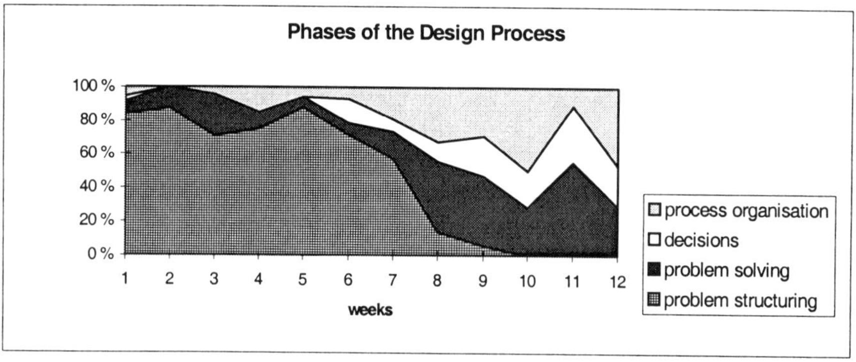

Figure 38.3. The design phases during the 12-weeks period

Finally, we examined whether the learning environment would facilitate consideration of design constraints in the participants' designing. The analysis revealed that design constraints played a significant role in the students' designing: 24% (f=81) of the design statements dealt with design constraints. The composition design composed 22% (f=75) of the statements in which they considered the visual aspects of the clothing (*i.e.*, colour, size). The construction design (*i.e.*, technical design aspects of the clothing) were handled in 19% (f=64) of the design statements. These findings indicated that the student used FLE-tools quite actively to co-ordinate the design process as well as to construct and share knowledge of the design constraints related premature neonates and special aspects related to their clothing.

38.4 Discussion

We examined how the FLE-Tools environment supported collaborative designing through organising a pilot course in which teams of first-year students of textiles solved an authentic and complex design task. The study focused on providing support for further development of FLE's design tools through detailed analyses of a limited number of cases. Since the study involved intensive investigation of a small number of students the data does not support the drawing of comparative conclusions. In particular there is no direct evidence permitting the assessment of the degree to which students' performance in the design task was better than in some alternate context.

The study indicates that all of the students analysed in this study actively provided and acquired information in the shared space and engaged in reciprocal interaction by answering and asking questions. An examination of the role of new information during problem structuring indicates that the students expanded their understanding of the constraints of designing clothing for premature neonates and shared their cognitive achievement through the database. The team members appeared to work in a very close collaboration.

The qualitative content analysis of the students' productions posted to the FLE's database indicates that knowledge produced by the participants represented prototypical elements of design process. The participants' design activity for instance, closely resembled corresponding processes revealed by protocol-analysis studies (see Akin, 1986, Akin and Lin, 1995; Goel, 1995; Seitamaa-Hakkarainen, 1997). The analysis suggested, further, that the iterative nature of designing was typical for the participants who generated their solutions through a sustained process of developing and testing design ideas.

Working in the FLE-Tools environment appeared to facilitate expert-like designing in the sense of engaging students in intensive problem structuring and in-depth analysis of design constraints. The results of this study suggest that by working in a networked learning environment for solving an authentic design task, students' engagement in problem structuring may be considerably facilitated. In the case of the students examined, collaborative technology appeared to facilitate the conceptual stage of designing through encouraging participants to engage in a sustained process of knowledge acquisition and problem structuring focused on determining and responding to design constraints. Participating in knowledge-building discourse in the FLE-Tools environment appeared to guide the students to use and intensively search for various kind of information sources in order to determine design constraints and increase their understanding of premature neonates. To conclude, the present study indicated that the FLE-Tools environment may foster sharing in the design process among university-level student-teachers.

38.5 References

Akin Ö (1986) Psychology of architectural design. Pion, London

Akin Ö, Lin C (1995) Design protocol data and novel design decisions. Design Studies 16:211-236

Chi MTH (1997) Quantifying qualitative analyses of verbal data: A practical guide. Journal of the Learning Sciences 6:271-315

Cross N, Cross A (1995) Observations of teamwork and social processes in design. Design Studies 16:143-170

Ferguson ES (1992) Engineering and the mind's eye. The MIT Press, Cambridge MA.

Goel V, Pirolli P (1992) The structure of design problem space. Cognitive Science 16:395-429

Goldschmidt G (1995) The designer as a team of one. Design Studies 16:189-210

Glaser R, Chi MTH (1988) Overview. In: Chi HTM, Glaser R, Farr M (eds.) The nature of expertise. Erlbaum, Hillsdale NJ

Goel V (1995) Sketches of thought. MIT Press, Cambridge MA

Leinonen T, Mielonen S, Seitamaa-Hakkarainen P, Muukkonen H, Hakkarainen K (1999) FLE – tools prototype: a WWW-based learning environment for collaborative knowledge building. In: Proceedings of ENABLE99 (Enabling Networked-based Learning). Espoo Finland

Perry M, Sanderson D (1998) Coordinating joint design work: the role of communication and artefacts. Design Studies 19:273-288

Scardamalia M, Bereiter C (1994) Computer support for knowledge-building communities. The Journal of the Learning Sciences 3:265-283

Seitamaa-Hakkarainen P (1997) Composition and construction in novices' and experts' weaving design. Techne Serie Research in Sloyd. Education and Crafts Science A (2). Piima-Offset, Helsinki

Seitamaa-Hakkarainen P, Hakkarainen K (in press a) Construction and composition in experts' and novices' weaving design. Design Studies

Seitamaa-Hakkarainen P Hakkarainen K (in press b) Visualization and sketching in design process. Design Journal

Seitamaa-Hakkarainen P, Raunio A-M, Raami A, Muukkonen H, Hakkarainen K (in press c) Computer support for collaborative designing. International Journal of Technology and Design Education

Valkenberg R, Dorst K (1998) The reflective practice of design teams. Design Studies 19:249-271

Chapter 39

Co-designing over the Net

B. Kolarevic and E.Y.Y. Ng

39.1 Introduction

The Information Revolution, like the Industrial Revolution before it, is not only challenging what we are designing but also how we design. It is not only that architects have to work with consultants who are located in other cities or branches elsewhere in the world, but that they themselves are increasingly moving beyond their local and geographical strongholds and becoming global and international.

In structuring a global organisation, two models are commonly used (Lucas *et al.,* 1994). The vertical integration model is quite simple and is widely practiced. In this model, branch offices are set up and closed as projects come and go. Basically, each of the offices manages its own projects and offers a complete range of services. The horizontal integration model, however, is more sophisticated. It makes use of the geographical, economic and social advantages of different branch offices to maximize their throughput. Offices are specialized and the model relies on a concerted effort of almost all the branches to complete a project (Popova, 1997). It requires the support of an excellent and cost-effective communication protocol for collaboration among geographically distributed teams.

Apart from the communication problems, researchers have already identified the need to look into the social implications and consequences of the horizontal integration model (Winter and Taylor, 1996). Culture, language, social norms, time differences, values and practice are some of the issues identified. 'Empirical studies of design have demonstrated that design is a social activity as much as it is an individual practice' (Vervenne, 1994). Therefore, 'supporting the social communication is as important as supporting communication about the design problem'. Previous studies (Poh and Myers, 1994; Sasada, 1994; Wojtowicz, 1995; Mitchell www; Maher *et al.*, 1996; Cicognani and Maher, 1997; Schmitt, 1998) have cast some light in this direction. However, the findings are far from conclusive, as the interactive dynamics are complex and yet to be fully understood.

Last but not least, there is a need to look into issues related to the design process (Lawson, 1980; Schon and Wiggens, 1992; Schön, 1996). Design is 'a process which is in continual flux, moving between asynchronous and synchronous activity where group membership over the life-cycle of the design is dynamic and transitory' (Rutherford and Maher, 1993), meaning that a large number of the constituting parameters are changing during the development of the design. Different people are involved over limited periods. In order to follow the changes, the design information system has to be sufficiently open to incorporate new participants, including their preferred means of representation.

Ultimately, questions had to be answered as to whether or not the horizontal integration model could yield better design services. Many of the previous studies on virtual design studios focused on various aspects of enabling technology, communication, social dynamics and design process, but these are only means to an end; to clients and end-users, they have no tangible values (Maver, 1997).

39.2 Virtual Design Studio

The virtual design studio (VDS) described in this paper is an experiment in remote design collaboration that explored the issue of collective design authorship. Teachers and students in Hong Kong and Singapore worked on a common design project ('Place2Meet on the Water') using traditional and digital media, a central database, World Wide Web, and video-conferencing.

The studio was an attempt to examine some of the issues related to the use of the horizontal integration model in design, namely communication, social and design issues. In contrast to most of the previous works and virtual design studios, this studio was a semester-long program, with specific pedagogical, academic and design agendas. Throughout the studio, design issues were emphasized and technical requirements like structure, construction, detailing, material and siting were discussed. The studio tried to mimic, as much as possible, the design flow of an architectural practice assuming a horizontal integration model.

In addition, addressing the shortcomings of previous virtual studios, the studio was structured to limit possible variations and contingencies due to culture, time, language, and academic differences of the teams. The University of Hong Kong and the National University of Singapore made an almost ideal pair in that respect. Students of both Universities are Chinese in origin and from a similar cultural background. They are in the same time zone. Both universities use English as a medium of instruction and they have a similar academic structure. By relying on these similarities, the idea was to minimize the distracting elements (the 'noise') and to concentrate on the few identified critical issues.

In the first part of the studio, which was three weeks long, students were required to analyse some precedents and present their analyses as web pages. They had to study the following aspects of the selected buildings: design concept, structural behaviour, assembly process, details, building materials, building services, and the integration of architectural design and technology. Students worked in teams of five, made up of two students from Hong Kong and three students from Singapore. They had been advised to pay attention to and seek, on their own, efficient communication and information protocols to work with each other.

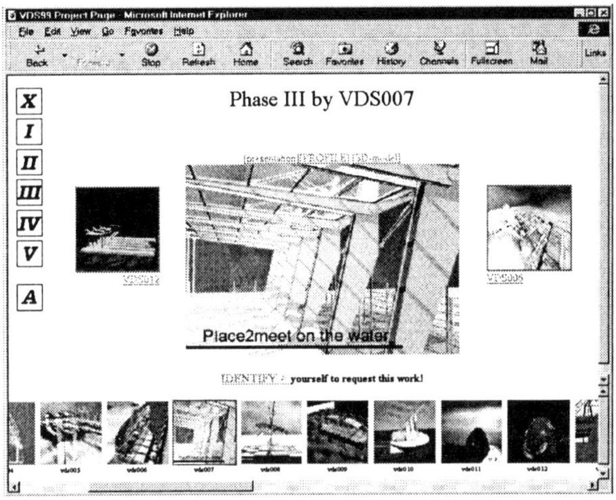

Figure 39.1. The VDS database (http://home.arch.hku.hk/vds99/)

In the second part of the studio, students were required to design a 'Place2Meet on the Water' over seven weeks in a collective, collaborative fashion. The collaboration, however, took a different form in this part of the studio. It was modelled after previous, weeklong experiments conducted at the University of Hong Kong with various partners around the world (Kolarevic *et al.,* 1998). The project was divided into five phases, each one or two weeks long. At the end of each design phase, students placed their work into a common design database, a 'digital pinup board' (Wojtowicz, 1994). The VDS database was accessible through a web browser. The browser window is split into three parts (frames). 'Buttons' in the frame along the left edge permit viewing of the projects produced in each phase, which are shown in iconic form in the frame along the bottom edge. The selected project can be then reviewed in the main frame (Figure 39.1)[1]. Then, at the beginning of each phase, students selected the best work they could find from the previous phases by browsing through the database. They were not allowed to continue with their own design. Since in each phase they had to select someone's design, they implicitly formed teams (Kolarevic *et al.,* 1998).

The final results were design projects with shared authorship that could be traced back to the contributing authors and co-authors using the database (Figure 39.2). In this process of 'evolutionary' design, authorship was not a question, as

---

[1] A web server (a 'digital pinup board') was established in Hong Kong for the project. Tim Yeung in Hong Kong especially developed the studio database for this project. It was modelled after the 'Phase[x]' database developed at ETH in Zurich (Hirschberg and Wenz, 1997), which was also used in 1997 and 1998 VDS experiments that involved several universities (Kolarevic *et al.,* 1998).

the contribution of all students involved in the design was recorded in the database (Kolarevic *et al.*, 1998).

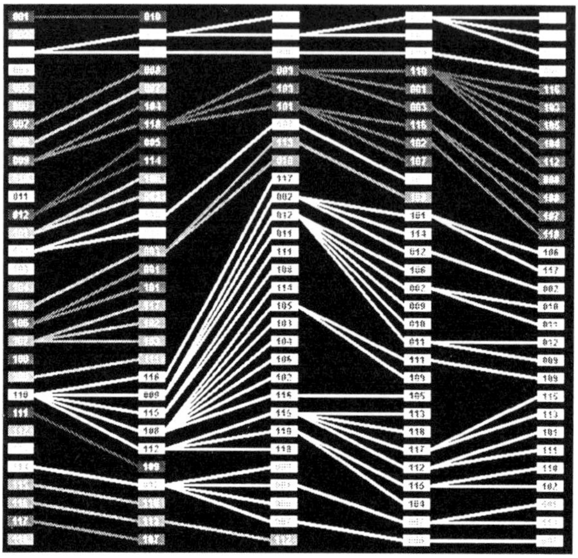

Figure 39.2. The genealogy of designs (note that there are four identifiable 'families' of designs at the end of the 'evolutionary' process)

39.3 Observations

Two surveys were designed to capture feedback related to the key issues that were investigated in this studio, namely modes of communication, collaboration, collective design, social processes, and pedagogy. Recognizing that the use of technology in the design process is a matter of critical judgement rather than a fact, the questionnaires included sections where respondents can explain the rationale behind their choices. Two general essay type questions were also included to capture some of the general observations about the virtual design studio. Apart from this more structured approach, informal inquiries were conducted during the studio sessions to collect views and feedback. We also monitored the design progress through the VDS database to formulate observations on the collaboration process as well as the design evolutions.

39.3.1 A Socio-technical Perspective

A socio-technical system perspective has been adopted in this studio. It is believed that it is essential to allow flexibility for individuals to seek means to construct their own communication protocols (Winter and Taylor, 1996). While it has always been assumed that higher bandwidth and resolution will lead to an improvement of communication (Chen and Maver, 1996; Chiu, 1997), our findings indicate otherwise. The availability of technology allows but does not necessarily change the established behaviours of the users; at times, it might be necessary to 'de-optimise' technology to accomplish the behaviour of the individuals (Yamaguchi www). A number of communication models or roles were identified:

- 'Philosophers' – They generally prefer slower, simpler and more direct communication methods. They use email to compose their thoughts and prefer a lag between dialogues. They use web pages to publish their thoughts. They participate in synchronous exchanges only when it is absolutely necessary.
- 'Misers' – They prefer using and maximizing a particular means of communication. They perceive communication as necessary but generally prefer to work and contribute quietly. For them, the meetings should be over quickly. The only thing they want to achieve is what needs to be done next.
- 'Party goers' – They like meeting people and being seen. Their outgoing characters mean that they also generate most of the exchanges on the network. They like experimenting with various means to communicate. To them, communication can sometimes be an end in itself.
- 'Followers' – They communicate to maximize their own work. They generally position themselves at the receiving end of the communication chain. They like to be on-line to ensure that they don't miss anything. They like asking questions so that they could get useful answers and information.
- 'Initiators' – They are the opposite of followers. They are at the front end of the communication and always wish to broadcast their ideas and contributions. These would often provide help to ensure that everybody in the team gets their message. They like instant feedback so that they could contribute more.
- 'Loners' - These people have their own agenda and do not normally wish to be bothered. Communication has little purpose except to put various 'individual' contributions together. For them, high bandwidth and resolution is a waste of time and energy.

The choice of technology at any one time of the design process is not entirely a technical decision. ICQ ('real-time chat' software) has been the most preferred communication protocol among the different user styles (for 83% of the survey respondents). Two thirds of surveyed students (68%) thought it essential for VDS collaboration and communication. Students found it 'the most convenient way to communicate', 'informal', 'fast', 'instantaneous' and welcomed its low system overhead, autosave and built in file transfer features. However its text-based interface poses limits to support-design activities. The Whiteboard feature in NetMeeting was used extensively to supplement ICQ because of its ability to share graphical information and, more importantly, to exchange sketches. One third (32%) thought that the whiteboard is essential for VDS. Email was the least preferred technology for two thirds of surveyed students, as it was perceived to be 'formal', 'only good for giving instructions' and 'slow'. Surprisingly, half of the students who were surveyed thought that it was barely needed. It was used mostly when it was understood that other party was not there. The use of web technology was, on the other hand, enthusiastically endorsed. Sharing and obtaining information were the main motives behind its support. Beyond the technological need to exchange information, the existence of a common design database somewhere provided a psychologically comforting proof of the existence of the teams. The mixed feeling expressed about whether or not it is important to 'see' the face(s) of the party was illuminating. Contrary to popular belief, our findings indicate that seeing the others is not an essential feature needed for effective collaboration. According to our survey, only 4% found it essential, while 32% though it is not important (16% thought it was often needed, 28% sometimes, and 20% barely). There are some concerns about privacy here, but the main reason given is that seeing the work is more important than seeing the face(s). However, at this point, it must be pointed out the students 'prefer' seeing the face(s) of their tutor(s) during tutorial and review sessions.

39.3.2 A Design Perspective

Whether or not design could be improved with the horizontal integration model was another inquiry of this studio experiment. Design is a response to a situation, thus it is difficult to develop objective measurements to quantify its quality (Schön, 1996). What this study investigated instead is whether or not the horizontal integration model could be 'perceived' to provide a more conducive environment for better design.

Design has been traditionally seen as a continuous and integrated process (Schön and Wiggens, 1992). In an idealized situation, the designer, like an artist, would wish to have control of the entire process. A good design has always been seen as one that has a coherent concept throughout. It is therefore not difficult to understand that some students would express misgivings about the idea of shared authorship. Two thirds of the students surveyed stated that they could accept some 'sharing' but not all. For some students it was 'difficult to work on others' ideas while they still had their own in mind', *i.e.*, to 'shift from one's initial design framework into somebody else's'. On the other hand, some noted that collective design provides an opportunity to try out different approaches to the project, and to discover potentially interesting ideas that might have emerged had one been working alone.

Discussion with the students and surveys revealed that most of them tended to choose designs that were 'similar' to theirs for further development. Only a few of them admitted that they chose the design because it was different and 'better'. One assumption of the shared authorship concept is that designs would evolve for the better and that the fittest would survive. However, underlying that is another assumption that the rules and principles of what could be considered good design have been established and understood. This was not the case in this studio. It would therefore be interesting to see if the evolution theory of design holds in a more objectively defined design situation.

On the positive side, feedback on whether or not the virtual studio could yield a richer and more diverse design and learning environment has been reasonably positive. For example, more than two thirds of the students thought that having to work on somebody else's idea was a good learning experience; one quarter could imagine being taught design in that fashion alone. In summary, an enlarged knowledge pool and the availability of a large database of different ideas were seen as valuable learning resources. This observation is in agreement with Schmitt's and Hirschberg's findings in their 'Phase(X)' and 'fake.space' experiments (Schmitt, 1998; Hirschberg, 1998).

39.3.3 A Teaching Perspective

The studio also examined the issue of teaching pedagogy. During the review sessions, tutors, or the clients, do not normally draw (Schön and Wiggens, 1992). The way information is transferred is mainly through verbal means and facial and body expressions. And more importantly, feedback is spontaneous. This means that the bulk of the information flow is intangible, perishable and unrecorded. To better understand this dynamic of information flow, four styles of teaching/tutorial were experimented with:

- 'Real' - tutors tutoring physically in a well-experienced and understood process.
- 'Real/virtual' - tutors tutoring a team of real and virtual students in a single session.
- 'Virtual' - tutors tutoring virtually over the Internet.
- 'Reverse virtual' - tutor from Singapore virtually tutors students in Singapore from Hong Kong.

On the whole, and as expected, students preferred 'real' tutors. Some concerns were expressed during 'real/virtual' tutorials. The problem is more to do with the fact that while it is not difficult to tutor either virtually or physically, it has not been easy finding ways to teach in both ways simultaneously. For example, for the benefit of the virtual tutee, a tutor might wish to speak slowly and use no body language. This restricted communication mode, however, is not very suitable for the 'physical' tutee. On the other hand, it was difficult to hold the attention of the virtual tutee while the tutor talked to the physical tutee.

'Virtual' tutors were welcomed, as they were perceived to have brought with them additional expertise to the studio. However, students in general find them 'remote' and more difficult to 'engage'. Spontaneity has been the major concern. Students generally feel that one of the main differences between real and virtual tutors is the lack of spontaneity. Even with the aid of high quality PictureTel-based video conferencing, it has been felt that there is a lack of intimacy. This perceived distance, entirely psychological, deterred students from attempting to build a relationship with the tutor. It has been felt that while video conferencing is useful in a lecturing environment, a tutoring environment is much more personal than that. Tutorials work the best when the students know the tutor personally; time is required to build that relationship.

Comments expressed by students about 'virtual' tutors shed some light on how 'reverse virtual' tutors were received. It was of no surprise to note that students regarded the 'reverse virtual' tutor almost as real as the 'real' tutor. Previous relationships built between the tutor and the tutee obviously help here. On further examination and discussion with students, it was revealed that the missing body language was inferred, through past experience with the tutor, by the tone and rhythm of his verbal expression.

39.4 Conclusions

The net-enabled collective design authorship is an instance of a wider concern for validating the effectiveness of the emerging horizontal integration model in architectural practice. Although the model has been used with some success by other, mainly engineering-based, design professionals, its usefulness in supporting the architectural design process is yet to be proven.

Architectural design process is solution-led and highly 'wicked' (Lawson, 1980). As such, there is a need for information to loop-back to the previous stages of the design process. A lot of these loops are unstructured and unpredictable. Our studio experiment demonstrated that flexibility in the provision of technology is more important than its optimisation. It demonstrated that to facilitate and encourage the loop-back mechanism, it is important to ensure that the information system is designed to cater for, or tolerate, minor and informal channels to exist within the more formal and structured framework. At present the two are rather dissociated. The challenge for future technological development will surely be in seeking ways to integrate them.

Coupled with the provision of technology is the social dimension of engaging technology for a purpose. People communicate for a need, and they will seek the best means to accomplish that need. This studio has identified some of the styles and needs of the individuals in the virtual world. However, this socio-technical interface is a dynamic one, and behaviours will be affected by the development of new technologies. The challenge here is to continue monitoring to capture where the balance points are in the rapidly changing world of information technology.

The encouraging news is that the studio is perceived to have provided a more enriching environment for design. Unfortunately, it is still inconclusive as to whether or not this studio arrangement could yield better design. Design quality is judgmental and is experience-based. The problem with this studio is that the designers are mostly inexperienced and may not fully appreciate the nature of architectural design process. To address the problem more adequately, a similarly structured studio should be conducted with more advanced students or even with design professionals.

The roles and styles of the teachers in the information age has been the subject of some debates by education theorists (Penz, 1992; McCullough, 1996; Popova, 1997). Virtual design studios add another level of complication. Apart from some of the concerns expressed so far, this studio highlighted the need to 'know the tutor' at a more personal level as a pretext to the need for a more spontaneous exchange. If virtual tutoring is ever to be promoted as a replacement for real tutoring, solving the contextual issue of what goes on outside the tutorial session is as important as providing higher technology to enable the session itself.

39.5 References

Chen Y, Maver T (1996) The Design Implementation of a Virtual Studio Environment. In: Proceedings of 2nd East-West Conference on IT in Design. Moscow

Chiu ML (1997) Representations Communication Channels in Collaborative Architectural Design. In: Proceedings of IFIP Conference. Sydney

Cicognani A, Maher ML (1997) Models of collaboration of designers in computers in computer supported environment. In:Proceedings of IFIP 97 Conference. Sydney

Hirschberg U, Wenz F (1997) PhaseX: Memetic Engineering for Architeture. In: Proceedings of ECAADE 97 Conference. Vienna

Hirschberg U (1998) Fake Space - An Online CAAD Community - a Joint Enquiry into the nature of Space. In: Proceedings of CADDRIA 98. Conference Osaka

Kolarevic B, Schmitt G, Hirschberg U, Kurmann D, Johnson B (1998) An Experiment in Design Collaboration. In: Proceedings of ACADIA '98 Conference. Quebec Canada

Lawson B (1980) How Designers Think. The Architectural Press, London

Lucas HC, Roitman DB, Roskies E (1994) The Role of Information Tehnology in Organisation Design. Journal of Management Information System 10(4)

Maher ML, Simoff S, Cicognani A (1996) The Potential Current Limitations in a Virtual Design Studio. University of Sydney

Maver T (1997) Added Value for CAAD in Education. In: Proceedings of AVOCAAD Conference. Brussels Belgium

McCullough M (1996) Abstrating Craft: The practiced digital hand. MIT Press, Cambridge

Mitchell W (www) Creative Design in the Computer Era: The Design Studio of the Future. MIT, http://sap mit edu/dsof/research/creative_design html

Penz E (ed.) (1992) Computers in Architecture: tools for design. Longman, London

Popova M (1997) IT the Architect's role. Division of Design Computing, Chalmers University of Technology, CTH-A CADLAB, Gothenburg

Poh L, Myers L (1994) A Distributed Co-operative Model for Architectural Design. Automation in Construction 3 (2/3)

Rutherford JH, Maher ML (1993) Video Conferencing CAD. In: Proceedings of the 27th Annual Conference. Australian and New Zealand Architectural Science Association

Sasada T (1994) Open Design Environment and Collaborative Design. In: Proceedings of ECAADE 94 Conference. Glasgow Scotland

Schmitt G (1998) Design Construction as Computer-Augmented Intelligence Process. In: Proceedings of CAADRIA 98 Conference. Osaka Japan

Schön D (1996) The Reflective Practitioner: How Professionals Think in Action. Basic Books, New York

Schön D, Wiggens G (1992) Kinds of seeing and their functions in designing. Design Studies 13 (2)

Vervenne D (1994) Interactive annotation processing for extraction of strategic knowledge. In: Proceedings of Management of Industrial Corporate Knowledge Conference. Compiegne France

Winter S J, Taylor SL (1996) The Role of IT in the Transformation of Works: A Comparison of Post-Industrial, Industrial and Proto-Industrial Organisation. Information Systems Research 7 (1)

Wojtowicz J (ed.) (1995) Virtual Design Studio. Hong Kong University Press

Yamaguchi S (www) The Next-Generation Architectural Design Environment http://archigraf kit ac jp/INT/Res

Chapter 40

Writing and Sharing of Rule-text Heuristics to Support Dispersed Design Communities

P.A. Rodgers, N.H.M. Caldwell and A. Huxor

40.1 Introduction

In recent years design has evolved from an individual activity to one where multiple participants with diverse skills collaborate on a project. Simultaneously, emerging advances in networking technologies now offer greater support to dispersed design teams. As a consequence of these social, cultural and technological developments, modern design is a highly social activity carried out between communities of individuals with disparate skills and perspectives (Bucciarelli, 1988). This paper presents a novel theoretical approach, 'AI as writing', to the development of network-based support tools for dispersed design teams. This approach views these tools as a form of, and extension to, writing, with the tool as a 'knowledge medium' (Stefik, 1986) for expressing and communicating knowledge. This paradigm is particularly appropriate in the domain of consumer products which relies heavily upon context-specific knowledge (regional, cultural, environmental, *etc*.,) applicable to the intended market (Harel and Prabbu, 1999; Hoft and Ito, 1999), and where there is unlikely to be a single 'correct' design. The AI as writing approach has been implemented in WebCADET – a web-based conceptual design support tool. This incorporates a knowledge capture facility allowing designers to 'author' and 'publish' new rule-texts by amending existing rule-texts or 'writing' completely new rule-texts. The eventual aim is to create a networked 'community' of WebCADET knowledge servers supporting dispersed designers with context-specific design knowledge.

40.2 Emerging Design Practice

Current, emerging styles of design working typically vast amounts of information and involve large numbers of individuals who have a varied range of education, skills and experiences all collaborating towards the development of the project.

Moreover, recent advances in information and computing technologies offer ever-increasing support for designers and design teams that may be dispersed throughout the world. Therefore, it is common for design teams to communicate and collaborate during design projects without actually meeting one another face-to-face. Design is, however, an intrinsically social process (Bucciarelli, 1994). As the well-known Italian designer Vico Magistretti (Pasca, 1991) indicates, design objects are not born on the drawing board, but are produced during a number of exchanges and suggestions between the designer and the manufacturer. Obviously, the various members of the design team can be expected to possess divergent views on elements such as the aesthetic nature of the product, the costs involved in the project development schedule and the technical functionality of the product scheme. Thus, informal reports, discussions and banter will have as much impact on the successful outcome or otherwise of the design project as the role the technical documents, guidelines and formal meetings play during the process (Ferguson, 1992). Recent developments in technology facilitate the exchange of information in a wide variety of communication and collaboration media, including email and face-to-face communication via video and audio conferencing (Court *et al.*, 1997). Methods currently used to communicate and collaborate with others in dispersed design projects include: email-based discussion lists such as design-research@mailbase.ac.uk (Figure 40.1) and idforum@yorku.ca; news groups such as alt.cad.autocad and comp.cad.pro-engineer; and Internet-based communication and collaboration tools such as NetMeeting and CUSeeME (Rodgers *et al.*, 1999). These tools and techniques provide effective and efficient mechanisms for enabling designers and other design project stakeholders, from very different geographic locations, to work together.

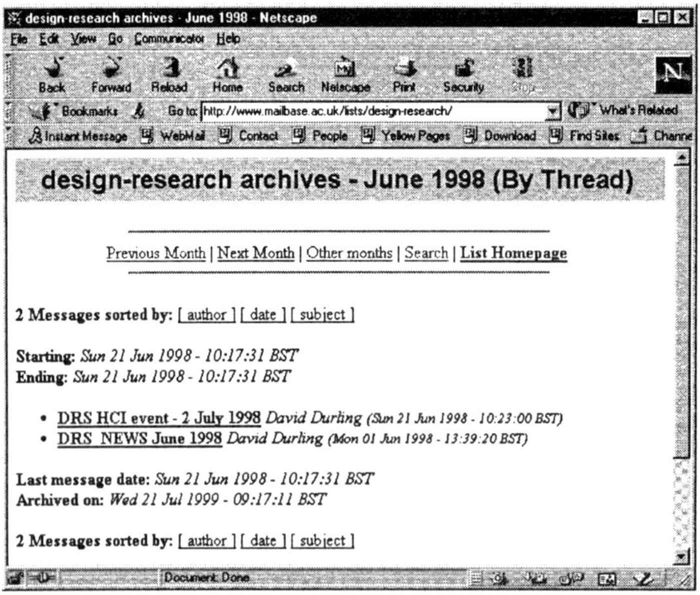

Figure 40.1. Design-research discussion list (web-based archives)

40.3 AI as Writing Paradigm ('Rule-texts')

The approach described in this paper is based on an alternative view of Artificial Intelligence (AI) in that symbolic AI is seen as an extension to writing (Ong, 1986). That is, symbolic AI systems take the written word and give it new powers. This approach also recognises the importance of transparency and modifiability of knowledge-base content. That is, the author, nature and provenance of the knowledge embodied in the system must be made accessible to the user (reader), who can contest and modify the knowledge-base contents in the light of their own experience, skills and requirements (Figure 40.2).

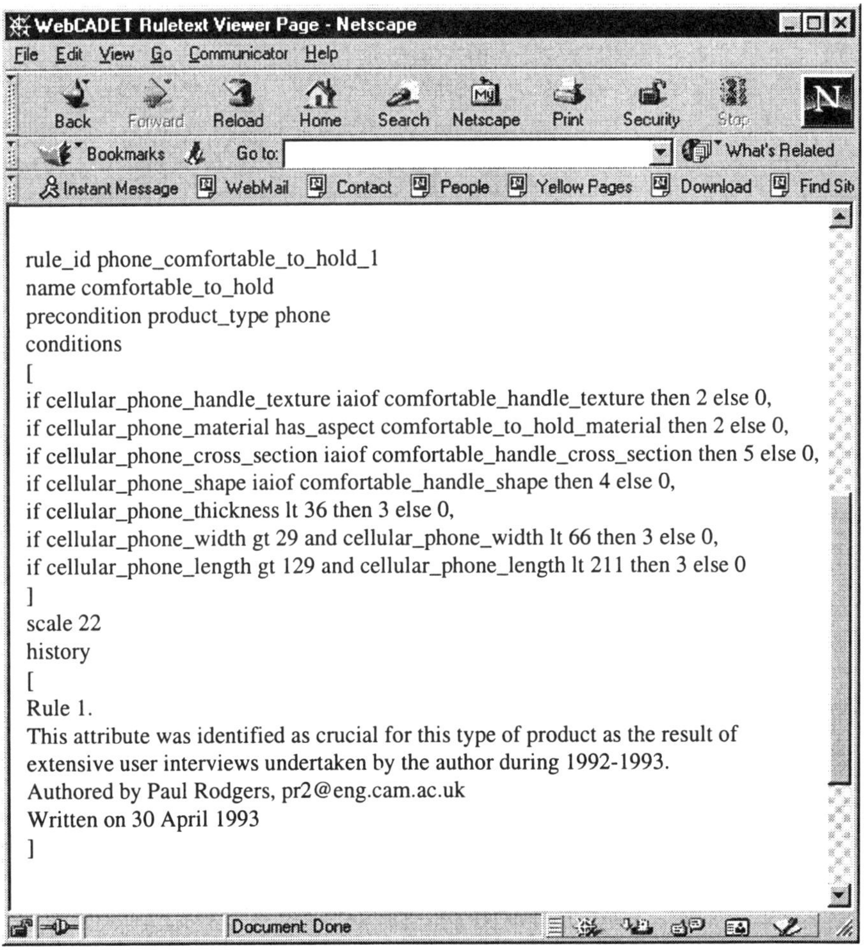

Figure 40.2. Rule-text heuristic example

It is widely accepted that design involves a range of different types of knowledge (Yoshikawa, 1993; Nonaka and Takeuchi, 1995; Rodgers and Clarkson,

1999). For example, design involves a kind of procedural knowledge (*i.e.*, both tactical understanding and know-how) and a kind of substantive knowledge outside the procedures themselves. For instance, a carpenter has a 'know-how' understanding of procedures such as sawing and joining as well as 'knowing-that' understanding of when such tactics should be applied (Schön, 1983). Designers, it has been claimed, take a 'heuristic' as opposed to an 'algorithmic' approach in their design activities (Stauffer and Ullman, 1988). Thus, the focus of the knowledge used in the 'AI as writing' approach is heuristic-based. Heuristics are general rules of thumb that can often help in solving a problem, but a solution is not guaranteed (Perkins, 1981).

From Figure 40.2, the original 'rule-text' heuristic example shown on the left concerning the design of ergonomic handles initially authored by P.A. Rodgers has been read by other users who have considered the initial rule to be inappropriate for their specific task. Edited versions of the initial rule-text have been created by different users (authors), such as that of A.P. Huxor. This may, for example, be due to the different market niches in which that user is working, such as for children or the elderly who have specific requirements. In the 'AI as writing' paradigm, users (authors) are able to submit their own versions for publication, review and use. Thus, it is envisaged, that a market in authored rule-texts will develop as users (readers) seek and apply rule-texts that are appropriate to their particular design scenarios.

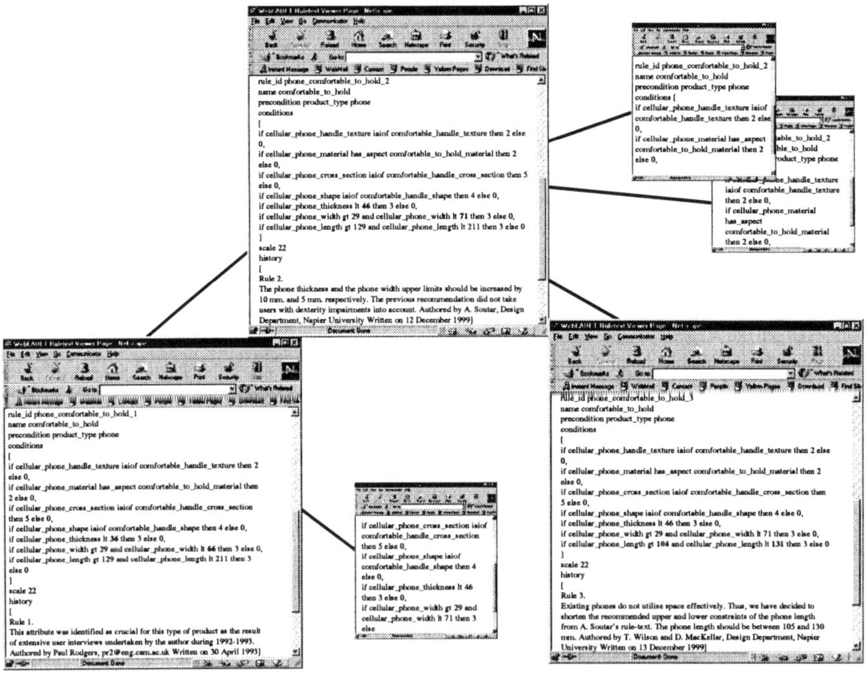

Figure 40.3. AI as writing 'rule-text' evolution example

In summary, in the AI as writing approach, participants in the process of knowledge externalisation (Sinding-Larsen, 1988) can be viewed in terms of their roles in a publishing model. That is, as editors, authors and readers.

40.4 WebCADET

WebCADET is an Internet-based design support tool. It provides support to designers in two main ways. First, it acts as support for designers during the concept evaluation and selection stage. Designers input details of their concept schemes, and the heuristics contained within WebCADET's knowledge base are used to return feedback. The feedback is justified by the inclusion of online links to published evidence from anywhere on the web as well as references to the design literature, both of which reinforce WebCADET's evaluation scores (Figure 40.4).

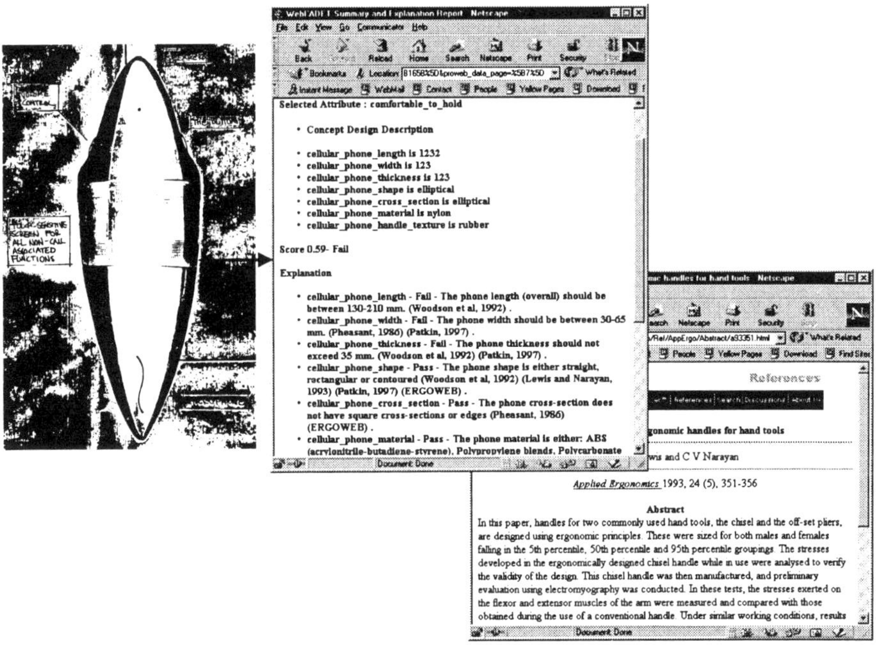

Figure 40.4 WebCADET concept evaluation schema

Second, WebCADET facilitates design knowledge capture. That is, any designer (user) with access to a web browser can interrogate the content within WebCADET's knowledge base and modify it in the light of their own views. New rule-texts can be created from scratch or existing rule-texts can be altered through the addition, removal or adjustment of the constituent content. Such revised rule-texts do not replace the original rule-texts within the knowledge base; rather the revised rule-texts supplement the original versions, and in this way, the knowledge base grows by the sharing of differing perspectives. Additionally as rule-texts on a specific class of products accumulate, a shared lexicon of terms for describing aspects of concept schemes for that product class develops, and this assists designers to submit their expertise more easily.

The rule-text formalism implemented within WebCADET supports transparency in terms of the origin, nature and provenance of content, by recording who submitted the content and the contributor's affiliation, when it was submitted, and why it was submitted. When an existing rule-text is modified, the equivalent details concerning the revising author and the rationale for making the changes is added to the details concerning the original author. Thus credit and responsibility is apportioned justly to originators and revisers. Moreover when WebCADET is used to support designers in concept evaluation and selection, the origin and rationales of applicable rule-texts are made visible to the designer, so that an informed decision may be made by the designer as to which rule-text is most relevant in the circumstances of the current design project.

40.5 Conclusions and Future Work

The paper has outlined the 'AI as writing' approach and its implementation detail in network-based support tools for dispersed design teams. This approach views AI and KBS technologies as a form of writing, with the tool itself seen as a 'knowledge medium' capable of expressing and communicating knowledge. This approach is particularly appropriate in the design domain of consumer products that have a high degree of user interaction and rely heavily upon market-specific knowledge, be it at a local or national level. In this type of design it is highly unlikely for there to be any single 'correct' design. Thus, it is important that the nature and provenance of the knowledge embodied in the system be made accessible to the reader or user, who can contest or modify the content in the light of their own background and requirements.

To date, the 'AI as writing' approach has been tested in a couple of preliminary studies. This has involved participants using the tool, WebCADET, in a number of concept design tasks. The first of these studies was carried out within the design offices of one of the research team's industrial partners in an effort to make the test as real as possible. This particular test adopted a think-aloud protocol similar to that described in Ericsson and Simon's work (1996) as a method for evaluating the usability of WebCADET. In this test, the think aloud protocol involved each selected participant of the company (*i.e.*, one industrial designer and two engineering designers) to talk about what they were doing and what they were thinking when using the WebCADET system to aid in a common design task. This test identified several weaknesses of the WebCADET system, most notably problems with the interface and the 'non-interruptible' nature of the system. Further details of this work can be found elsewhere (Rodgers *et al.*, 2000).

More recently, a number of postgraduate engineering students at Cambridge University were asked to 'author' one new rule using WebCADET's information capture modes. Each of the students provided significant feedback and a number of important suggestions on how WebCADET might be improved. In terms of the rule capture mode, there were few difficulties with navigating the user interface, although a couple of the students indicated a preference for increasing the height and width of the forms to be more consistent with the rest of the page. Generally speaking, however, the students did not encounter as many difficulties with the WebCADET system as the designers had in the previous study. It is planned to develop further the 'AI as writing' approach and the WebCADET system itself in a series of extensive investigations within both academic and industrial settings.

Future extensions to WebCADET are likely to include the provision of support for other forms of design knowledge, such as case-based representations of successful and unsuccessful product concepts, and structured representations of the significant content within design documents. It is envisaged that individual WebCADET knowledge servers will support distinct communities of designers and users, with each community having a different emphasis, either in terms of the design sector or target market. To facilitate the dissemination and sharing of knowledge among the distinct communities, future versions of the WebCADET servers will be able to communicate with each other so that rule-texts and other design knowledge from multiple knowledge bases can be brought to bear upon a design decision. It is anticipated that this facility will be particularly effective in providing insights from across product boundaries.

40.6 References

Bucciarelli LL (1988) Engineering design process. In: Dubinskas F (ed.) Making time: ethnographies of high-technology organizations. Temple University Press, Philadelphi PA, pp 96-97

Bucciarelli LL (1994) Designing engineers. The MIT Press, Cambridge MA

Court AW Culley SJ McMahon CA (1997) The influence of information technology in new product development: observations of an empirical study of the access of engineering design information. International Journal of Information Management, 17 (5):359-375

Ericsson KA, Simon HA (1996) Protocol analysis. The MIT Press, Cambridge MA

Ferguson ES (1992) Engineering and the mind's eye. The MIT Press, Cambridge MA

Harel D, Prabhu G (1999) Designing for other cultures: a strategic approach. Design Management Journal 10 (4):60-68

Hoft K, Ito Y (1999) A method for culture - and mindset-harmonised design. In: Lindemann U, Birkhofer H, Meerkamm H, Vajna S (eds.) Proceedings of the 12th international conference on engineering design (ICED'99). Technical University Munich, Germany, (vol. 1) pp 313-316

Nonaka I, Takeuchi H (1995) The knowledge creating company. Oxford University Press, New York NY

Ong WJ (1986) Writing is a technology that restructures thought. In: Baumann G (ed.) The written word: literacy in transition. Clarendon Press, Oxford, pp 23-50

Pasca V (1991) Vico Magistretti: elegance and innovation in postwar Italian design. Thames and Hudson, London

Perkins PN (1981) The mind's best work. Harvard University Press, Cambridge, MA

Rodgers PA and Clarkson PJ (1999) An investigation and review of the knowledge needs of designers in SMEs. The Design Journal, 1 (3):16-29

Rodgers PA Huxor AP Caldwell NHM (1999) DESRES: towards a virtual design space. In: Press M, Cooper R (eds.) Proceedings of the third European academy of design conference, European Academy of Design, Salford, Manchester, UK, (vol. 2) pp 235-252

Rodgers PA Caldwell NHM Clarkson PJ Huxor AP (2000) Managing knowledge in dispersed design companies. In: Gero JS (ed.) Artificial intelligence in design 2000 (AID'00). Kluwer Academic Publishers, Dordrecht, The Netherlands

Schön DA (1983) The reflective practitioner: how professionals think in action. Basic Books, New York NY

Sinding-Larsen H (1988) Information technology and the externalisation of knowledge. In: Sinding-Larsen H (ed.) Artificial intelligence and language: old questions in a new key. Tano, Oslo, pp 77-89

Stauffer LA, Ullman DG (1988) A comparison of the results of the empirical studies into the mechanical design process. Design Studies, 9 (2):107-114

Stefik M (1986) The next knowledge medium. AI Magazine, 7 (1):34-46

Yoshikawa H (1993) Systematization of design knowledge. Annals of the CIRP, 42 (1):131-134

Chapter 41

A Collaborative Design Studio: Architecture and Prototype

V. Zaychik, J. Sevy and W.C. Regli

41.1 Introduction

In this paper we discuss the Drexel Collaborative Design Studio (DCDS), an environment for distributed engineering under development at the Geometric and Intelligent Computing Laboratory at Drexel University. DCDS is intended to enhance collaborative design work, focusing in particular on design of electro-mechanical artifacts. Our goal is to provide transparent collaborative work and archival services to document the design process. The DCDS environment consists of IP-based communication and collaborative tools, a design knowledge repository, and a design-specific message model for capturing communication during the design process.

This research seeks to enhance our understanding of how to make designers more effective in networked engineering environments. The new design process consists of teams of decentralized designers, collaborating over networks using communication and engineering software tools. This convergence of the network with engineering practice creates new research challenges:

- **Integration of Collaborative Work and Computer-Aided Design (CAD).** Advances in Computer-Supported Collaborative Work (CSCW), coupled with the ubiquity of the Internet as a communications medium, create new opportunities to facilitate *context-aware* interaction among decentralized design teams by coupling tightly design and collaboration tools.
- **Representing Collaboration via Message Modeling.** Previous research has viewed systems integration as a problem of achieving shared semantics for data sharing. This research builds on past approaches to create extensible *message models* that provide frameworks for describing the interactions among human and software agents in the design process.
- **Capturing Design Processes to Build Design Repositories.** We view this new digital workspace as an authoring environment for the entire design process. Design repositories represent the next generation of engineering databases

(Szykman, *et al.*, 2000), a union of knowledge-based systems, Product Data Management (PDM), workflow tools and groupware.

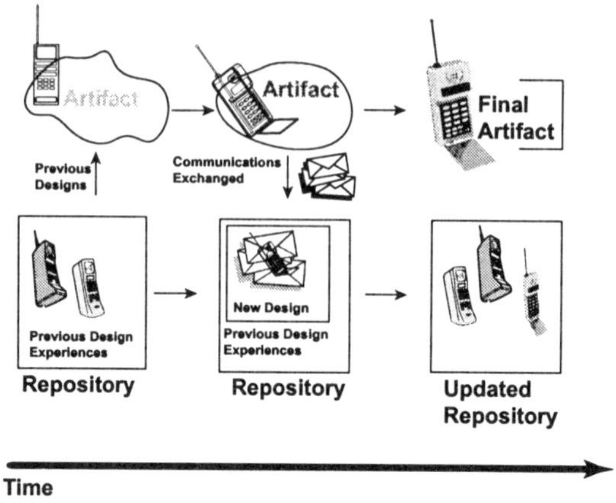

Figure 41.1. The design process viewed as collaborative work evolving over time

Figure 41.1 shows the evolution of a design project over time, emphasizing communications among design team members and software agents. DCDS will enable distributed engineering enterprises to archive communications exchanged during the design process. Our goal is to create intelligent collaboration and design systems to help engineers archive the digital design process and manage design repositories – thus providing leverage on previous design experience. Design repositories can also provide useful insights into the design rationale employed by experienced designers.

We view the Collaborative Design Studio as an enabling mechanism for the capture of design intent and design rationale. Integration of collaborative work tools with the worker's activity context, as provided by CAD and other engineering software systems, will both improve productivity gains for CSCW tools and provide an unparalleled opportunity to record the design process in a Design Repository. Lastly, the design repository creates a digital filing cabinet for storing design process and collaboration information for traditional and distributed engineering enterprises. The availability of this information should prove to be especially beneficial for product redesign and design reuse.

41.2 Related Research

The literature most relevant to our efforts comes from two areas: CSCW and design repositories.

41.2.1 Computer-supported Collaborative Work

Considerable research exists describing general-purpose collaboration systems, interaction modalities and evaluation of such environments (Grudin and Poltrock, 1997; Mark, Grudin and Poltrock, 1999; Greenberg and Roseman, 1998; and Herlea and Greenberg, 1998). The Collaborative Design Studio project is most interested in the integration of general-purpose CSCW tools with design tools. In addition, recent academic and commercial systems for group work have emerged (Eroom, 2000 and Teamware, 2000). Our present research is based on MITRE Corporation's Collaborative Virtual Workspace (CVW) (Collaborative Virtual Workspace, 2000), an open-source multi-platform, multi-user collaboration environment that is highly customizable and based on widely accepted Internet protocols.

41.2.2 Design Repositories

Reuse of design artifacts is a standard practice; it is also generally believed that most design projects are those of redesign (Ullman, 1992). In such projects, reusing accumulated design knowledge is much more cost- and time-effective than rediscovering it, yet current CAD tools are inadequate for recording design knowledge for future reuse (Baya *et al.*, 1992). A number of research efforts that attempt to record and archive information generated in the process of design in a structured manner are underway. These efforts can generally be divided into two groups based on the source of design information: *formal* (documented engineering reports, CAD drawings) or *informal* (design notebooks, e-mails, memoranda). Recent study confirms our belief that concept generation and development in engineering design occur mostly in informal media, such as e-mail and video conferencing, where capture tools are lacking (Yen *et al.*, 1999). Informal design information is difficult to capture. The SHARE project (Toye *et al.*, 1993) (later replaced by the Design Space Colonization project) at the Center for Design Research at Stanford University approaches the problem by providing an electronic design notebook in which designers can express ideas, rationales and any other relevant information. Another system, Recall, facilitates and records sketch activity along with video conferencing sessions in the conceptual stage of design (Yen *et al.*, 1999).

41.3 Technical Approach

By archiving informal communications, a significant percentage of previously lost design experience can be preserved and made available for later reuse. The communication tools, however, need to be tightly coupled with CAD tools to ground the context of communication to a particular design issue. Thus in DCDS, e-mail and video conferencing are integrated with a CAD system. Communication is structured according to a *message model* (Section 41.3.2) and archived in the design repository.

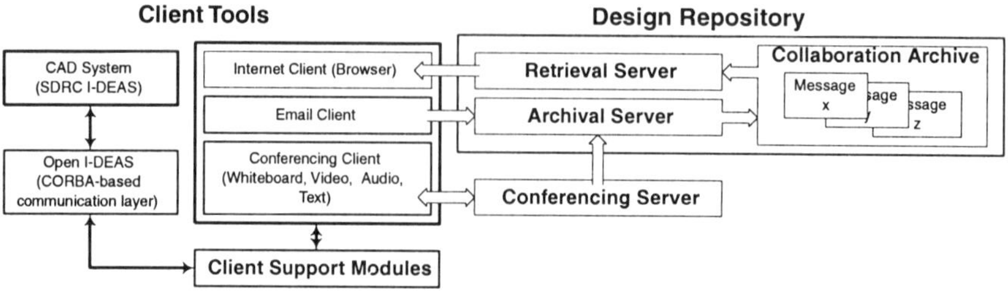

Figure 41.2. Architecture of the Collaborative Design Studio

Our approach builds on insights gained from the SHARE project (Toye *et al.*, 1993), with a focus on network-oriented distributed collaboration in the process of concurrent design. Both projects concentrate on archiving a combination of informal and formal design knowledge, with different focuses. SHARE is *process-oriented*, centering on a personal design notebook in which a designer enters relevant information by choice; thus a designer chooses to include e-mail he considers relevant. The Collaborative Design Studio is *communication-centered*: the focus is on messages exchanged in the process of design. All communications are automatically archived, requiring no action on the part of the designer. Elements of our DCDS also draw on Stanford's Interdisciplinary Communication Medium (ICM) (Fruchter *et al.,* 1995). Like DCDS, the ICM provides archival services for design change notifications. However, the systems differ in the manner in which communications are archived. In ICM, change notifications are intimately associated and linked with particular CAD files while in the DCDS, they are archived in a general-purpose database, allowing efficient searching of communications associated with distinct projects.

41.3.1 Architecture

The DCDS architecture has five components, illustrated in Figure 41.2:

1. *client tools* for communication and archiving support;
2. a *collaboration server* that facilitates communication;
3. a *design repository*, including:
   a. a *collaboration database* to store design process information;
   b. an *archival server* that indexes design knowledge and collaborative exchanges for storage in the collaboration database;
   c. a *retrieval server* to access information from the database.

The client communication tools provide enhancements to standard communication software. Their central function is to encapsulate information from a communication using the message model, and deliver this to the archiving server. These tools automatically extract information about a design from the SDRC I-DEAS CAD suite, using the Open I-DEAS programming interface. This interface allows client software to query and send commands to the CAD software, using the CORBA architecture for interprocess communication.

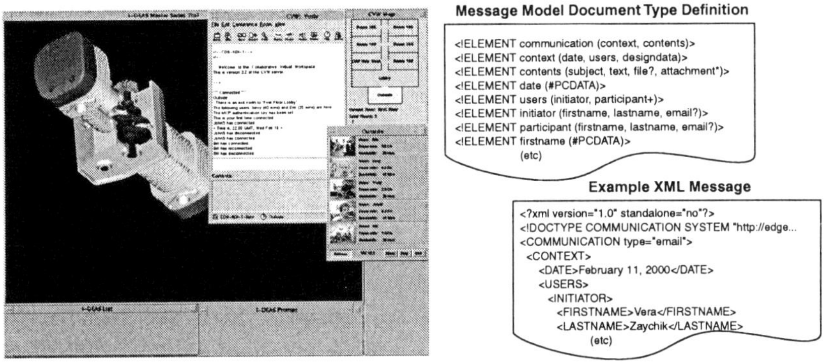

a b

Figure 41.3. A collaborative exchange with CVW: a. an XML-based document type definition for the message model; b. an example message for the exchange

In our environment, the CAD package is queried to extract design context, including: project and model file; specific parts within the model; features within the part; design history and recent operations; *etc*. This design context is included in the message, allowing the message to indicate unambiguously which design feature is being discussed.

Our design repository includes both formal and informal design information. Informal information is stored in the collaboration archive; this is populated by the archival server, which receives information from client communication tools. The retrieval server can search design messages in the archive (via a Web browser), using the message model to structure returned information. The archive database has been designed to be extensible to permit the inclusion of additional types of communications, preserving the object-oriented approach used in the message model. In addition, the architecture permits a message to be stored in multiple formats. Thus, an audio teleconference can be stored as both a digital audio file and a text-based transcript. The digital audio file is able to preserve information such as emphasis and tone of the participants, while the transcript can be used to provide search capabilities.

41.3.2 Message Model

Archiving design communications requires the development of an effective messaging model which can capture the information needed for efficient storage and retrieval. A message model can be viewed as an ontology to represent the semantics of information exchange among design agents. Our message model has been defined with the eXtensible Markup Language (XML) (World Wide Web Consortium, 1999) and encapsulates information used both to drive database storage of communications and to provide contextual information to client tools receiving the communications. In this way, the message model represents the context and content of communication. This, for example, can include pointers to the design project being discussed (as extracted from a design authoring tool, such as SDRC I-DEAS). All messages share a common set of fields that include participants in the communication, date, subject, and the actual content of the communication. We distinguish between several types of subject, such as *inquiry*, *problem*, and *change request*, to help classify the role of a communication In the design process. Our model can be extended with XML to accommodate other communications paradigms and design contexts. For example, one could extract structure-behavior-function from the CAD environment (Szykman, *et al.*, 2000) and associate it with the collaboration. An example is shown in Figure 41.3b.

41.3.3 Implementation and Example

We are using MITRE's CVW (Collaborative Virtual Workspace, 2000) as the collaboration server and to provide client-side video conferencing, audio conferencing and shared whiteboard. Netscape Messenger has been selected as our e-mail client. The collaborative tools have been integrated with the I-DEAS Master Series CAD software suite from Structural Dynamics Research Corporation (SDRC). Our client communication tools have been designed as modules that work in concert with standard communication software, so that users will not be required to switch from those they are already using to gain the benefits of archiving. The e-mail support module extracts project information from the CAD package, encapsulates it as a partial XML message (to be completed by the archiving server, as discussed below), and starts a chosen e-mail client with this message specified as an attachment and the archiving server as one of the recipients. This sequence can be initiated by an entry in I-DEAS' Application Launcher, making this functionality immediately available from within the CAD environment. The communication support module at the receiver functions as a MIME content type handler that can be registered with the receiver's e-mail package to handle attachments corresponding to the XML-based message model. When invoked, this module instructs the CAD package automatically to display the part referred to in the communication.

The archival server stores received communications in the design repository database according to information in the attached XML message and communication headers. The partial XML representation containing design information extracted from the CAD package is first completed by the archiving server with information extracted from the received communication, and is then used to direct the storage of the data in the various tables in the database. Conversely, the retrieval server generates and returns XML messages to structure the information as specified in the message model. Search access to the communication archive is obtained through a standard Web browser; the MIME content handling module can be registered with the Web browser to provide automated display of design data.

Presently, we are implementing multi-modal collaboration agents with CVW's Intelligent Multi-modal Participant interface. These agents will be able to archive different conversation threads and collaborative exchanges (other than e-mail) and extract design context from the CAD environment.

41.3.4 Scenarios for Use of the DCDS

We have chosen to examine design teams working on *mechatronic systems*: electro-mechanical systems that combine electronics and information technology. These designs include mechanical, electronic and software components and often involve many multi-disciplinary trade-offs among differing design domains (mechanical, software and electrical). We present three scenarios in which the DCDS can assist in the design process. The scenarios address distinct aspects of design efforts: communication; design information repository access; and design reuse and adaptation.

41.3.4.1 Scenario 1: Tools to Enhance Communication During Design

The tools described above have features that should make them attractive and useful to designers in addition to their archiving abilities, particularly regarding their ability to enhance communication at a distance. In this scenario, we envision several designers working on a mechatronic project, in the detailed (CAD) design phase, and using our communications tools together with the SDRC I-DEAS design suite. We imagine the designers as being physically separated, though the capabilities to be discussed below would be advantageous even for engineers working at the same facility.

In any detailed design process, there is often a need for engineers to communicate about parts in a project, and specific features of parts in the design In question. To communicate with another designer, an engineer must determine and enter specific file system or data installation information to identify the project, model file and part. In addition, she must find a way to refer to a specific feature of the part. Virtually all CAD packages provide means for designers to name features and dimensions to facilitate such communication. However, this places two burdens on designers: first, they must provide (unique) names for the features and dimensions about which they wish to exchange information; and second, they must either name all features and dimensions, or decide at the outset which are likely to require reference in the future. Neither of the latter two possibilities is attractive (or perhaps even feasible). CAD packages will often provide automatic naming of features (as in SDRC I-DEAS), but these names are generally very generic and unsuggestive (e.g., 'dimension 13'). Thus, an engineer sending a communication must determine and specify precisely the feature name to which she wishes to refer, in addition to the specific object and model-file-data-installation information and part name. This places a burden on the initiator of the communication.

An engineer receiving a communication about a specific feature or dimension, referred to either by an assigned or automatic name, must either be familiar enough with the detailed design to know to which feature/dimension the name refers, or must use tools provided by the CAD system to search for the named feature in the model. In addition, the communication must contain enough information for him to be able to locate unambiguously the project and model files, and the part within these being discussed; even armed with this information, it will be up to the engineer to take whatever steps are necessary to be able to view the part under consideration using his CAD system (e.g., start up the CAD package, browse for the project/model/part-file information, search the displayed part for the feature or dimension in question). This constitutes an unnecessary burden for the recipient.

In the Collaborative Studio, we have integrated communication tools with CAD tools to automate and facilitate reference to projects, models, parts and features. Our principal goal was to enable efficient and complete archiving of the design process; however, the capabilities incorporated into the tools also enhance communication during design. For example, the client support tools are designed to communicate with the I-DEAS CAD application automatically to include reference information about the project, model and part that is currently displayed on the designer's workstation, as well as any selected (highlighted) features of displayed parts. The initiator of an e-mail message will have this information automatically incorporated into the message, alleviating the burden of having to enter this information manually. The designer will have the freedom to refer to the feature using whatever name she chooses ('the pad next to the casting riser', perhaps); the information needed to identify it unambiguously will have been embedded into the message and can be used by client support tools automatically to display on the recipient's screen the specific part(s) and selected features referred to in the message.

While these are relatively minor benefits when compared with the advantages offered by the archiving capabilities of the tools in the suite, these enhancements to communication should make the tools attractive to members of the design team.

41.3.4.2 Scenario 2: Utilizing Currently Archived Communications During Design

In complex design projects, such as in the aerospace and automotive industries, the time between conceptual design and production of an artifact may be measured in years. In such a case, members of the design team will be unable to recall all of the trade-offs and rationales used in developing the design, owing to individual limitations and personnel changes. It may be essential for team members to have access to documentation indicating reasons for specific design decisions, so that, as the design evolves, they will be aware of important assumptions made in original decisions when they consider modifying previous choices (e.g., the re-routing of electrical or hydraulic lines). Formal documentation produced during the design process may indicate the rationale for a decision and conditions that are considered to be essential to support this decision but it may not include a discussion of alternatives considered acceptable but not chosen. Efficient access to the original communications that took place during the decision-making process can make the team aware of both alternatives and hidden assumptions. The Collaborative Studio is designed to make this information readily available. A designer looking at a particular hydraulic subsystem in an aircraft may wish to be aware of any previous change notifications in the routing of the lines before re-routing, to be certain that all conditions that need to be satisfied are maintained. He could then use the Studio's search tools to search all communications in the archive for those that pertain to the system in question and are indicated as change notifications. In addition, he may wish to search either the subject field or the text of the communication for specific terms or phrases such as 'routing' or 'hydraulic lines'. Retrieved e-mail communications that pertain to CAD model files could then be viewed in the e-mail client, which could be configured automatically to open the relevant model file and highlight the part or feature in question, as discussed above. Such information could inform a designer rapidly about important considerations to keep in mind when considering modifications.

41.3.4.3 Scenario 3: Design Reuse and Adaptation

In this scenario, a team of designers starts work on a new project. After goals and specifications have been defined, the designers will be interested in determining whether any design project in the past resembles the new one, for possible reuse or adaptation purposes. A search for relevant previous designs could, for example, be performed by representing the current design as a conceptual model (Anthony *et al.*, 2000) and performing a search on the design repository. The search interface to DCDS is then used to find communications associated with similar projects relating to their conceptual design, to provide guidance to the team and alert designers to problems and pitfalls that arose in previous efforts. At a later stage in the design process, during detailed design, designers can search communications from closely related projects for information pertaining to specific parts corresponding to ones in the current model. A designer can look at the CAD models for a previous project and find out why particular design decisions were made by looking at appropriate archived communications. For example, a design team may have identified a particular related project and a designer may be interested in finding out how and why a certain part appearing in both the current and previous models was changed during the design process. The project data can be searched for communications referring to the name of that particular part, retrieving any communications where the part in question was discussed. This can reveal information regarding important factors to consider in the design of the new part as well as the potential suitability of the part previously designed for use in the current project. For example, assumptions about the part's operating environment (expected stresses, temperatures, *etc.*) may suggest or preclude direct reuse, or indicate areas in which changes are likely to be needed (such as material or section thickness), based on previous testing. A search might also be restricted to communications involving a specific designer known to have had an important role in resolving a particular issue, or those having a specific subject, keyword, message type (inquiry, request for change, *etc.*), or other attribute. Additionally, searches may be made for parts that are not necessarily functionally related to a part in question but that have similar Model Dependency Graphs (MDG), by searching over MDGs in the repository (Cicirello and Regli, 1999). An MDG indicates the sequence of solid modeling operations to create a CAD model of a part; as such, the MDG is closely related to the sequence of manufacturing operations that can be used to produce the part. Communications about the production of parts with similar MDGs to one being designed may thus provide information on issues related to part manufacture.

41.4 Discussion

This paper has presented our architecture and initial prototype for a Collaborative Design Studio. We believe this research will create a unique environment to address multi-disciplinary issues in networked engineering:

- **How Will Networked Engineering Environments be Evaluated?** Past research in engineering design has yielded many successful tools for performing the detailed drafting and data management central to the modern engineering enterprise. As design is increasingly performed by distributed teams working over information networks, the design process becomes one of both engineering and computer-supported collaboration. We are interested in building and evaluating environments for collaborative and distributed engineering.
- **What Technologies are Required to Create and Manage Large-scale Engineering Digital Libraries?** Increasingly, engineers depend on PDM systems, large-scale engineering digital libraries, and knowledge-bases of CAD solid models (Regli and Cicirello, 2000). Current usage patterns often involve searching through vast amounts of corporate legacy data and navigating manufacturers' catalogs to retrieve precisely the right components to assemble a new product. Such 'design repositories' are limited in their ability to capture design intent and to reason about CAD knowledge, often relying on the textual annotations of the designer, or the part's file-name.
- **Can Models of the Collaborative Process be Used to Build Tools that Capture Design Rationale?** Design rationale is an explanation of why an artifact, or some part of it, is designed the way it is (Hu *et al.*, 2000) – including all the background deliberating, reasoning, trade-offs and decision-making in the design process. This information can be valuable, even critical, to various people who deal with the artifact. The DCDS will create a testbed for exploring how to capture rationale from IP-based collaborative work sessions.
- **How can Design Knowledge be Indexed for Reuse?** We are exploring graph and semantic-network-based knowledge representations for modeling collaboration networks. Indexing, retrieval and reuse from the repository can then be approached as a problem of structure matching and graph matching – problems for which there is a wide body of domain-independent heuristic and approximation techniques.

Our future work will include: development of a general CAD/communication interface architecture; ontologies for structure of design information repositories and the evolution of the design communication message model. We believe that DCDS will enhance our understanding of the design process by providing an architecture for archiving communications from the entire design process and tools to support the transparent storage and ready retrieval of communications. We feel that collaboration software and network technologies have matured and that many new research questions can now be addressed via integration with real design tools and tasks.

41.6 Acknowledgements

This work was supported in part by National Science Foundation (NSF), Grants CISE/IIS-9873005 and CISE/IRIS-9733545. Additional support was provided by AT&T Labs. Any opinions, findings and conclusions or recommendations expressed in this material are those of the author(s) and do not necessarily reflect the views of the NSF or the other supporting government and corporate organizations.

41.7 References

Collaborative Virtual Workspace (2000) Mitre Corporation http://cvw.mitre.org

Eroom (2000) http://www/eroom.com

Teamwave (2000) http://www.teamwave.com

Anthony LPA, John JE, Lombeyda SV, Regli WC (2000) Cup: A computer-aided conceptual design environment for assembly modeling. IEEE Computer Graphics and Applications (under review)

Baya V, Gevins J, Baudin C, Mabogunje A, Toye G, Leifer L (1992) An experimental study of design information reuse. In: ASME Design Engineering Technical Conferences, 4th International Conference on Design Theory and Methodology, New York NY, USA, September 13-16, Scottsdale, Arizona 1992. ASME Press, pp141-147

Cicirello V, Regli WC (1999) Resolving non-uniqueness in design feature histories. In: Anderson D, Bronsvoort W (eds.) Fifth Symposium on Solid Modeling and Applications, New York, NY, USA, June 8-11 1999. ACM, ACM Press, Ann Arbor, MI

World Wide Web Consortium (1999) XML. http://www.w3c.org

Fruchter R, Reiner KA, Toye G, Leifer LJ (1995) Collaborative mechatronic system design. In: 2nd International Conference on Concurrent Engineering, Research and Applications, August 23-25 1995

Greenberg S, Roseman M (1998) Groupware toolkits for synchronous work. In: Beaudoin-Lafon M (ed.) Computer-Supported Cooperative Work, Trends in Software Series, John Wiley and Sons Ltd., pp135-168

Grudin J, Poltrock SE (1997) Computer supported cooperative work and groupware. Advances in Computers 45:269-320

Herlea D, Greenberg S (1998) Using a groupware space for distributed requirements engineering. In: WET ICE '98: IEEE Seventh International Workshops on Enabling Technologies: Coordinating Distributed Software Development Projects, Stanford University, California, USA, June 17-19 1998

Hu X, Pang J, Pang Y, Atwood M, Regli W, Sun W (2000) Design Rationale: A survey of representation, capture and reuse. (in preparation)

Mark G, Grudin J, Poltrock SE (1999) Meeting at the desktop: An empirical study of virtually collocated teams. In: 6th European Conference of Computer Supported Cooperative Work, September 12-16, Copenhagen, Denmark 1999

Regli WC, Cicirello V (2000) Managing digital libraries for computer-aided design. International Journal of Computer Aided Design, 32(2):119-132, Special issue: CAD After 2000, Rezayat M (guest ed.)

Szykman S, Bochenek C, Racz JW, Senfaute J, Sriram RD (2000) Design repositories: Next-generation engineering design databases. IEEE Intelligent Systems (in press)

Toye G, Cutkosky MR, Leifer LJ, Tenenbaum JM, Glicksman J (1993) Share: A methodology and environment for collaborative product development. In: IEEE Workshop on Infrastructure for Collaborative Technologies, also available as Stanford CDR-TR #19930507

Ullman DG (1992) The mechanical design process. McGraw Hill, Inc.

Yen SJ, Fruchter R, Leifer L (1999) Facilitating tacit knowledge capture and reuse in conceptual design activities. In: ASME Design Engineering Technical Conferences, 11th International Conference on Design Theory and Methodology, New York, NY, USA, September 12-16, Las Vegas, NV 1999. ASME Press DETC99/DTM-8781

Chapter 42

Consumer-designer Interaction Through a Generative Design Medium

M. Kälviäinen and M.S. Pontecorvo

42.1 Introduction

In the past, the linkage between designer and consumer was often indirect, mediated primarily by marketing departments and ultimately by a product's performance in the marketplace. With advances in technologies and methodologies aimed at creating a more efficient, flexible, and responsive product development process, new possibilities for the sharing of information between consumer and designer are becoming available. New consumer-centric, participatory models arise from these possibilities, incorporating near real-time design research and the direct involvement of the consumer in the product design process. The product design itself becomes the central medium of communication between the stakeholders (designer, fabrication, marketing, and the consumer) in the development process.

This paper introduces a generative software system for consumer interaction and product image development in the design process. A suitable way of testing the consumer's holistic product perception for design purposes is to have available a wide set of product choices from which the consumer can make their preference distinctions. One possibility of realizing this set of choices in the design process is to create a vast amount of new product images through generation.

42.2 Pleasurable Distinction in Consumer Experience

The emphasis on aesthetics in everyday life, individuality, and the rise of the inner-orientated, self-searching consumer are evident in many contemporary lifestyle descriptions. Consumers seek to create their own lifestyle that focuses on authentic concerns of identity expression (Dant, 1999 p30; Chaney, 1996 p29, p122). The comsumer's desire for self-expression raises questions about the customization of deep experiences in products. The current discussion about 'experiential' design concentrates on these questions (Margolin, 1997; Designing the Experience, 1996; Jantzen and Ostergaard, 1998 pp126-128).

Useful theories for understanding product perception, meanings and pleasure can be found in consumer research, psychology, anthropology, semantics, the cognitive sciences, the sociology of taste, aesthetics and usability studies. The basic issue is that pleasurable product experience is based on the human capacity to make distinctions between physical objects and to favour some of them over others.

The product judgments we make reflect the inner map that we used unconsciously to organize and understand sensory experiences. Our mental maps are made up of iconic imagery that is not easy to describe in words. Most human meaning is shared non-verbally through intuitive emotional understanding that involves interdependencies between cognitive, physiological, expressive, and phenomenological components. Verbal information scales cannot grasp all these components of emotional experience. Visually or sensually delivered information offers experiential and emotional knowledge that prompts us to consider, and perhaps vicariously feel a part of, a particular person, situation, or event (Holbrook and Hischman, 1993 p20; Belk 1998 pp294-295; Zaltman, 1997 pp427-428). In the design task, visual and other sensory information helps designers to be empathic through emotional understanding.

The same words can be interpreted in many different visual ways and they may carry very different value-based meanings for different people. So it is possible to talk in words about a certain product image whilst having a quite different understanding about it and to talk with different words about the same image as you can appreciate it for different reasons. It is essential to have knowledge other than mere words from the consumer about the images and feelings that correspond to consumer preferences.

Categorization and prototypicality have a profound influence on our object perception and judgments. Categorization is the mental activity of grouping like things together in mental conceptual categories. The boundaries for certain categories are fuzzy and dependent on context. Also, experiences, familiarity and frequency of encounter with the species of the category produce differences in categorization. The conceptual representation of a certain category is lodged in prototypes that combine the characteristics of the typical members of the category. Typical category members will possess more typical features in common than atypical ones, producing a family resemblance (Roth and Bruce, 1995 pp19-54, pp73-95, pp111-130). The acceptable level of deviation from typical features and layout forms the possible range of variation in a category member. Designers have to have knowledge of how far from the prototypical image of the product they can drive the design.

A product has perceivable qualities called affordances from which we can understand its scenario of use; this reveals to us the possible action and motion associated with products. It is important to visualise categories as objects, because the designer's mental category for a certain product may be different from the one in the consumer's mind. If a product is categorized on first impressions as suitable for meeting the consumer needs, it is likely to have to go through further stages of consideration. Even though a product might be functionally ideal, it might fail because it conveys the wrong associations (Shackleton, 1996 p35). Categorization goes as far as identifying suitable user ages, genders and roles for the product.

Preference through typicality arises because a more typical is easier to recognize and to process. It demands expertise, involvement, and a higher focus of attention to become interested enough to engage with unfamiliar objects (Bloch, 1995 p20; Crozier, 1994 pp69-70). Yet there are conflicting and differing results in research about typicality and preferences. The relationship between preference and typicality differs between product groups (Shackleton *et al*, 1999). The consumer segment group in the context of the product's real-life situation also affects typicality preferences.

People differ in their expressed interest in stimuli. Sources of arousal in certain physical properties of environmental stimuli include ambiguity, complexity, novelty and potential for surprise. A more ambiguous or novel object will produce greater arousal than a less ambiguous or a familiar one (Crozier, 1994 p62). Pleasure arises from the reward-value of novelty and change, but too high a degree of novelty can be perceived as distressing and threatening. Novelty, change, and variety - within recognizable boundaries - are pleasant (Bianchi, 1998 p74-76). The consumer's pleasurable stimulation level and their acceptable range of object variation should be studied in design tasks.

There are two approaches to product-related beliefs: holistic visual perceptions of the product form, or linear processing of one design element at a time. It is possible to assume that both occur. The product may first be perceived as a whole. Individual elements are considered on further processing (Bloch, 1995 p19). Categorization and novelty perceptions are examples of how our brain organizes visual input into perceptual wholes, or Gestalts.

According to the Gestalt position, the whole has an organization that has its own dynamic properties that cannot be reduced or broken down to their parts. In product, the Gestalt produces the intensity, complexity, coherence, and order that have been recognized as important factors of pleasure in product perceptions (Crozier, 1994 p41, p64-66). The perception of the whole serves a useful purpose in revealing the whole flavor of the product and the emotions and experiences it evokes. The designer needs information about the general feel of the ground from the detailed features of the product and its brand image. The general feel and flavor can be represented in different visual compositions that are either acceptable or not.

Style discrimination is an essential part of aesthetic enjoyment. Style differentiates products through their distinctive quality or form, their holistic composition of color, shape, line, pattern in visual style, sound, and scent. Style causes intellectual and emotional associations because it is a multi-sensory device easily triggering our emotional understanding (Schmitt and Simonson, 1999 p84-85). It is thus important for our embodied experience. Different bipolar style traits such as romanticism/classicism form part of the personality variables (Bloch, 1995 p23). Style deals in surface impressions, yet it forms a corridor between the world of things and human consciousness (Ewen, 1990 p43). Style perception can be conveyed to the designer in holistic product images.

Especially in stylistic evaluation, certain people with design acumen have quicker sensory connections and exhibit more sophisticated preferences regarding the design of things than those with little design acumen. A person with expertise learns what to look for in certain product styles and what the important determinants of attractiveness are. Levels of design acumen and experience in the target segment are important guides for design. There is also no point in producing marginal variations, which are not perceived as different (Jones, 1991 p49-51; Bloch, 1995, 22). These nuances in holistic differentiation could be studied through wide sets of slightly varied images.

Our interpretation reflects an inner map that is used unconsciously to organize and understand sensory experiences. We interpret intangible product meanings with the help of associations, or 'Stimulus Chaining' that occurs in human cognition. One thought leads literally to another. Thoughts are not language but often images. A secondary stimulus is paired with a discriminative stimulus resulting in the secondary stimulus becoming a condition reinforced in its own right. In this way, stylistic attributes can become strongly associated with other product features. Low, sleek cars might be imagined to be fast (Shackleton, 1996 p15; Zaltman, 1997 pp427-428). The product meaning interpreted through associations has a profound effect on how pleasurable our experience of it is. We associate and feel pleasure in products through their metaphoric associations (similarity to other domains), narratives, and even object animation or personification. These factors are all connected to holistic product perceptions.

The warmth, cuteness, sincerity, or other such character qualities in the product Gestalt can be intuitively interpreted because of the associations of meanings. Product character is often described in words in the design brief. With the help of a large set of images it is possible to research these words, which often derive from the brand image-to-image connections in the consumer's mind.

Meaning is interpreted both from denotative and connotative messages in the product. In denotation an object conveys information about its functions and what it stands for. Connotation refers to an aesthetic dimension, which conveys a subjective impression and emotion about the product. There is the possibility of conflict in these different messages: the shape of the object can indicate its use but also adds an aesthetic dimension, which may contribute to a re-definition of the object or relate to its function (Heilbrunn, 1998 p196). As the nature of our lives and understanding is conflicting and complex, this offers possibilities for conflicts and openness in the meaning of the product. But there has to be enough consistency to give cues for understanding.

Emotive and cognitive elements interrelate in the structure of thought. Any judgment task is likely to contain an interactive mixture of both constructs, most of it unconscious (Zaltman, 1997 pp426-427). Product preferences are formed on the basis of aesthetic pleasure and socially determined preferences. They rely on form and content that is inherently combined in a product. Perception and semiosis are interrelated. It is difficult to link them to separate product attributes or features. Also our relation to product meanings seems to encourage the use of holistic Gestalts in information acquisition about pleasurable products.

Visuals can be used in many ways to aid the study of consumer experience. They can serve to authenticate details about consumers and their groups. They can make the consumer's experience visible. Visuals can also be a collaborative support to the results expressed in words. Design deploys visual messages that the brain is able to grasp without conscious analysis (Hine, 1995 p237). The design process is intuitive and holistic and dwells in a strongly non-verbal culture. The styling process is holistic, concerned with envisaging the overall design solution as a visual entity (Tovey, 1997). As designers think intuitively and visually it might be easier for them to apply consumer information gained in images than in verbal information. The consumer information elicited in consumer-centered design has often been in the form of words describing and translating the product concept. The research might, however, be done with more holistic and visual approaches using , for example, associations with mental images.

The software presented in this paper is based on the idea of holistic Gestalt distinction of products, because it seems to offer the most practical and comprehensive way of inquiring about intuitive product perceptions. The information can be used quantitatively and qualitatively; it can be combined with words representing the images, and especially it can work inspirationally and intuitively in visual design tasks. Through a vast amount of images the results can reveal the consumers' mental images, clarifying nuances in holistic understanding of products.

42.3. Product Generation

Using a generative software-based system it is possible to synthesize images of as yet non-existent products to match the consumer's mental images and future aspirations.

This type of system can support an unprecedented level of interaction between the stakeholders in the product design process (designer, manufacturer, marketeer, consumer).

42.3.1 Generative Software

The product image generation system is based on the GDS (Generative Design System) framework described in Pontecorvo (1998) (see Figure 42.1). A biological metaphor of 'breeding' is used by this system, allowing the user to rapidly gain an internal 'theory of operation' for the generation process.

Initially, the designer defines the design instance representation as a set of parameters and constraints. In this biological metaphor, the representation can be thought of as the 'DNA' of a 'species' of product. These parameters and relationships embody aesthetic, material, structural, and cost factors involved in fabrication, packaging, and distribution of the target product.

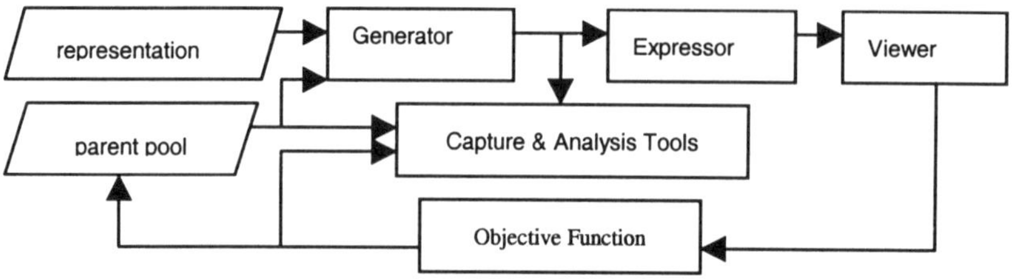

Figure 42.1. Generative image synthesis system flow

Operationally, the consumer is presented with a set of design instance variations generated in a generate-review-select cycle. In each successive generation the consumer selects instances to be included in the parent pool for subsequent generations. This generate-review-select cycle can be repeated as required to meet experimental objectives.

The capture and analysis of consumer session activity enables the designer to observe and understand the trends and patterns of the consumer's exploration of the space of designs. This information forms an essential element of exchange between the consumer and designer. With it the designer can rapidly react to consumer activity.

42.3.2 Selection Session Preparatory Activity

To aid the user's understanding of the system's operation and their own personal objectives in the selection process, the basic generative selection system can be enhanced by adding a preparatory session.

By allowing the consumer to engage in the activity of expressing internal goals for the selection process, the consumer can start to build a set of selection criteria. Whether through free-form narrative or through selection of properly crafted inventories of categories and questions, this activity can help put the consumer in the frame of mind to be receptive to subtle relationships between visual cues and 'extra-geometric' properties of the product's design.

42.4 Mass Customization and the Consumer

The generative product image system described here for consumer product taste research with slight modification is being applied to drive the serial manufacturing process utilizing computer-controlled machines to form and assemble entire products. This lets the consumer guide the process of creating products that reflect their own personal needs, desires, and lifestyle.

Mass customization, the ability to create products that are tailored to an individual consumer's needs and desires at a cost approaching a mass-produced product, is attractive both to consumer and to manufacturer (Pine, 1993). Adding generative design techniques extends mass customization beyond the configuration of standard elements (e.g., Ikea's furniture configuration, Levi's designer jeans, *etc.*) to the creation of unique component forms with standard operational interfaces.

The type of software presented here is already taking the experiential product qualities into consideration more than marketing or usability research has previously. This generative interaction system is an improvement in the move towards developing holistic product image information. With the continued addition of research into consumer experiences and object perception, it can be enhanced to better serve the consumer-centred design processes. The next step would be to increase the product constellation, use environment, and multi-sensory perception possibilities in the alternatives offered by the generation process.

42.5 References

Belk RW (1998) Multimedia approaches to qualitative data. In: Stern BB (eds.) Representing consumers. Voices, views and visions. Routledge, London New York

Bianchi M. (1998) Taste for novelty and novel tastes. The role of human agency in consumption. In Bianchi M (eds.) The active consumer: novelty and surprise in consumer choice. Routledge, London

Bloch PH (1995) Seeking the ideal form: product design and consumer response. Journal of Marketing 59:16-29

Chaney D (1996) Lifestyles. Routledge, London New York

Crozier R (1994) Manufactured pleasures: psychological responses to design. Manchester University Press, Manchester

Dant T (1999) Material culture in the social world: values, activities, lifestyles. Open University Press, Buckingham

Designing the Experience (1996) Strategies for developing consumer products and services. Design Management Journal 7 (4)

Ewen S (1990) Marketing dreams. The political elements of style. In: Tomlison A (ed.) Consumption, identity, and style: marketing, meanings, and the packaging of pleasure. Routledge, London New York

Heilbrunn B (1998) In search of the lost aura: the object in the age of marketing romanticism. In: Brown S, Doherty AM, Clarke B (eds.) Romancing the market. Routledge, London New York

Hine T (1995) The total package: the evolution and secret meaning of boxes, bottles, cans and tubes. Little Brown, Boston

Holbrook MB, Hirschman EC (1993) The semiotics of consumption: interpreting symbolic consumer behaviour in popular culture and works of art. Mouton de Gruyter, Berlin New York

Jantzen C, Ostergaard P (1998) The rationality of 'irrational' behaviour: George Bataille on consumer extremities. In: Brown S, Doherty AM, Clarke B (eds.) Romancing the market. Routledge, London New York

Jones PL (1991) Taste today: the role of appreciation in consumerism and design. Pergamon Press, Oxford Seoul New York Tokyo

Margolin V (1997) Getting to know the user. Design Studies 18:227-236

Pine BJ (1993) Mass customization: the new frontier in business competition. Harvard

Business School Press, Boston Mass
Pontecorvo MS (1998) Designing the undesigned: emergence as a tool for design. In: Generative Art Proceedings of the 1st International Conference Generative Arts '98. Editrice Librerie Dedalo, Rome
Pontecorvo MS, Elzenga RN (1999) Exploring designer/consumer dialog in evolutionary design systems. In: AISB '99 Conference, Society for the Study of Artificial Intelligence and Simulation of Behaviour. Edinburgh, Scotland
Roth I, Bruce V (1995) Perception and representation: current issues. Open University Press, Buckingham
Schmitt BH, Simonson A (1999) Marketing aesthetics: the strategic management of brands, identity, and image. The Free Press, New York
Shackleton JP (1996) The application of a 'prototype theory' framework to the modeling of product perception and the emergence of new product groups. Chiba University, Graduate School of Science and Technology
Shackleton J, Jung-Wan H, Kazuo S (1999) 'Typicality' and consumer preference: prototype theory in product design. In: Design cultures: proceedings of the third European academy of design conference (vol 2). Sheffield Hallam University, 30th March - 1 st April 1999
Zaltman G (1997) Rethinking market research: putting people back. Journal of Marketing Research Nov 4:24-437

Part V Computer-mediated Collaborative Design

Summary

The research presented in this Part attests to the vital role that computer-based digital technology can play in supporting the activities of collaborating design groups, where the effective exchange of *knowledge* is of primary importance. All of the contributions succeed in demonstrating how the fluent communication of knowledge between designers can readily be supported through carefully constructed computer-based systems and tools. For example, Seitamaa-Hakkarainen, Lahti, Muukkonen and Hakkarainen's networked learning environment (FLE) incorporates tools that are seen to facilitate collaborative knowledge building in relation to conceptual and visual ideas. The virtual design studio (VDS) described in Kolarevic and Ng's paper successfully enabled collective design authorship and evolutionary idea-development via a central database augmented with video-conferencing and web technology. Rodgers, Caldwell and Huxor similarly demonstrate how their web-based, design-support system (WebCADET) allows designers to generate, amend and share knowledge in a transparent manner that facilitates the adaptation of existing consumer products to meet specialised markets. In addition to the topic of knowledge interchange in collaborative design, two further themes appear to dominate the contributions in this Part.

One such theme relates to the growing realisation within the CAD-research community that digital support for early, *conceptual* design is a vital component of software-support for collaborating designers. The computer-based support of conceptual design is associated with some unique challenges, including those relating to the need to accommodate the vital processes of collaborative problem structuring, idea sketching and imprecise concept modelling. Zaychik, Sevy and Regli's DCDS system, for example, allows users to discuss early design ideas and generate high-level conceptual models for them without having to perform detailed three-dimensional CAD work. These conceptual models can then be used to constrain a search of an associated repository of previous design information in order to find communications related to similar previous projects. Retrieved information can thence provide early guidance to the team about the potential problems and pitfalls that arose in previous, related design efforts. Seitamaa-Hakkarainen *et al.*'s FLE environment incorporates tools for supporting the determination and structuring of design requirements and constraints, and for facilitating the generation, explication and externalisation of high-level, conceptual ideas. Data analysis indicated that the FLE environment was readily able to support the students' conceptual design activities - such as problem structuring - at levels comparable to those typically displayed by expert designers.

The second recurrent theme that emerges from the contributions in this Part concerns the unique power of computer-based technology to facilitate access to vast repositories of potentially useful design knowledge (e.g., in the form of previous design cases and CAD models, manufacturers' catalogues and component libraries). As already noted, for example, the DCDS system of Zaychik *et al.* enables the designer to use a conceptual design that they have generated as either a starting point for further refinement or as a query to an intricate design knowledge-base that enables the retrieval of potentially relevant design information. Kälviäinen and Pontecorvo's system enables the creation of a vast repository of product-image variants that can be presented to consumers in order to explore their intuitive, stylistic preferences. Seitamaa-Hakkarainen *et al.*'s FLE environment enabled student designers to access valuable information from internet sources about the typical needs of premature babies for whom they were designing clothes. An eventual aim for Rodgers *et al.* is to extend their WebCADET system to include additional repositories of design knowledge, such as case-based representations of successful and unsuccessful product concepts and structured representations of the significant content present within design documents.

The common ambitions expressed in the contributions in this Part – for example, to facilitate effective computer-mediated knowledge exchange and to integrate computer-based, collaborative design environments with a multiplicity of design repositories and knowledge bases - are timely and well-motivated. Clearly the value for collaborative design of achieving such ambitions is potentially great. However, the realisation of such goals also foregrounds issues relating to the management of informational complexity. For example, access to a range of intricate knowledge repositories for the storage and retrieval of design information raises problems of optimal indexing and data management, and effective navigation and search. The rigorous empirical investigation of such problems and their potential solutions represents a challenging avenue for future research on computer-mediated collaborative design.

Part VI

Educating for Collaborative Design

Part VI Educating for Collaborative Design

Introduction

Collaborative design may still be considered a relatively new process. It generates additional problems for designers and non-designers. Many of these have been documented in the case studies presented at this conference. In this, the last part, attention is focused on the challenges these new methods of working present for the education of designers and sometimes non-designers as well. The remit of design education is broad and the papers in this Part reflect this, with attention focusing not just on degree level students but at sixth-formers; on changing the philosophy of design education, and case studies of undergraduate courses; on looking at social networking in design studios and the impact of technology (videoconferencing) on design students. Two themes run through the section; firstly the myth of the hero, or lone designer. Design remains exciting, innovative, creative and challenging, yet it cannot be undertaken by one individual, in isolation - it is a team process, requiring social, verbal, and interdisciplinary skills and commitment from those who wish to practice it. The second theme relates to the utility of design research, individual experience and reflective practice in informing design education.

The first two papers focus on aspects of the lone, hero designer and ways in which the barriers associated with this idealised view of 'the designer' might be broken down. In the first contribution, Yair, Press and Tomes consider the relationship between design cognition and verbalisation. Through a literature review and interviews they evaluate the detrimental effects that the marginalisation of verbalisation has on the personal and professional growth and effectiveness of designers. They use their experiences to design an educational tool which will introduce verbal discourse and negotiation in to an holistic pedagogy in which contextual understanding, professional skills and studio-based practice may be developed as an integrated whole. As Yair *et al.* focus on the inarticulacy of designers, so Brown and Moreau-Yates, in their paper on architectural design, focus on another issue seen as being a crucial impediment to co-designing, that of the disparity between the models of designers and non-designers. Architects and non-architects view the built environment in different ways - the architect focusing on perceptual values and representational meanings, the non-architects emphasising associational and responsive meanings. Such disparity becomes manifest in both linguistic problems and solutions which are a compromise rather than a synthesis. Environmental psychologists have pinpointed education as being the primary source of these differences in architecture. Education is seen as having a dual role, firstly in informing the non-designer, in this case the non-architect, and secondly by educating the designer to better interact with others so as to become cognisant of and learn to value disparities in perception. The latter is the focus of this paper.

Felton moves the discussion to the arena of product design and team working in his paper on the development of a BSc in Computer Aided Product Design (CAPD). He reviews the team support required by 'recent lone inventors' and surveys engineering companies, revealing the need for graduates to possess greater knowledge of team working and the place of design in the company. The need for graduates to gain such experience (for example, on sandwich courses on industrial placements) is discussed with reference to the added financial burden this places on the students. Professional practice is again the topic of Hargreaves' paper, which uses detailed case studies to develop a new course model on professional practice to form the basis of an industrial design programme course. The findings suggest that courses must deliver graduates with sufficient expertise to engage in professional practice. These include, the ability to listen, consider and work to a brief, deliver contracts on time and generate 'powerful, creative manufacturable product concepts which exceed the customer's expectations'.

McDonagh-Philp also concentrates on undergraduate courses, this time in relation to the need for providing opportunities for undergraduate designers to gain experience of user-centred design techniques (such as focus groups) and collaboration with product manufacturers to provide an appreciation of commercial context. Her case study documents both the benefits and the costs of student projects having a high degree of user involvement.

Molyneux moves the discussion into a new arena, that of secondary education, by providing an account of one College's twenty years' experience in collaborative, user-centred, communal design. The student-centred approach adopted to the tuition of sixth formers provides a foundation for the personal development of all students, whilst enhancing vision, creativity, skill transfer, design capacity and ability.

In focusing on the attitudes and behaviour of students towards the final exhibition, Molyneux shows the way in which students spark off each other. Ashton also considers this topic. Drawing on social science methodologies, she studies undergraduate design studios, in a formal analysis of studio usage and the emergence of socio-emotional leaders and task experts who create group norms which in turn contribute to the wider learning experiences of the cohort.

MacLachlan addresses studio culture of a different kind in the last paper of this Part. Increasingly designers will find themselves working remotely with clients as partners. In this paper, the experiences of Glasgow engineering students who undertook a project brief with their counterparts in Brataslava are considered with especial emphasis on cultural differences and the impact of technology.

Chapter 43

Look Who's Talking: Developing Communication and Negotiation Skills Through Design Education

K. Yair and M. Press

43.1 Introduction

Both the educational tool described here and the research underpinning it were funded by the UK Design Council, in order to implement the findings of previous research undertaken into communication in designer - client relationships. This research had identified the importance of the designer's capabilities in articulating verbal meanings associated with visual design, and conversely, in interpreting verbal messages in visual terms (Oates *et al.* 1997, Tomes *et al.* 1998). This process was, however, found to be inhibited by designers' lack of confidence in verbal discourse, reinforced, it was argued, by an undervaluing of verbal articulacy in design education.

Design Council support was secured for the development of a teaching tool to address this problem. Its fundamental objectives were to demonstrate to undergraduate design students the relationship between conceptual and practical thinking, and to develop the skills necessary for their communication and translation. Learning outcomes were defined as:

- understanding the relationship between visual and verbal communication in the design process;
- appreciating the nature of communication between designers and clients in professional design practice;
- demonstrating the skills necessary in negotiating design through the verbal - visual translation of design propositions.

Because the original research had focused on specific issues and cases, there was a need to broaden the research base to include pedagogic practice and design process theory. This informed a development programme which refined and tested a set of integrated teaching tools applicable across a wide range of design disciplines.

43.2 Literature Review

43.2.1 Verbal Discourse and Design Cognition

Design is recognised as a complex and distinctive form of intelligence, assimilating reasoned and intuitive thought, and visual and verbal forms of articulation (Coyne, 1991; Thomas-Mitchell, 1993; Cross, 1994). Perhaps uniquely, it draws upon the full range of spatial, linguistic, kinaesthetic and logical modes of thinking identified by the theory of multiple intelligences (Gardner, 1983), in order to apply both tacit and explicit forms of knowledge to the development of product solutions.

Despite wide acceptance of this paradigm, perceptions of design activity as intuitive, mysterious and predominantly visual persist amongst some theorists, who maintain the need for a suspension of verbal and reasoned thought if products are to be truly innovative or to make profound psychological connections with users (Ward, 1984).

Such assertions are, however, based on the premise that the location of design cognition in the brain's right hemisphere connects it with aesthetic perception, visuo-spacial awareness and intuition, whilst separating it from speech, analysis and logic (see Press, 1998). The complex picture of brain functioning revealed by more recent research may therefore challenge this argument. This redefines right hemisphere activities more broadly, to include the management of information and symbols and the processing of complex, multi-representational information (Cross, 1986). It also reveals an interdependency of right and left hemisphere activity, in which drawing is seen as supportive (Cross, 1994) rather than formative (Herbert, 1988), and 'intuition' as a process of integrating imaginative thinking with experiential knowledge (Cross, 1994). In this context, the verbal articulation of design thinking is understood to play a crucial role in bridging intuitive and reasoned thought, thereby stimulating creativity and facilitating evaluation, decision-making and problem-solving processes (Fleming, 1998).

This viewpoint is substantiated by linguistic theories asserting the fundamental importance of verbal language across human cognitive activities, as a means of stabilising and categorising thought in words and phrases (Vygotsky, 1978; Fielding, 1994). Such theories indicate a crucial role for speech in accessing these coded memories and perceptions, enabling them to be accessed and applied appropriately in planning and guidance activities (Vygotsky, 1978). It may be argued, therefore, that verbal articulation enables designers to reflect on experience, and to assimilate subsequent learning into the conceptual and methodological frameworks which guide professional practice (Mitchell, 1996). Only by combining verbal and visual forms of cognition and articulation can product development benefit from the full breadth of the designer's knowledge, and can design become an on-going learning process.

43.2.2 Verbal Discourse and Negotiative Design

Verbal discourse and reasoned thinking - whilst contributing to all design activities - may be considered particularly important in a negotiative design situation, where designers are required to work with others, or to respond to a complex, multi-faceted problem. The development of a hybrid verbal - visual language, where words and visual forms of representation play a mutually supportive role, appears to contribute significantly to its success in several respects. Although it has obvious applications in identifying parameters, requirements and evaluation criteria (Byrne, 1990) and in explaining proposals to non-design colleagues (Lawson, 1990), its true value is as a aid to cognition as well as communication.

Firstly, the verbal-visual language assists in averting the 'icon trap' (Lawson, 1990), a tendency for the 'design-by-drawing' approach to prioritise aesthetic considerations over those relating to a product's use and its impact on the physical and social environment (Jones, 1980). Ill-informed decisions with harmful implications can result, especially when the directness and apparent simplicity of renderings or models (Lawson, 1990) are combined with the high degree of realism achievable through product simulation technologies. Verbal articulation explicates and challenges the rationale behind the image, identifying hidden and potential problems (Lawson, 1990) and focusing on processes rather than their tangible outcomes (VanGundy, 1988; Lumsdaine and Lumsdaine, 1995).

Secondly, the verbal - visual language can clarify understanding of the proposed product's subjective aspects: its character or 'feel', its target market, and its potential users' values and preferences. Such considerations resist definition (Sack, 1998), yet contribute significantly to the coherence between consumer needs and product attributes (Zaccai, 1995) demanded by the increasing emphasis in design on product signification (Dant, 1999) and consumer experience (Rhea, 1992). Often embodied in visual 'mood boards' (Cooper, 1995), their encapsulation in adjectives and metaphors is advocated by semiologists as a means of bridging verbal and visual thinking through evocative description (Byrne, 1990). This form of word play assists in translating marketing knowledge into actual product attributes (Dumas, 1994) and can lead directly to product solutions through a process of 'verbal - visual punning' (Tomes *et al.*, 1998). It has also been found to contribute significantly to design coherence in complex products such as cars, enabling a single vision to guide the development of features from interior trim to engine noise (Clark and Fujimoto, 1990).

Thirdly, verbal - visual discourse is essential in involving non-designers in the product development process, allowing individual expertise to be explicated and synthesized (see Peters and Waterman, 1995). In the emerging knowledge-based economy (Drucker, 1993), product development values knowledge creation (Nonaka, 1990) and organisational learning (Edmondson and Moingeon, 1996) as well as actual product outcomes. Because innovation resulting from unique combinations of expertise is valued for its resistance to imitation (Rhodes, 1995) designers are collaborating increasingly with specialists as diverse as sociologists, psychologists, linguists (Thomas-Mitchell, 1993) and technologists (Armstrong *et al.*, 1999).

The verbal - visual language assists product development teams in overcoming the cultural and methodological differences between these functional specialists (see Walker, 1990), thereby encouraging the empathy, communication and social adroitness understood to be crucial to all business and organisational activities (Goleman, 1995, 1998). This allows designers to undertake the strategic roles (Corbin, 1996; Ruekert, 1995) and broad responsibilities (Zaccai and Badler, 1996) demanded by negotiative design, and to develop the on-going alliances (Bruce, 1998, Jevnaker, 1998) characterising it. As such, it is essential in the shift in emphasis from products to processes and systems (Thomas-Mitchell, 1993).

Designers need both verbal and visual articulacy, if they are to involve their colleagues in the 'reflective conversation' characterising design (Schön, 1983; Forester, 1989; Cuff, 1991). They need to challenge the Modernist perception of the visual as a universal language (Metcalfe, 1993; Itami and Roehn, 1987), and instead to develop a hybrid verbal - visual language capable of explicating complex design thinking and involving colleagues in the design process.

Negotiative design requires particular attitudes as well as skills: designers and their colleagues need to value collaboration as a creative catalyst rather than a constraint, and design as the embodiment of diverse expertise in product form, rather than an individualistic act of creation. They need to accept the product development team's diversity as its greatest creative asset, but also a potential source of friction, as individuals bring particular value systems, working methodologies and objectives as well as expertise (Sonnenwald, 1996; Bucciarelli, 1988). They need to value conflict as chaotic yet ultimately productive (Kernan, 1997), and to develop the communication and negotiation skills which prevent it from becoming personalised and therefore detrimental to creativity (Lumsdaine and Lumsdaine, 1995).

43.3 Talking Design

43.3.1 The Problem

The skills and attitudes required for negotiative design practice are said not to be developed during designers' education, but through professional experience (Sonnenwald, 1996; Press and Cusworth, 1998; Partridge, 1997; Ball and Price, 1999). This theory is confirmed by interviews held with design tutors, who criticise students' verbal articulacy and their reluctance to contextualise and inform design through discussion:

> *'...They can't give a rationale for their idea: they'll say "it just came to me"...... they can't reason about the choices they make: they describe it as 'messing about' ......And that reflects in the poor standard of presentations they make to clients at the end of the live project.'*

> *'...Students go to see the technicians and end up with something totally inappropriate, because they've explained it wrong to the technician who's trying to help them.'*

'...The way that they gather information is very shallow: it's a case of asking one person's opinion or reading one article, then seeing that as the solution to the problem, whether it's studio work or an essay. They'll take that one thing as gospel, and it produces very one-sided results.'

The conceptual and methodological approaches evident in these comments suggest that, despite the negotiative paradigm prevalent in theoretical discourse, an outmoded view of design as individualistic, intuitive and predominantly visual persists amongst design students. Mitchell (1996) suggests that such a stance can be implicitly embedded in an institution's culture and endorsed by tutors, even when analytical engagement with creative work is officially endorsed. This may explain why, despite their criticisms, some tutors describe a wariness of design briefs potential to 'stifle creativity' and question the validity of 'trying to articulate how inspiration strikes'.

Understanding of negotiative design - and the communication and negotiation skills that it requires - are often assumed to be developed through live projects, team projects, tutorials and crits. However, both the interviews and literature review indicate that the validity of this assumption is questionable. Assessment by critique, for example, encourages students to adopt a defensive stance which precludes meaningful dialogue whilst emphasising product outcomes rather than cognitive processes (Hall-Jones, 1996). As one design tutor states,

'...They (students) don't like arguing in tutorials and crits. They don't want to risk being tripped up.'

Applications for such discourse beyond the design studio appear in any case to be limited (Hall-Jones, 1996). Because tutorials and critiques occur solely between designers, verbal explanation plays a subordinate role to visual forms of representation (Fleming, 1998) creating a specialist language which effectively excludes non-designers from discussion (Lloyd, 1996) and reinforces professional and cultural divisions. Live projects, according to design tutors, are inherently problematic, particularly when discrepancies arise between client and course objectives, or when clients lack time or genuine commitment. This means that in terms of both dialogue and design context, they often differ little from the standard studio project. Paradoxically, it appears from staff interviews that a project's capabilities to raise communication issues and to provide a critical context for design activity may be mutually exclusive. As one design tutor explains,

'...Some clients never use design and they are so easily impressed, they love everything they (the students) do. And there are others who work with designers a lot and are far more critical, but with them the language is easy, so the students might as well be talking to us.'

Projects involving student teams assembled from different academic disciplines (business studies, engineering and design, for example) appear to offer one potential solution to these problems. However, reference to international exemplars (Shida, 1995; Smith, 1994; Blackburn, 1994) indicates the need for a degree of co-ordination, planning, support and evaluation beyond the scope of many UK institutions. Without this continual intensive support, teams tend to remain multi-disciplinary rather than inter-disciplinary, combining their expertise in a linear manner reminiscent of traditional product development methods (Buchanan, 1994).

In practice, verbal articulation is often marginalised as a 'professional skill', perceived as useful in business administration and as a support for visual presentations, but essentially valued as a means of gathering and disseminating objective information (Fleming, 1998), rather than a creative process in its own right. This segregation, whilst further reinforcing perceptions of design as intuitive, visual and individualistic, may also be considered detrimental to learning. Firstly, its categorisation of verbal articulation as a vocational skill devalues it when the provision of such skills is feared to undermine academic credibility. Secondly, its segregation of studio practice and contextual studies impedes the dialogue between theoretical and practical learning known to consolidate both subject-specific understanding and learning capabilities (Morris, 1996, Gibbs, 1988). Finally, its contravention of design students' preference for holistic, project-based learning with tangible outcomes (Ball and Price, 1999; Morris, 1996) can lower attendance and motivation. As one design studies lecturer states,

> *'...The problem with any contextual studies is that design students tend to resist any sort of teaching not directly related to studio practice.'*

By resisting contextual awareness as a creative constraint, by separating theoretical and practical learning, and by overestimating the potential of traditional learning methods, design education appears to be perpetuating the myths of the ultimate design solution, the deified 'design hero' and the intuitive, visual design process. Guided only by the outmoded model presented by studio practice and by cultural stereotypes of the autonomous 'design hero', the conceptual frameworks and professional self-images developed by students are becoming increasingly inconsistent with the realities of professional practice (see Fisher, 1997). This approach may also be criticised for limiting the transferability of design expertise to other situations and occupations. As Ball and Price (1999) state, when graduates are able to value and articulate their thinking and working methodologies, they become more self-aware and confident, and are therefore better equipped for a diversity of career options. In addition, the individualistic approach encouraged by traditional design pedagogy may be considered less appropriate to many occupations than the creative collaboration characterising negotiative design. This is an issue of increasing importance, in view of by the pressure on resources and need for accountability facing design education, the emphasis on transferable skills highlighted by the Dearing Report (NCEHE, 1997) and the trend towards flexible, multi-track career patterns (Ball and Price, 1999; Partridge, 1997; Press and Cusworth, 1998). Developing the skills and attitudes demanded by negotiative design may therefore be considered an appropriate preparation not only for professional design practice, but for a broader perspective on life after graduation.

43.3.2 The Response

The literature review and interviews outlined above suggest that if graduates are to practise negotiative design, they need to develop both contextual awareness and professional skills. The challenge for the *Talking Design* project was to encompass both aspects, in a format which explicated their relevance to design practice for students and overcame the shortcomings identified with traditional teaching methods. This suggested an holistic pedagogy, with contextual understanding, professional skills and studio practice to be developed as an integrated whole.

The *Talking Design* project offers a suggested structure for this integrated approach, developed from the concept of 'externalised design' (Yatt, 1994). It guides tutors in reformatting the traditional, one-dimensional project brief into a complex 'design scenario', incorporating the potentially conflicting viewpoints and objectives of a range of 'stakeholders'. Tutorials are replaced or supplemented with a 'mid-project meeting', at which students discuss alternative proposals with the 'client' (a role undertaken by the tutor). Similarly, critiques become 'end of project meetings', whilst assessment incorporates students' design rationales and capabilities in communication and negotiation, as well as their final proposals.

The design scenario is not an objective problem with a single, ultimate solution, but a rich context for design thinking which encourages a diversity of individual interpretations and an intelligent, creative response. One sample project concerns a ceramic tableware manufacturer diversifying into the casual dining market. Students are asked to use an evaluation of consumer and social trends, manufacturing capabilities and company strategy, as a basis for their work and for considering their roles as designers.

The development of the contextual awareness and skills required by students in order to respond to the design scenario is supported by a range of learning resources:

- an introductory video, presenting negotiative design as a dynamic, creative process;
- two study guides, providing overviews of the contextual issues and skills summarised in Figure 1, together with annotated bibliographies and written exercises;
- video case studies, presenting two exemplars of negotiative design in action, and a detailed description of and reflection on a successful negotiative design project.

Although these resources may be used independently or adapted in response to individual requirements, their learning potential is optimised by the integrated approach detailed in an accompanying tutor guide.

Firstly, the relevance of communication and collaboration for professional practice is explicated, thereby providing the rationale for learning which motivates proactive engagement with learning activities amongst design students (Ball and Price, 1999).

Secondly, the consolidation of theoretical and practical learning is encouraged throughout the project. Initial written exercises require a summary of contextual issues, identification of their implications for studio practice, and an evaluation of personal capabilities in teamwork, communication and negotiation. Seminar discussion centred on the video case studies then assists in defining criteria for good negotiative design practice, a closer evaluation of personal capabilities and the identification of potential areas for improvement. As the project progresses, further exercises guide students in formulating a written rationale for design thinking that encompasses a redefinition of the brief, a mid-project evaluation of alternative proposals, an explanation of the product outcome and its implications, and a record of 'client discussions'.

This rationale is a crucial aspect of the *Talking Design* project. It develops skills in verbal articulation, and encourages the integration of reasoned and intuitive thought. It assists students in preparing for the 'client meetings', by providing an opportunity to clarify design thinking and formulate it in words prior to discussion. By emphasising design process rather than tangible outcomes, it encourages students to reassess personal working methodologies and professional self-image. Furthermore, it facilitates tutors' evaluation of negotiative design thinking, by explicating context-related reasoning in a written format which may be submitted for assessment.

Finally, reflection, self-evaluation and refinement are emphasised, in recognition of their importance to lifelong learning and personal development, as well as in professional design practice (Rogers, 1983; Ball and Price, 1999). In relation to contextual awareness, these skills are developed through the process of formulating the design rationale and discussing it in 'client meetings'. In relation to communication and negotiation skills, they are encouraged by the provision of a 'log', where performance in 'meetings' can be self-evaluated according to the criteria for good practice identified in seminar discussions. The pedagogy adopted by Talking Design embraces the defining characteristics of an holistic approach, advocated as a means of ensuring quality in learning (Ball and Price, 1999). It can be summarised as:

- the opportunity to develop self-evaluation capabilities, as an aid to independent critical judgemen;
- the provision of a rationale for engaging in the project;
- The provision of a progressive development model which encourages students to build on experience;
- the opportunity to put theory into practice in a safe environment;
- the opportunity to reflect on learning and articulate its benefits.

43.4 Conclusion

The discrepancy between the methodological and conceptual frameworks developed through design education and those applicable to life after graduation is an issue of increasing importance. There is a clear need for new pedagogic theories and methods, developed in response to diversifying graduate career patterns and the current redefinition of designers' roles and responsibilities.

This paper suggests that in contrast to current practice, such new developments should reflect the uniquely holistic nature of design thinking. It proposes that only by amalgamating intuitive and reasoned thought, verbal and visual forms of articulation and individual and collaborative activities, can design graduates fulfil their potential as professionals and life-long learners.

The teaching materials described offer one potential solution, centred on the introduction of negotiative, context-specific design as a dynamic and creative process, and communication and collaboration skills as intrinsic to design thinking. It is expected, however, that their underpinning theoretical framework will inform the development of alternative methods and tools. In addition, it is anticipated that further projects will apply a similar process of formulating educational tools with practical application from combined theoretical and empirical research.

43. 5 References

Armstrong P, Tomes A, Erol R, Dunmur D (1999) Surfing Applied Science: The Scientist-Designer in Action. Paper presented at the Design Cultures conference, Sheffield, UK, April 1999

Ball L, Price E (1999) Graduate Into Business: Rethinking Business Start Up. University of Brighton, Brighton / DfEE, London

Blackburn RS, Chapin D (1994) Design and Management Education: Collaboration Between Practice and Academia. Design Management Journal Fall: 47-54

Bruce M, Morris B (1998) In-House, Outsourced or a Mixed Approach to Design. In: Bruce M, Jevnaker B (eds.) Managing Design Alliances . John Wiley, Chichester

Bucciarelli L (1988) An Ethnographic Perspective on Engineering Design. Design Studies 9 (3): 159 - 168

Buchanan R, Vogel C (1994) Design in the Learning Organization: Educating for the New Culture of Product Development. Design Management Journal Fall: 39-46.

Byrne K (1990) A 'Semantics' of Visual Design: the Care and Feeding of Studio Projects Within a Communication Theory Context. Design Studies 11 (3):14-163

Clark K, Fujimoto T (1990) The Power of Product Integrity. Harvard Business Review 68 (6):107-118

Cooper R, Press M (1995) The Design Agenda. John Wiley, Chichester

Corbin J (1996) The Design Consultant as Strategic Resource. Design Management Journal 7 (2):38-42

Coyne R, Snodgrass A (1991) Is Designing Mysterious? Challenging the Dual Knowledge Thesis. Design Studies 12 (3):124-131

Cross A (1986) Design Intelligence: the Use of Codes and Language Systems in Design. Design Studies 7 (1)

Cross N, Cristiaans H, Dorst K (1994) Design Expertise Amongst Student Designers. Journal of Art and Design Education 13 (2):39-56

Cuff D (1991) Architecture: the Story of Practice. MIT Press, Cambridge Massachussets

Dant T (1999) Material Culture in the Social World. Open University Press, London

Drucker P F (1993) Post-Capitalist Society. Butterworth-Heinemann, London

Dumas A (1994) Building Totems: Metaphor-Making in Product Development. Design Management Journal Winter:71-82

Edmondson A, Moingeon B (1996) When to Learn How and When to Learn Why: Appropriate Organizational Learning Processes as a Source of Competitive Advantage. In: Moingeon B, Edmondson A (eds.) Organizational Learning and Competitive Advantage. Sage, London

Fielding R (1994) Human Language and Drawing Development: Their Productive Iterations. Journal of Art and Design Education 13 (2):145-150

Fisher T (1997) The Designer's Self-Identity: Myths of Creativity and the Management of Teams. Creativity and Innovation Management 6 (1):10-18

Fleming D (1998) Design Talk: Constructing the Object in Studio Conversations. Design Issues 14 (2)
Forester J (1989) Planning in the Face of Power. Berkeley University of California Press, California
Gardner H (1983) Frames of Mind: the Theory of Human Intelligence. Fontana Press, London
Gibbs G (1988) Learning By Doing: A Guide to Teaching and Learning Methods. Further Education Unit, London
Goleman D (1995) Emotional Intelligence: Why it Can Matter More Than IQ. Bloomsbury Press, London
Goleman D (1998) Working with Emotional Intelligence. Bantum Press, London
Hall-Jones S (1996) Crits - An Examination. Journal of Art and Design Education 15 (2):133-141
Herbert D M (1988) Study Drawings in Architectural Design: Their Properties as a Graphic Medium. Journal of Architectural Education 41 (2):26-38
Itami H, Roehn TW (1987) Mobilizing Invisible Assets. Harvard University Press, Harvard
Jevnaker B (1998) Building up Organizational Capabilities in Design. In Bruce M, Birgit J (eds.) Management of Design Alliances: Sustaining Competitive Advantage. John Wiley, Chichester
Jones JC (1980) Design Methods: Seeds of Human Futures. (2nd ed.). John Wiley and Sons, New York
Kernan S (1997) Two Scripts. Design Issues September / October:22-28
Lawson B (1990) How Designers Think: The Design Process Demystified. Butterworth – Heinemann, Oxford
Lloyd P, Lawson B, Scott P (1996) Can Concurrent Verbalization Reveal Design Cognition? In: Cross N, Christiaans H, Dorst, K (eds.) Analysing Design Activity. John Wiley, Chichester
Lumsdaine E, Lumsdaine M (1995) Creative Problem Solving: Thinking Skills for a Changing World. McGraw-Hill, New York
Metcalfe B (1993) Replacing the Myth of Modernism. American Craft April / May:40-47
Mitchell S (1996) Institutions, Individuals and Talk: the Construction of Identity in Fine Art. Journal of Art and Design Education 15 (2):143-154
Morris P (1996) Modular or Integrated? A Choice for Courses in Art and Design. Journal of Art and Design Education 11 (3):327-334
National Committee of Enquiry into Higher Education (1997) Higher Education in the Learning Society. Hayes, UK
Nonaka I (1990) The Knowledge-Creating Company. Harvard Business Review, December:96 - 104
Oates C, Tomes A, Armstrong P (1997) Negotiating Design: the Case of Charity Affinity Credit Cards. Paper presented at the Second European Academy of Design conference, Stockholm, Sweden.
Partridge S (1997) The Art of Getting Started: Graduate Skills in a Fragmented Labour Market. The London Institute / DfEE: London
Peters T,Waterman R (1995) In Search of Excellence: Lessons from America's Best Loved Companies. Harper Collins, London
Press M, Cusworth A (1998) New Lives in the Making: The Value of Craft in the Information Age. Sheffield Hallam University, Sheffield / Crafts Council, London
Rhea D (1992) A New Perspective on Design: Focusing on Customer Experience. Design Management Journal 3 (4)

Rhodes E, Carter R (1995) Emerging Corporate Strategies. Co-Design 1 (3): 6-13
Rogers C (1983) Freedom to Learn for the 80s. Merrill Publishing, London
Ruekert RW (1995) Cross Functional Interactions in Product Development and their Impact on Project Performance. Design Management Journal 6 (3):389-402
Sack M (1998) Using Research to Create Visual and Verbal Agreement. Design Management Journal Fall:59-63
Schön DA (1983) The Reflective Practitioner: How Professionals Think in Action. Basic Books, Aldershot
Shida T (1995) Innovations in Corporate Sponsorship of Education. Design Management Journal, Spring: 2 -35
Smith G, Loose E, Duane M (1994) Marketing, Design and Education: Cross-Functional Integration and Design Communication. Design Management Journal Fall:75-81
Sonnenwald D (1996) Communication Roles that Support Collaboration During the Design Process. Design Studies 17:277-301
Thomas-Mitchell C (1993) Redefining Designing: From Form to Experience. Van Nostrand Reinhold, New York
Tomes A, Oates C, Armstrong P (1998) Talking Design: Negotiating the Verbal - Visual Translation. Design Studies 15 (4):124-142
VanGundy AB (1988) Techniques of Structured Problem Solving. Van Nostrand Reinhold, New York
Vygotsky LS (1978) Mind in Society. Harvard University Press, Harvard
Walker D (1990) Managers and Designers: Two Tribes at War? In: Oakley M (ed.) Design Management: a Handbook of Issues and Methods. Basil Blackwell, Oxford
Ward A (1984) Design Cosmologies and Brain Research. Design Studies 5 (4):229-238
Yatt B (1994) Externalised Design. Design Management Journal Fall
Zaccai G (1995) Art and Technology: Aesthetics Redefined. In: Buchanan R, Margolin V (eds): Discovering Design: Explorations in Design Studies. University of Chicago Press, Chicago
Zaccai G, Badler G (1996) New Directions for Design. Design Management Journal 7 (2):54-60

Chapter 44

The Search for a New Paradigm for Practice and Education

R. Brown and D. Moreau-Yates

'Partnering' is the latest buzz word in architectural circles today, one manifestation of this being co-designing. The value of the variety in knowledge and experience that can, theoretically, be obtained from this experience has inspired a growing trend of collaboration in design between architects and other professionals of related areas (e.g., engineers, contractors, *etc*.), and, similarly, of participatory design, where clients and end-users are invited to take an active part in the design process. However, this growing trend, which tries to address the complex nature of architecture and draw all the parties involved out of the isolation of their own concerns, has encountered ever-growing difficulties in establishing itself within the professional circles of architecture and the building industry.

But why is this happening and what can we do about it? In this paper we will identify and review the nature of what we believe is one of the fundamental obstacles that currently challenges co-design. We will focus on the implications it poses for the development of architectural practice within the environment of co-design and present our work as a model for exploring and developing a possible way forward.

While architecture as a profession has traditionally sought to distinguish itself from others sectors of the building industry, and so preserve for itself the special role of 'the designer', recognition is growing amongst practitioners that architecture is indeed a collaborative process which needs to engage with others. Calls for this change of approach are not only emerging from within the profession, but from the building industry as a whole, and are being reinforced by the architectural press (e.g., 'Meeting the Challenge', 1999; Prasad, 1999). With these declarations on the imperativeness of co-designing have also come recommendations on how to go about it, and accounts of how it is currently being implemented in practice.

Along with of all this encouraging news and advice is also a clear recognition of how problematic the process can be. Problems in communicating and achieving satisfactory results are difficult to overcome, and seem to be solved, more often than not, through compromise rather than synthesis. Moreover, even though in the particular case of architecture the success of co-design is vital, (as the built environment affects us all), in practice the models, procedures and strategies which have been developed to guide architecture tend to embody only the values and concerns of those responsible for design and construction. This only reinforces the sense of isolation of the architect, the consultants and/or the end-users which gives rise to fairly evident consequences that pose serious implications for co-designing and for the viability of the profession as a whole.

With this thought in mind, it is important to point out that much of what has been written on co-designing in architecture has focused mainly on the need to establish effective communication between parties by researching and providing the best techniques and protocols to facilate it. Central to this has been the notion of identifying a common vocabulary, shared between architect and non-architect. Even so, problems and disputes in co-designing projects are still being experienced in practice with disputes still arising, and the architect and non-architect appearing to be going in different directions and pursuing different aims. Whilst communication is an important issue in the development of co-design, its focus on shared 'words' does not recognise or address more profound obstacles to the exchange of ideas and concepts between people. In the light of this experience it becomes clear that what is overlooked in these initiatives is any reference to fundamental disparities between architects and non-architects in how each experiences, perceives and relates to the built environment. As long as this disparity remains unaddressed, it will continue to inhibit co-designing. Thus, an important step towards successful co-designing practice is to identify and explore the nature of this disparity and utilise its potential as a resource in design.

Research in the field of environmental psychology has demonstrated that the way in which architects and non-architects relate to the built environment is indeed different. While architects choose to emphasise the perceptual values and representational meaning of the built environment, non-architects place more emphasis on its associational and responsive meaning (Devlin, 1990; Hershberger, 1988, Rapoport 1977; Rapoport 1982). How this plays out in practice is that architects in designing the built environment place emphasis on visual qualities such as the use of materials, geometric proportions and the concepts these can embody. On the other hand, non-architects place more emphasis on the way of life they associate with the built environment (*i.e.,* the image of the built environment is seen as a symbol for a way of life) and the values associated with that way of life. For the purpose of our work, what is most relevant in this observation is that while both architects and non-architects are exposed to similar experiences of the built environment, they consciously (or sub-consciously) choose to emphasise and focus on different aspects of it. Thus, the architect and non-architect are each bringing to the process of co-design their own way of relating to the environment, which in turn informs how each approaches and envisions its design.

Since the initial reference points and aims are very dissimilar, the assumption that they are both talking about the same thing is inevitably mistaken, and leads to a discussion where each tries to impose on the other their own subjective values. This failure to recognize and discuss the difference in the nature of their relation to the built environment in turn leads (in the existing model of practice) to an inevitable breakdown in communication that is, on both sides, only solved through a compromising of aims and objectives. Needless to say, this result is far from satisfactory for either the architect or non-architect.

This state of affairs raises the question of whether architects need to rethink how they practice. There are of course exceptions, but on the whole they have been disinclined to think and practice in way which are conducive to the approach and attitudes needed for co-designing. By the time the architect enters practice a mindset has been created which subconsciously works against co-designing. It is our contention that what is needed therefore is a new paradigm for educating architects for co-designing.

To pursue this further, we first need to address how architects currently practice and what beliefs inform that practice. The idealized model that persists today is one founded over 500 years ago, going back to the days of Sir Christopher Wren: namely that of the architect as a uniquely gifted universal man. In this context, the architect is seen as, above everyone else involved in a building project, responsible for controlling all aspects of the job. In subsequent years, the profession found its pre-eminent position challenged, first by contractors, then later by surveyors, planners, engineers, *etc*. So what once had been a building industry incorporating various crafts organized loosely under the architect evolved into a highly competitive system often pitting the architect against others within the industry (Finch, 1999).

Faced with this threat, the profession has increasingly sought to defend its position of control over design, the one thing that it feels distinguishes it from the other factions. Inherent in this mission is the belief that architects are uniquely gifted creatively and have a right to this position of control over design, even to the extent of overriding the concerns of the client. This attitude is advanced by leaders within the profession, for whom 'the idealized professional holds the power to decide what is best...' (Cuff, 1992).

These beliefs are supported by and reflected in the architectural press. Here, the architect is portrayed not as a collaborating professional, but is instead held up as a lone hero, justly pitting their creative genius against those with limited vision and ambition. This promotion of the architect as sole designer is perhaps best personified by the character of Howard Roark in 'The Fountainhead' (Saint, 1983). These attitudes, together with the focus on the architect's perceptual and representational values, persist even in the light of recent articles on the value of co-designing. Little, if any discussion is given to the non-architects' role, perspective or values in the review and accreditation of projects, and only limited consideration is given to architectural design as a collaborative process.

Within this context, developing a new model would be an extremely difficult undertaking (though, we recognise, not impossible, as well-documented examples from practice do exist). A new model would require a re-thinking of how we practice and even of how we think about and define architecture. Such a radical evolution would not be readily accepted within the status quo of practice. Even were there openness to this change, the day-to-day demands and economic constraints of practice generally would not be able to afford the considerable time such evolution would require. If those in practice cannot implement this much-needed change, where is this new paradigm to be generated?

The previously noted research in environmental psychology has identified a number of factors which contribute to the disparity between architects and non-architects, including background, economic status *etc.*, but identifies its primary source as differences in experience and, most particularly, education. From the time the future architect enters architectural education, she is exposed to and indoctrinated into a distinct culture. Within this culture, emphasis is placed on (perceptually- and representationally-based) form-making. This traditional appraoch (Hellman, 1987) persists today. Indeed, a review of recent discourse on architectural education in the UK notes and critiques its emphasis on architecture as a primarily visual subject (Hyett, 1997; Morrow, 1997).

It is also within architectural education that the primacy of the architect as sole designer is imprinted. In most schools of architecture it is the design studio, in which students generally work on projects individually and free of interaction with others (such as clients or non-design professionals, e.g., engineering students), that is paramount. The emphasis within this system reinforces the belief in 'the primacy of the individual...' emphasizing singular '...personality, talent, creativity, and convictions.' (Cuff, 1992). In this context, there is little room (or even concern) for co-design. Indeed, architects have fought to defend this approach to architectural education. (Schools Face The Great Divide, 1989) A recent report form the Ove Arup Foundation, titled 'Interdisciplinary Skills for Built Environment Professionals', comes as no surprise in finding that '...architectural education is not doing enough to foster a multi-disciplinary approach...', with architecture performing the worst in this respect when measured against other construction-related studies. (Schools Miss The Bigger Picture, 1999).

There are of course calls for addressing the disparity between architects and non-architects by increasing the level of education on the built environment provided to non-architects. Currently there are a number of proposals, as well as attempts being undertaken on a one-off/ad-hoc basis, for introducing education on the built environment in to the National Curriculum (e.g., 'Architecture on the Curriculum', 1999). Simultaneously, the architectural profession is calling for the need to do a better job of marketing, so that non-architects will better understand what it is that architects have to offer (Long, 1999).

While these efforts are valuable, it is debatable whether approaching this disparity simply by 'educating the non-architects' is either morally or objectively valid. Inherent in this intent is the maintenance of a hierarchy where the ideas and values of architects are placed at the top. Such an approach inevitably tends to dismiss the relevance and value of the non-architects' knowledge and experience.

In questioning the appropriateness of this one-sided approach, we would suggest that a better way forward can only be found in combining the learning experience of both architects and non-architects. Indeed, it is important for non-architects to gain a more critical appreciation of the built environment and how it affects them. But it is equally important for architects to develop a greater understanding of how others perceive and relate to the built environment. Recognising this as a positive source, architects can employ their skills and trained knowledge to develop new and viable forms that reflect the non-architects values as well as their own.

So, if we are to achieve a new paradigm for co-designing by re-thinking how we practice and think of architecture, the circumstances outlined above clearly indicate that the most viable place to address this condition is within (architectural) education. It is here that this mind-set first emerges distinctly, and it is here where a re-thinking of architecture can take place free of the constraints of practice and of well-entrenched attitudes.

Our concern with the issues outlined above has led us to investigate possible ways of addressing them within architectural education. With the collaboration of the University of East London School of Architecture and the support of the RIBA, we are developing research that aims at identifying teaching methods that are effective in raising an awareness of this disparity and encouraging exploration of its intrinsic potential as a positive source of inspiration in the development of architectural design. We believe addressing this issue at an early stage is of great value as it can contribute to the development of a more holistic outlook in future professionals and help them gain some of the skills necessary for the successful development of co-design and communication.

In the last year, we have been exploring the potential and value of having architecture students interact directly with different groups of non-architects such as school children, students from other disciplines, and members of the general public. These interactions allow the students to gain first-hand experience of other people's views, values and forms of communication. At the same time, they help raise an awareness of architectural concepts in non-architects. In both instances, the focus of these interactions is centred on the act of communicating and establishing a process of cross-fertilisation between the two groups where views and knowledge are the main element of exchange. We have purposefully avoided engaging in 'live' projects which may result in an actual building proposal, in order to circumvent the constraints such projects can impose on the learning process (e.g., timing of 'live projects' conflicting with academic calendar and students' development (for a descriptive example, see Rawson, 1998).

During these interactions students are prompted to look beyond the simple programmatic or functional interests that non-architects might initially express, and try to investigate the experience and values on which they are basing their ideas. To achieve this, they are encouraged to explore different forms of dialogue and interaction that free them from sophisticated architectural jargon and a reliance on representation, so enabling the non-architects to engage in a constructive discussion with them. By being exposed to the underlying ideas of the non-architects' point of view and gaining a better understanding of the way they express themselves, the students are then able to gain an insight into the architectural implications that these statements entail.

Considering that most of the preconceptions and barriers that keep us from establishing a constructive dialogue are based on personal experience and education, we have based our teaching method on 'heuristic' learning. That is, learning by doing. From their experience with non-architects, the architecture students are asked to develop their own methods of communication and find ways of 'conveying' to the non-architects the architectural issues in which they are personally interested. This allows the student not only to gain feedback on the development of their work, but also to develop their own line of thought in a way that is relevant to this particular learning process.

An example that illustrates the dynamics of these interactions can be found in an activity carried out by one of our students who worked with children from the Vicarage Primary School in the summer of 1998. This interaction programme took place over a period of three weeks during which the students and the children met one day a week to carry out different activities. During his first session, the architecture student walked with a small group of children through the school prompting them to talk about the things they liked/disliked. As they took him to their favourite places in the school he noticed a particular sensitivity within the children to the size of the rooms in relation to themselves and to smaller children. Again the issue of size caught his attention in his second interaction where he was in turn questioned by the children about the reasons he liked some examples of architecture he presented to them. Time and again comments and questions on the size of the spaces and buildings he showed them were a persistent issue in the dialogue. For his third interaction he was asked to develop a didactic activity that would help the children to explore and become explicitly aware of an architectural issue he felt they would be interested in. Following his observations, the student chose to develop an activity that focused on the direct relationship that can be established between the size of buildings and the size of other objects in the room including their own bodies. He introduced this as a form of measurement and a way of establishing references that illustrate the relativity of scale. While the children became familiar with these concepts, the architecture student gained a valuable insight into how they felt in different spaces, and where and in what circumstances they felt most comfortable. Through this interaction it became clear to him that the idea of scale the children were most interested in was not one that compared one object with another, but one that compared the objects with their own perceptions of different sizes. This information heavily influenced his design proposal for a space in the school, and prompted him to consider spatial relationships to which he would not have otherwise paid much attention.

The interactions also provide the non-architects with greater insight into architectural concepts, and a better understanding of the built environment and the way it affects them, than would be the case without them. They become more familiar with the nature and relevance of the architect's ideas and concepts and learn to articulate their ideas with more accuracy and clarity. Through the interactions they gain a better understanding of the processes that inform their own experience, which enables them to visualise ways and possibilities of improving it, and to contribute in a more direct way to the process of design. Following the example above, by engaging with the classroom activities the architecture student set them, children were able to appreciate the relationship of scale between themselves and the spaces around them, and they started to discuss how the sizes of these spaces affected their perception and sense of well being. While a spacious room could be appropriate for one of the 10-year-olds, it might seem to be too big for a smaller child who could feel uncomfortable in it (this observation was made by one of the children participating in the workshop mentioned above).

The experience gained from these interactions is complemented by the reading of supporting texts and seminar discussions, where students have the opportunity to share their thoughts and discuss their ideas. During these sessions the architecture students are prompted to review and define their findings in terms of the challenges and/or implications they pose for their role as architects and the development of their work as designers of the built environment.

These exercises help students' question the self-referential patterns widely encouraged by the nature of contemporary architectural education, the profession and the media. They also help students to develop skills in graphic and verbal communication, interpretation and critical analysis, and to listen to and recognise the architectural value intrinsic in other peoples experience and ideas. Finally, and most importantly, they can contribute to the exploration of establishing a platform for dialogue where both architects and non-architects can communicate in a constructive way.

Assessing these interactions in terms of measuring how much the students or the non-architects have learnt is a very difficult task. The learning is not linear nor is it based on the acquisition of a specific body of knowledge or information that can be 'measured' with more traditional methods of assessment. At the same time, evaluating the designs as 'end products' (in accordance with the traditional way of evaluating architectural design) would not be, in our view, a fair and accurate judgement of what the students have gained through the process of the interactions. Thus moving away from the visual and formal aspects of the work produced, we have so far based our evaluations on the students' ability to become self critical of their own processes and gains. As part of the programme of activities that accompany the interactions, students have been asked not only to keep records of the development and evolution of their work and designs, but to write a paper in which they review from a critical perspective the impact these activities have had on the work they developed during the course or workshop. However, to fully assess the value of this initiative, its implications and potential design needs to be explored within the context of a longer timeframe that will allow the students to develop their design proposals fully. Our next step will be to investigate the ways in which the knowledge gained from this experience can be transformed into built forms.

The attempt to develop a new model for architectural education and practice which reflects the insights and attitudes required for co-designing, and in essence to transform both, is an ambitious one. It is at the same time an all-together necessary one. With the growing emphasis on partnering in architecture today, effectiveness in co-designing is paramount. To replace the idea of compromise with the pursuit of synthesis as the predicted outcome of co-design is, in our view, a fundamental issue for the success of co-design. This in turn can only be achieved by recognizing the fundamental disparity in how architects and non-architects interpret, value and communicate about the built environment. As long as this disparity remains unaddressed it will continue to impede those engaged in co-designing in truly achieving their goals.

It is our contention that while it is questionable whether effective reconciliation of this disparity can be fully achieved in practice, it can (and should) be addressed at its primary source within architectural education. It is here that the future architect's distinct approach to the built environment, and thus design, is nourished. Examination of this divergence will raise future architects' awareness of this issue, and open up for exploration possibilities for this disparity to become a positive source for inspiration and collaboration. Ultimately such learning experiences will form in students and future professionals a strong foundation for the continued exploration of co-design later in practice.

References

Architecture on the Curriculum (1999). Building Design, 29 October, pp1

Cuff, D (1992) Architecture: The Story of Practice. MIT Press, Cambridge Massachusetts

Devlin, K (1990) An Examination of Architectural Interpretation: Architects Versus Non-Architects. The Journal of Architectural and Planning Research 7 (3):235-243

Finch, P (1999) How Did We Get to Where We are Today. Lecture at University of East London, 10 February

Hershherger, R (1988) A Study of Meaning and Architecture. In: Nasar J (ed.) Environmental Aesthetics. Cambridge University Press, Cambridge

Hellman, L (1987) Two-way Teach-in. The Architects' Journal, 2 December:27-29

Hyett, P (1997) Lessons From the Andes School. The Architects' Journal, 3 April:27

Long, K (1999) President Marco. Building Design, 5 March:1

Meeting the Challenge (1999) RIBA Journal, November:105-114

Morrow, R (1997) People-centred Design. Streetwise, 32:17-18

Prasad, S (1999) Rethinking Architecture. RIBA Journal, August:68

Rapoport, A (1977) Human Aspects of Urban Forum. Pergamon Press, Oxford

Rapoport, A (1982) The Meaning of the Built Environment., Sage Publications, Beverley Hills CA

Rawson, J (1998) Learning to Listen. The Architects' Journal, 15 October: 40-47

Saint, A (1983) The Image of the Architect. Yale University Press, New Haven

Schools Face The Great Divide (1989) The Architects' Journal, 31 May:9

Schools Miss The Bigger Picture (1999) Building Design, 13 August:1

Chapter 45

Developing an Integrated Approach to Product Design

A.J. Felton

45.1 Introduction

The BSc. in Computer Aided Product Design (CAPD) at the University of Wolverhampton is ten years old.

Conceived as a collaborative venture between the School of Engineering and the School of Art and Design it has integrated engineering design from the former with industrial design from the latter developing a mixture of technical and design skills using the computer.

The field of product design is rapidly changing due to the increasing use of computer based technologies. Design for manufacture and design for assembly have resulted in the development of new approaches to design, including parametric solid modelling and concurrent/simultaneous engineering concepts.

The adoption of new methodologies to develop these new concepts indicates a need for the academic world to foster the culture of the integrative philosophy to meet industrial requirements for product designers.

This paper discusses the findings of part of the authors research into this area and reviews the BSc. CAPD's integrated approach to group design in terms of:

- curriculum development to promote and foster the integrative philosophy;
- company survey on recruitment criteria for product designers;
- team working skills;
- group projects;
- meeting industrial needs.

45.2 Curriculum Development to Promote and Foster the Integrative Philosophy

It was decided at the outset that the curriculum would have to develop a broad range of transferable skills, in order to equip the graduate to enter the fast changing job market that did not offer a traditional safe career structure, but required flexibility and team participation. The new BSc. CAPD course transcended traditional institutional boundaries and brought together a blend of skills in design, engineering, manufacture and computing in what was one of the first inter-school collaborative ventures in the institution.

The inter-school staff team felt that five main skill areas needed to be addressed by the course:

- design;
- computing;
- visual awareness and communication;
- manufacture;
- working within a design team.

These five areas it was felt needed an integrating vehicle to bring them together and test the delivery of teaching and learning in product design. This integrating vehicle was the design project.

45.3 Company Survey on Recruitment Criteria for Product Designers

The questionnaire was designed to establish recruitment criteria and indicate the level of design skills companies look for, or adopt, when employing newly qualified product designers. The aim of the research was to highlight any specific shortfall in the BSc. CAPD design curriculum for product design.

Twenty-five companies were targeted out of which seventeen returned the completed questionnaire, thus giving a response rate of 68% (17/25).

The companies in question were selected because of their diverse product range when compared with one another rather than different companies producing similar products. The reasons for selecting companies in this manner were based upon the following factors:

1. The BSc CAPD curriculum was designed and developed for general product design to maximise employment chances for students in a diverse range of product design areas.
2. Students wishing to specialise in a particular design field more suited to their intended career vocation had the facility of module options made available.
3. The diverse range of industrial placements secured by BSc CAPD students both in the UK and abroad.
4. Students final year design projects could be tailored to their appropriate design field vocation.
5. Students graduated from the BSc CAPD course at the University of Wolverhampton and gained employment in various design disciplines in a number of companies that produced all types of products and a varied range of services.

One of the questions put to the companies was: 'What skills do you feel new designers are lacking?' Many companies responded to this question and their comments were categorised under the headings of:

- practical experience, manufacturing;
- communication, drawing skills, CAD-related skills;
- business;
- creative/innovative ability;
- team working skills.

45.4 Team Working Skills

Some of the company's comments from the questionnaire regarding the skills that newly qualified product designers are quoted below:

> *'Team working skills and operating in multi-functional environment'*
>
> *'Team working skills, rounded ability'*
>
> *'Many students lack appreciation of the role of design and its relationship with other departments'*
>
> *'Designers need to be fully integrated into the manufacture and production engineering environment'*
>
> *'On-the- job experience'*
>
> *'Experience of working in a non-academic environment'*
>
> *'More work experience within the University course years would help the design of practical products that can also be manufactured'*

It can be seen from the comments made above that some of the companies identified students as lacking a number of specific team-working skills. Comments made, such as 'many students lack appreciation of the role of design and its relationship with other departments', 'operating in multi-functional environment' and 'designers need to be fully integrated into the manufacture and production engineering environment', would suggest that designers of today need to adopt the 'integrative philosophy' of concurrent, simultaneous and parallel engineering concepts. Group design is more often the case in industry rather than individual design.

45.4.1 The Team Approach to Innovation

The Professional Engineering journal published an article: 'The innovative imperative' by Pullin (1999) which makes the statement that 'the biggest spur to innovation is competition, and the delivery of innovation is more and more dependent on collaboration'

Lone inventors like Mike Burrows, Trevor Baylis and James Dyson who have the original design idea often still require a highly skilled team to develop, manufacture and market the product. Mike Burrows, the inventor of the bicycle with a one-piece, carbon fibre frame, worked with Lotus. Lotus provided the technical backup with their extensive knowledge of moulding composites and wind tunnel testing to improve the aerodynamics of Mike Burrows, composite bicycle, on which Chris Boardman won his 1992 Olympic gold medal.

In the case of Trevor Baylis, the inventor of the clockwork radio, the invention was only the start of a long process of innovation transforming the prototype into the aesthetic, technical specification to meet the commercial markets. The British Council's (1999) article on the inventor highlights how he had to collaborate with others in a partnership of technical skills, market research, commercial, finance and intellectual property rights. The project team needed to understand engineering, design and commercial issues to develop the product from the initial idea. The chosen team was skilled in engineering, even if they were marketing or design people. A materials engineer designed the material for the spring, a radio engineer the radio board, an electrical engineer helped to develop the right generator and a mechanical engineer put it together. In essence they had to work together to produce a radio that was robust for the market place, and energy-efficient.

With reference to an article by Howell (1999), the importance of innovation to companies' success cannot be over-emphasised. Dyson states in the article 'We cannot afford not to strive continuously for innovation'. Dyson's products provide an example of how to brilliantly exploit innovation in cleaners from one of the fastest growing manufacturing companies in the UK.

Developing improvements in product innovation can be achieved through the early involvement of wide-ranging functions in the product design process. These include techniques of design for manufacture, design for assembly, structured approaches such as quality function deployment, risk assessment, cost planning and control, but more importantly, getting multifunction teams involved as early as possible, *i.e.*, making use of concurrent/simultaneous engineering concepts.

Pullin's (1999) article 'Innovative rethinking' opens by saying British manufacturing industry is disturbingly complacent about the process of innovation and few were managing the process at all well. With as many as seventy per cent of the surveyed companies relying on customers for new product ideas and fifty per cent of companies scanning their competitors for inspiration, with little evidence that company staff, academia or independent researchers were used to identify opportunities or generate ideas.

Pullin goes on to say that the concept of forming multifunctional project teams to drive product innovation is accepted more in theory than in practice, and is often allocated to an engineering department rather than a team. Innovation needs to be treated seriously by the management within a company, and a structure created to support projects. Innovation is seen as a way of gaining competitive advantage especially if the company can react quickly to market demands.

45.4.2 Group Projects

Companies with specialised personnel attached to the product team, *i.e.*, concept designer, engineering designer, manufacturing, prototyping, tooling, marketing and sales, can adopt and introduce this 'integrative philosophy' for new products. However, this is often a difficult environment to simulate or replicate at University.

First year BSc CAPD students are introduced to the 'integrative philosophy' approach to product design in the module Product Design and Development through concurrent/parallel engineering concepts and its important role and relationship with other departments.

The curriculum has been designed to enable students to work in groups on a number of design projects throughout the course to help foster a team-working approach, in particular in modules such as Graphical Modelling and Design Practice in the second year, and Design Enterprise Project in the final year.

Graphical modelling applies modelling techniques to product design using a 3-dimensional, parametric solid modeller. This enables students to work as design teams on a group project, e.g., a food mixer. Each individual group member takes responsibility for modelling various components of the product and the group then assemble the individual components to produce an assembly of the complete artefact to include rendered solid models, 2-dimensional working drawings and various mass properties of the product. Design Practice takes the process one stage further whereby the group of students design and then make a physical prototype of the product.

Design Enterprise Project facilitates the full product design process. It enables students to work in teams to set up a company to produce a business plan, design, manufacture and market a given product.

45.5 Meeting Industrial Needs

Industrial-based assignments and projects have played an increasing role in achieving innovation and bringing integrated approach to product design for the BSc CAPD course. Companies have provided sponsorship through prize monies to students for the most appropriate designs and money to the department for equipment or software. Projects set up with companies by the author have included the following:

- design of 'Tote' boxes for both stacking and nesting;
- the external design of a lorry cab;
- design of tower-boots for builders, electricians, *etc.*, a type of stilts, up to a metre in height, that are fixed to the feet to enable work overhead at an elevated level;
- design of a locking mechanism for UPVC windows;
- design of top-of-the-range bathroom accessories for four- or five-star hotels and exclusive department stores.

Working on this type of 'live' project enables the students to acquire skills in concept design, analysis, manufacturing, prototyping, marketing, *etc.*, thus allowing students to take on these roles whilst at University. This gives students more flexibility for employment and equips them with some of the necessary skills to be able to communicate with the specialised personnel involved in a company's design team when developing a new product.

King and Preston (1995) discuss the Integration of Undergraduate Engineering Through Total Product Design (TPD), a team-based activity, and its benefits. The philosophy of the TPD final year module put forward by King is 'to get students involved in real, open-ended, industrially based projects. These projects are intended to graduate the students with the necessary skills to work as effective professional engineers upon their first appointment. This emulates the real-world engineering environment where team activity is the norm as opposed to individual academic exercises, designed for ease of assessment'.

45.5.1 Links with Industry Using Video Conferencing

The field of product design is changing due to the increasing use of computer-based technologies. Felton (1995; 1996) discusses the role which video conferencing can play in a distance learning environment. Video conferencing can also provide the vehicle for co-designing where the design team may be spread out over different sites or even different countries.

The technique of sharing CAD data between centres using videoconferencing techniques and shareware software is starting to be accepted by industry. Often known as Collaborative Engineering this offers a quicker response time by designers/engineers to 2-dimensional and 3-dimensional design data and reduced lead times for new products being introduced. This opens up new avenues for industrially-based projects providing a link between University courses and industry.

45.5.2 Work Experience

A large percentage of companies from the survey (nearly sixty five per cent) stated that work experience was essential, with the issue of work experience also playing a significant part in selection of prospective employees.

In the past, Industrial placements have played an important role for BSc. CAPD sandwich students' who have gained valuable experience in companies, and many obtain positions in the company when graduating. However, the ending of ERASMUS and COMETT grants have seen the number of BSc CAPD students studying abroad diminish even though these schemes have been replaced by SOCRATES and LEONARDO funding.

The introduction of student fees in 1998 may now make it more difficult for many students to take the opportunity of a sandwich placement and thus gain valuable experience in a company within the UK or abroad because of the financial implications. An increasing number of students now wish to complete their degree in the shortest time with minimal cost. It would appear more students are living at home to reduce costs. In addition, the payment of fees during the placement year is seen as yet another financial burden unless the placement company is willing to pay them on the student's behalf.

Early indications from the BSc. CAPD course show the lowest number of second year students entering work placements in October 1999.

Exposing students to 'live' company-based projects and National Competitions in modules may be one way of alleviating this problem; this practice has been encouraged throughout the BSc. CAPD curriculum but is by no means a substitute for industrial experience. Some graduates find themselves in a 'Catch 22' situation: they need work experience to get employment, but they need employment to get the work experience.

45.6 Conclusion

Companies are expected to develop successful new generations of products whilst reducing costs and shortening development time. Therefore, the importance cannot be over-emphasised of forming multifunctional project teams to drive product innovation in product design and development: many companies use such teams to gain a competitive edge over their rivals. However, developing this environment and becoming innovative requires effort by both staff and students in academia and product designers and managers in industry.

The possible reduction in the number of design students obtaining practical experience during the course through work placements, and its associated problems requires Universities to strive to include as much practical experience in the curriculum as possible. Liaisons with companies for 'live' projects and national design competitions can play a vital role in achieving a more practical solution. Co-designing, which involves students working together on group projects in particular 'live' projects, can provide a good foundation in developing the integrated philosophy of concurrent and parallel engineering concepts.

Skill levels and knowledge bases are changing: the academic world needs to recognise this and continue to develop new teaching and learning strategies to promote and foster the culture of the integrative philosophy.

45.7 References

Felton AJ (1995) The valuation of video conferencing in a distance learning environment. International Conference on Hypermedia. University of Sheffield, 3rd-5th July 1995, pp 225-233

Felton AJ (1996) The evaluation of video conferencing in a distance learning environment for product design. Thirteenth Conference of the Irish Manufacturing Committee, (Re-Engineering for World Class Manufacturing). University of Limerick, 4th-6th September 1996, pp 639-646

Howel D (1999) Innovation: ideas in action. UK: Professional Engineering 12 (15):32

King PD, Preston ME (1995) Integration of undergraduate engineering through total product design. Proceedings of the 2nd National Conference on Product Design Education. Coventry University, 10th-11th July 1995, UK

Pullin J (1999) The innovative imperative: accelerating product development time and information systems put yet more pressure on firms to innovate. UK: Professional Engineering (MX2000-Manufacturing Excellence) 12 (17):24

Pullin J. (1999) Innovative rethinking. UK: Professional Engineering 12 (18):14-16
The British Council (1999) Trevor Baylis – inventor of the clockwork radio; innovation–invention was just the start…England:
http://www.britcoun.org/science/science/personalities/text/ukperson/baylis2.htm

Chapter 46

Undergraduate Design Project Utilising Users as a Design Resource: A Case Study

D. McDonagh-Philp[1]

46.1 Introduction

The Nuffield Foundation supports student research through Undergraduate Research Bursaries (URB). The aim of this scheme is to provide the student with experience of research, which will foster design research. This paper will discuss a ten-week project conducted by a student at the end of their second year on the Industrial Design and Technology programme at Loughborough University under the direct supervision of the author. Collaborating with Morphy Richards, a leading consumer product manufacturer, provided the commercial context. User-centred design approaches were employed (e.g., focus groups, questionnaires, product handling, interviews) to provide the student with experience of evidence-based design decision-making (McDonagh-Philp *et al.,* 1999).

The aim of the project was to apply focus group methods, currently being developed by the author, to ascertain if such approaches would be appropriate for undergraduate level training. This paper will discuss the design research project, the methods and techniques employed, and present a summary of the outcomes. Particular emphasis will be placed upon the advantages and disadvantages of conducting such projects at undergraduate level.

The URB scheme is aimed at supporting students conducting research projects during the summer vacation period, before the final year of their degree programme, in this case the Industrial Design and Technology degree. The project concentrated on applying user-centred design approaches. A key feature of this project was that it involved *concentrated study* (Parlett and King, 1971), which resulted in the student focusing solely on one design project, instead of conducting several projects concurrently.

---

[1] The author would like to thank the Nuffield Foundation (Grant number: AT/100/99/0039) and Morphy Richards for their financial support. In addition, Chris Tickle, who undertook the brief with diligence and professionalism.

The project involved the investigation of the ironing process (e.g., ironing garments), with the manufacturing company providing a user profile. This consisted of female homeowners, in full time employment, and between 25-55 years of age. The research subjects were contacted within the University campus from a range of professions (e.g., lecturer, domestic assistant, and administrator). The collaboration with Morphy Richards provided additional funding to the URB by funding the focus group activities (e.g., research participant attendance fees and hospitality), supplying product samples, and providing feedback on completion of the project from a commercial view point.

The student organised and conducted a series of three sessions over the ten-week period, involving a core team of eight research participants. Focus group discussions, product handling, and visual questionnaire exercises were integrated into the 3-hour research sessions. The activities of each session were recorded (via photographs, audiotape, and videotape) for further analysis. The research sessions were strategically planned to provide the student with access to a core team of 8 users at three key stages in the designing process: (i) prior to designing; (ii) after concept generation to assist in concept selection; and (iii) concept development and refinement. To widen the empathic horizon of the student (McDonagh-Philp and Denton, 1999), access to the user group took place within the first 2 weeks of the project. This took the form of intensive focus group discussion and user evaluation of existing products, Figure 46.1. There is evidence to suggest that designers and design students, on receiving a design brief, immediately commence the concept generation stage. For this particular project the student resisted concept generation until user access had been gained. This was to ensure the design decisions were based on actual understanding and appreciation of the users' needs, rather than generalised assumptions.

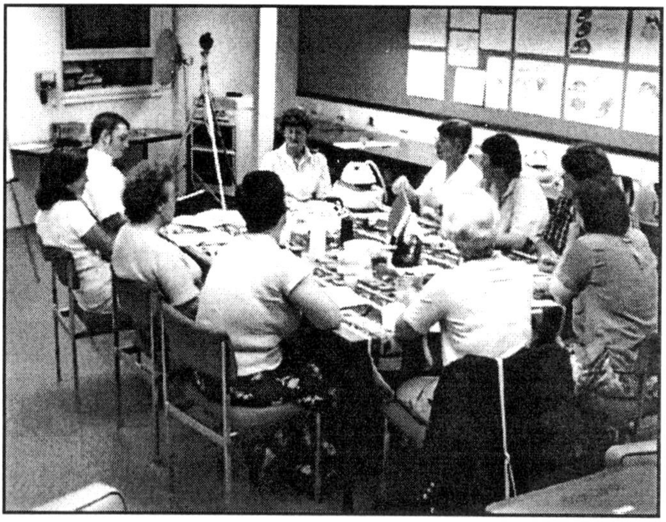

Figure 46.1. Focus group discussion

46.2 The Case Study

The project concentrated on ironing as an activity, and irons as the product area for investigation. Female users had been identified as the primary purchasers and consumers of this product type. The student conducted a review of current products on the market and further conducted intensive product handling of a sample of products. The project relied upon three meetings with the users during the project. An overview of the project plan is illustrated in Table 46.1.

Table 46.1. Project overview

Week	Activity	Outcomes
1	Literature review Construction of questionnaire(s)	Understanding of research method, preparation of research subject feedback booklet
2	Design review of existing products	Market awareness Product Specification
3	Preparation of Research Session	Protocol and schedule planned Feedback booklets produced
4	Research Session 1 Concept Generation	Feedback, visuals, video evidence and sketch work
5	Concept Generation 3D Modelling	2D and 3D sketches CAD renderings
6	Research Session 2 Concept selection	Feedback, visuals, video evidence
7	Concept Development 3D modelling	CAD animation Full-scale model
8	Research Session 3	Feedback, visuals, video evidence
9	Concept Refinement 3D Modelling	CAD animation Full-scale model
10	Design refinements Final company presentation	PowerPoint presentation

Three research sessions, each with the group of 8 subjects, were conducted in weeks 4, 6 and 8. Each research meeting integrated a range of exercises and tasks (product handling, concept selection, product personality profiling), with particular emphasis on focus group activity. Focus groups are a meeting of individuals that are brought together to discuss a particular topic, area or concern. A moderator chairs the group discussion through a prepared agenda and aims to promote discussion between the group of individuals. This approach relies upon the synergy of the group discussion (Kitzinger, 1994) and provides the opportunity for the user group to express their thoughts, fears and aspirations (Krueger, 1988). The distinctive aspect of this project was the involvement of the student as both the focus group moderator and designer.

During the focus group sessions mood boards were utilised. These boards represent a collection of visual stimuli that have been brought together by the student in direct response to the design brief. Such collections aim to capture the essence (mood, feeling, atmosphere) of the design project, design context and problem to be solved. Within this first research session the student presented the user group with a group of visuals that explored abstractly, the activity of ironing, domestic environment and the iron as a product. Though caution needs to be taken when introducing such unfamiliar material to users, the group responded favourably on presentation of the mood board material. This led to rich and fruitful discussion. Introducing such designerly exercises and materials are intended to assist in promoting and identifying design opportunities. A researcher who is not a designer (Cross and Clayburn-Cross, 1995) may not identify these. Such material also provides visual cues (e.g., colour, texture, and form) that can later support the designer from concept generation through to concept refinement.

On completing the first research session, the collected data was translated into information to assist and support the concept generation. The videotapes and notes taken during the sessions, as well as written feedback from users, provided the stimulus for the conceptual design phase. The initial session provided the opportunity for the student to become more familiar with the user experience generally, and more specifically user perceptions of existing products currently on the market. Some of these, such as steam stations, were unfamiliar to the user group (although they are widely used in parts of continental Europe). Though such products were unfamiliar to this particular user group, they stimulated discussion and provided additional visual stimuli. In preparation for the first session the student had conducted a design audit (assessment of the use of design and materials), literature search and had become familiar with the research methods to be applied.

The second research session concentrated on generating user responses to a series of 10 design concepts, presented two-dimensionally (see Figure 46.2). This range was highly conceptual and aimed to stimulate discussion amongst the user group. User feedback assisted the student in making key design decisions through formal (e.g., feedback forms) and informal (e.g., verbal, body language) communication. It is important to appreciate that users do not replace the designer, but provide evidence on which to base the decision-making.

A technique that has proved valuable is *product personality profiling* (PPP). This exercise involves subjects viewing a product, either in sketch or model form, as if it were a person. This *person* has certain characteristics (e.g., job, house, car, and gender). This exercise provides an insight into *who* the user perceives the product to be targeted at. For example, if a product were perceived as a professional who drives a high performance car, earns £100k as a merchant banker for a major City bank, who studied at Oxford and so on, this would not relate to a product aimed at the mainstream market. The user may aspire to own such products but in reality the user does not relate to that profile and therefore is highly unlikely to purchase the product. In this case the manufacturer could revisit the product semantics, colour application, form *etc.* The PPP exercise assisted in the concept selection process. Existing products and sketch models provided additional visual stimuli and further discussion among the user group (Figure

46.3). An ironing station was conceived, and is currently being developed into a full working prototype, whilst focus group activities continue to support the design decision-making process.

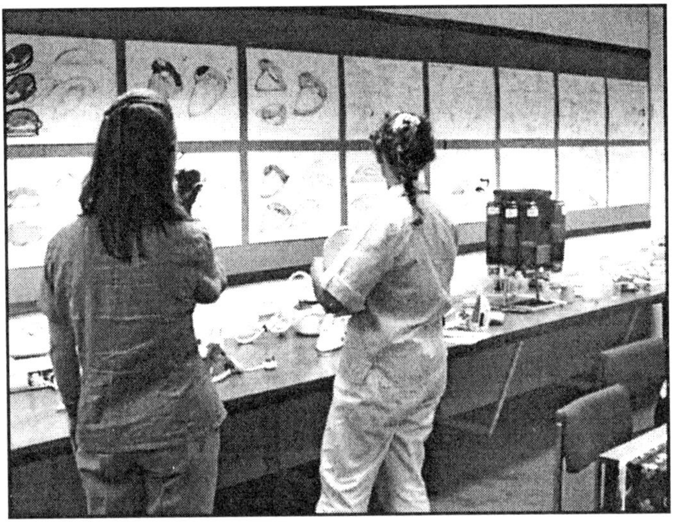

Figure 46.2. Users evaluating the conceptual design proposals

Figure 46.3. Student conducting personality-profiling exercise with user group

The final research session concentrated on one design concept, an ironing station that incorporated an iron, a board, and a steam generating system. A full-scale model of the ironing board was produced. Users were invited to conduct task analyses to ascertain its ergonomic appropriateness. Tasks included holding, pushing and pulling an iron across a surface to simulate ironing. In addition, the convenience of moving the ironing station was examined. Subjects represented a varied range of percentile heights, hand sizes and statures, which assisted the design student in identifying the concept's most appropriate dimensions.

From this project, pursuing the application of research methods, a nine-month project has developed involving three final year degree students and increased manufacturing company involvement. In addition, the new design team is exploring domestic environments for their appropriateness for running focus groups to enable comparison with experience gained from hosting research sessions within the Department of Design and Technology on the University campus.

46.3 Outcomes: The Student Experience and the Design Project

The main issues raised in relation to the process and the product under investigation, within this project, were as follows:

Issues in relation to *ironing*:

- user approaches varied from ironing at a set time and place (the *ritual*) to the more casual (*as* and *when* required);
- safety (care is taken to carry out such activity away from young children);
- generally users viewed ironing as an essential activity;
- the user group forecast that non-iron fabrics will dominate in future years.

Issues in relation to *irons*:

- cost and life expectancy (irons were perceived to be expensive and to have a short life expectancy);
- safety concerns (e.g., when used in the home near young children);
- weight of the iron over an average ironing period of 30 minutes, led to fatigue
- storage (users tend not to display iron when not in use);
- cordless irons are less efficient due to heat loss when detached from the energy source;
- functions and features (e.g., steam was perceived as essential by some users, left handed considerations important).

The user group was less knowledgeable about the retail cost of this product area in comparison to other small domestic appliances, in keeping with earlier research findings (McDonagh-Philp, 1999). This is probably due to the low turnover of this product type. Participants provided feedback from each session, reporting a high level of satisfaction, gained through involvement with the focus group and associated activities (product handling, visual evaluation). Overall, participants viewed the discussion favourably, but the filling in of questionnaires (evaluation forms, subject payment forms *etc.*) consistently gained a low rating. As with previous studies conducted by the author, users have expressed concern regarding the estimated life of small domestic appliance products. Polymer kettles for example are perceived only to last 2-3 years, whilst irons are less frequently purchased and are expected to last between 5-15 years. The user group valued their continued involvement during the ten-week project, which resulted in the eight subjects investing 9 hours of their time, each session lasting three hours. The indications from the student are that valuable insight and empathy was gained to support the project at critical stages (e.g., pre-concept generation) through contact with the user group over the duration of the project.

46.4 Conclusions

This project provided the opportunity for an undergraduate design student to explore a design project in collaboration with a manufacturing company whilst applying methods that concentrated on user involvement from the initial stages of the designing process. The main constraints regarding this method involve the cost of hosting and running such sessions. This can be overcome by a less formal approach. The student can conduct sessions at low cost by utilising their student networks, family and friends, and conducting such sessions within domestic environments. However, care must be taken to introduce a degree of informality, as the perceived importance and value of the user contact must be protected. This method offers immediate contact with users and as it provides a flexible framework, the student can further develop their own unique exercises in response to the unique problem being solved. The following points represent a range of disadvantages and advantages experienced during this design project.

Disadvantages:

- time consuming for the supervisor of the student over the summer vacation period (company visits, meetings);
- focus group session are resource consuming (subject fees, hospitality);
- if the student operates without the support of fellow peers it can lead to feelings of isolation and de-motivation;
- time management becomes a crucial factor for the student.

Advantages:

- provides a unique opportunity for the student to become involved in design research at an early stage in their career and training;
- the holistic nature of simultaneous research and development provides a qualitative improvement to a student's design education;
- student develops their design 'toolkit' of skills, experiences and expertise over the summer period in preparation for final year design practice;
- supports the development of a design research culture within a relatively new discipline;
- in this example the student has chosen to adopt the method(s) for a major design project which will involve a further two students.

To conclude, there are advantages in a user-centred design approach for participants, students and design departments. The student, supervisor and manufacturing company perceived the project to be successful on the grounds that it contributed to knowledge, it supported the development of design research methods, and the project was undertaken within a live design brief scenario. Applying design research methods at undergraduate level is perceived to be feasible, though further work is currently being undertaken to monitor students integrating these methods independently for three major projects in the final year, based on the experience gained by the student during this study.

The non-designer, when presented with visual data and abstract problem solving exercises can be potentially very creative. It is the responsibility of the designer to tap into this resource to assist and support the designing process. Creativity is not exclusive and through such contact with users. Teaching and developing the methods of user-centred design should be integrated more fully into the undergraduate syllabus. Professional and undergraduate designers can begin to utilise users as a rich design resource.

References

Cross N, Clayburn-Cross A (1995) Observation of teamwork and social processes in design. Design Studies 16:143-170

Kitzinger J (1994) The methodology of focus groups: the importance of interaction between research participants. Sociology of Health and Illness 16:103-21

Krueger J (1988) Focus groups: a practical guide for applied research. Sage, London

McDonagh-Philp D (1999) User evaluation techniques applied to small domestic appliances: a pilot study. In: Lindemann U, Birkhofer H, Meerkamm H, Vajna S (eds.) Proceedings of the international conference of engineering design (Volume 3). Munich, pp. 1525-1546

McDonagh-Philp D, Denton H (1999) Using focus groups to support the designer in the evaluation of existing products: a case study. The Design Journal 2(2):20-31

McDonagh-Philp D, Lebbon C, Torrens GE (1999) An evidence based design method within a user-centred approach. Proceedings of the 4th Asian design conference international symposium on design science, Japan, October. Program committee of the 4th Asian Design Conference: Japan. (CD Rom) (ISBN4-9980776-0-0)

Parlett MR, King JG (1971) Concentrated study: a pedagogic innovation observed. Research into higher education monograph, 14. Society for Research into Higher Education, London

Chapter 47

Co-designing among Sixth-formers

R.W. Molyneux

47.1 Introduction

I wish to present an empirical study of the development and use of collective and collaborative teaching among sixth form design students in the period 1979 - 1999 at Haywards Heath College, West Sussex, UK, and to examine the subsequent benefits and outcomes when entering higher or further education, adult life and the world of work. I have used the terms 'Sixth Form' and 'Sixth Formers' because for most of you these are readily understood descriptions; the term includes for my purposes a number of older students, perhaps three per cent, who were returning to gain qualifications. I believe that their *vision, creativity* and *capacity for design* were significantly enhanced by the collaborative, user-centred and communal nature of their collective experience; many learned valuable and necessary lessons for social and economic success in life. These students and their progress are my chief resource for this paper; there have been more than eight hundred of them in the Design Department at the College during this period, many of them maintaining subsequent contact, some for almost twenty years. Based on this experience, I would like to suggest to the Conference some important educational insights and lessons that can be embraced; one example would be the need, deeply felt in my case, for parallel, co-operative development pathways in enhancing personal creativity-intending artists, designers, engineers and business people working and learning together, through teams and groups, to both personal and externally-set goals, some of which would be self-determined.

Firstly, some working definitions are needed. My perception of these has deepened over the years; I am grateful to all who helped in this process. I referred just now to vision, creativity and capacity for designing and, whatever the medium of expression, these qualities are vital for any contribution of worth to take place.

- **Vision.** For me, suggests both desirable direction and a destination. It bridges past, present and future; it contemplates success and failure; it sees what could be within a context. It looks with confidence towards the future; it is often bold and optimistic, rarely dystopian. The 'distance' between a vision and reality inspires creativity, or 'creative tension'; keeping one ahead of the other maintains that 'stretch' and leads to constant improvement. Collaborative work sustains and magnifies this process, in my view.

- **Creativity.** The National Advisory Committee on Creative and Cultural Education (1999) sees creativity as 'imaginative activity fashioned so as to produce outcomes that are both original and of value'. The Oxford Shorter English Dictionary (p.544) considers creativity as 'being imaginative and inventive, bringing into existence, making, originating'. I would be reasonably happy with both of these within a student context, adding only that problem finding, the perception and formulation of a problem, is as important a feature of creativity as problem-solving. The 'problem-solving' that one so often meets in education and training does not go far enough; the 'problem' is often presented in a form which guarantees a solution in the presenter's mind.
- **Design Capacity And Designing.** NADE (The National Association for Design Education) presents a view that I would subscribe to: that the capacity for design is a human fundamental, its development a proper part of general education, as well as a necessary part of the education and training of professionals whose work is founded in design (and many others, I should think). For many students and professionals, designing represents a complex and collaborative working out of needs, ideas, opportunities, choice, materials, aesthetics, value, cost, ease of manufacture, marketing and quality, among many others; they are present in both design capacity and designing.
- All the above are non-territorial; they can (and do) overlap in some degree and this was certainly so in the experience I shall describe.

47.2 Finding Out

I was a late convert to teaching; twenty years earlier, I had abandoned an Art Teacher's Certificate offer and made a career in manufacturing industry, working as designer, marketer and manufacturing manager for international companies in the UK, USA and Europe. It had been tough on a young family, though, and I needed an opportunity to reflect; a one-year course in teaching design would give me that chance, also a return to academia before commencing a fresh venture.

I was right about the venture; completely wrong about the kind of venue. September 1978 found me, newly-qualified, about to teach design in a West Sussex Comprehensive converting to a Sixth Form College. Or so I thought; publicity for the course had been negligible, no takers could be found. Until the process of recruitment could produce an intake for design, I busied myself with whatever came to hand; vital skills and knowledge development for the future.

I also gained that opportunity to reflect. An enlightened Principal, Ted Winter, had told me at interview that he wanted to see design used as 'a bridge between subjects'; it was plain that, once in being, support and encouragement would be forthcoming , including those valuable cross-curricular opportunities. The college also accepted some contentions:

- that design and designing are important human skills; good design is never easy;
- that design is an ever-changing and evolving process, based often in need and also linked to artistic, scientific and technological achievements;
- that it can be our imaginative world made real; a form of visible thinking.

There was scope too for programme research; many of the likely recruits were already in College and some were in my classes. I encouraged them to come and chat about 'design', some brought their girl or boy friends and discussion ranged widely. Some expressed surprised that a designer had 'gone slumming' (as they saw it) in teaching; a greater source of wonder was how widely design could range. Some were very specific: 'Sir, I'd like to be a car designer' (subsequently, Kevin did just that with Giugiaro); others were interested, but vague in their intention: they felt 'creative' but did not see an outlet for their talents, real or imaginary. Just one thing was certain; there was interest across the whole ability range.

47.3 The Beginnings of Sharing and Co-operation

At the start of this year (1979), I had made contact with NADE; I found its values stimulating and a member's book, 'About Design' by Ken Baynes, was to prove fundamental to the envisaged course. His espousal of the so-called 'Oxford' Design examination (A83 - later 9883) made the decision for us. The 'clincher' came when he quoted Malcolm Deere, a principal examiner. An extract reflects on collaborative endeavour:

> *'...The quality of the people that it produces, as individuals, ...I have not found this to the same extent in any other area of the curriculum. This reflects in part, I think, a very important feature. It is the very fruitful relationship that is apparent in design education. It is an approach based on the teacher learning with students, on recognition of fallibility, on a kind of mutual respect. This, too, is not seen elsewhere; our task is to develop it, to improve it, above all to preserve it.......its potential contribution is quite unique.'*

This confirmed that we should adopt the Oxford syllabus; we were to find that, handled imaginatively, it conferred immense freedom of approach on both student and teacher. Given, though, the wide range of interest shown in discussion with students, it would also be a brave choice. Would we be able, for instance, to sustain the depth of enquiry that some plainly wanted, seeing their course as maybe an opening to a lifetime's interest and employment? Deere's recognition of co-operation, of learning collaboratively, but based in mutual respect, was in large measure to provide an answer, at least for the time being. It would be difficult, though, not least in the commitment suggested by such student-centred learning. Each was being promised a programme of considerable individuality, working ultimately towards a moment of high endeavor and personal expression. Wonderful learning, perhaps, but also wonderfully time-consuming. No computers to help in 1979, very little in the way of direct learning resources and just six 'official' periods a week. I think we must all have been a bit mad.

Paradoxically, perhaps, working together and in cells and groups provided constant demonstrations of highly-personal development. Within the relatively large groups that were to develop, sometimes up to forty or more, there was potential for interest groups to develop, sometimes a group of intending architects, engineers or textile designers, sometimes a 'buddy group' from a previous school or class. The sum of the whole was almost always greater than the parts; classes took on sometimes, and to the consternation of colleagues, something of the character, if not the appearance, of their former primary schools. The educational wheel had indeed turned full circle.

One learned, sometimes rapidly, to facilitate these 'growth points'; groups of sixteen or seventeen year olds can run very quickly with an idea. Those less able (or willing) were still caught up in the general enquiry; very few to my knowledge stumbled disastrously. One learned, too, that the ideas sparking this 'running conversation' on design are often deceptively simply; a brief of six or seven words can sometimes sustain several fertile weeks of work.

What else did one learn through this adventure in learning and facilitating? Thoughts which come to mind developed over a period of several years; we tried to integrate them in the ongoing work as a kind of re-cycling process:

- for the age group, the activities and experiences should be full of promise for the future; students should see that it's about what they do, rather than how we, the teachers and facilitators, think;
- students should experience a genuine situation, preferably discovered by themselves, and for which they have made proposals and suggested solutions for which they will be ultimately responsible;
- a form of group, or self-initiated, learning involves the whole group and individual; it is lasting, has depth and involves feelings as well as intellectual and practical activity;
- all kinds of learning are helped, sometimes dramatically, when there is seen to be full participation by everyone in the activity; much of this significant learning can be achieved by hands-on experience in doing and making;
- the activity has to have adequate relevance and meaning, at least for the majority; even the most apt stimulus may have to be 'sold' to a minority, although consistent experience of success may make this much less needful;
- in the process of learning, all should become acutely aware of how and why innovation and change has been effected;
- this awareness should continue into self-evaluation, which is highly important; evaluation by others, working in groups, is needful but should be seen by the individual as less important;
- there should always be ample opportunity for reflection, whether in groups or on a one-to-one basis.

Apart from these more wide-ranging considerations, we found that there were others that we should see in student participation, such as:

- students should feel they had participated on an equal footing, whatever their role in the venture;
- they should be aware of, and accept, differing points of view; diversity is vitally necessary in an ideas culture;

- in particular, each should be prepared to listen and to be moderate when 'attacking' others; mutual demolition was not a worthy objective (competitive students did not always see it this way);
- honesty and truthfulness were vital in every situation;
- there had to be a degree of confidentiality; students should appreciate why disclosure was not always appropriate, particularly with half-formed ideas.

47.4 Group Collaboration and Performance

Group work, in varying forms and contexts, was introduced within our teaching groups, so that small projects were developed as forms of practice towards ' the big one' itself: the Major Project, worth 40% of the available marks. This took almost a year and encouraged a broad view of design, as well as giving students opportunity to experiment within more detailed contexts in which satisfying personal objectives might be found. Two aspects were vital for everyone:

- that all had group and inter-personal experience which would illuminate the assessment process, the final exhibition and the value of the outcome;
- that the outcome, or 'Product', was not only a contribution in professional terms towards the student's chosen sphere of enquiry, but that it was also continually, and finally, assessed on the way the student justified the actions taken and subsequently evaluated them.

It can be seen from the above how a very wide range of student interest could be both accommodated and subsequently celebrated; how it was, too, that students with varying interests could share, perhaps for the first time, a collaborative experience of creativity in their personal development.

The process whereby the Exhibition was used as a common tool of learning is equally interesting. By the end of the Autumn Term in their second year, most students could present a clear idea of what their project entailed. An analysis of a typical student response at this stage is shown below as Figure 47.1.

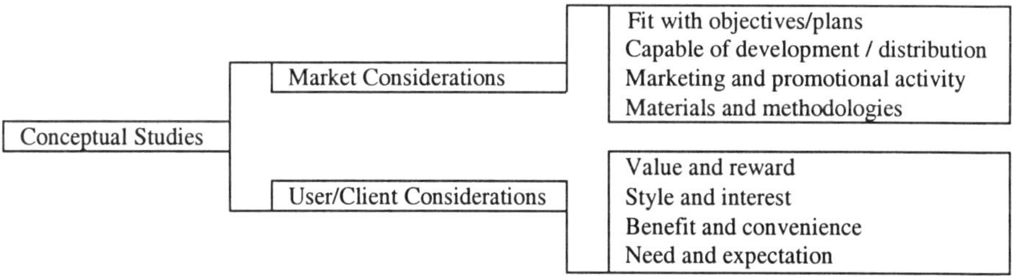

Figure 47.1.

In discussion, it was then possible to allocate spaces within the planned area of the exhibition. With some concepts the students were thus able to conduct their work or build their projects within the allocated space; they could also test their allocation, as obviously space requirements differed greatly. As an example, in 1992 a group of five students, who were designing various aspects of a restaurant car for British Rail, were able to construct, in tape, the boundaries and parameters to which they would work. As the project evolved, they constructed, tested and evaluated all aspects of their work: including uniforms, house style and cooking/serving methodologies within the original taped 'shell'. Another student, an aspiring professional golfer, constructed a large scale model in sand of his proposed golf course within the given space; a process taking approximately three months. Many students used their allocated space as their workshop for a period of time, gradually changing the space into their final exhibition stand or presentation; others of course, would work elsewhere, perhaps out of College, and only fully use their space in the latter stages of their work. Very large projects, like a full-scale lorry cab for a trucking magazine or a special effects 'pod' for making sci-fi movies, might mysteriously appear behind the curtains on the College stage.

This form of preparation, while demanding for all concerned, conferred important benefits; many of these related to the desirable qualities already set out. Perhaps the most important was that of being able to compare and evaluate work with that of one's neighbours (or even those further afield; the process could take place in up to three adjoining rooms). Comparison of this kind is highly valuable; many changes were made as the result of being able to reflect on what others were doing, but rarely did this degenerate into a form of 'me-tooism'. Valuable social skills were also learned; not least the ability to work with one's neighbours to mutual benefit (most students would help others, particularly when in possession of specialised skills or knowledge). A mutual and pleasing concern emerged in many instances: the boundaries between stands and projects were often carefully negotiated to mutual benefit; colour and content were frequently manipulated in order to give a better chance to others and to ensure that sight lines and the eventual overall look of that part of the exhibition were felicitous and successful.

Not the least advantage, of course, was the spectacular effect of 'growing the Exhibition' on some less active students; their dawning realisation that they would be seen as not in the same league unless they rapidly upgraded their approach could be quietly enjoyed, by staff and by other students alike. In practice, very few would not readily raise their game; partly in consequence, examination results nearly always reflected added value, even at a time when this was not an issue of quality performance. The Exhibition became in consequence an upwelling of pride and concern that often went far beyond the requirements of assessment; there was a palpable sense of occasion which, to my mind, often encouraged students to go well beyond what one might have reasonably expected. Mutual uplift in performance may now be taken for granted; what was astonishing was the lengths to which the students might go in presenting their work. Young women, never otherwise seen out of trousers, would attend their Viva in a skirted outfit colour-matching their exhibit; young men, never normally seen out of jeans would somehow find a suit and afterwards crumple in a corner. To some extent it was theatrical, but it was too a constant reminder of the burgeoning group ethos; each successive year truly believed through their preparation that they were the best, taking off to the worlds of university and work a daunting confidence in their ability to succeed. Theatre had become in fact reality.

In terms of development in design thinking, the effect on the first-year group could be quite spectacular. In general, we did not allow them much sight of what their elders were up to, although one's guess is that rumour worked to our advantage. In the wake of the examiner/moderator's visit, exhibits were quickly refreshed and the First Year came to look and learn, with the instructions that they were to:

- find an exhibit that they felt an affinity with; sit, read and absorb, then write their own critique on the success, or otherwise, of what had taken place, together with their own recommendations for improvement;
- develop this critique in conjunction with at least one other student.

In this way, each student had the guarantee of not only being the recipient of 'token-passing', receiving the standards of the year that was past, but also of being involved in at least two developmental works of their own (the critiques) at a time when the form of their own Major Project was beginning to be of concern to them. Conscientiously undertaken, this was to be a powerful influence for good in the development of their own work and also, patently, to the long-term success of the course itself. Continuity of this kind cannot be emphasised too highly in a creative situation, where the written or spoken word is too frequently inadequate; the group had thus re-invented itself for another year, and perhaps another, particularly as the intending students of the course would also see the show.

47.5 Group Work in Contexts

The Major Project and its associated Exhibition (which also included, for some years, so-called 'supporting work' achieved in a range of minor projects, many self-determined) was, for us, always the prime example of collective and collaborative teaching and learning among our sixth-form students; it was often, as it should have been, the most successful, embodying much of what had gone before and building from those experiences. It was particularly important that the timespan of the Project took place within a stimulating context of discovery, sometimes of self, and the acquisition and management of knowledge, materially helped by the reality of discussion and reflection among one's peers and, of course, tutors and 'expert witnesses', who could be (and frequently were, as the process caught their imagination) staff of other disciplines, or 'real' experts in the field of inquiry. Some reasons have already been offered as to why this should be so, but one might add that, in my view, rarely was there an innovative approach, or release of a student's true potential for creativity, without a genuine experience of group contextual working within the learning field. The process whereby students with a (sometimes carefully seeded) variety of interests, talents and skills worked together, sometimes for several weeks, to investigate, propose and synthesise, which not only benefited them as participants but gave a decided 'buzz' to the whole group. I might also add, it was an on-going effect that kept those who taught, or otherwise became involved, constantly on their toes.

The process of wide-ranging working developed over many years, but did not become fixed. The advent of an Advanced GNVQ (General National Vocation Qualification) in Art and Design in September 1993 added fresh interest and possibilities to group contextual working; care and co-operation with timetabling and blocking ensured that teaching sets from both A Level and GNVQ were able to work together for approximately two-fifths of their taught time, although it should be borne in mind that GNVQ Advanced had a much greater allocation of time, being seen as equivalent to at least two A Levels.

By this time, perhaps regretfully, there was a narrowing-down in Project possibilities due to the greater emphasis on Technology adopted by the examining board (still Oxford), so opportunity to regain breadth in a joint process of contextual working was most welcome. It was highly effective for developing vision and creativity; what follows is an example of first-year joint action:

- Groups of (normally) five students were asked to undertake a joint enquiry. This might be (for instance) an integrated study of aspects of the local town centre (Haywards Heath) and its future suitability for a changing, younger population. Each group would concentrate on no more than two or three considerations. Although guidance was provided, a significant degree of choice of topic was built in to the collaborative work, so that the interests and abilities of each group were properly reflected. The findings, views and proposals of each team were then presented in turn to the rest (perhaps forty students in total), who were complemented by the presence of, say, an architect, a civil engineer, a member of the District Council and a working artist, as well as the staff teaching. It was always encouraging to witness the degree of support which students received from their peers; significant in-depth knowledge and the ability to empathise with an audience were achieved.
- Students were then required to undertake a detailed study, which for A Level students would contain a proposal involving a detailed specification and construct; GNVQ Advanced students were asked to consider forms of 'Public Art' which could be a stimulating feature of a redevelopment – this could be highly focused as a single artifact, or be more general in character, suffused across an area; it would serve as a complete outcome for one of the Mandatory Units. The considerable joint awareness which by now had been built up led, in a natural and unforced manner, to extremely sophisticated and effective interpretations of the brief(s), and for those who were currently contemplating careers in architecture, town-planning or the three-dimensional arts it constituted an excellent and thought-provoking demonstration of that flow and counter-flow of ideas, needs, performance (and sometimes prejudice) that are so much part of these situations. Particularly significant was the constant reminder that proposers of change, such as artists, designers and engineers, have to lobby and fight constantly for their ideas; this was one of the most valuable lessons of the specific group experience.

We have always been fortunate in keeping in touch with former students; and an analysis of destinations of the three cohorts 1987 – 89 is shown in Figure 47.2 – this is regarded as being typical over many years.

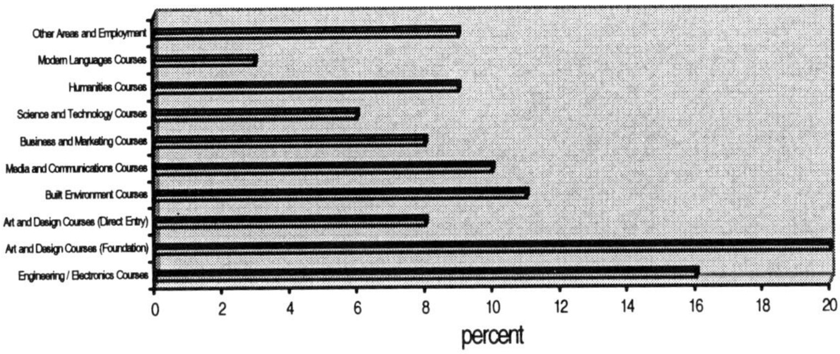

Figure 47.2. Student destinations, 1987-1989

Significant numbers of them have confirmed how much this formative experience was in helping them to prepare for group and public tasks, often undertaken in HE and FE environments, which demanded developed skills of integration and co-operation. Some felt this was a good introduction to work and to life-long learning.

Communication in groups thus came to be seen by students as a natural part of the design process. Students developed, because they had to (or risk peer comment), the power to communicate effectively, to associate and to express freely the relationships between their ideas and those of form, function, marketing, cost and the many commercial and cultural factors which contribute to today's often complex projects. Constant experimentation and testing was also seen as vital and necessary to this end. A further benefit was that almost all of them, from time to time, gained useful experience of the problems of leadership. This was not only important to personal development, but also recognised that many 'design' processes take considerable periods of time. Students should become aware of, if not conversant with, the many different contexts of design, if, like some, they are eventually destined to lead and to be involved from the very beginning to the very end of a large project (The Millennium Dome?). They must be acutely aware of how and when to call for expertise, in design or other disciplines, and always try to regard their involvement as a collaborative, multi-disciplinary experience.

47.6 Looking to the Future

It may now be seen that, by encouraging our teenage students to work in co-operative and collaborative ways, we were able to develop and sustain a design programme which helped them to gain personal authority and maturity, sometimes beyond their years. Developing their ability to integrate, and to carefully weigh the hierarchical importance and significance of ideas, processes and events would, we felt, confer the potential to become successful participants in a changing world. In the process, a variety of student personal developmental needs and career intentions were also addressed; stimulus for self-discovery came as a bonus.

The process, one may say, not only benefited those with a designing career in mind; it also helped others in a wide-ranging cohort of students to apply different strands of their advanced studies to a variety of career-orientated tasks or opportunities. It should perhaps be argued that such a formative experience ought to be available to the majority, at least, of all sixth formers, and, particularly for its contribution towards the multi-skilled portfolio, perhaps digitally expressed, a requirement which is now seen by some as part of a personal and career strategy for the 21st Century (Vandevelde, 1999). The concept of the 'digital portfolio' may also reinforce the arguments put forward for developing a broad range of skills (Hodgson and Spours, 1999) alongside particular opportunities for integration, contrasting, as we have seen, with too-early a specialisation. Educational developments such as the International Baccalaureate, or IB, are also complemented. Teachers and practitioners of design thus have an important role to play in devising or refining strategies, programmes and courses that will encourage *vision, creativity* and *capacity for design*, across the widest range of ability and interest. While considering the needs of so many, one may accept that specialised design (and technology) A Levels, or Advanced GNVQs, can do much to cultivate the talent and vocation of a minority if presented and taught in a collaborative manner. A greater need, therefore, exists for a focused experience of designing, constrained only by time and available resources, which can be shared and valued by those who are currently disbarred, by lack of opportunity or provision.

Bentley (1998) praises the value of solutions and outcomes achieved through the interaction of individuals in a system of learning where collaboration ensures that experience is shared. Hills (1994) acknowledges, too, the collaborative, comprehensive and integrating nature of design, arguing for a new model in post-graduate design education which would persuade schools and colleges to increase respect for design as a serious A Level proposition for sixth formers. There are past instances (I would include Haywards Heath) where this has been achieved but these have been relatively few and much work remains to be done – design institutions can continue to play a major and valuable role in this. Thus, Hills' concept of a generic design course available to both post-graduate and undergraduate students could have its precursor in a course, available at the sixth-form level, which would allow students to build bridges between their varying subjects, in the manner that design does (and did, in the experience I have set out). It might not be necessary to insist on physical outcomes, as has been the case, but the generation of a virtual outcome would be a principal and beneficial concern. The course would also require some understanding of the following:

- innovation and change;
- human factors (ergonomics);
- aesthetic judgement and the scope of technology;
- evaluating success and failure;
- marketing (including research);
- IT and the development of awareness of materials and value, risk and reward;
- social volition and collaboration.

Most of these 'topics' feature strongly in other subject areas and it would thus be possible for cross-curricular bridging to be both intuitive and easily facilitated.

The presence of these areas of knowledge and understanding, associated with formal subject areas in an active context of designing, may not only stimulate real creativity but also encourage, firstly, a positive attitude to the transfer of skills and, secondly, a real use of reflection by the student on the nature and relevance of his or her learning. As Bentley (1998) also points out, such reflection, and its connection to other knowledge and situations that may be encountered, often in practical and work place situations, is crucially important to personal development.

47.7 Acknowledgements

I forget whom it was who said that in order to find one's way, one must first get lost. I did that, most successfully, half-way through a working life and I shall not forget how, with the great help of others, particularly my students, I found a pathway to collaborative teaching. I would like to thank all of my guides, young and not-so-young, for their generosity, honesty and understanding over these twenty years. Many are shown below; if I fail to give credit, I apologise now.

The Principal, Vice-Principals and staff, teaching, technical and administrative, of Haywards Heath College and in particular Des Roue, Malcolm Smith and Alice Tyrrell, who were colleagues during that time, and to the Physics and Art Departments.

The students, more than eight hundred of them, who are the true heroes and heroines of this story, sometimes just for one day (or even fifteen minutes).

Their long-suffering parents, guardians and friends, who provided an endless stream of advise, coffee and sympathy,

The examiners, moderators and others who had wisdom to provide teaching materials capable of such wide interpretation, in particular Bernard Aylward, Malcolm Deere, Richard Kimbell, Ken Baynes and Laurie Latheron.

The HMIs, advisers and members of the Design Council's education service who gave so much by way of encouragement, advice and direct help to students, in particular John Price, Michael Pullee, Doug Marshall, Frank Whiteman, Steve Burroughs, Freda Hooberman and Ghislaine Kidd.

The seemingly endless procession of willing professionals, tradespeople, servicemen, senior citizens and children of all ages who freely gave their time as guinea pigs, test marketers, critics, doubters and cheerleaders to all these young people.

47.8 References

Baynes K (1976) About design. 1st edn. Design Council Publications, London
Bentley T (1998) Learning beyond the classroom. 1st edn. Routledge, London
Eggleston J (1976) Developments in design education. Open Books Publishing, London
Hills G (1994) A renaissance in design education. Open Eye Publishing, Horsham Hodgson
A, Spours K (1991) New Labour's educational agenda. Kogan Page, London
Hubel J, Lusson DB (1984) Focus on designing. 1st edn. McGraw-Hill Ryerson, Canada

Man and his Technology (1971) Engineering concepts project. State University of New York
Mayall WH (1979) Principles in design. Design Council Publications. London
Morrison J, Twyford J (1994) Design – capability and awareness. Longman Group, Harlow
NADE Journal (1985) NADE conference on innovation and change, Haywards Heath College
Rogers R (1997) Cities for a small planet. Faber & Faber, London
Vandevelde H (1997) Beyond the CV – securing a lifetime of work. Butterworth-Heinemann, Oxford

Chapter 48

Collective Action for Design Learning: An Investigation in the Design School Studio

P. Ashton

48.1 Introduction

Pressure on resources has been a major feature of the management of design education in the UK for several years. One impact of this has been that the use of the studio environment has been questioned. In fact design educators know very little about the benefits and role of the studio. One key property of the studio is that it allows social interaction to take place between the learners. Indeed it is very much a group learning experience. Membership of groups has been identified as an important influence in the development of our identities and skills. This paper will explore one aspect of the studio - the interaction between students in this environment and how membership of the group plays a role in learning. Using data drawn from studies in two Universities, it will describe studio usage, the position of the students within a social network and the role of some students as leaders or role models. It will explore the hierarchies and networks that exist and in particular will show the relationship between social groups and the students' positive rating of peers as task experts. The paper will show that this collective action is an important feature of the learning experiences of design students.

48.2 Groups, Identity and Social Comparison

The recent changes in the funding and nature of Higher Education have caused educationalists to constantly question the effective use of resources. Great emphasis has been placed on the notion, for instance, of independent learning, (GLAD, 1987) which could tend to ignore the idea that learning is a highly social process (Ashton, 1999). It is the premise of this paper that learning in the Design School studio takes place within a group, so to establish the context for the research presented here, it is necessary to consider the nature of groups and the consequences of being part of a group for individual participants. The Design

School year group or course is rarely considered to be a group with dynamics, structures and rules, even though it is known that any crowd of people who have not met before, quickly adopts a kind of dynamic. Goals and purposes are shared and agreed, people will assume or are assigned roles according to their status or personalities, socio-emotional leaders and task-orientated leaders emerge which others support and follow. These structures and guiding principles are described as the norms of the group. In a learning community like a Design School year group these kinds of tasks, roles and structures are of course not assigned or explicit as they might be in a formal team. However as Cooper and Alderfer point out:

'...It is one thing to say that experiential learning groups do not have structured activities, and another to pretend that they do not have structure to them - a set of assumptions, values, goals. For it is our contention here that such groups are highly structured and the practitioners and members both act in accordance with this structure' (1978, p3)

The Design School studio provides considerable opportunities for the development of hierarchies and networks of influence and communication through which norms will be formed and disseminated. The experience of a design student in Higher Education is, however, one of working within a group in a shared informal learning space where it is possible to forge relationships and develop a sense of the working practices of others.

The role of the social group in this context is all the more powerful because of the influence that it has the particular stage of life most students have reached. Eighteen to twenty one-year olds are still the largest age group represented in undergraduate design education. This is a time when individuals establish new identities and when peer groups exert particular influence: individuals are motivated to comply with group norms - they become socialised into different roles (Breakwell, 1992).

It is believed that individuals undergo two 'layers' of socialisation in their lives. Primary socialisation, which takes place during childhood, enables an individual to become a member of society. Secondary socialisation is the process that inducts us into new sub-groups within society as a whole (White and Mufti, 1979). Secondary socialisation into societal groups becomes an important part of our later lives and we define ourselves and who we are by our membership of and position within a variety of groups. As Brown observes:

'... our social identity - our sense of who we are and what we are worth - is intimately bound up with our group memberships. Thus, one of the first consequences of becoming a member of a group is a change in the way we see ourselves. Joining a group often requires us to redefine who we are which, in turn, may have implications for our self-esteem.', (Brown, 1988)

One of the important functions of group membership is to allow social comparison. In other words we make judgements about our own worth, abilities and personalities by comparing ourselves with others. Knowing how good we are in any field of endeavour by making comparisons with others, allows us to identify how close we are to group norms. Brown says:

'...social comparison takes place within - and often gives rise to - status hierarchies, and it is suggested that this strongly implicates self-evaluation as the

motive underlying them. By locating ourselves in the group's status structure we gain insight into our abilities relative to our peers (Brown, 1988)

Furthermore Festinger's exploration of social comparison theory suggests that we have a need to try to change ourselves so that we compare favourably with our peers, and thereby become more highly valued members of the group. The role of leaders, or those identified by members of the group as embodying valued attributes, is critical in providing role models against which others can compare themselves (Festinger, 1954). Thus writes Breakwell: 'On the basis of Festinger's social comparison theory, it is assumed that we have an upward directional drive which leads us to compare ourselves with others who are similar to (or slightly better than) ourselves on relevant dimensions', (1992).

It was felt that the idea of social comparison within groups may have particular relevance in design learning. It has been stated that reflection plays a particularly critical role within design education. (Schön, 1984; Ashton, 1995) Experiential learning - learning from action - requires a period of reflection when the learner evaluates his or her actions. It is likely that social comparison in the Design School studio could be an important means of reflection.

48.3 Constructing Group Norms

If our identities are constructed by group membership, the norms with which we aspire to conform are also a product of the group and most notably of those who are able to exert influence or leadership. Those who emerge as leaders play a fundamental role in shaping the norms of the group as others compare themselves to them. From the literature on leadership (Northouse, 1997) we can see that those who generate the best ideas are often assigned this role; however, those who are strong in socio-emotional traits - helping and supporting - are also valued. Certain members of the group may be prevented from assuming these roles because of societies' perceptions about obvious characteristics like gender and ethnicity (Farr, 1993; White and Mufti, 1997). Group norms may also be influenced by intergroup relationships. For instance, notions about the role of a designer may be influenced by the expectations of those who are not themselves designers. In this way, the idea that who we are and what we 'know' are the product of our interactions with others is defined as social constructivism. Burr says that, 'It is through the daily interactions between people in the course of social life that our versions of knowledge are fabricated.' (1995, p79). It is for this reason that the understanding of a student about the role of a designer, namely, what they do and how they do it, will be a product of the individual's interaction with others - their peers, teachers and also influences outside the University environment. If what we should learn and how we learn is socially constructed, the networks of social interaction within which we exist assume particular significance and need to be studied and further understood.

48.4 The Research Method

The research is based on the view that social interaction amongst design students enables learning, and also mediates learning and one of the contexts in which this may take place is within the Design School studio. To establish the nature of social interaction, an initial study was undertaken which took a broad overview of the features of undergraduate design education in four different Design Schools. The resulting qualitative data reinforced the view that social interaction takes place and plays a role in design learning (Ashton, 1999).

The subsequent research, which directly informs this paper, is the result of two studies undertaken in different University Design Schools. The research attempts to discover the communication networks that exist in the studio and how students are positioned within this framework. The resulting data are specific to each of their contexts and do not attempt to represent the totality of undergraduate design education but to provide 'snap shots' which, because of the common methods of research and analysis, are comparable. The two situations have been compared to reveal similarities and differences.

As a representative sample of Design Schools was not required for this study, two Schools in England were selected randomly. The method used in the research which is described in this paper is informed by the pioneering work of Moreno who developed ways of investigating and describing structures of influence and communication within social groups (Moreno, 1954). The data were gathered by asking all students in one, second-year group in each institution, to complete a questionnaire. There were 19 students in one group and 28 in the other. There was an adequate balance of female and male students in both groups to allow an analysis of possible differences between the genders. However, there was only one representative from an ethnic minority in each group and therefore an analysis of differences based on ethnicity was not conclusive. In the questionnaire, students were asked to name two peers whom they believed were strong in a range of characteristics. The criteria are divided into two broad areas - task abilities, such as, drawing, making and ideas, and socio-emotional traits, such as, supporting, advising and confidence. These characteristics were arrived at by drawing on the results of the previous study. This was further corroborated by a literature search which looked for research into the skills valued in the task of design and characteristics required for group leadership (Northouse, 1997; Cross *et al.*, 1994) Students were also asked about their patterns of usage of the studio, their satisfaction with their course and their allegiance to different groups in their lives. In addition staff provided the marks they had awarded the students for the last two projects.

The data were analysed by establishing the mean profile for both groups on a number of dimensions. This enabled an individual profile for each student to be developed based on deviation from the mean. In this way it was possible to compare the data from both groups. A socio-gram was also constructed for each student who responded to the questionnaire, identifying who they networked with and who networked with them. It was hoped that analysis would allow the building of a further socio-gram which described the total network of communication within the year group but this was more difficult than originally envisaged and further work is required to achieve this. It was possible, however, to cross reference the data in a number of ways, by, for instance, comparing the level of peer rating with satisfaction with the course and level of usage of the studio. From this analysis six students who exemplified particular profiles were selected. A period of observation in the studio followed where the six selected students were observed to validate the data in the questionnaires.

48.5 Research Findings

48.5.1 Membership of Groups

Table 48.1. Membership of groups (GA and GB – Group A and B, R – Rating, MR and FR- Male and Female rating)

Q.22	GA	R	GB	R	FR	MR
I am a member of a family (parent, sister, partner, *etc.*)	34	1	82	1	1	2
I am a member of a religious community (Muslim, Christian, *etc.*)	13	6	9	9	10	8
I am a graphic/product design student	33	2	73	2	3	1
I am a part of a group of friends	32	3	69	3	2	2
I am a member of a team or society (football, quiz team, *etc.*)	24	4	34	5	6	5
I am a University student	19	5	51	4	4	4
I am part of an ethnic community (Greek, Pakistani, *etc.*)	3	10	6	10	9	10
I am a particular class in society (working class, middle class, *etc.*)	7	9	27	7	7	6
I come from a particular part of the country (Northerner, Londoner, *etc.*)	8	8	33	6	5	7
I am an employee	12	7	18	8	8	8
If there is a group or community not mentioned above to which you feel you belong, please add it here	1	11	0	11	10	11

It was firstly important to discover the relative importance of the students' membership of their design group. The respondents were asked to select five groups from a list of ten and rate the importance of these in their lives. The results of both studies were consistent (See Table 48.1). Each group rated membership of their family as being most important and being a graphic/product design student as second most important to them, with their group of friends third. The total scores for these three categories were over twice as high as for the next most highly rated group. This adds weight to the idea that they are, therefore, motivated to conform to the norms of these groups rather than any others. There was some difference between the genders, however, with female students rating their friends marginally more important than the design student group.

48.5.2 Studio Usage

In the initial study prior to the two case studies described here, substantial variation in the amount of time students attended the studio had been observed. The second study provided the opportunity to substantiate and quantify this pattern. In both groups the median distribution point falls below the mean and this is accounted for by a small number of students who attend considerably more than average. There was no difference between the attendance of male students and female students in either group.

The two groups were different in terms of the balance of time spent in the studio, at home and elsewhere whilst working on a project. (see Table 48.2) In Group A students spent more time at home on their project than in the studio whilst this was reversed in Group B. Both groups spent 18% and 15% in places other than the studio or at home. There was some evidence that female students spent less time at home than male students but more time in places other than home or the studio. The most common reason for being in 'other places' was to undertake research to support the project.

Table 48.2. Balance of time spent in different places during the project

Q5 On this project, what percentage of your time did you spend in the following places?	**Group A** **Mean**	**Group B** **Mean**	Female Mean	Male Mean
Time at home	**52%**	**36%**	26%	45%
Time in studio	**30%**	**49%**	44%	44%
Time elsewhere	**18%**	**15%**	24%	11%

Both the initial study and the subsequent study identified that students who studied at home preferred to do so because they had their own equipment at home - mainly computers - although pressures of work and family were also mentioned. However, another clue which may explain the pattern of usage was the fact that of those students who rated the teaching below average, sixty three percent of these attended the studio below the average for that group. Students tend to spend more time out of the studio at the beginning of a project - the need to do research being cited as the main reason. They spend more time in the studio at the end of a project - the need to get feedback and use equipment being cited as the reasons for this.

48.6 Who Rates Whom

48.6.1 Task Ratings

In both groups studied, students could clearly identify a group of peers as task experts. Table 48.3 shows task ratings for one group. In both groups a few students stood out as being rated very highly in task characteristics. These students had ratings that were at least 3 times and up to 5 times above the average for that group. In the task ratings most students frequently mentioned the same one or two peers. The most clearly defined rating was that for high marks and this was most apparent amongst Group B. Different practices for informing students of their marks may account for this difference between the groups. These students were awarded marks that were above average for that group in the two projects prior to the study.

Table 48.3. Task ratings for Group B

Students who rated	Rated Student
C C I I	A
I M M M	B
I C	C
	D
	E
	F
C	G
	H
A A A B C C C F F F J K K K M M M N N N N N O O P R R R	I
A A C H M N P	J
O	K
	L
A C F F F H H J J J K K K N N N O O O P P	M
A A B C C F G H J J K K M O P P P R R S S S S	N
P	O
F G J L L R S S S S S	P
	Q
I	R
	S

	T
F	U
	V
	W
A B E E E F H H H H H J J K K M M N O P P	X
E E E I I I I L	Y
G H I	Z
	AA
	AB

Mean for ratings - 5.0 **Male Mean - 6.9** **Female Mean - 3.1**

Males were more likely to be rated highly as task leaders than females and females more likely to be rated highly as socio-emotional leaders in both groups. There was no evidence, however, of students being more likely to rate those of the same gender. This may be a result of inter-group rather than intra-group influence. It is likely that the views of society about the appropriate roles and abilities of the genders has had an impact. This view is, of course, socially constructed and has not been proved to be a real reflection of the abilities of females. A number of students in each group were not mentioned at all by their peers in the task rating. Significantly, these students were also much more likely not to have mentioned any of their peers within their own questionnaires. When questioned they said that they either did not know who were the 'good' students or did not know their names. The students from ethnic minority groups were within this sector. This fact is potentially interesting although a sample of one in each group is not large enough to prove statistically that it is significant. There was no correlation between those rated high in task dimensions and these students' satisfaction with either the course or the teaching they received. There was also no correlation between those rated high in task dimensions and high studio usage. This potentially means that students in both groups do not see attending the studio a lot as being a way of achieving good task results.

48.6.2 Socio-emotional Rating

In both groups more students were mentioned as being strong in socio-emotional characteristics than in the task dimensions. So it seems that socio-emotional leadership is more equally shared than task leadership. There was a high incidence of task-rated students also being mentioned as those who were strong in socio-emotional characteristics. In both groups, over half the students identified as being above average in task abilities were also rated above average in these dimensions. Again the majority of students mentioned at least two students several times on these dimensions. Also prevalent was the occurrence of mutual support between two or three students. As is illustrated in Figure 48.1, two peers would rate each other several times as being supportive and these often included students who sit together in the studio.

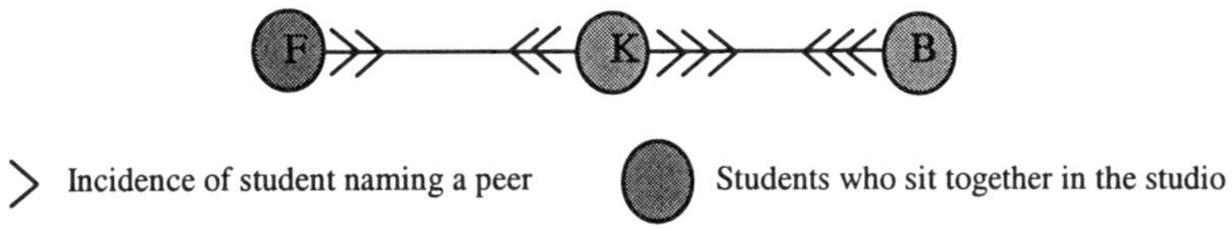

Figure 48.1. A social support network from Group A

As for the task ratings, a significant number of students were not mentioned by any of their peers and neither could many of these students mention peers who they found supportive. Indeed, students rated below average on the task dimensions were more than seventy five percent less likely to be rated above average on socio-emotional dimensions than other peers were. The socio-emotional ratings charted more closely who students spent time with than the task rating. So for instance, each student in both groups, mentioned amongst the peers who supported them, those students with whom they sat in the studio and those with whom they lived. By charting the networks of social support between peers in both groups, three specific patterns of social activity have been identified. The three types were:
Below average-sized network, high incidence of mutual support between two students (as exemplified in Figure 48.1); Larger than average network, widely distributed support (as exemplified in Figure 48.2); No discernible network either in terms of supporting or being supported.

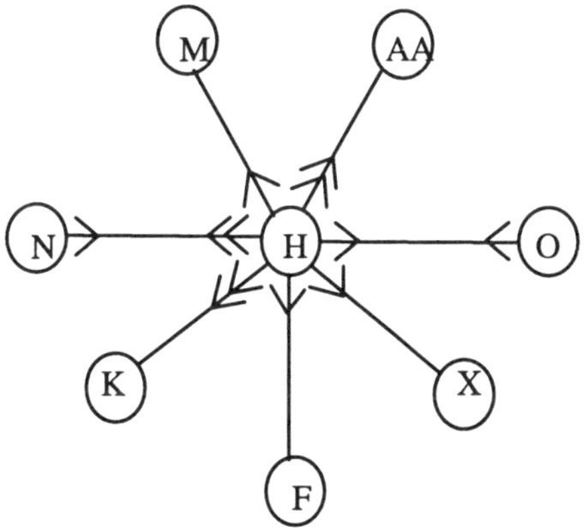

Figure 48.2. Larger than average network, widely distributed support from Group B

48.6.3 The Relationship between Social Support and Task Rating

The most significant results of the task and socio-emotional leadership ratings was obtained by studying the socio-gram profile of high task-rated students to find those peers with whom these students associated. To understand the relationship between the socially supportive networks and the peer task rating, the socio-grams of those students who were above the average in the task rating were examined. Here it was seen that in both groups those students with a higher than average task rating had within their socially supportive network a high incidence of other students who were also highly rated on this dimension. This was more marked where the relationship was mutually supportive. An example is shown in Figure 48.4.

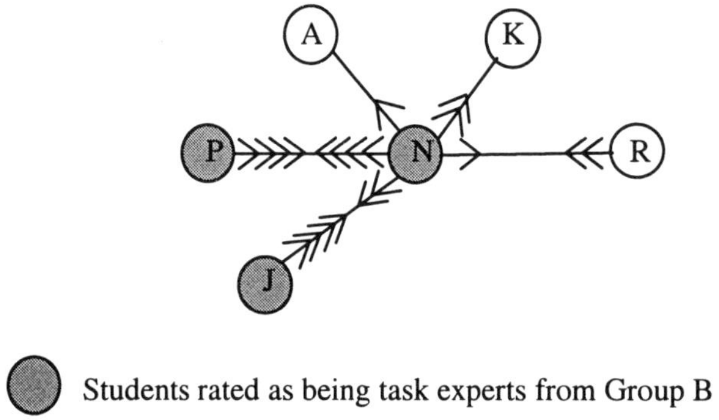

Figure 48.3. Socio-gram for high task-rated students

The inference here is that students identified as being task experts by the group are closely associated. If processes of social comparisons are taking place it is likely that these students are more likely to compare themselves with students who are also task experts. The high proportion of task experts who are also socio-emotional leaders also makes it likely that these students are in a stronger position to influence norms through the positive contact they have within social networks. This will mean that they could be considered to be on an 'upward spiral' as their comparisons will lead them to wish to be compared favourably against those students who rate well against the norms of the group. In contrast, those students who do not have task experts within their social network could be said to be on a 'downward spiral' as their points of comparison do not conform favourably to group norms. Those who are both socially isolated and not rated for their task abilities, may have fewer channels of communication and less opportunities to both influence and be influenced by group norms.

48.7 Summary of Findings

In summary, the key findings from the analysis of the data were:

- students felt that membership of the design year group and their group of friends was comparatively important to them;
- use of the studio varies considerably, with small numbers using it far more than average, but over half using it below the average for that group;
- the balance of time spent in the studio, at home and elsewhere differed for both groups;
- students are able to identify clearly who are task experts;
- male students are more likely to be rated as task-experts than females were;
- task experts are also likely to be socio-emotional leaders;
- task experts exist within networks with other task experts;
- a significant proportion of students are not rated as either task or socio-emotional leaders and cannot identify peers who are either task or socio-emotional leaders.

48.8 Conclusions

The social networks and the identifying of particular students as task experts and socio-emotional leaders is a natural and integral part of the social interaction within the Design School studio. Students appear to be very able to locate themselves and others within the social hierarchy. Breakwell and others believe the impact of this social structuring is an example of collective action: 'More fundamentally it has been argued by Barnes and others that group identity itself is an example of collective action... in the real world, social identities are themselves the product of collective action.', (1992).

Processes of reflection enabled by modelling and social comparison are in a very real sense examples of collective action between students. Working alongside each other and interacting socially either to support peers or to provide task targets are collaborative actions. The processes of socialisation into group norms are fundamentally group-enabled. The informality of the Design School year group means that norms are, however, implicit rather than explicitly stated and if we accept the social contructivist view, the groups' actions enable the construction of group norms. This suggests that the approaches and skills of a designer, to which the students aspire, is open to interpretation in different groups and by each individual, and is subject to influence by group leaders.

Whilst there are some illuminating aspects in the research described here and it has provided empirical evidence where little such data has existed before, further study is clearly required. Two aspects in particular are ripe for further investigation. Some of the patterns revealed are 'chicken or egg' situations. Do task experts group together socially because they see themselves as being similar or does associating closely with a task expert make students more likely to become like them? Similarly, are some students naturally socially limited or does being neither a task expert nor a socio-emotional leader cause them to be socially isolated. Furthermore, there is some evidence that ethnic minority students are more likely to be socially isolated and this needs further investigation for the impact this has on the equal opportunities claims that are a feature of many University Mission Statements.

48.8 References

Ashton P (1995) The Impact of Learning and Teaching Styles in Vocational Preparation. International Forum on Design Management. Paris

Ashton P (1999) Independent Learning in the Design School Studio. European Academy of Design, Sheffield

Breakwell G (1992) Social Psychology of Identity and the Self Concept. Surrey University Press, Surrey

Brockbank A, McGill I (1998) Facilitating Reflective Learning in Higher Education. Open University Press, Milton Keynes

Brown R (1988) Group Processes - Dynamics within and between Groups. Basil Blackwell, Oxford

Burr V (1995) An Introduction to Social Constructivism. Routledge, London

Cooper C, Alderfer (1978) Advances in Experiential Social Processes. Wiley

Cross. N, Christiaans. H, Dorst. K (1994) Design Expertise Amongst Student Designers. Art and Design Studies 13 (1)

Farr RM (1993) Theory and method in the study of Social Representation. In: Breakwell, GM, Canter D (eds.) Empirical Approaches to Social Representations. Clarendon Press, Oxford

Festinger L (1954) A Theory of Social Comparison. Human Relations 7

GLAD (1987) On Not Sitting with Nellie. CHEAD/CNAA

Moreno J L (1954) Who Shall Survive. Foundations of Sociometry, Group Psychotherapy and Sociodrama. Beacon House Inc., New York

Northouse P (1997) Leadership - theory and practice. Sage

Schön D (1987) Educating the Reflective Practitioner. Jossey-Bass

White G. Mufti, San Fransisco

Chapter 49

Disrupting Isolation: Exposing Distance Learning Students to Cultural Differences

I. Ali-MacLachlan

49.1 Introduction

The introduction of Communications and Information Technologies to students can enhance their learning experience (Dearing, 1997; Maher *et al.*, 1997) and introduce them to leading-edge industrial technologies where distributed resources are available to product development teams (Campbell *et al.*, 1999).

Employment of pedagogical techniques such as 'open and distance learning' will undoubtedly bring tertiary education to many students who cannot physically attend institutions on a regular basis. The provision of 'alternative' learning paradigms gives rise to a search for new methods for mentoring students. Adding live tutorial feedback for this type of education enables staff to mentor students in a way that is more effective than evaluating work asynchronously.

Distance learning students who embark on the type of course normally conducted within a studio are not only detached from the collaborative studio environment but also the ability to interact on a daily basis with other students following the same course of study. Being remote from the academic institution also makes it difficult for the student to experience the collaborative studio culture associated with their profession.

The challenge of the ICON (Institutional Collaborations Over Networks) (Thomson *et al.*, 1997) projects has been to introduce Product Design Engineering (PDE) students to remote collaborative working through a series of Internet-supported design projects. In particular ICON 2, completed in 1999, showed that a collaborative project between PDE students from institutions with differing teaching and learning cultures enabled the participants to gain a deeper understanding of their subject area.

The PDE course managed jointly by Glasgow School of Art (GSA) and Glasgow University has developed a working relationship with the Transportation Design course at the Academy of Fine Arts and Design in Bratislava over a period of approximately three years. The Slovak course has a parity with BA courses in the UK whilst the PDE course has a BEng/MEng curriculum and offers a major transportation project in Year 3 as part of the studio course.

The relationship has involved contributions to both institutions from visiting professors and a project where initial concepts developed in Glasgow have been further developed independently in the Slovak school. This method of totally asynchronous work was lacking insofar as the original concept designers had no input in the further development of these designs. Video Conferencing (VC) was seen to be a medium by which students in Glasgow and Bratislava could collaborate on their designs.

In anticipation of collaborative projects being introduced as part of the curriculum in session 2000/01, a consultation period was arranged where students undertaking a remote collaborative project based in Glasgow could discuss their ideas at the concept stage with students in Bratislava before moving on to more developed designs.

49.2 Objectives

The main objectives of this project can be outlined as follows:

- to ascertain the practicality of operating Internet-based remote collaborative design projects for students involved in remote learning courses;
- to assess the suitability of collaborations between remote courses of a similar nature as an exercise in heightening awareness of cultural differences;
- to study the impact of the technology on the remote teaming process;
- to investigate the performance and reliability of the technology.

49.3 Technology

The equipment used in the project consisted of medium specification PCs with VC cameras and a choice of headset (single user) or microphone and speakers (team use). Both institutions use PCs as their Computer Aided Design (CAD) machines as a matter of course. Though cross-platform conferences between PC and Apple have been tried as part of the ICON programme, it was concluded that this made the system even less stable.

All machines were installed with Microsoft NetMeeting. The Glasgow participants generally use Solidworks and 3d Studio Viz software for CAD whilst the Slovak course uses the Autodesk suite of products. A comprehensive guide to the Academy of Fine Arts and Design in Bratislava's computer systems and a brief history of the school can be found in Martin Sperka's paper written for Electronic Art in Universities (Sperka, 1999).

Both GSA machines were located in different rooms in the same building at Glasgow School of Art. Over the time of the collaborative project, the rate of data transfer between them was found to be approximately 44kbit/s. The rate of data transfer between Glasgow and Bratislava was not found to be significantly different and the participants did not perceive any major change in the experience of communicating with either.

This rate of transfer is an approximation of what a distance learning student working from home may expect in two or three years' time. (Ali-MacLachlan, 2000)

49.4 Project Background

The design project took place at GSA over a period of ten days and initially involved two students, who had completed year three of BEng/MEng Postgraduate Diploma in Education and one member of academic staff. The Slovak students became involved on day six after the initial concepts had been completed and the project was entering the development phase. The project personnel diagram is shown in Figure 49.1.

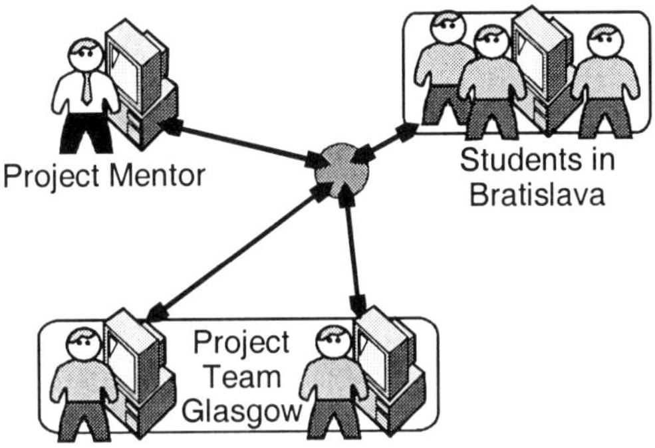

Figure 49.1. Project personnel diagram

One of the students had also been involved in ICON 2 and both felt they were confident with the available tools. Both students also showed strong evidence of ability in troubleshooting IT problems. Both students and staff were known to each other but had not met the team in Bratislava. Research does suggest that VC (Video Conferencing) is more effective when the participants have previously met face-to-face (MacFarlane, 1999).

The project brief was designed around subject matter that would be easily accessible on the Internet. Students were asked to consider the security of bicycles and belongings for a chosen target market. The brief also indicated that an urban or rural setting could be chosen and indicated that each would have a quite different set of design constraints. The solution could be either *in situ* or portable giving the designers a wide range of possible outcomes.

In this way, the brief enabled the Glasgow students to adopt a systematic approach to problem solving without the time constraints normally associated with project research of this type.

The students in Bratislava had recently completed a project looking at two-wheeled transport and were able to provide images of their completed designs as a visual stimulus for discussions, and well considered thoughts based on their own research.

49.5 Methodology

The adopted methodology is summarized in Figure 49.2.

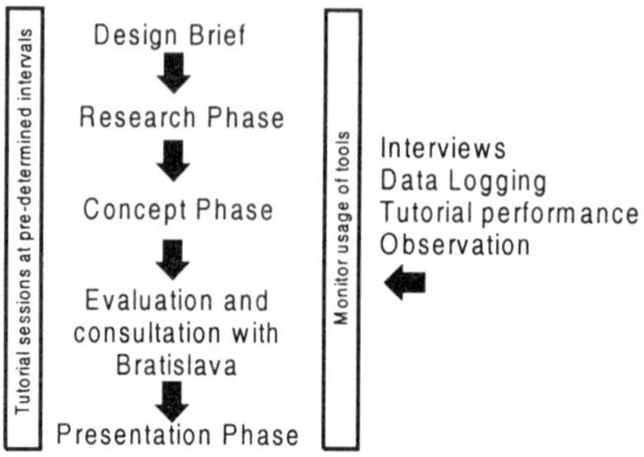

Figure 49.2. Project methodology

On the first day of the project the participants at GSA were introduced to their workstations and asked to complete questionnaires about their knowledge of IT and software packages. The software available was:

- operating system - Windows NT;
- communication - MS NetMeeting;
- CAD - Solidworks, 3d Studio Viz;
- graphic – PhotoShop.

In addition to a mouse and keyboard, A6 drawing tablets were also supplied. The design brief was delivered and a timetable was agreed for VC tutorial sessions. Participants were asked to attempt to solve any technical problems themselves and only to request external technical help when unable to solve problems themselves. In case of total breakdown of communications, a telephone was available to each participant.

The project was monitored in a number of ways:

1. **Project Log.** Both participants and tutorial staff kept a log of any problems and feelings about the technology, making entries approximately every two hours.
2. **Interviews.** Participants were interviewed before, during and after the project, either by VC or in person.
3. **Observation.** Carried out at both participant locations by both technical and design staff.

Before the collaboration began with the students in Bratislava, the participants were asked to research the project and prepare a concept which could be tabled to invoke a discussion and further development.

49.6 Findings

The Head of the Design Department at the Academy of Fine Arts and Design, Prof. Stefan Klein said of the project, 'The discussion has revealed that each of us is keen to preserve the original creations typical of individual regions and that each nation should maintain its traditions and specific features. Yet it is necessary to disrupt the isolation and create a state when both sides will be kept informed about each other and will maintain lively communication and cooperation.' (Demovicova, 1999)

This 'disruption of the isolation' is one of the ways in which students can widen awareness of their chosen subject area. Due to the nature of Glasgow's BEng/MEng curriculum, students are required to consider physical, mechanical and manufacturing constraints with a higher priority than the Slovak students, whose strengths lie in aesthetics and form. Both sets of students come from a culture of human-centred design, and have strong skills in ergonomics and human factors.

49.6.1 Methods of Working

In the first phase of the project, the GSA students felt that non-electronic methods of communication were faster and more suitable when working both synchronously and asynchronously up to concept stage. It was felt that the camera was of more use when trained on the user's work, and both team members modified their camera mounting so the image appeared in the correct orientation on their partner's screen.

A technique was developed for mounting a still view of a sketchbook page on the whiteboard so both parties using pen tools could discuss details. An example of a typical whiteboard layout is shown in Figure 49.3.

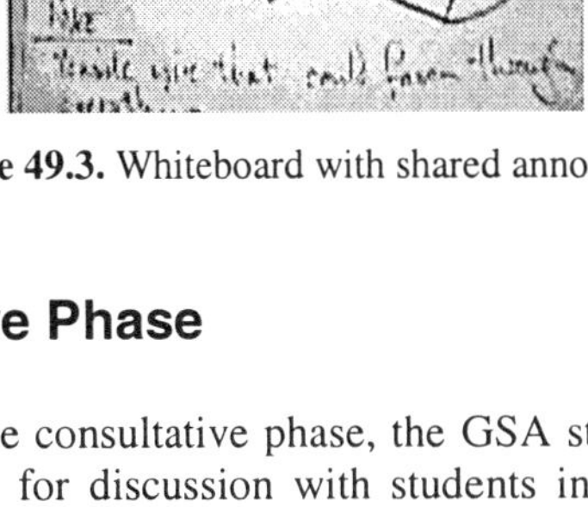

Figure 49.3. Whiteboard with shared annotation

49.6.2 Consultative Phase

As the project entered the consultative phase, the GSA students felt apprehensive about tabling their work for discussion with students involved in a course with much more visual content. The phenomenon of judging visual work (and in particular highly finished computer modelling) to be of more importance than engineering work seems prevalent amongst GSA's PDE cohort, and this inherent competitiveness has been evident in all ICON projects.

Students began by discussing their respective courses but as they became more acquainted and the technology started to become more transparent, the conversation turned towards a more social dimension. Results of past ICON projects have shown the importance of audio above all other means of communication but in this particular case, due to the spoken language differences, text-based chat very quickly became the norm.

The main effect of the collaboration with Bratislava was to encourage the GSA students to present their work in a much stronger way whilst holding on to the engineering integrity which has become a mainstay of their course. The final design presentation showed a change of working practice and references to wider cultural knowledge such as refinement of form and aesthetic.

49.7 Conclusions

The project objectives are presented with their conclusions as follows:

- *To ascertain the practicality of operating Internet based remote collaborative design projects for students involved in remote learning courses.* The ICON projects have illustrated that it is feasible to facilitate remote collaborative design projects supported through Internet technology. Quality of learning experience is somewhat dependent upon guaranteed minimum bandwidth or greater speed of operation. Evidence suggests that with the support of emerging technologies, these should both be available to the home user within the next three years.
- *To assess the suitability of collaborations between remote courses of a similar nature as an exercise in heightening awareness of cultural differences.* Students have an inherent interest in related courses and can benefit greatly from exposure to them. The study has shown that these influences help participants to gain a stronger and more varied skill set. Collaborative projects expose the student to alternative processes and solutions, whilst helping to minimize any feelings of insularity. It will be necessary to foster a strong feeling of belonging amongst students who do not have the opportunity for regular social interaction with other members of their cohort.
- *To study the impact of the technology on the remote teaming process.* Evidence shows that participants learned to adapt their working methods to take account of the technology. Students at this stage on a degree programme realise that a computer is 'just another tool' and are aware of its limitations.

- *To investigate the performance and reliability of the technology.* Reliability of the system and software depends upon the perception of the individual. Students who are able to solve basic hardware and software problems are less likely to expect a system to run fault-free and are more confident in initial troubleshooting. As bandwidth increases, Internet group conferences will be possible but experience of design studio practice has shown that one-on-one tutorials yield the best overall results.

49.8 References

Ali-MacLachlan I (2000) A Synchronous Feedback System for Distance Collaborative Projects. In: Proceedings of the 3rd Working Conference on Engineering Education. Sheffield, April 17-20

Campbell L, Ali-MacLachlan I, Thomson A, Ion W, Macdonald A (1999) Institutional Collaboration over Networks, ICON, A Comparison of Two Collaborative Design Projects, In: Proceedings of the 3rd conference on Computers in Art & Design Education (CADE 99). University of Teesside, April 7-9

Dearing R (1997) Higher Summary Report in Education in the Learning Society. National Committee of Enquiry into Higher Education, http://www.leeds.ac.uk/educol/ncihe, para 65.

Demovicova M (1999) Video Conference Glasgow-Bratislava. Designrum, the Journal of the Slovak Design Council 4:33-34 (English Summary)

MacFarlane A (1999) Towards a Learning Strategy for the University of the Highlands and Islands [Report of the Learning Environments & Technology Working Group] sect. 4.3.2 Computer-mediated communication CMC. http://www.uhi.ac.uk/LET_Report.html

Maher ML, Cicognani A, Simoff S (1997) An Experimental Study of Computer Mediated Collaborative Design..http://www.arch.usyd.edu.au/kcdc/cmcd/paper/index.html

Sclater N, Sclater M, Campbell LM (1997) ICON -Evaluating collaborative technologies. Active Learning 7:41-46

Sperka M (1999) Electronic Media and Computers at the Academy of Fine Arts and Design in Bratislava. http://www.uiah.fi/bookshop/isea_proc/eaiu/j/15.html

Thomson A, Campbell LM, Ali-MacLachlan I (1999) Institutional Collaboration over Networks, ICON2, A Remote Design Project. In: Proceedings of the 12th International Conference on Engineering Design (ICED 99). Munich, August 24-26

Chapter 50

Design Practice in New Zealand: Seeking the Basis for a Teaching Programme Course

G. Hargreaves

50.1 Introduction

As a lecturer in a design degree programme with need to learn more about the professional practice prospects facing new graduate industrial designers once they have taken up employment, the author of this paper selected some New Zealand product design projects for *post facto* case study.

This paper raises some issues and topics for discussion from a broader research aim that was to develop a new course model on professional practice. The study focused on professional practice seeking a basis for an industrial design programme course, taking a qualitative research approach. The study objectives included piloting and implementing appropriate inquiry methods to explore group design process, its context and content. The final discussion in this paper considers the outcomes of five product innovation case studies, the co-design process and the part played by the early-career and new graduate industrial designers.

From a literature search on enquiry methods (Bessant and McMahon, 1979, Hales, 1987; Frake, 1962; Spradley, 1979), published reports of some New Zealand studies on innovation (Frater *et.al.*, 1995; Slappendale, 1994),and studies elsewhere on multifunctional design group process (Davies-Cooper and Jones, 1995), the following criteria was used to select projects for enquiry.

Regarding operations, the following project content wasneeded: a collaborative multifunctional design group; an early career industrial designer; broad ranging organisational aspirations for innovation in design policy, strategy, project management, and product outcomes; difference regarding organisational type, size, capability, specialisation of expertise, and product design context.

Regarding research, the following project content was needed: researcher distance between project participant/informant (PI) practice; PI's informed consent; only academic research involvement; researcher intervention to each design group by networking through the industrial designer, (new graduate industrial designers taking this option when finding employment and getting established in professional practice); freedom to investigate consistency of agreement and disagreement amongst PI's reporting, allowing triangulation in the method (Hammersley, 1993); scope limited to researcher's personal time and funding; project records to be made accessible by PI's and colleague personnel; PI reporting over a three-year period (1997– 2000) allowing study of innovation/NPD process impact on organisational change.

The eventual selection of case studies (CS) included these:

- CS.1 - An up/down light in optional standard or wall-mounted form for commercial and domestic use. A low-technology product, it was inhouse funded, design/originated developed, batch manufactured to order and distributed for sale by a local branch office of a major international corporate lighting distributor. The project has been running about two years. Tooling cost was limited to NZ$10,000. At present the product is only available for sale in New Zealand.
- CS.2 - A stadium seat for either outdoor or indoor installation. Consultant designed by a small size manufacturer who successfully tendered, becoming a design/originator subcontractor to supply and install for a main building contractor employed by a client board of trustees. As sponsor, the client up-front funded the subcontractor to have the moulds built. The product is now being produced to order (in thousands) by high-tech manufacturing for home and export. From idea to installation/release the project took two years. Tooling cost was about NZ $250,000.
- CS.3 - A range of street receivers into which the public deposit mail for collection by service van drivers. The product range was industrial design consultant originated for a new postal organisation (client/sponsor) shortly after the service was deregulated in the mid-1980's. A successful tenderer from the building industry was contracted to self-fund, project-manage product development and low-technology manufacture to order in batches of 100. The manufacturer flat-pack component supplied the product back to the client, the latter effecting on site product assembly and installation. Retaining an intellectual property right to the product, the contractor has since pursued export opportunities. From idea to mark-one product installation, the project took twelve months. Tooling cost is not known.
- CS.4 - An automatic toll dialler (ATD). Following deregulation of telecommunications in the late 1980's, a transnational corporate (TC) entered the market to compete with a newly privatised company which monopolised the network. The TC's company called tenders for design/origination, manufacture, and supply of a miniature ATD for attachment to their subscribers' line phones. The ATD simultaneously alters digits dialled by the customers, re-routing calls and opening access into New Zealand and global telecommunications networks. A very small company that designed modems made the successful tender and was contracted to self-fund design/originate the product hardware, software and high-technology manufacture in batches, supplying to the TC's company (the sponsors) whose contract service installed and maintained connections. In other forms the product is now exported by the contractor. From tender success to product commissioning, the project took six months. Tooling cost is unknown.

- CS.5 - A domestic dishwasher. An in-house-funded, designed/originated, developed, manufactured and marketed product for distribution to home and international markets. Following the 1987 global economic recession, an established medium-size whitegoods manufacturer decided that competing, in an unbounded home market left no option but to innovate products for international markets. The domestic dishwasher was a first step towards this achievement, the product and NDP process becoming a catalyst, which over five years, transformed the organisation's core competencies (Prahalad and Hamel, 1990), corporate culture and product range. From idea to launch the project took nine years. Tooling cost is estimated at $40 million. Manufacturing is high-technology automated. Volumes are 50,000 p.a.

50.2 Co-design

To some extent each CS analysis showed organisational attributes of 'co-design' as described in Duncan's typologies of research methods. He writes: 'This entire category centres around designing (and all that entails) with end-users.' (Duncan, 1994). The serial numbering of each CS coincides with the author's impression of core participants' commitment to co-design in the early stages of the project, 1 rating low and 5 rating high.

Figure 50.1, proposes the author's, CS inspired, model of multifunctional group competence, access and approach to an iterative 'means to ends' design process during product innovation and NPD. The central focus is 'product innovation'. The concept of product innovation has been defined by Souder as: 'a high risk idea that is new to the sponsoring organisation, and which the organisation believes has high profit potential or other favourable commercial impacts for them.' Souder (1987). He draws attention to a 'sponsoring organisation' that might not necessarily 'originate' the idea. The author's model positions contributing functions as either dominant 'sponsor', e.g., directors, and funding; or dominant 'originator', e.g., operations, engineers, ergonomists and industrial designers. An 'originator/sponsor' mix is suggested in operations, sales marketing, and end-user functions. This positioning of contributing functions was broadly evidenced during each CS, as represented in Figure 50.1

Across the CSs, blurred functional role interaction between industrial designers and other groups was concentrated in the following areas: creative research, concept generation, design and development, manufacturing engineering, form-giving (aesthetics), ergonomics, sales and marketing, and product concept evaluation meetings with the end-user/customer. Operations' and technical managers' interest in product innovation and design was occasionally hands-on, yet at times distanced when mentoring originators or championing project funding at board level.

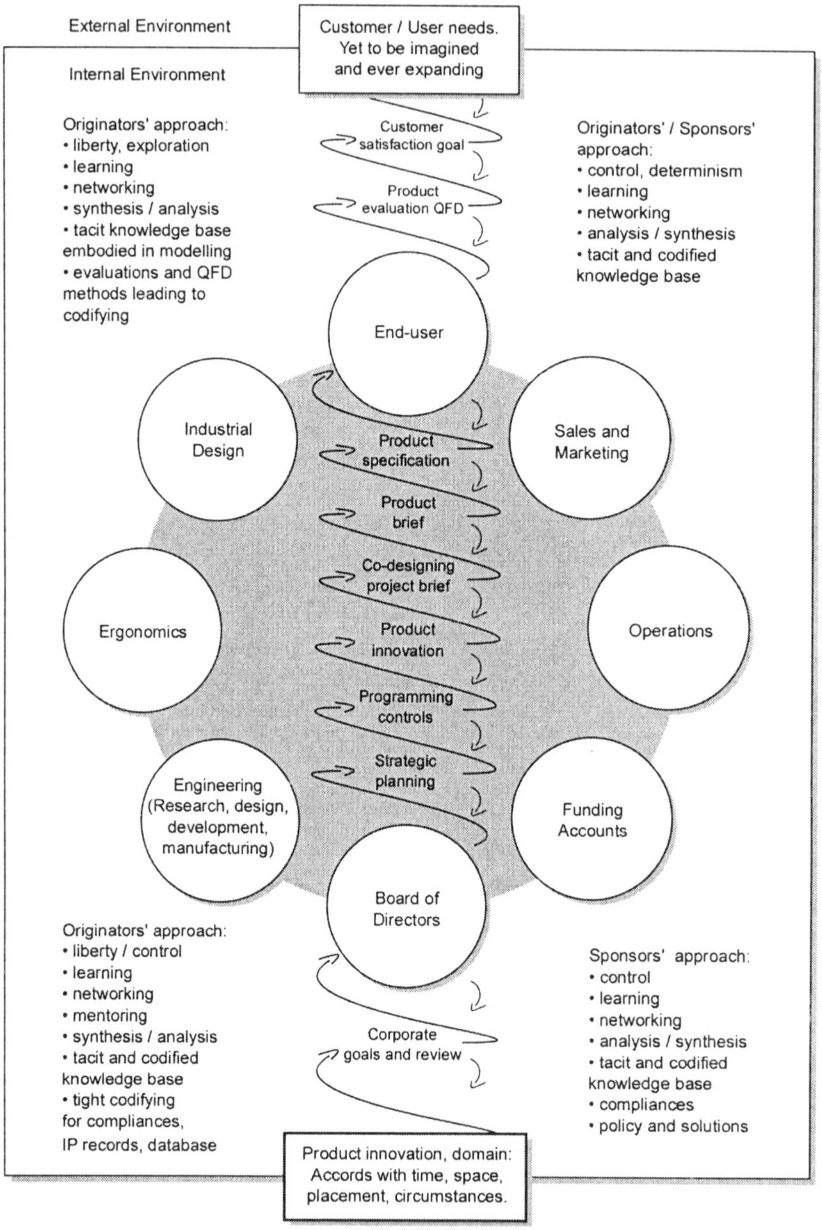

Figure 50.1. Multifunctional group competences and access to an iterative 'means to ends' design process

50.3 Method of Enquiry

Planning the method of enquiry ran parallel to a literature search and review. During, 1997 projects were reconnoitred for enquiry with a minimum of five PIs. Later that year, CS.2 was piloted. PIs were contacted by telephone and informed-consent research conditions were arranged. For a year during these occasional telephone and, later, work-place conversations the researcher built a professional relationship with the PIs. Over 1998-99 PIs were interviewed at their places of work. Throughout each interview project records, many of these three-dimensional constructions, sometimes sketches, spreadsheets and the like, were available as reminders. A narrative tour of the project was sketch-annotated by the PI and researcher on multiple overlaid A1 sheets of tracing paper to enable broad exploration of dimensionality. PIs preferred this approach to audio tape and video recording. Selected interview settings usually had a pleasant view to look out onto, the PI and researcher sitting side by side to encourage facilitation and to avoid confrontation. The interview time was usually forty five minutes. A1 sheets were set up in loose matrix form. About four major events were elicited and listed down the left vertical margin. Along the top horizontal margin, milestone dates were recorded. From this came mutual sketching of flow diagrams describing: program tasks, decision points, budgets, costs, means, ways and ends; functional and actual roles, and relationships of core project participants; ideation of three-dimensional product concepts and their development. PIs set up the project record, structured in generative task stages with decision pressure points where concepts were evaluated and screened, with the generation/evaluation process either iterated or further progressed toward a prototype and specification. PI descriptions of process tended to accord with other research findings, (Cooper, 1990). This influenced the enquiry later in the study. Pilot data was transcribed and returned to respective PIs for their interest and confirmation as an accurate record of the interview. Ethnographic in approach, interviews considered and explored ‘cultural themes’, (Frake, 1962), *i.e.*, in response to PI’s referencing, use of metaphor, vocabulary, and sketch annotation preferences, the researcher learns to converse with PIs on their own terms. The qualitative ethnographic objective was to enrich descriptive data with authentic narrative depicting PI’s individual way of communicating their perception of a ‘means to ends’ design process (Topalian,1980). Piloting alerted the author to the subsequent CS-interview need for a semi-structured questionnaire with flexibility for measured assessment of issues that PIs felt were important to the success of the project and the product outcome. A ‘design process’-based questionnaire was developed. Through 1998-99 each CS was set up, run and checked with PIs, then data tabulated for cross-comparative analysis.

50.4 Findings

With the CS research focus seeking a basis of professional practice for a teaching programme course in industrial design, co-design context in the early stages of product innovation and NPD is now discussed.

All PIs affirmed that their projects from idea through to product launch or release consisted of a 'means to ends' design process. 'Means' refers to their context-related resources of expertise *i.e.,* knowledge and skill in handling: ideation, research, design and development, manufacturing, marketing, financing, liabilities and management, obtaining funds, and programming events. 'Ends' refers to the satisfying of either sponsors' or originators' needs, and the satisfying (sometimes exceeding expectations) of client and customer or end-user needs. PIs held differing and changing perceptions of design process, with project sponsors and originators constantly experiencing phases of conflict and reconciliation, not only over interpretations of appropriate fit regarding 'means to ends', but also in their searching, testing and validation of 'ways' for effecting best fit. In the CS's innovation groups PI's self assessments and the assessment of others' functional contributions in the early stages of projects, both sponsors and originators assigned high rating to effective collaboration and communication. Sponsors rated next highest effective programming and product engineering. Originators rated next highest identifying opportunities and relating these to a champion product concept. Financial management and total quality management (TQM) tended to be relaxed. Industrial designers and creative engineers as originators felt a need for liberty to play with and explore ideation early on. CS.5 group leaders felt that an early TQM induction or immersion helped the innovation group to generate simple concepts.

Research studies report that design quality calamities are traceable to errors in human judgement during the early stages of projects. The need for TQM emphasis at the low-cost front end of projects with a 'go' or 'no-go' screen decision (Cooper, 1990) can help avoid difficulties at the high-cost end nearer production (Hollins and Pugh, 1990). In the early part of CS2 the subcontractor underrated designing and evaluating the on-site securing of seats to stadium terraces. A number of seats were installed before the error was rectified and securing made good. In the absence of a standard, a performance test was invented to prove repairs. Sponsors and originators attributed the outcome resolution to co-design attitude, concurrent management, and open lines of communication.

By the specification stage all CSs showed the advantage of multiple prototype modelling early in the project, a flexible design process and the modelling of an anticipated product life cycle. PIs were evaluating their early assumption-based concepts by eliciting product end-user reaction. All CSs engaged end-users in judgements regarding aesthetics (emotive qualities of form and materials), ergonomics, engineering, and choice of materials affecting production processes. The insights gained improved product utility and helped in getting the human/techno interface to work and feel good. CS.1 involved customers/end-users in the development of a technically similar product, immediately before embarking on their current NPD mission.

In CS.3 the client wanted the product to improve postal service for the end-user, incorporate the launch of a new corporate identity, and almost eliminate maintenance cost. Product evaluations were staged around user trials by able and disabled persons. Misuse and vandal abuse scenarios were generated and implemented by the core innovation group, to forecast or test various community interests and concerns. Representation included regulation enforcement specialists, e.g., health and safety at work, pedestrian and vehicular traffic safety, with workers' unions also taking part. Co-design networking made it possible for the consultant industrial designer and a creative mechanical engineer to both fault-analyse the existing street receiver designs, and generate/test a benchmark specification for the new designs.

From their winning of tender, the CS.4 contractor originator engaged a consultant industrial designer, and together with his electronics engineers linked in with the client/sponsor's innovation group. The latter consisted of in-house telecommunications expertise, product field-testing specialists and business consultants in customer liaison. The group networked deeply into the community, inventing and effecting measured tests of the product as first specified. From the test results, technology opportunities unexpected emerged. The contractor's design response introduced new novel services and branding options that placed the product ahead of its competition. The contractor's firm and the industrial designer consultant are now positioned to actually pick their customers in international markets.

In CS.5 the operations management encouraged their innovation group to forget about dishwashers as 'whitegoods' and start regarding them as 'life-style fashion goods'. Liberation and idea sharing started a 'different' concept which was rigorously multiple-prototype tested and customer evaluated using quality function deployment (QFD) techniques (Griffin and Hauser, 1992). A short list of no-compromise design attributes resulted. For the innovation group, this structured co-design exercise greatly benefited the product concept in exceeding customer satisfaction as current international commercialisation indicates.

50.5 Discussion

The findings suggested that design education must graduate industrial designers to a level of expertise ready for professional practice. As employees, new graduates will be expected to listen, consider and work to a brief and deliver on contracts for reward. Some first principles of their expertise will include knowledge and skill in creatively visualising and modelling artistically elegant form in highly functional products that are ergonomic and safe in use. Environmentally friendly, and consisting of appropriate materials, their product designs will need to be manufacturable and affordable. Product concepts may need to be integrated with a company's strategic design policy and so forth. For the customer and individual user, now and in the future new products will need to feel good at the techno/human interface and be emotionally appealing. The products need to develop a relationship of trust between designer, vendor and user. To generate this function in a group design process the new graduate entering professional practice will need to become familiar with the cultures (norms, values), core competencies and the approach to tasks of co-design colleagues. Plenty of mentoring and encouragement will be needed whilst learning about the contingent nature of the design process, *i.e.,* taking strategic action as things and events unfold in real time.

Among New Zealand design educators teaching professional practice it is likely that various assumptions exist about what the complexities of a 'design group process' entail. This paper describes some regional case studies, whose features invite further research questions and dialogue on perceptions of design group process (see Dorst and Dijkhuis, 1996). In the domain of professional practice each product innovation project journey contained its own particular set of process circumstances. Questions of time, space, placement, society and culture, economic politics, technologies and markets were all under pressure because of short lead times, CS.5 being the only exception. From the CS outcomes it seemed likely that under such conditions the social, cultural and environmental values associated with the co-designing process became imbued in the product. It appeared important for the product customer/end-user to have an ongoing involvement in design, research and development rom an early stage in the project. This is especially essential if imaginative and sensitive conceptions of needs, wants or demands were to be discovered and ethically integrated among product attributes. CSs 3, 4 and 5 clearly demonstrated such advantages.

Of the industrial designers participating in the CSs, four out of five had recently begun their careers in research, design and development of new products. In this regard in CSs 1 and 5, co-design group mentors appeared to take a nurturing approach. However, across the co-design groups there was a high expectation for the young designers to generate powerful, creative, 'manufacturable' product concepts to exceed the customer's expectation, if not their imagination. In the generation of ideas and their evaluation, PIs rated highly the industrial designer's capability in effectively sharing, networking and communicating their design concepts. This is an ongoing requirement which design educators and teachers ought to be responding to, exploring in research and developing in courses.

A downside to the research method used by the author was the *post facto* limitation. Whereas colleague co-design functions tend to operate from a cumulative (analytic/synthetic), codified knowledge base, much of the knowledge and skills in the industrial designers' operations are apparently of a (synthetic/analytic) tacit nature, and embodied in the context of process (Senker, 1998). To access this aspect of operations, teachers and design researchers need the benefit of creating appropriate *in vivo* research methods. For instance, the intuitive artistry and gestures in sketch modelling, articulated expression and other designerly communications might be better assessed through real-time researching into contingent aspects of the process in specific contexts and environments.

50.6 Conclusion

From the viewpoint of a teacher/design researcher, the reported study set out to seek the basis for a teaching programme course in the professional practice of industrial design. Concentration on design group process (in co-designing) yielded a rich source of ideas, analogy and metaphor useful for conceptual integration into course structures, teaching and discovery-learning activities. It was valuable to learn how the core expertise of young industrial designers and their leadership roles became integrated with broader systems, communications and environments.

The *post facto* semi-structured ethnographic interview method had flexibility and, with respect to the research goal, worked well enough. The limitations were mentioned earlier in the discussion.

For students nearing graduation, some project work, research, internships and their first employment in professional practice can all offer some great learning opportunities in co-designing. To help them with this transition from student to professional, teachers and design researchers could be playing a stronger role by monitoring the processes and fostering partnerships between education and professional practice.

50.7 References

Bessant J, McMahon B (1979) Participant Observation of a Major Design Decision in Industry. Design Studies 1 (1):21-26

Cooper RG (1990) Stage-Gate Systems: A New Tool for Managing New Products. Business Horizons. May/June:44-54

Davies-Cooper R, Jones T (1995) The interfaces between design and other key functions in product development. In: Bruce M, Biemans W (eds.) Product Development: Meeting the Challenge of Design-Marketing Interface. John Wiley & Sons Ltd, Chichester

Dorst K, Dijkhuis J (1996) Comparing Paradigms for Describing Design Activity. In: Cross N, Christiaans H, Dorst K, (eds.) Analysing Design Activity. John Wiley and Sons, Chichester

Duncan E (1994) User Research, IDEO draft document for discussion purposes. San Francisco, IDEO. In: Stanton N (ed.) Human Factors in Consumer Products. Taylor & Francis, London

Frater P,Stuart G, Rose D, Andrews G (1995) The New Zealand Innovation Environment. The BERL Foundation, Wellington

Frake C (1962) The ethnographic study of cognitive systems. In: Gladwin, T,Sturtevant, W (eds.) Anthropology and Human Behaviour. Anthropological Society of Washington, Washington

Griffin A, Hauser J (1992) The Voice of the Customer. Report No 92-106, Marketing Science Institute, Cambridge Massachusetts

Hales C (1987) Analysis of the Engineering Design Process in an Industrial Context. Grants Hill Publications, Hampshire

Hammersley M (1992) What's wrong with Ethnography? Routledge, London

Hollins B, Pugh S (1990) Successful Product Design. Butterworths, London

Prahalad C, Hamel G (1990) The Core Competence of the Corporation. Harvard Business Review May-June:79-91
Senker J (1998) The contribution of tacit knowledge to innovation. In: Williams R, Faulkner W, Fleck J, (eds.) Exploring Expertise. MacMillian, London
Slappendale C (1994) Ergonomics capability in product design and development: an organisational analysis. Applied Ergonomics 25 (5): 266-274
Spradley J (1979) The Ethnographic Interview. Holt, Rinehart and Winston, New York
Souder W (1987) Managing New Product Innovations. Lexington Books, Massachusetts
Topalian A (1980) The Management of Design Projects. Associated Business Press, London

Part VI Educating for Collaborative Design

Summary

Co-design represents a challenge for both design practitioners and educators. Maximum benefits from this practice can only be derived if designers are able to work effectively with end-users and other stakeholders at all stages in the design process. This may be considered a two-way process. However, the papers in this Part have placed the onus on design educators to develop courses and methods which will both encourage and train designers to work in this manner.

The contributions to this Part have tackled a diverse range of issues. Attention has been given to those issues which are seen as hindering co-design (for example, in terms of verbal communication skills, team-working ability and lack of empathy with other people's user models). Solutions have been proposed which seek to increase the understanding and practice of user-centred design, team-working, verbalisation and empathic skills. The level of solutions has ranged from the design of course components, courses themselves through to overall course philosophy. Consideration has also been given to the design student experience itself - to what it means to engage in a remote course, the influence of social networks in design studios and the enhancement of personal development through working with and near others.

Although many of the contributions have commenced with highlighting the weaknesses of designers, the primacy of the designer as the creative force in development is not challenged by this. Rather, the acknowledgement of this and the willingness to transform current educational practice (at all levels) to take into account changes in the societal, technological and cultural milieu, demonstrates a level of maturity within the profession which will produce designers who are adaptable, flexible and sensitive to the needs of their fellow team members and customer requirements.

Author Index